AP®

BIOLOGY 1

Student Workbook

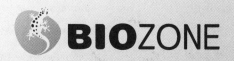

BIOZONE

AP® Biology
Student Workbook

First edition 2012

ISBN 978-1-927173-11-4

Copyright © **2012** Richard Allan
Published by **BIOZONE International Ltd**

Printed by REPLIKA PRESS PVT LTD using paper
produced from renewable and waste materials

About the Writing Team

Tracey Greenwood joined the staff of Biozone at the beginning of 1993. She has a Ph.D in biology, specialising in lake ecology, and taught undergraduate and graduate biology at the University of Waikato for four years.

Lissa Bainbridge-Smith worked in industry in a research and development capacity for eight years before joining Biozone in 2006. Lissa has an M.Sc from Waikato University.

Kent Pryor has a BSc from Massey University majoring in zoology and ecology. He was a secondary school teacher in biology and chemistry for 9 years before joining Biozone as an author in 2009.

Richard Allan has had 11 years experience teaching senior biology at Hillcrest High School in Hamilton, New Zealand. He attained a Masters degree in biology at Waikato University, New Zealand.

Purchases of this workbook may be made direct from the publisher:

www.the**BIOZONE**.com

USA, CANADA & REST OF WORLD:

BIOZONE International Ltd.
P.O. Box 13-034, Hamilton 3251, New Zealand
Telephone: +64 7-856 8104
Fax: +64 7-856 9243
Toll FREE phone: 1-866-556-2710 (USA-Canada only)
Toll FREE fax: 1 800 717 8751 (USA-Canada only)
Email: sales@biozone.co.nz
Website: www.the**BIOZONE**.com

UNITED KINGDOM & EUROPE:

BIOZONE Learning Media (UK) Ltd.
Burton upon Trent, United Kingdom
Email: sales@biozone.co.uk
Website: www.**BIOZONE**.co.uk

AUSTRALIA:

BIOZONE Learning Media Australia
Burleigh BC, QLD, Australia
Email: sales@biozone.com.au
Website: www.**BIOZONE**.com.au

Preface to the First Edition

This first edition of AP Biology 1 has been specifically structured and written, with its companion title AP Biology 2, to meet the content and skills requirements of the new AP Biology curriculum framework. The four big ideas form a thematic framework for presenting a wealth of illustrative examples to support the required content. Enduring understandings are clearly identified and developed in the key points of the chapter introductions and their supporting activities. AP Biology 1 and its companion AP Biology 2 emphasize inquiry-based learning and provide a flexible concept-based structure that accommodates diverse learning styles and allows for multiple approaches to teaching essential knowledge. Key features include:

▶ A contextual approach. We encourage students to become thinkers through the application of their knowledge in appropriate contexts. Many chapters include an account examining a 'biological story' related to the theme of the chapter. This approach provides a context for the material to follow and an opportunity to focus on comprehension and the synthesis of ideas. Throughout the workbook, there are many examples of applying knowledge within context.

▶ Concept maps introduce each main part of the workbook, integrating the content across chapters to encourage linking of ideas. A concept map also introduces the entire AP® Biology curriculum framework, identifying where material relating to each component of the course can be located.

▶ An easy-to-use chapter introduction summarizing essential knowledge in a bulleted list to be completed by the student. Chapter introductions also include a list of key terms and a summary of key concepts.

▶ An emphasis on acquiring skills in scientific literacy. Each chapter includes a comprehension and/or literacy activity, and the appendix includes references for works cited throughout the text.

▶ *Web links* and *Related Activities* support the material provided on each activity page.

A Note to the Teacher

This workbook is a student-centered resource, and benefits students by facilitating independent learning and critical thinking. This workbook is just that; a place for your answers notes, asides, and corrections. It is **not a textbook** and annual revisions are our commitment to providing a current, flexible, and engaging resource. The low price is a reflection of this commitment. Please **do not photocopy** the activities. If you think it is worth using, then we recommend that the students themselves own this resource and keep it for their own use. I thank you for your support.
Richard Allan

Acknowledgements

We would like to thank those who have contributed to this edition
• Sue FitzGerald, Gwen Gilbert, and Mary McDougall for their efficient handling of the office • Denise Fort, Gemma Conn, and Edith Woischin for graphics support • Paolo Curray for IT support • Dr Meredith Ross for editorial comment • Ben Lowe, University of Minnesota for advice and input on *Ensatina* subspecies and *Canis* distribution • Dr John Stencel, Olney Central College, Illinois for providing the data for the squirrel gene pool exercise • TechPool Studios, for their clipart collection of human anatomy: Copyright ©1994, TechPool Studios Corp. USA (some of these images have been modified) • Totem Graphics, for clipart • Corel Corporation, for vector clipart from the Corel MEGAGALLERY collection.

Photo Credits

Royalty free images, purchased by Biozone International Ltd, are used throughout this workbook and have been obtained from the following sources: **Corel** Corporation from various titles in their Professional Photos CD-ROM collection; **IMSI** (International Microcomputer Software Inc.) images from IMSI's MasterClips® and MasterPhotosTM Collection, 1895 Francisco Blvd. East, San Rafael, CA 94901-5506, USA; ©1996 **Digital Stock**, Medicine and Health Care collection; ©**Hemera** Technologies Inc, 1997-2001; © 2005 JupiterImages Corporation www.clipart.com; ©1994., ©**Digital Vision**; Gazelle Technologies Inc.; **PhotoDisc®**, Inc. USA, www.photodisc.com • 3D modeling software, Poser IV (Curious Labs) and Bryce.

The writing team would like to thank the following individuals and institutions who kindly provided photographs: • Graham Colm • Wadsworth Centre (NYSDH) for the photo of the cell undergoing cytokinesis • Alan Sheldon, Sheldon's Nature Photography, Wisconsin for the photo of the lizard without its tail • Image of Theodosius Dobzhansky used under fair use licence from the Theodosius Dobzhansky papers held by the APS • Louisa Howard and Chuck Daghlian Dartmouth College for image of freeze fracture of a cell membrane • Rita Willaert, Flickr, for the photograph

of the Nuba woman • Aptychus, Flickr for use of the photograph of the Tamil girl • Dept. of Natural Resources, Illinois, for the photograph of the threatened prairie chicken • Rocky Mountain Laboratories, NIAID, NIH • Aviceda for photos of treecreepers • Chad Lane, Flickr for the *Ensatina* photo • California Academy of Sciences for the photo of the ground finch • Alex Wild for his photograph of swollen thorn *Acacia* • Missouri Botanical Gardens for their photograph of egg mimicry in *Passiflora* • The late Ron Lind for his photograph of stromatolites • Cytogenetics Department, Waikato Hospital (NZ) for karyotype photographs • Ed Uthman for the image of the nine week human embryo • Alison Roberts for the image of the plasmodesmata • Dartmouth College for TEMs of cell structures.

Contributors identified by coded credits are as follows:

BF: Brian Finerran (Uni. of Canterbury), **BH**: Brendan Hicks (Uni. of Waikato), **BOB**: Barry O'Brien (Uni. of Waikato), **CDC**: Centers for Disease Control and Prevention, Atlanta, USA, **DS**: Digital Stock **EII**: Education Interactive Imaging, **JDG**: John Green (Uni. of Waikato) **NASA**: National Aeronautics and Space Administration, **NIH**: National Institutes of Health, **RA**: Richard Allan, **RCN**: Ralph Cocklin, **RM-DOC**: Rod Morris (DOC), **TG**: Tracey Greenwood, **USDA**: United States Department of Agriculture **VMRCVM**: Virginia-Maryland Regional College of Veterinary Medicine **UWMU**: Waikato Microscope Unit

We also acknowledge the photographers that have made their images available through **Wikimedia Commons** under Creative Commons Licences 2.5. or 3.0: • Andrew Dunn www.andrewdunnphoto.com • Dual Freq • Fritz Geller-Grimm • GDallimore • Kathy Chapman online • PLoS • Romain Behar • Xiangyux (public domain) • Barfooz and Josh Grosse • and3k and caper437 • Steve Lonhart (SIMoN / MBNMS) • Velela • Karl Magnacca • mbz1 • Laitche • Alan & Elaine Wilson • UtahCamera • Onno Zweers • Bruce Marlin • Lorax • AKA • Dirk Bayer • Omasz G. Sienicki • JM Garg • Khalid Mhmood • KTBN • Matthias Zepper • Ltshears • ART G it:Utente:Cits • Y tambe • MNOLF • Kaylaya • The Wednesday Island • John Hayman • Alice Wiegand • Opabinia regalis

Special thanks to all the partners of the Biozone staff

Cover Photograph

Mshindi, a silverback Western lowland gorilla (*Gorilla gorilla gorilla*). Mshindi is one of family group of gorillas at Louisville Zoo in Kentucky. Western lowland gorillas are a subspecies of the western gorilla found in montane, primary and secondary forests, and lowland swamps in the Congo Basin. They are the subspecies most commonly represented in zoos and are critically endangered in the wild, threatened by habitat loss, poaching, and the impacts of civil war. PHOTO: iStock © BonuraPhoto (www.bonuraphoto.com)

Contents

Activity is marked: ▣ to be done; ☑ when completed

Contents

Activity is marked: ☑ to be done; ✓ when completed

Contents

Activity is marked: ☐ to be done; ☑ when completed

Getting The Most From This Resource

This workbook is designed as a resource that will help to increase your understanding of the content and skills requirements of **AP Biology**, and reinforce and extend the ideas developed by your teacher. This workbook includes many useful features to help you locate activities and information relating to each of the big ideas and their enduring understandings.

Constructing New Ideas: The Five Es

Engage: Object, event, or question used to engage students.

Explore: Objects and phenomena are explored.

Explain: Student explains their understanding of concepts and processes.

Elaborate: Student can apply concepts in contexts, and build on or extend their understanding and skills.

Evaluate: Students assess their knowledge, skills and abilities.

Features of the Section Concept Map

Chapter panels identify and summarize the essential knowledge covered within each chapter.

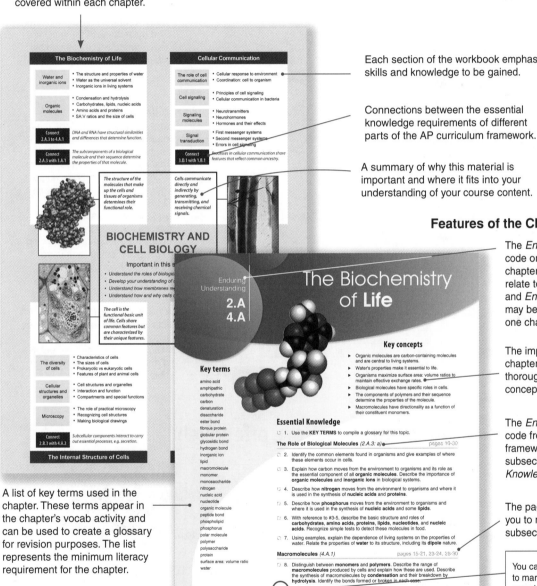

Each section of the workbook emphasizes skills and knowledge to be gained.

Connections between the essential knowledge requirements of different parts of the AP curriculum framework.

A summary of why this material is important and where it fits into your understanding of your course content.

Features of the Chapter Topic Page

The *Enduring Understanding* code or codes to which this chapter applies. Some chapters relate to more than one big idea and *Enduring Understandings* may be covered over more than one chapter.

The important key ideas in this chapter. You should have a thorough understanding of the concepts summarized here.

The *Enduring Understanding* code from the AP curriculum framework is indicated for each subsection of the *Essential Knowledge* key points.

The page numbers direct you to material related to this subsection of work.

A list of key terms used in the chapter. These terms appear in the chapter's vocab activity and can be used to create a glossary for revision purposes. The list represents the minimum literacy requirement for the chapter.

You can use the check boxes to mark objectives to be completed (a **dot** to be done; a **tick** when completed).

Periodicals of interest are identified by title on a tab on the activity page to which they are relevant. The full citation appears in the **Appendix** on the page indicated.

Essential knowledge provides a point by point summary of what you should have achieved by the end of the chapter.

The Weblinks cited on many of the activity pages can be accessed through the web links page at: *www.thebiozone.com/weblink/AP1-3114.html*
See page 4 for more details.

Student Review Series provide color review slides for purchase. Download via the free BIOZONE App, available on the App Store.

Using the Activities

The activities make up most of the content of this book. Your teacher may use the activity pages to introduce a topic for the first time or you may use them to revise ideas already covered by other means. They are excellent for use in the classroom, as homework exercises and topic revision, and for self-directed study and personal reference. You may wish to read the related material in your textbook before you attempt the activities or use simpler activities as an introduction to your texbook reading.

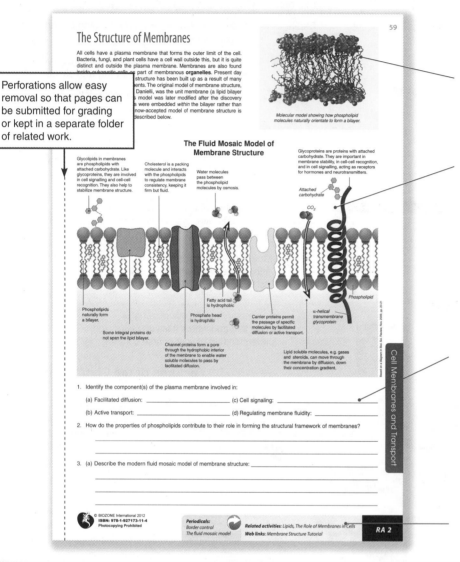

Perforations allow easy removal so that pages can be submitted for grading or kept in a separate folder of related work.

Introductory paragraph:
The introductory paragraph provides essential background and provides the focus of the page. Note words that appear in bold, as they are 'key words' worthy of including in a glossary of terms for the topic.

Easy to understand diagrams:
The main ideas of the topic are represented and explained by clear, informative diagrams.

Write-on format:
Your understanding of the main ideas of the topic is tested by asking questions and providing spaces for your answers. Where indicated by the space available, your answers should be concise. Questions requiring more explanation or discussion are spaced accordingly. Answer the questions adequately according to the questioning term used (see the introduction).

A tab system at the base of each activity page identifies resources associated with the activity on that page. Use the guide below to help you use the tab system most effectively.

Using page tabs more effectively

Periodicals:
Border control
The fluid mosaic model

Related activities: The Role of Membranes in Cells
Weblinks: Membrane Structure Tutorial

RA 2

Students (and teachers) who would like to know more about this topic area are encouraged to locate the periodical cited on the **Periodicals** tab. Articles of interest directly relevant to the topic content are cited. The full citation appears in the Appendix as indicated at the beginning of the topic chapter.

Related activities
Other activities in the workbook cover related topics or may help answer the questions on the page. **In most cases, extra information for activities that are coded R can be found on the pages indicated here.**

Weblinks
This citation indicates a valuable video clip or animation that can be accessed from the Weblinks page specifically for this workbook.
www.thebiozone.com/weblink/AP1-3114.html

INTERPRETING THE ACTIVITY CODING SYSTEM

Type of Activity
D = includes some data handling or interpretation
P = includes a paper practical
R = *may* require extra reading (e.g. text or other activity)
A = includes application of knowledge to solve a problem
E = extension material

Level of Activity
1 = generally simpler, including mostly describe questions
2 = more challenging, including explain questions
3 = challenging content and/or questions, including discuss

Resources Information

Your set textbook should be a starting point for information about the content of your course. There are also many other resources available, including journals, magazines, supplementary texts, dictionaries, computer software, and the internet. Your teacher will have some prescribed resources for your use, but a few of the readily available periodicals are listed here for quick reference. The titles of relevant articles are listed with the activity to which they relate and are cited in the appendix. Please note that listing any product in this workbook does not, in any way, denote BIOZONE's endorsement of that product and BIOZONE does not have any business affiliation with the publishers listed herein.

Comprehensive Textbooks

Comprehensive texts cover an entire program of work. These title represent a sample of those appropriate for AP Biology. Curriculum matches are found on the College Board website, AP Central. For further details or to make purchases, link to the publisher via BIOZONE's website: **www.thebiozone.com**

Mader, S.S & M. Windelspecht, 11th edn 2013
Biology AP® edition, 1040 p
Publisher: Glencoe/McGraw-Hill
ISBN: 0-076-62004-2
Comments: *This new edition addresses the redesigned AP biology curriculum. New features include chapter and part openers to introduce and connect the big ideas.*

Raven, P.H., G.B. Johnson, K.A. Mason, J.B. Losos, & S.S. Singer 9th edn 2011
Biology
Publisher: McGraw Hill
ISBN: 0-073-53222-3
Comments: *Incorporating chapter outlines, learning outcomes, inquiry questions, reviews and summaries. Logically organized and written for the new AP biology curriculum.*

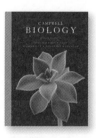

Reece, J.B, L.A. Urry, M.L. Cain, & S.A. Wasserman 9th edn 2010
Campbell Biology AP* Edition
Publisher: Pearson
ISBN: 0-13-137504-0
Comments: *This revised AP edition of Campbell's Biology includes updated content and several new features designed to help students made connections between different areas of biology.*

Sadava, D., Hillis, D.M. Heller, H.C. & M.R. Berenbaum
Life: The Science of Biology, 9th edn 2011
Publisher: Sinauer Assoc. & W.H. Freeman
ISBN: 1-42-921962-9
Comments: *Reorganized on the basis of the new AP biology curriculum framework. Experimental content has been introduced to illustrate practical applications of biological research and techniques.*

Starr, C., R. Taggart, C. Evers, & L. Starr 13th edn 2013
Biology: The Unity and Diversity of Life, 1024 pp. **Publisher**: Cengage Learning
ISBN: 1-11-142569-8
Comments: *Designed to help students easily locate material and make connections between key concepts. Provides 'real world' examples to help students put the learning material into relevant contexts.*

Periodicals, Magazines and Journals

Details of the periodicals referenced in this workbook are listed below. For enquiries and further details regarding subscriptions, link to the relevant publisher via Biozone's website: **www.thebiozone.com**

Biological Sciences Review (Biol. Sci. Rev.) *An excellent quarterly publication for teachers and students of biology. The content is current and the language is accessible.* Subscriptions available from Philip Allan Publishers, Market Place, Deddington, Oxfordshire OX 15 OSE.
Tel. 01869 338652
Fax: 01869 338803
E-mail: sales@philipallan.co.uk

New Scientist: *Published weekly and found in many libraries. It often summarizes the findings published in other journals. Articles range from news releases to features.*
Subscription enquiries:
Tel. (UK and international): +44 (0)1444 475636. (US & Canada) 1 888 822 3242.
E-mail: ns.subs@qss-uk.com

Scientific American: *A monthly magazine containing mostly specialist feature articles. Articles range in level of reading difficulty and assumed knowledge.*
Subscription enquiries:
Tel. (US & Canada) 800-333-1199.
Tel. (outside North America): 515-247-7631
Web: www.sciam.com

The American Biology Teacher: *The official, peer-reviewed journal of the National Association of Biology Teachers. Published nine times a year and containing information and activities relevant to the teaching of biology in the US and elsewhere.* Enquiries: NABT, 12030 Sunrise Valley Drive, #110, Reston, VA 20191-3409
Web: www.nabt.org

Biology Dictionaries

Access to a good biology dictionary is of great value when dealing with the technical terms used in biology. Below are some biology dictionaries that you may wish to locate or purchase. They can usually be obtained directly from the publisher or they are all available (at the time of printing) from www.amazon.com. For further details of text content, or to make purchases, link to the relevant publisher via Biozone's website: **www.thebiozone.com**

Hale, W.G. **Collins: Dictionary of Biology** 4 ed. 2005, 528 pp. Collins.
ISBN: 0-00-720734-4.
Updated to take in the latest developments in biology and now internet-linked. (§ This latest edition is currently available only in the UK. The earlier edition, ISBN: 0-00-714709-0, is available though amazon.com in North America).

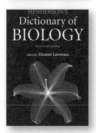

Henderson, E. Lawrence. **Henderson's Dictionary of Biological Terms**, 2008, 776 pp. Benjamin Cummings. **ISBN**: 978-0321505798
This edition has been updated, rewritten for clarity, and reorganised for ease of use. An essential reference and the dictionary of choice for many.

Using BIOZONE's Website

The current internet address (URL) for the web site is displayed here. You can type a new address directly into this space.

Use Google to search for web sites of interest. The more precise your search words are, the better the list of results. EXAMPLE: If you type in "biotechnology", your search will return an overwhelmingly large number of sites, many of which will not be useful to you. Be more specific, e.g. "biotechnology medicine DNA uses".

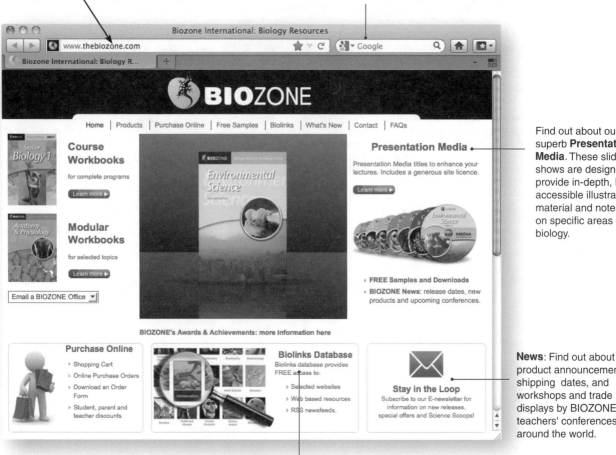

Find out about our superb **Presentation Media**. These slide shows are designed to provide in-depth, highly accessible illustrative material and notes on specific areas of biology.

News: Find out about product announcements, shipping dates, and workshops and trade displays by BIOZONE at teachers' conferences around the world.

Access the **Biolinks** database of web sites related to each major area of biology. It's a great way to quickly find out more on topics of interest.

Weblinks: www.thebiozone.com/weblink/AP1-3114.html

Throughout this workbook, some pages make reference to additional or alternative activities, as well as web sites and periodicals that have particular relevance to the activity. See example of page reference below:

| Periodicals:
Border control
The fluid mosaic model | | Related activities: The Role of Membranes in Cells
Weblinks: Membrane Structure Tutorial | RA 2 |

Periodicals: Full citations are provided in the Appendix for those that wish to read further on a topic.

Web Link: Provides a link to an **external web site** with supporting information for the activity.

Weblinks for AP Biology 1

Some of the activities in your BIOZONE workbook have references to specific websites, listed below under the relevant chapters. These websites (blue links) provide material, generally either animations or video clips, to help you visualize and understand the material presented on the relevant activity page.

For other activities, an extra worksheet, provided as a downloadable PDF file (orange links) can be printed and completed. These provide extension or review of material in the Workbook.

Chapter	Activity	Link
Chapter 1 The Biochemistry of Life	The Biochemical Nature of the Cell The Biochemical Nature of the Cell The Role of Water Nucleotides and Nucleic Acids Amino Acids	A Closer Look at Water The Structure of Water Water and pH DNA Anatomy Amino Acids and
	Proteins Biological Functions of Lipids Lipids Carbohydrate Chemistry Monosaccharides and Disaccharides	Peptide Bond Formation Amino Acids and Proteins Biomolecules: Lipids Formation of Triglycerides Condensation and Hydrolysis Biomolecules: Carbohydrates
Chapter 2 The Internal Structure of Cells	Prokaryotic Cells Animal Cells Prokaryotic Cells Plant Cells Plant Cells	Bacterial Conjugation Animation Eukaryotic Cells Interactive Animation Review of Cell Structure Eukaryotic Cells Interactive Animation Review of Cell Structure

AP Biology Guide

The AP biology curriculum is organized into four underlying principles, or big ideas. The guide below lists the enduring understandings for each big idea, and directs the student as to where to locate material supporting them in the **BIOZONE AP Biology** workbooks. Material located in **AP** Biology **1** is coded **blue**, and material located in **AP** Biology **2** is coded **black**.

Big Idea 1: The process of evolution drives the diversity & unity of life

Enduring Understanding 1A: Change in the genetic makeup of a population over time is evolution

1.A.1 Natural selection is a major mechanism of evolution	Genetic Change in Populations
1.A.2 Natural selection acts on phenotypic variations in populations	
1.A.3 Evolutionary change is also driven by random processes	
1.A.4 Biological evolution is supported by scientific evidence from many disciplines	Evidence for Biological Evolution

Enduring Understanding 1B: Organisms are linked by lines of descent from common ancestry

1.B.1 Organisms share many conserved core processes and features that have evolved	The Relatedness of Organisms
1.B.2 Phylogentic trees and cladograms are graphical models of evolutionary history	

Enduring Understanding 1C: Life continues to evolve within a changing environment

1.C.1 Speciation and extinction have occurred throughout the Earth's history	Speciation & Extinction
1.C.2 Speciation may occur when two populations become reproductively isolated	
1.C.3 Populations continue to evolve	

Enduring Understanding 1D: The origin of living systems is explained by natural processes

1.D.1 There are several scientific hypotheses about the natural origin of life on Earth	The Origin of Living Systems
1.D.2 Scientific evidence from many different disciplines supports models of the origin of life	

Big Idea 2: Biological systems utilize free energy and molecular building blocks to grow, to reproduce and to maintain dynamic homeostasis

Enduring Understanding 2A: Growth, reproduction and maintenance of the organization of living systems require free energy and matter

2.A.1 All living systems require energy	Energy in Living Systems, Homeostasis & Energy
2.A.2 Organisms capture and store free energy for use in biological processes	Energy in Living Systems, Energy Flow & Nutrient Cycles
2.A.3 Energy exchange maintains life processes	The Biochemistry of Life

Enduring Understanding 2B: Growth, reproduction and dynamic homeostasis require that cells create and maintain internal environments that are different from their external environments

2.B.1 Cell membranes are selectively permeable	Cellular Membranes & Transport
2.B.2 Growth and homeostasis are maintained by the movement of molecules across membranes	
2.B.3 Internal membranes in eukaryotic cells partition the cell into specialized regions	The Internal Structure of Cells

Enduring Understanding 2C: Organisms use feedback mechanisms to regulate growth and reproduction, and to maintain dynamic homeostasis

2.C.1 Organisms used feedback mechanisms to maintain their internal environments and respond to change in their environmental changes	Homeostasis & Energy Allocation
2.C.2 Organisms respond to change in their external environments	

Enduring Understanding 2D: Growth and dynamic homeostasis of a biological system are influenced by changes in the system's environment

2.D.1 Biotic and abiotic factors affect biological systems	The Nature of Ecosystems
2.D.2 Homeostatic mechanisms reflect both common ancestry and divergence due to adaptation in different environments	AP2 Part 2: The Physiology & Behavior of Organisms (all chapters)
2.D.3 Biological systems are affected by disruptions to their dynamic homeostasis	
2.D.4 Plants and animals have chemical defenses against infections that affect dynamic homeostasis	Internal Defense, Plant Structure & Adaptation

Enduring Understanding 2E: Many biological processes involved in growth, reproduction & dynamic homeostasis include temporal regulation & coordination

2.E.1 Timing and coordination of events are regulated, and necessary for development	Regulation of Gene Expression
2.E.2 Timing and coordination of physiological events are regulated by multiple mechanisms	Timing, Coordination, & Social Behavior
2.E.3 Timing and coordination of behavior are regulated and are important in natural selection	

Big Idea 3: Living systems store, retrieve, transmit & respond to information essential to life processes

Enduring Understanding 3A: Heritable information provides for continuity of life

3.A.1 DNA, and in some cases RNA, is the primary source of heritable information	DNA and RNA
3.A.2 In eurkaryotes, heritable information is passed to the next generation via the cell cycle and mitosis or meiosis plus fertilization	Chromosomes & Cell Division
3.A.3 The Chromosomal basis of inheritance provides an understanding of the transmission of genes from parent to offspring	Chromosomes & Cell Division, Chromosomal Basis of Inheritance
3.A.4 Inheritance pattern of many traits cannot be explained by simple Mendelian genetics	Chromosomal Basis of Inheritance

Enduring Understanding 3B: Expression of genetic information involves cellular and molecular mechanisms

3.B.1 Gene regulation results in differential gene expression, leading to cell specialization	Regulation of Gene Expression
3.B.2 Signal transmission mediates gene expression	

Enduring Understanding 3C: The processes of genetic information is imperfect and is a source of genetic variation

3.C.1 Genotype changes can alter phenotype	Sources of Variation
3.C.2 Multiple processes increase genetic variation	
3.C.3 Viral replication and genetic variation	

Enduring Understanding 3D: Cells communicate by generating, transmitting and receiving chemical signals

3.D.1 Common features of cell communication reflect shared evolutionary history	Cellular Communication Chromosomal Basis of Inheritance
3.D.2 Cells communicate by direct contact with other cells and by distance via chemical signalling	Cellular Communication
3.D.3 Signal transduction pathways link signal reception with cellular response	
3.D.4 Cell response to signal transduction pathways	

Enduring Understanding 3E: Transmission of information results in changes within and between biological systems

3.E.1 Communicating information with others	Timing, Coordination, & Social Behavior
3.E.2 Nervous systems and responses	Nervous Systems & Responses

---

Big Idea 4: Biological systems interact, and these systems and their interactions possess complex properties

Enduring Understanding 4A: Interactions within biological systems lead to complex properties

4.A.1 Properties of a molecule are determined by its molecular construction	The Biochemistry of Life
4.A.2 The structure and function of subcellular components, and their interactions, provide essential cellular processes	The Internal Structure of Cells
4.A.3 Gene expression results in specialization of cells, tissues and organs	Chromosomal Basis of Inheritance
4.A.4 Organisms exhibit complex properties due to interactions between their constituent parts	AP2 Part 2: The Physiology & Behavior of Organisms (all chapters)
4.A.5 Communities are composed of populations of organisms that interact in complex ways	The Nature of Ecosystems, Energy Flow & Nutrient Cycles, Populations
4.A.6 Matter and energy movements	

Enduring Understanding 4B: Competition and cooperation are important aspects of biological systems

4.B.1 Interactions between molecules affect their structure and function	Enzymes & Metabolism
4.B.2 Cooperative interactions within organisms promote efficiency	Enzymes & Metabolism AP2 Part 2: The Physiology & Behavior of Organisms
4.B.3 Population interactions influence species distribution and abundance	Populations, The Diversity & Stability of Ecosystems
4.B.4 Ecosystem distribution changes over time	The Diversity & Stability of Ecosystems

Enduring Understanding 4C: Naturally occurring diversity among and between components within biological systems affects interactions with the environment

4.C.1 Variation in molecular units provides cells with a wider range of functions	Internal Defense
4.C.2 Environmental factors influence the expression of the genotype in an organism	Chromosomal Basis of Inheritance
4.C.3 Variation in populations affects dynamics	
4.C.4 Species diversity within an ecosystem may influence the ecosystem stability	The Diversity & Stability of Ecosystems

---

Biochemistry and Cell Biology

The Biochemistry of Life
- The role of water
- Organic molecules
- Limitations to cell size

The Internal Structure of Cells
- The characteristics of cell types
- The organization of cells
- Microscopy

Cell Membranes and Transport
- The structure of membranes
- Passive transport processes
- Active transport processes

Cellular Communication
- The basis of cell communication
- Signaling molecules
- Signal transduction

CONNECT *The interaction of component parts of cells leads to the emergence of new properties.*

Evolution

Genetic Change in Populations
- Allele frequencies in gene pools
- Natural selection and evolution
- Special evens in gene pools
- Artificial selection

Evidence for Biological Evolution
- Fossil evidence of evolution
- Molecular evidence of evolution
- Current examples of evolution

The Relatedness of Organisms
- Descent and common ancestry
- Determining phylogeny
- Understanding taxonomy

Speciation and Extinction
- Isolation and speciation
- Sympatric speciation
- Adaptive radiation and extinction

The Origin of Living Systems
- An RNA world
- The origin of eukaryotes
- Tracing Earth's history

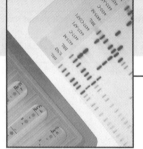

The cell is the basic functional unit of life. Cell communication mediates cell function and gene expression.

Diversity on Earth has arisen through evolution. Classification reflects this history.

Andrew Dunn

AP Biology 1

Evolution is responsible for diversity of life and the relatedness we see at all levels of biological organization.

Cells, organisms, and ecosystems are maintained by the input of matter and free energy.

Storing, retrieving, and transmitting information are essential to the continuation of life's processes.

Complexity and novel properties arise when living systems or their component parts interact.

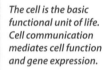

The genetic code is universal among living things. Genomes can be manipulated.

Meiosis creates new allele combinations in the offspring. Changes to DNA can create new alleles.

DNA and RNA
- Genetic code and DNA replication
- Protein synthesis
- Genetic manipulation

Chromosomes & Cell Division
- Mitosis and cell cycle
- Meiosis and genetic variation

Chromosomal Basis of Inheritance
- Mendelian inheritance
- Non-Mendelian inheritance
- Genes and environment

Regulation of Gene Expression
- Cellular differentiation
- Genes and development
- Regulation of transcription

Sources of Variation
- The nature and types of mutation
- Aneuploidy
- Viral replication

CONNECT *Gene expression can be mediated through signal transmission.*

CONNECT *Genetic variation provides the raw material for evolution.*

• Use representations and models • Use mathematics appropriately • Engage in scientific questioning • Plan and implement strategies to collect data • Analyze data and evaluate evidence • Work with scientific explanations and theories • Connect and relate knowledge.

Students can develop competency in these areas through the activities presented in Appendix A and the data handling activities in the workbook.

Molecular Genetics and Inheritance

Appendix A: Science Practices

Metabolism

Energy in Living Systems
- Maintaining order in living systems
- ATP in cells
- Photosynthesis
- Energy yielding pathways

Enzymes and Metabolism
- The characteristics of enzymes
- Enzyme reaction rates
- Cofactors and inhibitors
- Achieving functional efficiency

 RECALL *The internal structure of cells facilitates efficiency of metabolic functions.*

 CONNECT *Organisms use energy to maintain organization, grow, and reproduce.*

CONNECT *Energy availability is a major determinant of ecosystem function.*

Ecology

The Nature of Ecosystems
- Environmental gradients
- Patterns in ecosystems
- Adaptation and niche

Energy Flow and Nutrient Cycles
- Food chains and webs
- Production and trophic efficiency
- Nutrient cycles

Populations
- Features of populations
- Population growth and regulation
- The effects of species interactions

The Diversity and Stability of Ecosystems
- The nature of ecosystem stability
- The implications of disturbance
- Human impact: ecosystem responses and survival

 CONNECT *Feedback mechanisms operate at all levels of biological organization.*

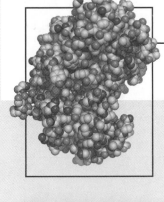

Metabolic pathways for capture and storage of free energy involve enzyme-catalyzed reactions.

Diversity and interaction in biological systems contribute to flexibility and resilience.

AP Biology 2

The diversity of life is a product of evolution. Diversity at all levels creates complexity and resilience in biological systems.

Dynamic homeostasis is maintained by the input of matter and free energy.

Individuals act on information and communicate it to others.

Interactions at the level of molecules, organisms, and ecosystems contribute to complexity, efficiency of function, and biological resilience.

Homeostasis is maintained through system interactions involving use of free energy.

Growth, reproduction, and homeostasis often involve temporal regulation and coordination.

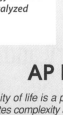

Homeostasis and Energy Allocation
- Principles of homeostasis
- Survival, fitness, and energy allocation

Plant Structure and Adaptation
- Plant structure and function
- Plant adaptation
- Plant defenses

Obtaining Nutrients and Eliminating Wastes
- Gas exchange
- Nutrition
- Osmoregulation
- Excretion

Interactions in Physiological Systems
- Blood: connecting the body's systems
- The role of the liver

Internal Defense
- Specific defenses
- Non-specific defenses

Coordination Timing, and Social Behavior
- Responses and rhythms
- Migration and homing
- The role of communication
- Cooperation and survival

Nervous Systems and Responses
- Neural communication
- The brain
- Sensory systems
- Nerves and muscles

 CONNECT *Homeostatic mechanisms reflect common ancestry and adaptation.*

Science practices are integrated throughout AP® Biology 2 in context by way of activities involving data handling, evaluation, and interpretation.

Physiology and Behavior

Science Practices

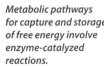

The Biochemistry of Life

Water and inorganic ions
- The structure and properties of water
- Water as the universal solvent
- Inorganic ions in living systems

Organic molecules
- Condensation and hydrolysis
- Carbohydrates, lipids, nucleic acids
- Amino acids and proteins
- SA:V ratios and the size of cells

Connect 2.A.3 to 4.A.1
DNA and RNA have structural similarities and differences that determine function.

Connect 2.A.3 with 3.A.1
The subcomponents of a biological molecule and their sequence determine the properties of that molecule.

Cellular Communication

The role of cell communication
- Cellular response to environment
- Coordination: cell to organism

Cell signaling
- Principles of cell signaling
- Cellular communication in bacteria

Signaling molecules
- Neurotransmitters
- Neurohormones
- Hormones and their effects

Signal transduction
- First messenger systems
- Second messenger systems
- Errors in cell signaling

Connect 3.D.1 with 1.B.1
Processes in cellular communication share features that reflect common ancestry.

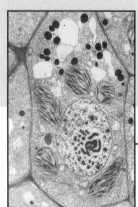

The structure of the molecules that make up the cells and tissues of organisms determines their functional role.

Cells communicate directly and indirectly by generating, transmitting, and receiving chemical signals.

Biochemistry and Cell Biology

Important in this section...

- *Understand the roles of biological molecules*
- *Develop your understanding of cellular structure*
- *Understand how membranes regulate cell transport*
- *Understand how and why cells communicate*

The cell is the functional basic unit of life. Cells share common features but are characterized by their unique features.

Membranes regulate exchange processes in cells. These exchanges are needed to maintain function.

The diversity of cells
- Characteristics of cells
- The sizes of cells
- Prokaryotic vs eukaryotic cells
- Features of plant and animal cells

Cellular structures and organelles
- Cell structures and organelles
- Interaction and function
- Compartments and special functions

Microscopy
- The role of practical microscopy
- Recognizing cell structures
- Making biological drawings

Connect 2.B.3 with 4.A.2
Subcellular components interact to carry out essential processes, e.g. secretion.

Membranes in cells
- How we know about membranes
- The structure of cellular membranes
- The roles of membranes in cells

Passive transport processes
- Concentration gradients
- Diffusion and facilitated diffusion
- Osmosis and water movements

Active transport processes
- Energy and transport processes
- Ion pumps and coupled transport
- Exocytosis and endocytosis

Connect 2.B.1 with 4.A.1
Selective permeability is a consequence of membrane structure, which is governed by the properties of its molecular components.

Connect 2.B.2 with 2.A.1
Active transport requires input of energy, which is provided by other processes.

The Internal Structure of Cells

Cell Membranes and Transport

The Biochemistry of **Life**

Key concepts

▶ Organic molecules are carbon-containing molecules and are central to living systems.

▶ Water's properties make it essential to life.

▶ Organisms maximize surface area: volume ratios to maintain effective exchange rates.

▶ Biological molecules have specific roles in cells.

▶ The components of bio-polymers and their sequence determine the properties of the molecule.

▶ Biological macromolecules have directionality as a function of their constituent monomers.

Key terms

amino acid
amphipathic
carbohydrate
carbon
denaturation
disaccharide
ester bond
fibrous protein
globular protein
glycosidic bond
hydrogen bond
inorganic ion
lipid
macromolecule
monomer
monosaccharide
nitrogen
nucleic acid
nucleotide
organic molecule
peptide bond
phospholipid
phosphorus
polar molecule
polymer
polysaccharide
protein
surface area: volume ratio
water

Essential Knowledge

☐ 1. Use the **KEY TERMS** to compile a glossary for this topic.

The Role of Biological Molecules *(2.A.3: a)*   pages 10-30

☐ 2. Identify the common elements found in organisms and give examples of where these elements occur in cells.

☐ 3. Explain how carbon moves from the environment to organisms and its role as the essential component of all **organic molecules**. Describe the importance of **organic molecules** and **inorganic ions** in biological systems.

☐ 4. Describe how **nitrogen** moves from the environment to organisms and where it is used in the synthesis of **nucleic acids** and **proteins**.

☐ 5. Describe how **phosphorus** moves from the environment to organisms and where it is used in the synthesis of **nucleic acids** and some **lipids**.

☐ 6. With reference to #3-5, describe the basic structure and roles of **carbohydrates, amino acids, proteins, lipids, nucleotides,** and **nucleic acids**. Recognize simple tests to detect these molecules in food.

☐ 7. Using examples, explain the dependence of living systems on the properties of **water**. Relate the properties of **water** to its structure, including its **dipole** nature.

Biological Macromolecules *(4.A.1)*   pages 15-21, 23-24, 28-30

☐ 8. Distinguish between **monomers** and **polymers**. Describe the range of **macromolecules** produced by cells and explain how these are used. Describe the synthesis of macromolecules by **condensation** and their breakdown by **hydrolysis**. Identify the bonds formed or broken in each case.

☐ 9. With reference to nucleic acids and proteins, explain how the properties of a polymer are determined by its component monomers.

☐ 10. Explain the biological significance of the **amphipathic** nature of some lipids, e.g. **phospholipids**, cholesterol.

☐ 11. Using examples, explain how the properties of a polysaccharide are determined by the monomers present and the nature of the bonds between them.

☐ 12. Explain how directionality in component monomers (e.g. nucleotides, amino acids, monosaccharides) influences the structure and function of a polymer.

Surface Area and Volume *(2.A.3: b)*   pages 31-33

☐ 13. Describe the relationship between surface area and volume in a unit structure, such as a cell. Explain how **surface area: volume ratio** affects exchanges in biological systems (e.g. between a cell and its environment).

☐ 14. Using examples, explain the adaptations of multicellular organisms to maintain exchanges (e.g. of gases or nutrients) with the environment.

Periodicals:
Listings for this chapter are on page 380

Weblinks:
www.thebiozone.com/
weblink/AP1-3114.html

BIOZONE APP:
Student Review Series
Molecules of Life

The Biochemical Nature of the Cell

The molecules that make up living things can be grouped into five broad classes: carbohydrates, lipids, proteins, nucleic acids, and water. Water is the main component of organisms and provides an environment in which metabolic reactions can occur. Apart from water, most other substances in cells are compounds of carbon, hydrogen, oxygen, nitrogen and phosphorus. These elements form strong, stable covalent bonds by sharing electrons. The combination of carbon atoms with the atoms of other elements provides a huge variety of molecular structures. Many of these **biological molecules**, e.g. DNA, are very large and contain millions of atoms. The role of these molecules in cells is outlined below on the diagram of a typical plant cell.

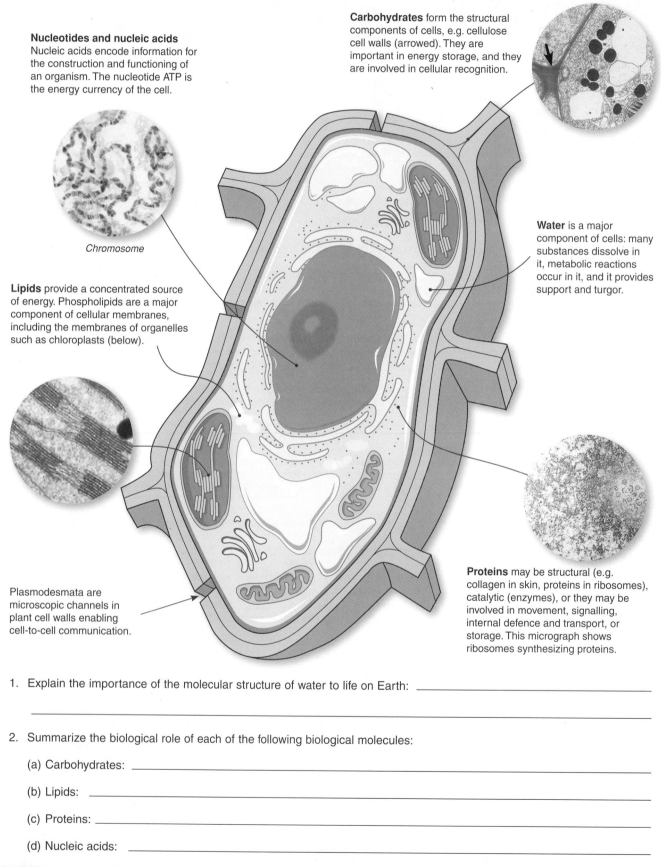

Nucleotides and nucleic acids
Nucleic acids encode information for the construction and functioning of an organism. The nucleotide ATP is the energy currency of the cell.

Chromosome

Carbohydrates form the structural components of cells, e.g. cellulose cell walls (arrowed). They are important in energy storage, and they are involved in cellular recognition.

Water is a major component of cells: many substances dissolve in it, metabolic reactions occur in it, and it provides support and turgor.

Lipids provide a concentrated source of energy. Phospholipids are a major component of cellular membranes, including the membranes of organelles such as chloroplasts (below).

Plasmodesmata are microscopic channels in plant cell walls enabling cell-to-cell communication.

Proteins may be structural (e.g. collagen in skin, proteins in ribosomes), catalytic (enzymes), or they may be involved in movement, signalling, internal defence and transport, or storage. This micrograph shows ribosomes synthesizing proteins.

1. Explain the importance of the molecular structure of water to life on Earth: _____

   _____

2. Summarize the biological role of each of the following biological molecules:

   (a) Carbohydrates: _____

   (b) Lipids: _____

   (c) Proteins: _____

   (d) Nucleic acids: _____

Related activities: The Role of Water, Organic Molecules
Weblinks: A Closer Look at Water

The Role of Water

Water is the most abundant of the smaller molecules making up living things, and typically makes up about two-thirds of any organism. Water is a liquid at room temperature and many substances dissolve in it. It is a medium inside cells and for aquatic life. Water takes part in, and is a common product of, many reactions. Water molecules are **polar** and have a weak attraction for each other and inorganic ions, forming large numbers of weak hydrogen bonds. It is this feature that gives water many of its unique properties, including its low viscosity and its chemical behavior as a **universal solvent**.

Important Properties of Water

A lot of energy is required before water will change state so aquatic environments are thermally stable and sweating and transpiration cause rapid cooling.

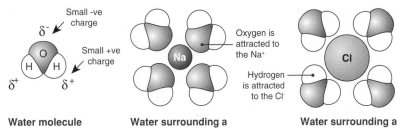

Water molecule
Formula: H_2O

Water surrounding a positive ion (Na⁺)

Oxygen is attracted to the Na⁺

Hydrogen is attracted to the Cl⁻

Water surrounding a negative ion (Cl⁻)

The most important feature of the chemical behaviour of water is its dipole nature. It has a small positive charge on each of the two hydrogens and a small negative charge on the oxygen.

Water is colorless, with a high transmission of visible light, so light penetrates tissue and aquatic environments.

Ice is less dense than water. Consequently ice floats, insulating the underlying water and providing valuable habitat.

Water has low viscosity, strong cohesive properties, and high surface tension. It can flow freely through small spaces.

1. On the diagram above, showing a positive and a negative ion surrounded by water molecules, indicate the polarity of the water molecules (as shown in the example provided).

2. Explain the importance of the **dipole nature** of water molecules to the chemistry of life: _____
_____
_____

3. For (a)-(f), identify the important property of water, and describe an example of that property's biological significance:

   (a) Property important in the clarity of seawater: _____

      Biological significance: _____

   (b) Property important in the transport of water in xylem: _____

      Biological significance: _____

   (c) Property important in the relatively stable temperature of bodies of water: _____

      Biological significance: _____

   (d) Property important in the transport of glucose around the body: _____

      Biological significance: _____

   (e) Property important in the cooling effect of evaporation: _____

      Biological significance: _____

   (f) Property important in ice floating: _____

      Biological significance: _____

Periodicals:
Water, life, & H bonding

Related activities: Biochemical Nature of the Cell, Organic Molecules
Weblinks: The Structure of Water, Water and pH

RA 2

Organic Molecules

Organic molecules are those chemical compounds containing carbon that are found in living things. Specific groups of atoms, called **functional groups**, attach to a carbon-hydrogen core and confer specific chemical properties on the molecule. Some organic molecules in organisms are small and simple, containing only one or a few functional groups, while others are large complex assemblies called **macromolecules**. The macromolecules that make up living things can be grouped into four classes: carbohydrates, lipids, proteins, and nucleic acids. An understanding of the structure and function of these molecules is necessary to many branches of biology, especially biochemistry, physiology, and molecular genetics. The diagram below illustrates some of the common ways in which biological molecules are portrayed. Note that the **molecular formula** expresses the number of atoms in a molecule, but does not convey its structure; this is indicated by the **structural formula**. Molecules can also be represented as **models**. A ball and stick model shows the arrangement and type of bonds while a space filling model gives a more realistic appearance of a molecule, showing how close the atoms really are.

Portraying Biological Molecules

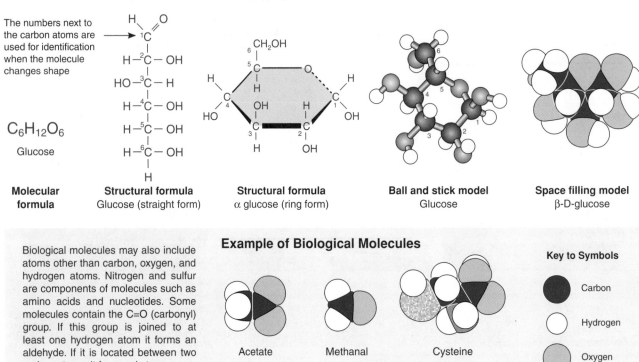

1. Which three main elements make up the structure of organic molecules? _____

2. Name two other elements that are also frequently part of organic molecules: _____

3. State how many covalent bonds a carbon atom can form with neighboring atoms: _____

4. Distinguish between molecular and structural formulae for a given molecule: _____

5. What is a functional group? _____

6. Classify methanal according to the position of the C=O group: _____

7. Identify a functional group always present in amino acids: _____

8. Identify the significance of cysteine in its formation of disulfide bonds: _____

Related activities: The Biochemical Nature of the Cell, Monosaccharides and Disaccharides

© BIOZONE International 2012
ISBN: 978-1-927173-11-4
Photocopying Prohibited

Building an Organism

Living organisms are very complex biological structures, containing many different components. The major components are the four macromolecules: proteins, nucleic acids, carbohydrates, and lipids. In addition, organisms also contain many smaller components, such as inorganic ions and water. In order to carry out life processes, organisms must obtain nutrients from the environment. In animals, this is achieved by the consumption of other organisms. In plants, nutrients are obtained from the soil (via roots) or from the atmosphere.

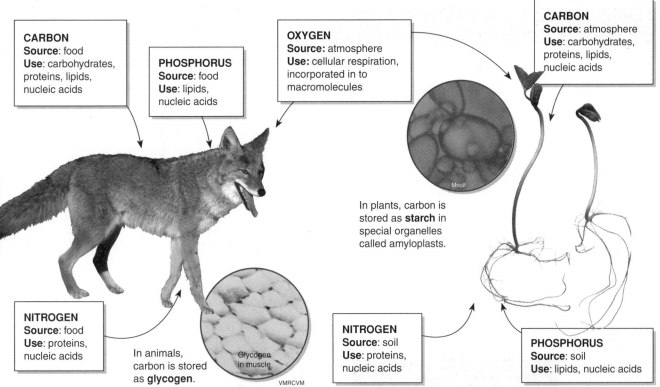

CARBON
Source: food
Use: carbohydrates, proteins, lipids, nucleic acids

PHOSPHORUS
Source: food
Use: lipids, nucleic acids

OXYGEN
Source: atmosphere
Use: cellular respiration, incorporated in to macromolecules

CARBON
Source: atmosphere
Use: carbohydrates, proteins, lipids, nucleic acids

Mnolf

In plants, carbon is stored as **starch** in special organelles called amyloplasts.

NITROGEN
Source: food
Use: proteins, nucleic acids

In animals, carbon is stored as **glycogen**.

Glycogen in muscle
VMRCVM

NITROGEN
Source: soil
Use: proteins, nucleic acids

PHOSPHORUS
Source: soil
Use: lipids, nucleic acids

Inorganic ions are important for the structure and metabolism of all living organisms. An ion is simply an atom (or group of atoms) that has gained or lost one or more electrons. Many of these ions are water soluble.

Some of the inorganic ions required by organisms and examples of their biological roles are described in this table (right). A deficiency in any of these ions can result in specific deficiency disorders.

Ion	Name	Example of Biological Roles
Ca^{2+}	Calcium	Component of bones and teeth, required for muscle contraction
Mg^{2+}	Magnesium	Component of chlorophyll, role in energy metabolism
Fe^{2+}	Iron (II)	Component of hemoglobin and cytochromes
NO_3^-	Nitrate	Component of amino acids
Na^+	Sodium	Component of extracellular fluid and the need for nerve function
K^+	Potassium	Important intracellular ion, needed for heart and nerve function
Cl^-	Chloride	Component of extracellular fluid in multicellular organisms

1. State the main source of carbon, phosphorus, and nitrogen for animals: _____

2. (a) State the main source of carbon for plants: _____

   (b) State the main source of phosphorus and nitrogen for plants: _____

3. List the four main macromolecule components of living organisms: _____

   _____

4. Explain why carbon is so important for building the molecular components of an organism: _____

   _____

   _____

Biochemical Tests

Biochemical tests are used to detect the presence of nutrients such as lipids, proteins, and carbohydrates (sugar and starch) in various foods. These simple tests are useful for detecting nutrients when large quantities are present. A more informative technique by which to separate a mixture of compounds involves **chromatography**. Chromatography is used when only a small sample is available or when you wish to distinguish between nutrients. Simple biochemical food tests will show whether sugar is present, whereas chromatography will distinguish between the different types of sugars (e.g. fructose or glucose).

Paper Chromatography

Set Up and Procedure

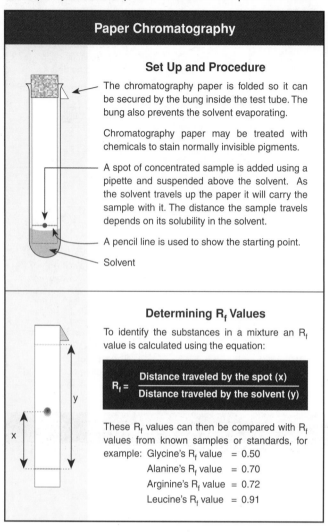

The chromatography paper is folded so it can be secured by the bung inside the test tube. The bung also prevents the solvent evaporating.

Chromatography paper may be treated with chemicals to stain normally invisible pigments.

A spot of concentrated sample is added using a pipette and suspended above the solvent. As the solvent travels up the paper it will carry the sample with it. The distance the sample travels depends on its solubility in the solvent.

A pencil line is used to show the starting point.

Solvent

Determining R_f Values

To identify the substances in a mixture an R_f value is calculated using the equation:

$$R_f = \frac{\text{Distance traveled by the spot } (x)}{\text{Distance traveled by the solvent } (y)}$$

These R_f values can then be compared with R_f values from known samples or standards, for example:

Glycine's R_f value = 0.50
Alanine's R_f value = 0.70
Arginine's R_f value = 0.72
Leucine's R_f value = 0.91

Simple Food Tests

Proteins: The Biuret Test

Reagent:	Biuret solution.
Procedure:	A sample is added to biuret solution and gently heated.
Positive result:	Solution turns from blue to lilac.

Starch: The Iodine Test

Reagent:	Iodine.
Procedure:	Iodine solution is added to the sample.
Positive result:	Blue-black staining occurs.

Lipids: The Emulsion Test

Reagent:	Ethanol.
Procedure:	The sample is shaken with ethanol. After settling, the liquid portion is distilled and mixed with water.
Positive result:	The solution turns into a cloudy-white emulsion of suspended lipid molecules.

Sugars: The Benedict's Test

Reagent:	Benedict's solution.
Procedure:	*Non reducing sugars*: The sample is boiled with dilute hydrochloric acid, then cooled and neutralized. A test for reducing sugars is then performed. *Reducing sugars*: Benedict's solution is added, and the sample is placed in a water bath.
Positive result:	Solution turns from blue to orange.

1. Calculate the R_f value for the example given in the diagram above (show your working): _____

_____

2. Explain why the R_f value of a substance is always less than 1: _____

_____

3. Discuss when it is appropriate to use chromatography instead of a simple food test: _____

_____

_____

4. Predict what would happen if a sample was immersed in the chromatography solvent, instead of suspended above it:

_____

_____

5. With reference to their R_f values, rank the four amino acids (listed above) in terms of their solubility in the solvent used:

_____

6. Outline why lipids must be mixed in ethanol before they will form an emulsion in water: _____

_____

Related activities: Monosaccharides and Disaccharides

© BIOZONE International 2012
ISBN: 978-1-927173-11-4
Photocopying Prohibited

Nucleotides and Nucleic Acids

Nucleic acids are a special group of chemicals in cells concerned with the transmission of inherited information. They have the capacity to store the information that controls cellular activity. The central nucleic acid is called **deoxyribonucleic acid** (DNA). DNA is a major component of chromosomes and is found primarily in the nucleus, although a small amount is found in mitochondria and chloroplasts. Other **ribonucleic acids** (RNA) are involved in the 'reading' of the DNA information. All nucleic acids are made up of simple repeating units called **nucleotides**, linked together to form chains or strands, often of great length (see the activity *DNA Molecules*). The strands vary in the sequence of the bases found on each nucleotide. It is this sequence which provides the 'genetic code' for the cell. In addition to nucleic acids, certain nucleotides and their derivatives are also important as suppliers of energy (**ATP**) or as hydrogen ion and electron carriers in respiration and photosynthesis (NAD, NADP, and FAD).

Chemical Structure of a Nucleotide

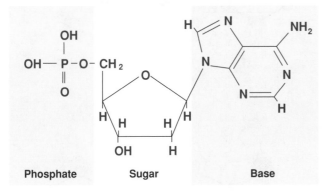

Phosphate   Sugar   Base

Symbolic Form of a Nucleotide

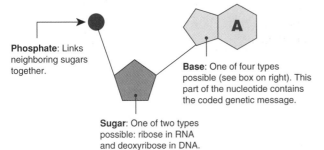

Phosphate: Links neighboring sugars together.

Base: One of four types possible (see box on right). This part of the nucleotide contains the coded genetic message.

Sugar: One of two types possible: ribose in RNA and deoxyribose in DNA.

Nucleotides are the building blocks of DNA. Their precise sequence in a DNA molecule provides the genetic instructions for the organism to which it governs. Accidental changes in nucleotide sequences are a cause of mutations, usually harming the organism, but occasionally providing benefits.

Bases

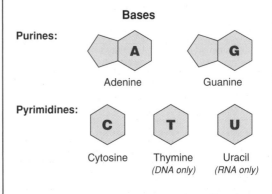

Purines: Adenine   Guanine

Pyrimidines: Cytosine   Thymine *(DNA only)*   Uracil *(RNA only)*

The two-ringed bases above are **purines**. The single-ringed bases are **pyrimidines**. Although only one of four kinds of base can be used in a nucleotide, **uracil** is found only in RNA, replacing **thymine**. DNA contains A, T, G, and C, while RNA contains A, U, G, and C.

Sugars

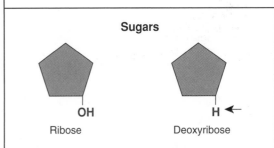

Ribose   Deoxyribose

Deoxyribose sugar is found only in DNA. It differs from **ribose** sugar, found in RNA, by the lack of a single oxygen atom (arrowed).

RNA Molecule

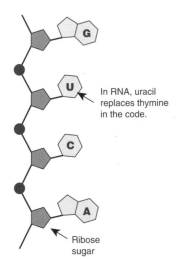

In RNA, uracil replaces thymine in the code.

Ribose sugar

DNA Molecule

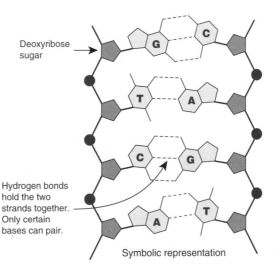

Deoxyribose sugar

Hydrogen bonds hold the two strands together. Only certain bases can pair.

Symbolic representation

DNA Molecule

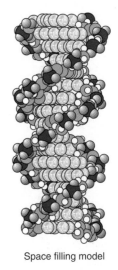

Space filling model

Ribonucleic acid (RNA) comprises a *single strand* of nucleotides linked together.

Deoxyribonucleic acid (DNA) comprises a *double strand* of nucleotides linked together. It is shown unwound in the symbolic representation (left). The DNA molecule takes on a twisted, double-helix shape as shown in the space filling model on the right.

Related activities: DNA Molecules, Creating a DNA Molecule
Weblinks: DNA Anatomy

Formation of a nucleotide

Condensation
(water removed)

A nucleotide is formed when phosphoric acid and a base are chemically bonded to a sugar molecule. In both cases, water is given off, and they are therefore condensation reactions. In the reverse reaction, a nucleotide is broken apart by the addition of water (**hydrolysis**).

Formation of a dinucleotide

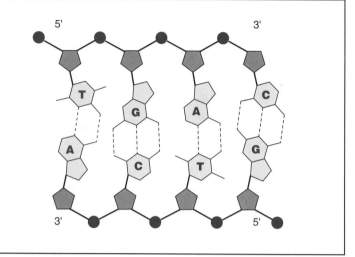

Two nucleotides are linked together by a condensation reaction between the phosphate of one nucleotide and the sugar of another.

Double-Stranded DNA

The double-helix structure of DNA is like a ladder twisted into a corkscrew shape around its longitudinal axis. It is 'unwound' here to show the relationships between the bases.

- The DNA backbone is made up from alternating phosphate and sugar molecules, giving the DNA molecule an asymmetrical structure.

- The asymmetrical structure gives a DNA strand a **direction**. Each strand runs in the opposite direction to the other.

- The ends of a DNA strand are labeled the 5' (five prime) and 3' (three prime) ends. The **5'** end has a terminal phosphate group (off carbon 5), the **3'** end has a terminal hydroxyl group (off carbon 3).

- The way the pairs of bases come together to form hydrogen bonds is determined by the number of bonds they can form and the configuration of the bases.

1. The diagram above depicts a double-stranded DNA molecule. Label the following parts on the diagram:
   - (a) **Sugar** (deoxyribose)
   - (b) **Phosphate**
   - (c) **Hydrogen bonds** (between bases)
   - (d) **Purine** bases
   - (e) **Pyrimidine** bases

2. (a) Explain the **base-pairing rule** that applies in double-stranded DNA: _____

   _____

   (b) How is the base-pairing rule for mRNA different? _____

   (c) What is the purpose of the hydrogen bonds in double-stranded DNA? _____

   _____

3. Describe the functional role of nucleotides: _____

   _____

4. (a) Why do the DNA strands have an asymmetrical structure? _____

   _____

   (b) What are the differences between the 5' and 3' ends of a DNA strand? _____

   _____

5. Complete the following table summarizing the differences between DNA and RNA molecules:

	DNA	RNA
Sugar present		
Bases present		
Number of strands		
Relative length		

Amino Acids

Amino acids are the basic units from which proteins are made. Plants can manufacture all the amino acids they require from simpler molecules, but animals must obtain a certain number of ready-made amino acids (called **essential amino acids**) from their diet. Which amino acids are essential varies from species to species, as different metabolisms are able to synthesize different substances. The distinction between essential and non-essential amino acids is somewhat unclear though, as some amino acids can be produced from others and some are interconvertible by the urea cycle. Amino acids can combine to form peptide chains in a **condensation reaction**. The reverse reaction, which breaks up peptide chains, uses water and is called **hydrolysis**.

Structure of Amino Acids

There are over 150 amino acids found in cells, but only 20 occur commonly in proteins. The remaining, non-protein amino acids have roles as intermediates in metabolic reactions, or as neurotransmitters and hormones. All amino acids have a common structure (see right). The only difference between the different types lies with the 'R' group in the general formula. This group is variable, which means that it is different in each kind of amino acid.

General structure of an amino acid

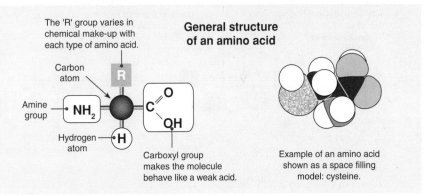

The 'R' group varies in chemical make-up with each type of amino acid.

Carbon atom

R

Amine group — NH_2

Hydrogen atom — H

C O OH

Carboxyl group makes the molecule behave like a weak acid.

Example of an amino acid shown as a space filling model: cysteine.

Properties of Amino Acids

Three examples of amino acids with different chemical properties are shown right, with their specific 'R' groups outlined. The 'R' groups can have quite diverse chemical properties.

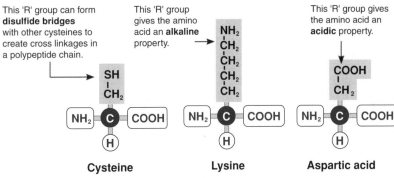

This 'R' group can form **disulfide bridges** with other cysteines to create cross linkages in a polypeptide chain.

SH | CH_2

NH_2 — C — COOH — H

Cysteine

This 'R' group gives the amino acid an **alkaline** property.

NH_2 | CH_2 | CH_2 | CH_2 | CH_2

NH_2 — C — COOH — H

Lysine

This 'R' group gives the amino acid an **acidic** property.

COOH | CH_2

NH_2 — C — COOH — H

Aspartic acid

A polypeptide chain

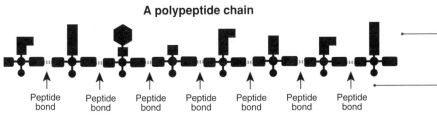

The order of amino acids in a protein is determined by the order of nucleotides in DNA and mRNA.

Peptide bonds link amino acids together in long polymers called polypeptide chains.

Peptide bond (×7)

The amino acids are linked together by peptide bonds to form long chains of up to several hundred amino acids (called polypeptide chains). These chains may be functional units (complete by themselves) or they may need to be joined to other polypeptide chains before they can carry out their function. In humans, not all amino acids can be manufactured by our body: ten must be taken in with our diet (eight in adults). These are the 'essential amino acids'. They are indicated by the symbol ♦ on the right. Those indicated with as asterisk (*) are also required by infants.

Amino acids occurring in proteins

Alanine	Glutamine	Leucine ♦	Serine
Arginine *	Glutamic acid	Lysine ♦	Threonine ♦
Asparagine	Glycine	Methionine ♦	Tryptophan ♦
Aspartic acid	Histidine *	Phenylalanine ♦	Tyrosine
Cysteine	Isoleucine ♦	Proline	Valine ♦

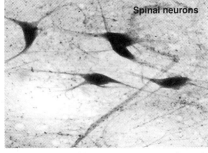

Spinal neurons

Several amino acids act as neurotransmitters in the central nervous system. Glutamic acid and GABA (gamma amino butyric acid) are the most common neurotransmitters in the brain. Others, such as glycine, are restricted to the spinal cord.

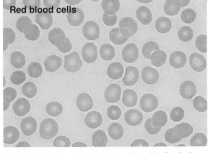

Red blood cells

Amino acids tend to stabilise the pH of solutions in which they are present (e.g. blood and tissue fluid) because they will remove excess H^+ or OH^- ions. They retain this buffer capacity even when incorporated into peptides and proteins.

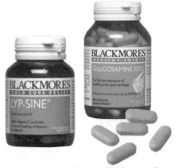

Amino acids are widely available as dietary supplements for specific purposes. Lysine is sold as relief for herpes infections and glucosamine supplements are used for alleviating the symptoms of arthritis and other joint disorders.

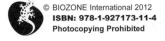
© BIOZONE International 2012
ISBN: 978-1-927173-11-4
Photocopying Prohibited

Related activities: Proteins, Translation
Web links: Amino Acids and Peptide Bond Formation

Condensation and Hydrolysis Reactions

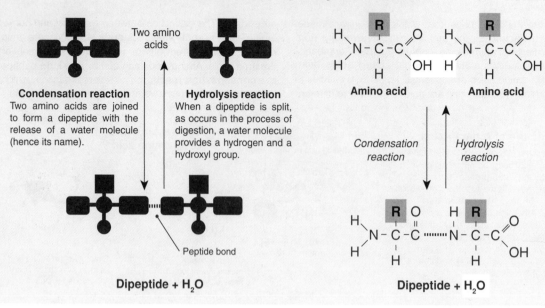

Condensation reaction
Two amino acids are joined to form a dipeptide with the release of a water molecule (hence its name).

Hydrolysis reaction
When a dipeptide is split, as occurs in the process of digestion, a water molecule provides a hydrogen and a hydroxyl group.

Peptide bond

Dipeptide + H₂O

Amino acid   Amino acid

Condensation reaction   Hydrolysis reaction

Dipeptide + H₂O

1. Discuss the various biological roles of amino acids: _____

_____

_____

_____

_____

_____

_____

2. Describe what makes each of the 20 amino acids found in proteins unique: _____

_____

_____

3. Describe the process that determines the sequence of amino acids in a polypeptide chain:_____

_____

_____

4. Explain how the chemistry of amino acids enables them to act as buffers in biological tissues: _____

_____

_____

5. Giving examples, explain what is meant by an **essential amino acid**: _____

_____

_____

6. Describe the processes by which amino acids are joined together and broken down: _____

_____

_____

_____

_____

Proteins

The precise folding of a protein into its **tertiary structure** creates a three dimensional arrangement of the active 'R' groups. It is this structure that gives a protein its unique chemical properties. If a protein loses this precise structure (through **denaturation**), it is usually unable to carry out its biological function. Proteins can be classified on the basis of structure (e.g. globular vs fibrous) or function, as described on the next page. The entire collection of proteins in a particular cell type is termed the **cellular proteome**. An organism's proteome is likely to be larger than its genome in the sense that there are more proteins than genes. This is the result of alternative splicing of genes (which can give rise to several proteins) and modifications made to proteins after they are translated, such as phosphorylation and glycosylation.

Primary Structure - 1° (amino acid sequence)
Strings of many amino acids link together with peptide bonds to form molecules called polypeptide chains. There are 20 different kinds of amino acids that can be linked together in a vast number of different combinations. This sequence is called the primary structure. It is the arrangement of attraction and repulsion points in the amino acid chain that determines the higher levels of organisation in the protein and its biological function.

Secondary Structure - 2° (α-helix or β-pleated sheet)
Polypeptides become folded in various ways, referred to as the secondary (2°) structure. The most common types of 2° structures are a coiled α-**helix** and a β-**pleated sheet**. Secondary structures are maintained with hydrogen bonds between neighbouring CO and NH groups. H-bonds, although individually weak, provide considerable strength when there are a large number of them. The example, right, shows the two main types of secondary structure. In both, the **'R' side groups** (not shown) project out from the structure. Most globular proteins contain regions of α-helices together with β-sheets. Keratin (a fibrous protein) is composed almost entirely of α-helices. Fibroin (silk protein), another fibrous protein, is almost entirely in β-sheet form.

Tertiary Structure - 3° (folding)
Every protein has a precise structure formed by the folding of the secondary structure into a complex shape called the tertiary structure. The protein folds up because various points on the secondary structure are attracted to one another. The strongest links are caused by bonding between cysteine amino acids which form disulfide bridges. Other interactions that are involved in folding include weak ionic and hydrogen bonds as well as hydrophobic interactions.

Quaternary Structure - 4°
Some proteins (such as enzymes) are complete and functional with a tertiary structure only. However, many complex proteins exist as aggregations of polypeptide chains. The arrangement of the polypeptide chains into a functional protein is termed the **quaternary structure**. The example (right) shows a molecule of hemoglobin, a globular protein composed of 4 polypeptide sub-units joined together; two identical **beta chains** and two identical **alpha chains**. Each has a heme (iron containing) group at the center of the chain, which binds oxygen. Proteins containing non-protein material are **conjugated proteins**. The non-protein part is the **prosthetic group**.

Denaturation of Proteins
Denaturation refers to the loss of the three-dimensional structure (and usually also the biological function) of a protein. Denaturation is often, although not always, permanent. It results from an alteration of the bonds that maintain the secondary and tertiary structure of the protein, even though the sequence of amino acids remains unchanged. Agents that cause denaturation are:
- **Strong acids and alkalis:** Disrupt ionic bonds and result in coagulation of the protein. Long exposure also breaks down the primary structure of the protein.
- **Heavy metals:** May disrupt ionic bonds, form strong bonds with the carboxyl groups of the R groups, and reduce protein charge. The general effect is to cause the precipitation of the protein.
- **Heat and radiation (e.g. UV):** Cause disruption of the bonds in the protein through increased energy provided to the atoms.
- **Detergents and solvents:** Form bonds with the non-polar groups in the protein, thereby disrupting hydrogen bonding.

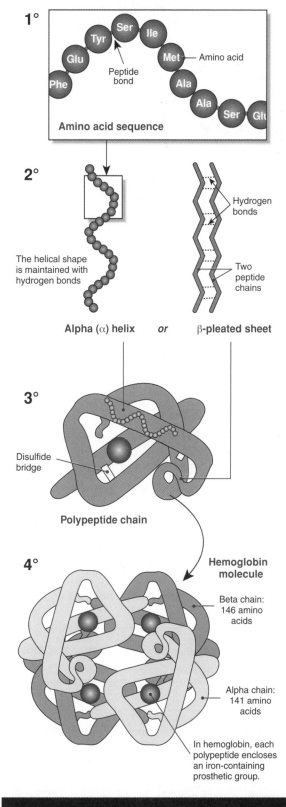

1°
Tyr · Ser · Ile · Glu · Met — Amino acid · Phe · Peptide bond · Ala · Ala · Ser · Glu

Amino acid sequence

2°
The helical shape is maintained with hydrogen bonds

Hydrogen bonds

Two peptide chains

Alpha (α) helix *or* **β-pleated sheet**

3°
Disulfide bridge

Polypeptide chain

4°
Hemoglobin molecule

Beta chain: 146 amino acids

Alpha chain: 141 amino acids

In hemoglobin, each polypeptide encloses an iron-containing prosthetic group.

Hemoglobin's Chemical Formula:

$$C_{3032}H_{4816}O_{872}N_{780}S_8Fe_4$$

The Biochemistry of Life

Periodicals:
What is tertiary structure?

Related activities: Amino Acids
Weblinks: Amino Acids and Proteins

RA 2

Structural Classification of Proteins

Fibrous Proteins	Globular Proteins
Properties • Water insoluble • Very tough physically; may be supple or stretchy • Parallel polypeptide chains in long fibers or sheets **Function** • Structural role in cells and organisms *e.g. collagen found in connective tissue, cartilage, bones, tendons, and blood vessel walls.* • Contractile *e.g. myosin, actin*	**Properties** • Easily water soluble • Tertiary structure critical to function • Polypeptide chains folded into a three-dimensional shape **Function** • Catalytic *e.g. enzymes* • Regulatory *e.g. hormones (insulin)* • Transport *e.g. hemoglobin* • Protective *e.g. antibodies*

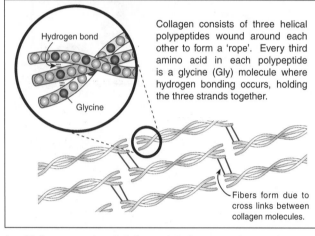

Collagen consists of three helical polypeptides wound around each other to form a 'rope'. Every third amino acid in each polypeptide is a glycine (Gly) molecule where hydrogen bonding occurs, holding the three strands together.

Fibers form due to cross links between collagen molecules.

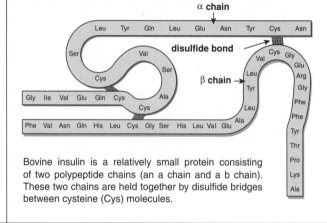

Bovine insulin is a relatively small protein consisting of two polypeptide chains (an a chain and a b chain). These two chains are held together by disulfide bridges between cysteine (Cys) molecules.

1. Giving examples, briefly explain how proteins are involved in the following functional roles:

   (a) Structural tissues of the body: _____

   _____

   (b) Regulating body processes: _____

   _____

   (c) Contractile elements: _____

   _____

   (d) Immunological response to pathogens: _____

   _____

   (e) Transporting molecules within cells and in the bloodstream: _____

   _____

   (f) Catalyzing metabolic reactions in cells: _____

   _____

2. Explain denaturation and its effect on protein function: _____

   _____

   _____

3. Why are fibrous proteins important as structural molecules both within cells and extracellularly? ____

   _____

   _____

4. Suggest why many globular proteins, in contrast to fibrous proteins, have a catalytic or regulatory role:

   _____

   _____

Modification of Proteins

Proteins may be modified after they have been produced by the ribosomes. This is termed **post translational modification**. Two important modifications are the addition of carbohydrates or lipids to the protein. **Glycoproteins** are formed by adding carbohydrates to proteins once they pass into the interior of rough endoplasmic reticulum. The carbohydrates may help position and orientate the glycoproteins in membranes, they may help guide a protein to its final destination, and they have roles in intercellular recognition and cell signalling. Other proteins may have fatty acids added to them to form **lipoproteins**. These modified proteins transport lipids in the plasma between various organs in the body (e.g. gut, liver, and adipose tissue). Other common post-translational modifications include degradation, cleavage, and phosphorylation (below).

Cleaving: Polypeptide chains may be cleaved to give smaller chains, which then fold or join to give the functional protein. An example is human insulin which is transcribed as one long polypeptide chain before being cleaved in two places to form two shorter chains that form the functional protein.

Glycosylation (adding carbohydrate groups): This is used to add an ID tag to the protein that will allow the cell to recognize its use and where it is to be transported (2a). The resulting glycoprotein may be used in the cell membrane or secreted. The carbohydrate tag may help position the glycoprotein within the membrane (2b).

Phosphorylation (the addition of phosphate groups). This may contribute to the protein's three dimensional structure, or help with cell signalling.

Lipid attachment: Proteins may have lipids attached to them which anchor the protein to the plasma membrane.

Degradation: Some polypeptide chains may be tagged for degradation when they are no longer useful and their amino acids reused in the formation of other proteins.

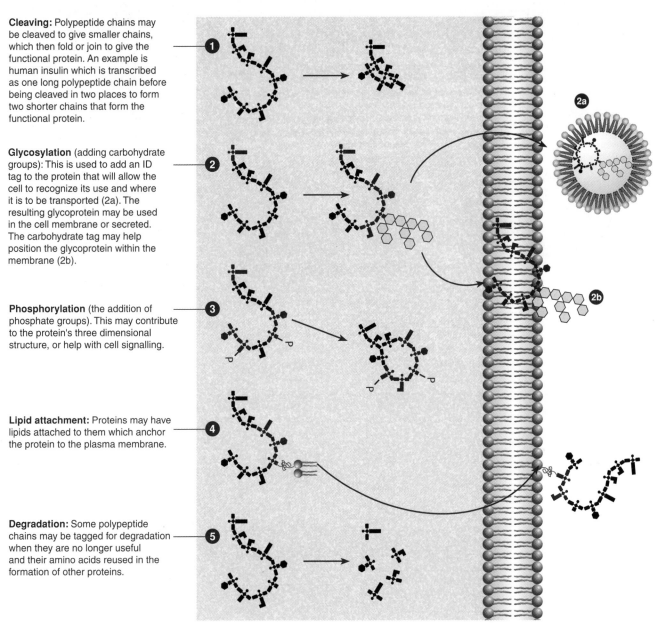

1. (a) Describe some of the modifications that polypeptide chains undergo before becoming functional proteins:

_____

_____

_____

(b) Explain why these changes are necessary: _____

_____

_____

2. Explain why the orientation of a protein in the plasma membrane might be important: _____

_____

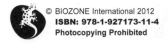

Related activities: The Role of Membranes in Cells, Golgi and the ER

A 2

The Biochemistry of Life

Biological Functions of Lipids

Lipids are a group of organic compounds with an oily, greasy, or waxy consistency. They are relatively insoluble in water and tend to be water-repelling (e.g. cuticle on leaf surfaces). Lipids are important biological fuels, some are hormones, and some serve as structural components in plasma membranes. Proteins and carbohydrates may be converted into fats by enzymes and stored within cells of adipose tissue. During times of plenty, this store is increased, to be used during times of food shortage.

Important Biological Functions of Lipids

Lipids are concentrated sources of energy and provide fuel for aerobic respiration.

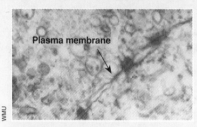

Phospholipids form the structural framework of cellular membranes.

Waxes and oils secreted onto surfaces provide waterproofing in plants and animals.

Fat absorbs shocks. Organs that are prone to bumps and shocks (e.g. kidneys) are cushioned with a relatively thick layer of fat.

Lipids are a source of metabolic water. During respiration stored lipids are metabolized for energy, producing water and carbon dioxide.

Stored lipids provide insulation. Increased body fat levels in winter reduce heat losses to the environment.

1. Explain how fats can provide an animal with:

   (a) Energy: _____

   _____

   (b) Water: _____

   _____

   (c) Insulation: _____

   _____

2. Explain why marine mammals (e.g. whales and seals) have thick layers of fat, or blubber, surrounding their bodies:

   _____

   _____

   _____

3. Oils and waxes are water repelling. Give two examples in animals or plants where this property would be useful:

   _____

   _____

   _____

4. Phospholipids have a polar head and non-polar tail. Explain how this allows them to spontaneously form the plasma membrane of a cell.

   _____

   _____

   _____

Related activities: Lipids
Weblinks: Biomolecules: Lipids

© BIOZONE International 2012
ISBN: 978-1-927173-11-4
Photocopying Prohibited

Lipids

Lipids are a diverse group of chemicals that lack an affinity for water, i.e. they are **hydrophobic**. They consist mainly of covalently bonded hydrogen and carbon molecules. Lipids can be divided into fats (comprising fatty acids and glycerol), phospholipids, and steroids. They can be solids or liquids at room temperature depending on the length of their carbon chains.

Neutral Fats and Oils

The most abundant lipids in living things are neutral fats. They make up the fats and oils found in plants and animals. Fats are an economical way to store fuel reserves because they yield more than twice as much energy as the same quantity of carbohydrate. **Neutral fats** are composed of a glycerol molecule attached to one (monoglyceride), two (diglyceride) or three (triglyceride) fatty acids. The fatty acid chains may be saturated or unsaturated (see below). Waxes are similar in structure to fats and oils, but they are formed with a complex alcohol instead of glycerol.

Glycerol   Fatty acids

Triglyceride: an example of a neutral fat

Condensation

Glycerol   Fatty acids

Triglyceride   Water

Triglycerides form when glycerol bonds with three fatty acids. Glycerol is an alcohol containing three carbons. Each of these carbons is bonded to a hydroxyl (-OH) group.

When glycerol bonds with the fatty acid, an ester bond is formed and water is released. Three separate condensation reactions are involved in producing a triglyceride.

Saturated and Unsaturated Fatty Acids

Fatty acids are a major component of **neutral fats** and **phospholipids**. About 30 different kinds are found in animal lipids. **Saturated fatty acids** contain the maximum number of hydrogen atoms. **Unsaturated fatty acids** contain some carbon atoms that are double-bonded with each other and are not fully saturated with hydrogens. Lipids containing a high proportion of saturated fatty acids tend to be solids at room temperature (e.g. butter). Lipids with a high proportion of unsaturated fatty acids are oils and tend to be liquid at room temperature. This is because the unsaturation causes kinks in the straight chains so that the fatty acids do not pack closely together. Regardless of their degree of saturation, fatty acids yield a large amount of energy when oxidised.

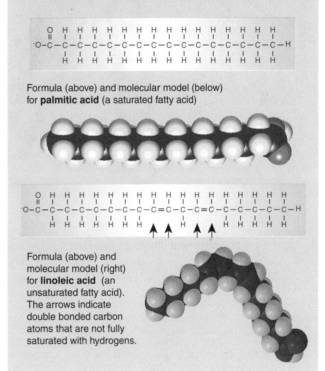

Formula (above) and molecular model (below) for **palmitic acid** (a saturated fatty acid)

Formula (above) and molecular model (right) for **linoleic acid** (an unsaturated fatty acid). The arrows indicate double bonded carbon atoms that are not fully saturated with hydrogens.

1. (a) Distinguish between saturated and unsaturated fatty acids: _____

_____

_____

(b) Explain how the type of fatty acid present in a neutral fat or phospholipid is related to that molecule's properties:

_____

_____

_____

2. Describe two examples of steroids. For each example, describe its physiological function:

(a) _____

(b) _____

Related activities: Biological Functions of Lipids
Weblinks: Formation of Triglycerides

A 2

Phospholipids

Phospholipids are the main component of cellular membranes. They consist of a glycerol attached to two fatty acid chains and a phosphate (PO_4^{3-}) group. The phosphate end of the molecule is attracted to water (it is hydrophilic) while the fatty acid end is repelled (hydrophobic). The hydrophobic ends turn inwards in the membrane to form a **phospholipid bilayer.**

Hydrophilic head

CH_2 — $N^+(CH_3)_3$
|
CH_2
|
O
|
$O = P — O^-$
|
O
|
H_2C — CH — CH_2
| |
O   O
| |
C=O   C=O

Hydrophobic tails

Steroids and Cholesterol

Although steroids are classified as lipids, their structure is quite different to that of other lipids. Steroids have a basic structure of three rings made of 6 carbon atoms each and a fourth ring containing 5 carbon atoms. Examples of steroids include the male and female sex hormones (testosterone and estrogen), and the hormones cortisol and aldosterone.

Cholesterol, while not a steroid itself, is a sterol lipid and is a precursor to several steroid hormones. It is present in the plasma membrane, where it regulates membrane fluidity by preventing the phospholipids from packing too closely together.

Like phospholipids, cholesterol is **amphipathic**. The hydroxyl (-OH) group on cholesterol interacts with the polar head groups of the membrane phospholipids, while the steroid ring and hydrocarbon chain tuck into the hydrophobic portion of the membrane. This helps to stabilise the outer surface of the membrane and reduce its permeability to small water-soluble molecules.

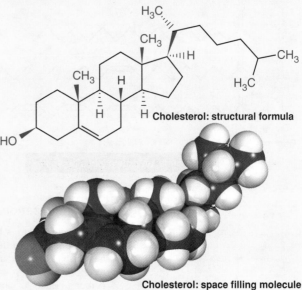

Cholesterol: structural formula

Cholesterol: space filling molecule

3. Outline the key **chemical** difference between a phospholipid and a triglyceride: _____

_____

_____

_____

4. Explain why saturated fats (e.g. lard) are solid at room temperature: _____

_____

5. (a) Relate the structure of phospholipids to their chemical properties and their functional role in cellular membranes:

_____

_____

_____

(b) Suggest how the cell membrane structure of an Arctic fish might differ from that of tropical fish species:

_____

_____

6. Explain how the structure of cholesterol enables it to perform structural and functional roles within membranes:

_____

_____

© BIOZONE International 2012
ISBN: 978-1-927173-11-4
Photocopying Prohibited

Carbohydrate Chemistry

Carbohydrates have the general formula $C_m(H_2O)_n$. Carbohydrate monomers are linked together in different ways to provide a variety of structurally and functionally different molecules. Monomers are linked together by **condensation reactions**, so called because in most condensation reactions a water molecule is produced. The reverse **hydrolysis reaction** splits polymers into smaller units by breaking the bond between two monomers. Hydrolysis literally means breaking with water, and so requires the addition of a water molecule. Disaccharides are two monomers joined together. Several disaccharides (lactose, sucrose and maltose) have important roles in human nutrition. Carbohydrates also exist as **isomers**. Isomers are compounds with the same molecular formula, but different structural formulae. Because of this they have different properties. For example, when α - glucose polymers are linked together they form starch, but linked β-glucose polymers form cellulose.

Isomerism

Compounds with the same chemical formula (same types and numbers of atoms) may differ in the arrangement of their atoms. Such variations in the arrangement of atoms in molecules are called isomers. In structural isomers (such as fructose and glucose, and the α-and β-glucose, right), the atoms are linked in different sequences. Optical isomers are identical in every way but are mirror images of each other.

Condensation and Hydrolysis Reactions

Monosaccharides can combine to form compound sugars in what is called a condensation reaction. Compound sugars can be broken down by hydrolysis to simple monosaccharides.

Condensation reaction

Two monosaccharides are joined together to form a disaccharide with the release of a water molecule. Energy is supplied by a nucleotide sugar (e.g. ADP-glucose).

Hydrolysis reaction

When a disaccharide is split, as in digestion, a water molecule is used as a source of hydrogen and a hydroxyl group. The reaction is catalysed by enzymes. For maltose (right), this is maltase.

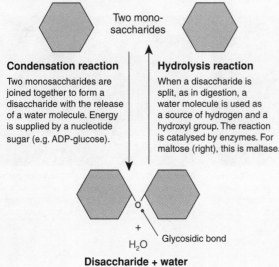

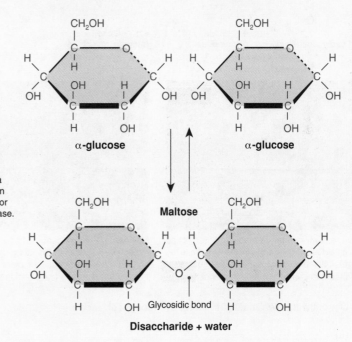

1. Distinguish between structural and optical isomers in carbohydrates, describing examples of each:

   _____

   _____

   _____

2. Explain briefly how compound sugars are formed and broken down: _____

   _____

   _____

3. Using examples, explain how the isomeric structure of a carbohydrate may affect its chemical behavior:

   _____

   _____

   _____

   _____

   _____

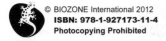
Related activities: Monosaccharides and Disaccharides
Web links: Condensation and Hydrolysis

A 2

Monosaccharides and Disaccharides

Sugars (monosaccharides and disaccharides) play a central role in cells, providing energy and, in some cells, contributing to support. They are the major component of most plants (60-90% of the dry weight) and are used by humans as a cheap food source, and a source of fuel, housing, and clothing. Disaccharides important in human nutrition include lactose, sucrose and maltose.

Monosaccharides

Monosaccharides are used as a primary energy source for fuelling cell metabolism. They are single-sugar molecules and include glucose (grape sugar and blood sugar) and fructose (honey and fruit juices). The commonly occurring monosaccharides contain between three and seven carbon atoms in their carbon chains and, of these, the 6C hexose sugars occur most frequently. All monosaccharides are classified as reducing sugars (i.e. they can participate in reduction reactions).

Single sugars (monosaccharides)

Triose
```
C
|
C
|
C
```
e.g. glyceraldehyde

Pentose
e.g. ribose, deoxyribose

Hexose
e.g. glucose, fructose, galactose

Disaccharides

Disaccharides are double-sugar molecules and are used as energy sources and as building blocks for larger molecules. The type of disaccharide formed depends on the monomers involved and whether they are in their α- or β- form. Only a few disaccharides (e.g. lactose) are classified as reducing sugars.

Sucrose = α-glucose + β-fructose (simple sugar in plant sap)
Maltose = α-glucose + α-glucose (a product of starch hydrolysis)
Lactose = β-glucose + β-galactose (milk sugar)
Cellobiose = β-glucose + β-glucose (from cellulose hydrolysis)

Double sugars (disaccharides)

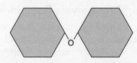

Examples sucrose, lactose, maltose, cellobiose

Lactose, a milk sugar, is made up of β-glucose + β-galactose. Milk contains 2-8% lactose by weight. It is the primary carbohydrate source for suckling mammalian infants.

Maltose is composed of two α-glucose molecules. These germinating wheat seeds contain maltose because the plant breaks down their starch stores to use it for food.

Sucrose (table sugar) is a simple sugar derived from plants such as sugar cane (above), sugar beet, or maple sap. It is composed of an α-glucose molecule and a β-fructose molecule.

1. Describe the two major functions of monosaccharides:

   (a) _____

   (b) _____

2. The breakdown of a disaccharide into its constituent monosaccharide units is an enzyme catalysed hydrolysis (see previous page). For each of the following common disaccharides, identify the enzyme responsible for the catalysis and the products of the hydrolysis, and describe an example of where this enzyme might naturally occur:

   (a) Lactose: Enzyme: _____ Products of hydrolysis: _____

   Found: _____

   (b) Maltose: Enzyme: _____ Products of hydrolysis: _____

   Found: _____

   (c) Sucrose: Enzyme: _____ Products of hydrolysis: _____

   Found: _____

3. Use your understanding of disaccharide chemistry to suggest how the digestive disorder lactose intolerance arises:

   _____

   _____

   _____

Related activities: Biochemical Tests, Carbohydrate Chemistry
Weblinks: Biomolecules: Carbohydrates
Periodicals: Glucose & glucose containing carbohydrates
© BIOZONE International 2012
ISBN: 978-1-927173-11-4
Photocopying Prohibited

Colorimetry

Colorimetric analysis is a simple quantitative technique used to determine the concentration of a specific substance in a solution. A specific reagent (e.g. Benedict's, which detects glucose) is added to the test solution where it reacts with the substance of interest to produce a color. After an appropriate reaction time, the samples are placed in a **colorimeter**, which measures the solution's absorbance at a specific wavelength. A dilution series can be used to produce a **calibration curve**, which can then be used to quantify that substance in samples of unknown concentration. This is illustrated for glucose in the example below.

1 Prepare glucose standards

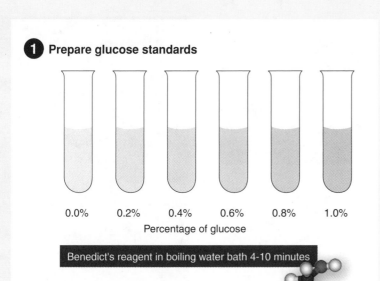

0.0%   0.2%   0.4%   0.6%   0.8%   1.0%

Percentage of glucose

Benedict's reagent in boiling water bath 4-10 minutes

Solutions containing a range of known glucose concentrations are prepared in test tubes.
A 1.0% glucose standard is made by dissolving 1 g of glucose in 100 cm^3 of water. The other dilutions are made by diluting the glucose solution with water.

Volume of 1% glucose standard	Volume of water	Percentage glucose
8 cm^3	2 cm^3	0.8
6 cm^3	4 cm^3	0.6
4 cm^3	6 cm^3	0.4
2 cm^3	8 cm^3	0.2

Benedict's reagent is added to each dilution, including the 0% blank. The test tubes are heated in a boiling waterbath for 4-10 minutes. At the end of the reaction time, samples containing glucose will have undergone a color change. The samples are cooled, then filtered or centrifuged to remove suspended particles.

Glucose

2 Produce the calibration curve

To produce a calibration curve, the prepared glucose standards are placed in a colorimeter, and their absorbance at 735 nm is recorded. These values are used to produce a calibration curve for glucose. For the best results, a new calibration curve should be generated for each new analysis. This accounts for any possible changes in the conditions of the reactants.

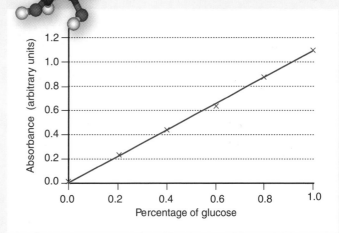

1. (a) How could you quantify the amount of glucose in a range of commercially available glucose drinks?

_____

_____

(b) What would you do if the absorbance values you obtained for most of your 'unknowns' were outside the range of your calibration curve?

_____

_____

_____

2. Why is it important to remove suspended solids from a sample before measuring its absorbance?

_____

_____

Related activities: Biochemical Tests

A 2

Polysaccharides

Polysaccharides or complex carbohydrates are straight or branched chains of many monosaccharides (sometimes many thousands) of the same or different types. The most common polysaccharides, cellulose, starch, and glycogen, contain only glucose, but their properties are very different. These differences are a function of the glucose isomer involved and the types of glycosidic linkages joining the glucose monomers. Different polysaccharides, based on the same sugar monomer, can thus be highly soluble and a source of readily available energy or a strong structural material that resists being digested.

Cellulose

Cellulose is a structural material in plants and is made up of unbranched chains of β-glucose molecules held together by 1,4 glycosidic links. As many as 10,000 glucose molecules may be linked together to form a straight chain. Parallel chains become cross-linked with hydrogen bonds and form bundles of 60-70 molecules called **microfibrils**. Cellulose microfibrils are very strong and are a major component of the structural components of plants, such as the cell wall (photo, right).

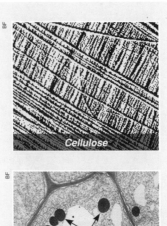

Cellulose

Starch

Starch is also a polymer of glucose, but it is made up of long chains of α-glucose molecules linked together. It contains a mixture of 25-30% amylose (unbranched chains linked by α-1,4 glycosidic bonds) and 70-75% amylopectin (branched chains with α-1,6 glycosidic bonds every 24-30 glucose units). Starch is an energy storage molecule in plants and is found concentrated in insoluble starch granules within plant cells (see photo, right). Starch can be easily hydrolysed by enzymes to soluble sugars when required.

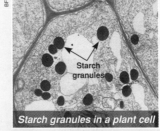

Starch granules in a plant cell

Glycogen

Glycogen, like starch, is a branched polysaccharide. It is chemically similar to amylopectin, being composed of α-glucose molecules, but there are more α-1,6 glycosidic links mixed with α-1,4 links. This makes it more highly branched and water-soluble than starch. Glycogen is a storage compound in animal tissues and is found mainly in liver and muscle cells (photo, right). It is readily hydrolysed by enzymes to form glucose.

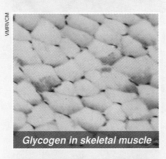

Glycogen in skeletal muscle

Chitin

Chitin is a tough modified polysaccharide made up of chains of β-glucose molecules. It is chemically similar to cellulose but each glucose has an amine group (–NH–) attached. After cellulose, chitin is the second most abundant carbohydrate. It is found in the cell walls of fungi and is the main component of the exoskeleton of insects (right) and other arthropods.

Chitinous insect exoskeleton

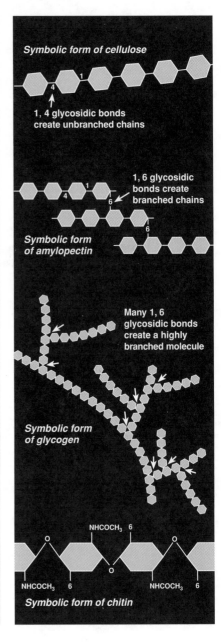

Symbolic form of cellulose

1, 4 glycosidic bonds create unbranched chains

1, 6 glycosidic bonds create branched chains

Symbolic form of amylopectin

Many 1, 6 glycosidic bonds create a highly branched molecule

Symbolic form of glycogen

Symbolic form of chitin

1. Why are polysaccharides such a good source of energy? _____

_____

2. Discuss the structural differences between the polysaccharides starch and glycogen, explaining how the differences in structure contribute to the functional properties of the molecule:

_____

_____

_____

_____

_____

Related activities: *Carbohydrate Chemistry*

Periodicals:
Designer starches

© BIOZONE International 2012
ISBN: 978-1-927173-11-4
Photocopying Prohibited

Cellulose and Starch

Cellulose is the most common molecule on Earth, making up one third of the volume of plants and one half of the volume of wood. Cellulose is made of thousands of β-glucose monomers arranged in a single, unbranched chain. In plants, cellulose makes up the bulk of the cell wall, providing both the strength required to keep the cell's shape and the support for the plant stem or trunk. As a material that can be exploited by humans, cellulose provides fibers which can be made into thread for textiles, and wood used for framing and cladding buildings. In contrast, **starch** is made of α-glucose monomers, and is a compact, branching molecule. It has no structural function, but can be hydrolysed to release soluble sugars for energy.

Cellulose Structure and Function

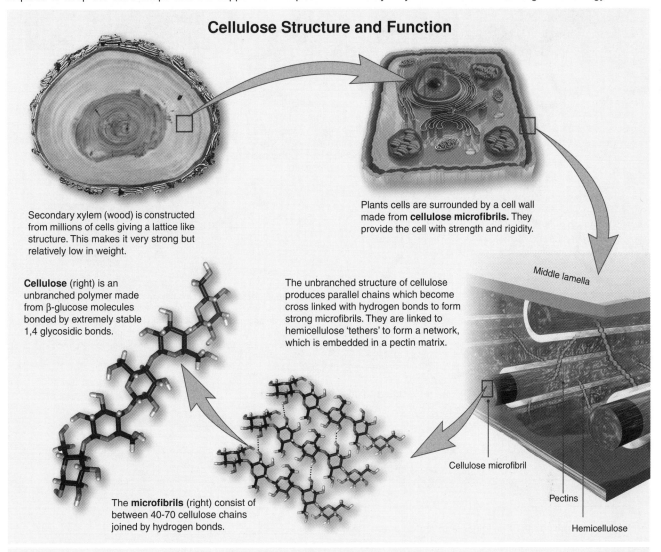

Secondary xylem (wood) is constructed from millions of cells giving a lattice like structure. This makes it very strong but relatively low in weight.

Plants cells are surrounded by a cell wall made from **cellulose microfibrils.** They provide the cell with strength and rigidity.

Cellulose (right) is an unbranched polymer made from β-glucose molecules bonded by extremely stable 1,4 glycosidic bonds.

The unbranched structure of cellulose produces parallel chains which become cross linked with hydrogen bonds to form strong microfibrils. They are linked to hemicellulose 'tethers' to form a network, which is embedded in a pectin matrix.

The **microfibrils** (right) consist of between 40-70 cellulose chains joined by hydrogen bonds.

Middle lamella

Cellulose microfibril

Pectins

Hemicellulose

Cellulose and Tensile Strength

The **tensile strength** of a material refers to its ability to resist the strain of a pulling force. The simplest way to test tensile strength is to hang the material from a support and increase the tension by adding weight at the other end until the material separates. The tensile strength is then expressed as the force applied per the area of material in cross section. Typically this is written as pascals (Pa) or Newtons per square metre ($N\,m^{-2}$), but is also often expressed as pounds per square inch (PSI). Cellulose fibrils show extremely high tensile strength (table below).

Material	Tensile strength (GPa)
Cellulose	17.8
Silk	1
Human hair	0.38
Rubber	0.15
Steel wire	2.2
Nylon (6/6)	0.75

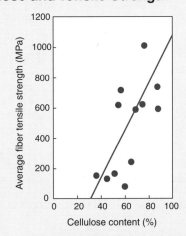
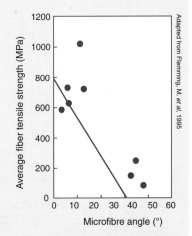

Adapted from Flemming, M. et al. 1995

Plants produce fibers of various strengths that can be used for textiles, ropes, and construction. In general, the greater the amount of cellulose in the fiber the greater its tensile strength (above left). The tensile strength of the plant fiber is also influenced by the orientation of the cellulose microfibrils. Microfibrils orientated parallel to the length of the fiber (0°) provide the greatest fiber strength (above right).

0

Starch Structure and Function

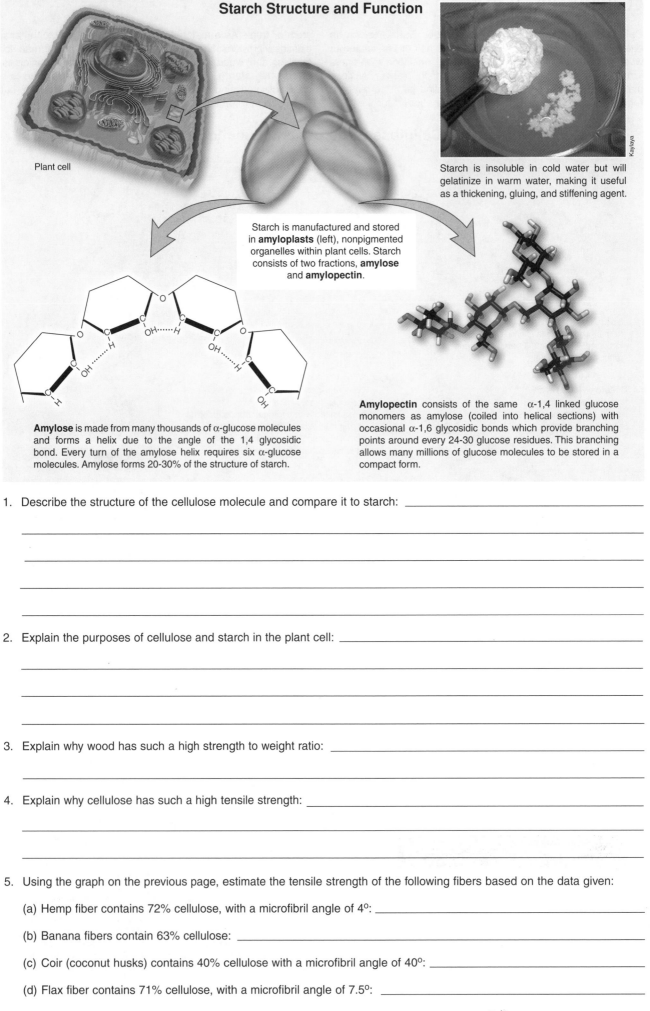

Plant cell

Starch is insoluble in cold water but will gelatinize in warm water, making it useful as a thickening, gluing, and stiffening agent.

Starch is manufactured and stored in **amyloplasts** (left), nonpigmented organelles within plant cells. Starch consists of two fractions, **amylose** and **amylopectin**.

Amylose is made from many thousands of α-glucose molecules and forms a helix due to the angle of the 1,4 glycosidic bond. Every turn of the amylose helix requires six α-glucose molecules. Amylose forms 20-30% of the structure of starch.

Amylopectin consists of the same α-1,4 linked glucose monomers as amylose (coiled into helical sections) with occasional α-1,6 glycosidic bonds which provide branching points around every 24-30 glucose residues. This branching allows many millions of glucose molecules to be stored in a compact form.

1. Describe the structure of the cellulose molecule and compare it to starch: _____

_____

_____

_____

2. Explain the purposes of cellulose and starch in the plant cell: _____

_____

_____

_____

3. Explain why wood has such a high strength to weight ratio: _____

_____

4. Explain why cellulose has such a high tensile strength: _____

_____

_____

5. Using the graph on the previous page, estimate the tensile strength of the following fibers based on the data given:

   (a) Hemp fiber contains 72% cellulose, with a microfibril angle of 4°: _____

   (b) Banana fibers contain 63% cellulose: _____

   (c) Coir (coconut husks) contains 40% cellulose with a microfibril angle of 40°: _____

   (d) Flax fiber contains 71% cellulose, with a microfibril angle of 7.5°: _____

Cell Size and Shape

Cells come in a wide range of types and forms, each adapted for a specific role. In humans, there are over 200 types of cell and they vary widely in their size and shape. The longest neurons can be over a metre long. In other animals, such as giraffes and giant squid, they can be very much longer (up to 12 m). Plants also have a large variety of cell types. Plants have a very different body form to animals, but many of the organelles inside the cells are the same. The variety of specialised cells in living organisms is a reflection of their specific roles in the organism. For example, animals require long, flexible neurons to carry signals about the body. Plants require specialised cells to transport manufactured food (sugar) around the plant body.

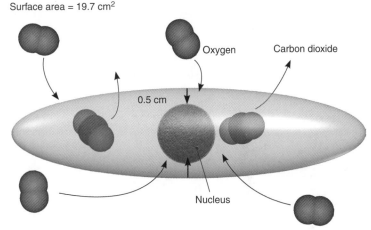

Cell A

Radius = 1 cm
Volume = 4.2 cm^3
Surface area = 12.5 cm^2

Oxygen
Carbon dioxide
1 cm
Nucleus

Cell B

Radius a = 0.5 cm, b = 0.5 cm, c = 4 cm
Volume = 4.2 cm^3
Surface area = 19.7 cm^2

Oxygen
Carbon dioxide
0.5 cm
Nucleus

Oxygen diffusing into a hypothetical spherical shaped cell of volume 4.2 cm^3 needs to cross 1 cm to reach the center of the cell when entering from any angle. There is a much lower surface area for diffusion than in an ellipsoid cell of the same volume: less oxygen per volume can diffuse into the spherical cell compared to an ellipsoid cell of the same volume.

The size and shape of a cell reflects its function and the need for essential molecules to diffuse in and out. The greater the spherical diameter of a cell, the more material it contains, and the further molecules have to move in order to reach the center. Molecules diffusing into the cell are used up faster than they can be supplied and may not reach the cell's center, leaving it starved of essential molecules (e.g. oxygen). This can be solved by reducing the diameter of the cell along at least one axis.

Oxygen diffusing into a hypothetical ellipsoid cell of volume 4.2 cm^3 needs to cross only 0.5 cm to reach the center of the cell when entering from the long axis. There is a greater surface area for diffusion than in a spherical cell of the same volume: more oxygen molecules diffuse into the cell per unit of volume than in a spherical cell of the same volume.

Root hairs are the elongated extensions of root-hair cells in the epidermis of plant roots. They increase the surface area available for absorption of water and nutrients.

Villi increase the surface area in the small intestine for absorption. They are not cells but protrusions of the small intestine wall. Their surface is lined with cells which are covered in microvilli.

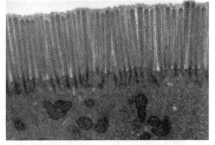

Microvilli are extensions of a cell's plasma membrane, greatly increasing its surface area. In the small intestine they increase the area for absorbing nutrient from food.

1. Why do cells need to be a variety of shapes and sizes? _____

_____

2. Use the dimensions of the hypothetical cells above to answer the following questions:

   (a) For both cell A and B, oxygen diffuses across the plasma membrane at a rate of 100 molecules per cm^2 per minute. Calculate the total number of oxygen molecules that enter the cells during a period of 5 minutes:

      Cell A: _____

      Cell B: _____

   (b) Calculate the number of oxygen molecules available per cm^3 of cell during the 5 minute period:

      Cell A: _____

      Cell B: _____

Limitations to Cell Size

When an object (e.g. a cell) is small it has a large surface area in comparison to its volume. In this case diffusion will be an effective way to transport materials (e.g. gases) into the cell. As an object becomes larger, its surface area compared to its volume is smaller. Diffusion is no longer an effective way to transport materials to the inside. For this reason, there is a physical limit for the size of a cell, with the effectiveness of diffusion being the controlling factor.

Diffusion in Organisms of Different Sizes

Respiratory gases and some other substances are exchanged with the surroundings by diffusion or active transport across the plasma membrane.

The **plasma membrane**, which surrounds every cell, functions as a selective barrier that regulates the cell's chemical composition. For each square micrometrer of membrane, only so much of a particular substance can cross per second.

The surface area of an elephant is increased, for radiating body heat, by large flat ears.

The nucleus can control a smaller cell more efficiently.

Oxygen

Food

A specialised gas exchange surface (lungs) and circulatory (blood) system are required to speed up the movement of substances through the body.

Carbon dioxide

Wastes

Respiratory gases cannot reach body tissues by diffusion alone.

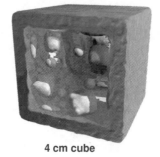

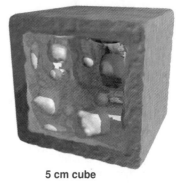

Amoeba: The small size of single-celled protists, such as Amoeba, provides a large surface area relative to the cell's volume. This is adequate for many materials to be moved into and out of the cell by diffusion or active transport.

Multicellular organisms: To overcome the problems of small cell size, plants and animals became multicellular. They provide a small surface area compared to their volume but have evolved various adaptive features to improve their effective surface area.

The diagram below shows four imaginary cells of different sizes (cells do not actually grow to this size, their large size is for the sake of the exercise). They range from a small 2 cm cube to a larger 5 cm cube. This exercise investigates the effect of cell size on the efficiency of diffusion.

| 2 cm cube | 3 cm cube | 4 cm cube | 5 cm cube |

1. Calculate the volume, surface area and the ratio of surface area to volume for each of the four cubes above (the first has been done for you). When completing the table below, show your calculations.

Cube size	Surface area	Volume	Surface area to volume ratio
2 cm cube	$2 \times 2 \times 6 = 24 \text{ cm}^2$ (2 cm x 2 cm x 6 sides)	$2 \times 2 \times 2 = 8 \text{ cm}^3$ (height x width x depth)	24 to 8 = 3:1
3 cm cube			
4 cm cube			
5 cm cube			

Related activities: *Passive Transport Processes, Cell Size and Shape*

Periodicals: *Getting in and out*

© BIOZONE International 2012
ISBN: 978-1-927173-11-4
Photocopying Prohibited

2. Create a graph, plotting the surface area against the volume of each cube, on the grid on the right. Draw a line connecting the points and label axes and units.

3. Which increases the fastest with increasing size: the **volume** or the **surface area**?

_____

4. Explain what happens to the ratio of surface area to volume with increasing size.

_____

_____

_____

_____

_____

_____

_____

5. The diffusion of molecules into a cell can be modelled by using agar cubes infused with phenolphthalein indicator and soaked in sodium hydroxide (NaOH). Phenolphthalein turns a pink color when in the presence of a base. As the NaOH diffuses into the agar, the phenolphthalein changes to pink and thus indicates how far the NaOH has diffused into the agar. By cutting an agar block into cubes of various sizes, it is possible to show the effect of cell size on diffusion.

(a) Use the information below to fill in the table on the right:

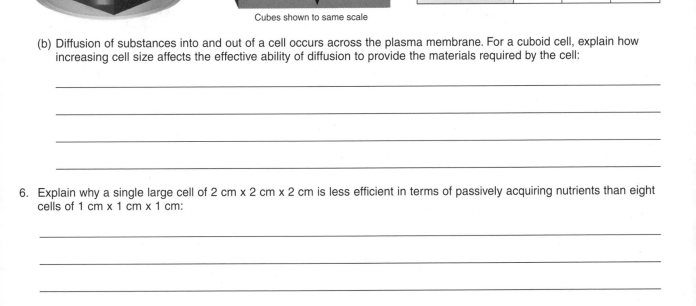

Cube	1	2	3
1. Total volume (cm³)			
2. Volume not pink (cm³)			
3. Diffused volume (1. – 2.) (cm³)			
4. Percentage diffusion			

Cube 1 — 1 cm
Cube 2 — 2 cm
Cube 3 — 4 cm
Region of no color change
Region of color change
Cubes shown to same scale

NaOH solution

Agar cubes infused with phenolphthalein

(b) Diffusion of substances into and out of a cell occurs across the plasma membrane. For a cuboid cell, explain how increasing cell size affects the effective ability of diffusion to provide the materials required by the cell:

_____

_____

_____

_____

6. Explain why a single large cell of 2 cm x 2 cm x 2 cm is less efficient in terms of passively acquiring nutrients than eight cells of 1 cm x 1 cm x 1 cm:

_____

_____

_____

_____

 © BIOZONE International 2012
ISBN: 978-1-927173-11-4
Photocopying Prohibited

KEY TERMS: Mix and Match

INSTRUCTIONS: Test your vocabulary by matching each term to its definition, as identified by its preceding letter code.

amino acid

amphipathic

carbohydrates

carbon

denaturation

disaccharide

ester bond

fibrous proteins

globular proteins

hydrogen bond

inorganic ion

lipids

macromolecule

monomer

monosaccharides

nucleic acids

nucleotide

organic molecule

peptide bond

phospholipids

polar

polysaccharides

proteins

surface area: volume ratio

water

A The building blocks of proteins.

B Organic compounds, usually linear polymers, made of amino acids linked together by peptide bonds.

C Organic molecules consisting only of carbon, hydrogen and oxygen that serve as structural components in cells and as energy sources.

D Carbohydrate monomers. Examples include fructose and glucose.

E The loss of a protein's three dimensional functional structure is called this.

F These proteins are very tough and often have a structural role in cells.

G The attractive interaction of a hydrogen atom with an electronegative atom, such as nitrogen, oxygen or fluorine.

H Complex carbohydrates with structural and energy storage roles in cells. Examples include cellulose, starch, and glycogen.

I A class of organic compounds with an oily, greasy, or waxy consistency. Important as energy storage molecules and as components of cellular membranes.

J A molecule in which the opposite ends are oppositely charged is called this.

K These proteins are water soluble and have catalytic and regulatory roles in cells.

L A molecule possessing both a hydrophobic and hydrophilic portion, e.g. phospholipids and fatty acids, is described as this.

M A double sugar molecule used as energy sources and building blocks of larger molecules. Examples are sucrose and lactose.

N A charged molecule that does not contain carbon.

O A very large complex molecule often made up of smaller units, e.g. protein or polysaccharide.

P A molecule made up of a single unit.

Q Molecules comprising a purine or pyrimidine bonded with a pentose sugar (ribose or deoxyribose) and a phosphate group.

R Polynucleotide molecules that occur in two forms, DNA and RNA.

S A molecule that contains carbon.

T A class of amphipathic lipids, they are a major component of all cell membranes as they can form lipid bilayers.

U Ratio of the surface area of a cell to its volume. The smaller the ratio the less effective the diffusion of material into the interior of the cell.

V Covalent bond formed when a carboxyl group and an amine group react and release a molecule of water. In biological systems this leads to the formation of proteins.

W The bond between an alcohol and an acid, e.g. glycerol and fatty acid.

X Atom around which life is based, mainly due to its ability to form four stable covalent bonds and long chained molecules.

Y Polar molecule consisting of two hydrogen atoms and one oxygen atom.

© BIOZONE International 2012
ISBN: 978-1-927173-11-4
Photocopying Prohibited

The Internal Structure of Cells

Key concepts

▶ Cells are the fundamental units of life.

▶ Cells share many of the same components but there are distinguishing differences between the cells of different kingdoms.

▶ In eukaryotic cells, specialized organelles localize reactions and promote functional efficiency.

▶ Microscopy can be used to understand cellular structure.

Key terms

apoptosis
cell wall
centrioles
chloroplast
cytoplasm
cytoskeleton
electron microscope
endoplasmic reticulum (ER)
eukaryotic cell
flagella (*sing*. flagellum)
Golgi apparatus
lysosome
magnification
membrane
mitochondrion
nuclear envelope
nuclear pore
nucleoid
nucleolus
nucleus
optical microscope
organelles
plasma membrane
prokaryotic cell
resolution
ribosome
rough ER
SEM
smooth ER
vacuole

Essential Knowledge

☐ 1. Use the **KEY TERMS** to compile a glossary for this topic.

Compartments in Cells *(2.B.3)*   pages 37-38, 43-47, 61-62

☐ 2. Using examples, e.g. **endoplasmic reticulum**, **chloroplast**, **mitochondria**, **Golgi apparatus**, or **nuclear envelope**, describe the role of **membranes** in localizing metabolic processes within the cell.

☐ 3. Describe the structure of a **prokaryotic cell**, recognizing the **nucleoid** region and **cell wall**. Explain where metabolic processes and enzymatic reactions take place in the absence of discrete membrane-bound cellular organelles.

Subcellular Components *(4.A.2)*   pages 36, 39-56

☐ 4. Describe the structure of **ribosomes** and relate this to function. Explain how the subunits of the ribosome interact to provide the site of protein synthesis.

☐ 5. Describe the structure of **endoplasmic reticulum** (ER), distinguishing between **rough ER** and **smooth ER**. Relate structure to function in each case.

☐ 6. Describe the structure of the **Golgi apparatus** (complex) and relate this to its function. Relate the orientation of the Golgi to export of its packaged products.

☐ 7. Describe the structure and function of **mitochondria** (*sing*. mitochondrion). Relate the internal structure of the mitochondrion to available surface area and localization of enzymatic reactions within the organelle.

☐ 8. Describe the structure and function of **lysosomes**, including their role in programmed cell death (**apoptosis**).

☐ 9. Describe the structure and function of **vacuoles** in animal and plant cells. Account for the differences in the role of vacuoles in these two cell types.

☐ 10. Describe the structure and function of **chloroplasts**. Relate the internal structure of the chloroplasts to available surface area and localization of enzymatic reactions within the organelle.

☐ 11. As required, describe the structure and function of other subcellular structures and organelles in eukaryotic cells, including **flagella**, **centrioles**, and **cytoskeleton**. Recognize the role of the cytoskeleton in intracellular transport.

☐ 12. Recognize the contribution of microscopy to our modern understanding of cell structure and function. If required, demonstrate an ability to use both stereo and compound **light microscopes** to locate material and focus images.

☐ 13. Interpret drawings and photomicrographs of typical plant and animal cells as seen using light microscopy and electron microscopy. Recognize features of **prokaryotic cells** in electron micrographs.

☐ 14. Demonstrate an ability to calculate linear magnification and make accurate biological drawings of material viewed with a light microscope.

Periodicals:
Listings for this chapter are on page 382

Weblinks:
www.thebiozone.com/
weblink/AP1-3114.html

BIOZONE APP:
Student Review Series
Cell Structure

Cell Sizes

Cells are extremely small and can only be seen properly when viewed through the magnifying lenses of a microscope. The diagrams below show a variety of cell types, together with a virus (non-cellular) and a multicellular microscopic animal (as a comparison). For each of these images, note the scale and relate this to the type of microscopy used.

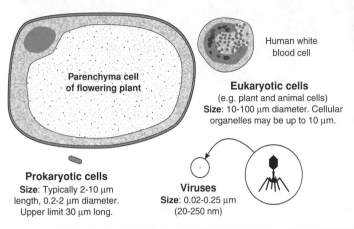

Human white blood cell

Parenchyma cell of flowering plant

Eukaryotic cells
(e.g. plant and animal cells)
Size: 10-100 µm diameter. Cellular organelles may be up to 10 µm.

Prokaryotic cells
Size: Typically 2-10 µm length, 0.2-2 µm diameter. Upper limit 30 µm long.

Viruses
Size: 0.02-0.25 µm (20-250 nm)

Unit of length (International System)		
Unit	**Meters**	**Equivalent**
1 meter (m)	1 m	= 1000 millimeters
1 millimeter (mm)	10^{-3} m	= 1000 micrometers
1 micrometer (µm)	10^{-6} m	= 1000 nanometers
1 nanometer (nm)	10^{-9} m	= 1000 picometers

Micrometers are sometime referred to as microns. Smaller structures are usually measured in nanometers (nm) e.g. molecules (1 nm) and plasma membrane thickness (10 nm).

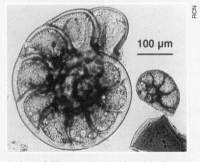

100 µm

An **Amoeba** showing extensions of the cytoplasm called pseudopodia. This protoctist changes its shape, exploring its environment.

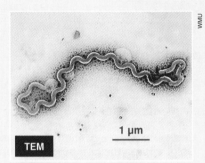

1 µm

TEM

A long thin cell of the spirochete bacterium **Leptospira pomona**, which causes the disease leptospirosis.

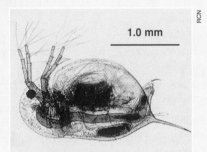

1.0 mm

Daphnia showing its internal organs. These freshwater microcrustaceans are part of the zooplankton found in lakes and ponds.

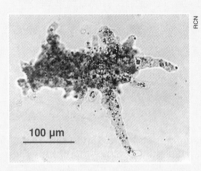

100 µm

A **foraminiferan** showing its chambered, calcified shell. These single-celled protozoans are marine planktonic amoebae.

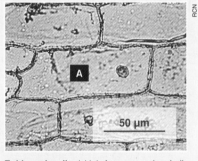

A

50 µm

Epidermal cells (skin) from an onion bulb showing the nucleus, cell walls and cytoplasm. Most organelles are not visible at this resolution.

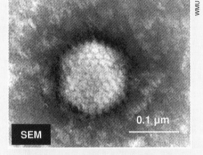

0.1 µm

SEM

Papillomavirus (human wart virus) showing its polyhedral protein coat (20 triangular faces, 12 corners) made of ball-shaped structures.

1. Using the measurement scales provided on each of the photographs above, determine the longest dimension (length or diameter) of the cell/animal/virus in µm and mm (choose the cell marked 'A' for epidermal cells):

   (a) *Amoeba*: _____ µm _____ mm   (d) Epidermis: _____ µm _____ mm

   (b) Foraminiferan: _____ µm _____ mm   (e) *Daphnia*: _____ µm _____ mm

   (c) *Leptospira*: _____ µm _____ mm   (f) *Papillomavirus*: _____ µm _____ mm

2. List these six organisms in order of size, from the smallest to the largest: _____

   _____

3. Study the scale of your ruler and state which of these six organisms you would be able to see with your unaided eye:

   _____

4. Calculate the equivalent length in millimeters (mm) of the following measurements:

   (a) 0.25 µm: _____   (b) 450 µm: _____   (c) 200 nm: _____

Related activities: Prokaryotic Cells, Optical Microscopes
Weblinks: Cell Size and Scale

Periodicals:
Size does matter

© BIOZONE International 2012
ISBN: 978-1-927173-11-4
Photocopying Prohibited

Prokaryotic Cells

Bacterial (prokaryotic) cells are much smaller and simpler than the cells of eukaryotes. They lack many eukaryotic features (e.g. a distinct nucleus and membrane-bound cellular organelles). The bacterial cell wall is an important feature. It is a complex, multi-layered structure and often has a role in virulence. These pages illustrate some features of bacterial structure and diversity.

Structure of a Generalized Bacterial Cell

Plasmids: Small, circular DNA molecules which can reproduce independently of the main chromosome. They can move between cells, and even between species, by **conjugation**. This property accounts for the transmission of antibiotic resistance between bacteria. Plasmids are also used as vectors in recombinant DNA technology.

Fimbriae: Hairlike structures that are shorter, straighter, and thinner than flagella. They are used for attachment, not movement. Pili are similar to fimbriae, but are longer and less numerous. They are involved in bacterial conjugation (below) and as phage receptors.

The cell lacks a nuclear membrane, so there is no distinct nucleus and the chromosome is in direct contact with the cytoplasm. It is possible for free ribosomes to attach to mRNA while the mRNA is still in the process of being transcribed from the DNA.

Cell surface membrane: Similar in composition to eukaryotic membranes, although less rigid.

Cell wall: A complex, semi-rigid structure that gives the cell shape, prevents rupture. The cell wall is composed of a macromolecule called **peptidoglycan**; repeating disaccharides attached by polypeptides to form a lattice. The wall also contains varying amounts of lipopolysaccharides and lipoproteins. The amount of peptidoglycan present in the wall forms the basis of the diagnostic **gram stain**. In many species, the cell wall contributes to their virulence (disease-causing ability).

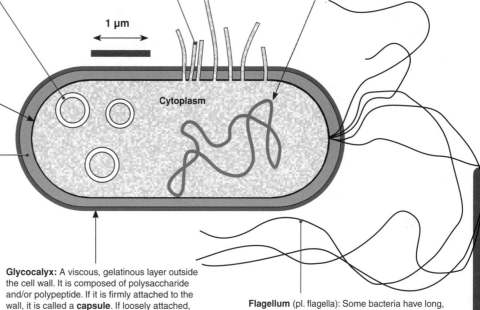

Cytoplasm

1 μm

Glycocalyx: A viscous, gelatinous layer outside the cell wall. It is composed of polysaccharide and/or polypeptide. If it is firmly attached to the wall, it is called a **capsule**. If loosely attached, it is called a **slime layer**. Capsules may contribute to virulence in pathogenic species, e.g. by protecting the bacteria from immune attack. In some species, the glycocalyx allows attachment to substrates.

Flagellum (pl. flagella): Some bacteria have long, filamentous appendages, called flagella, that are used for locomotion. There may be a single polar flagellum (monotrichous), one or more flagella at each end of the cell, or the flagella may be distributed over the entire cell (peritrichous).

Bacterial cell shapes

Most bacterial cells range between 0.20-2.0 μm in diameter and 2-10 μm length. Although they are a very diverse group, much of this diversity is in their metabolism. In terms of gross morphology, there are only a few basic shapes found (illustrated below). The way in which members of each group aggregate after division is often characteristic and is helpful in identifying certain species.

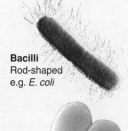

Bacilli
Rod-shaped
e.g. *E. coli*

Cocci
Ball-shaped
e.g. *Staphylococcus*

Spirilla
Spiral-shaped
e.g. *Leptospira*

Bacilli: Rod-shaped bacteria that divide only across their short axis. Most occur as single rods, although pairs and chains are also found. The term bacillus can refer (as here) to shape. It may also denote a genus.

Cocci: usually round, but sometimes oval or elongated. When they divide, the cells stay attached to each other and remain in aggregates e.g. pairs (diplococci) or clusters (staphylococci), that are usually a feature of the genus.

Spirilla and vibrio: Bacteria with one or more twists. Spirilla bacteria have a helical (corkscrew) shape which may be rigid or flexible (as in spirochetes). Bacteria that look like curved rods (comma shaped) are called vibrios. Vibrio may also denote a genus.

Bacterial conjugation

The two bacteria below are involved in conjugation: a one-way exchange of genetic information from a donor cell to a recipient cell. The plasmid, which must be of the 'conjugative' type, passes through a tube called a sex pilus to the other cell. Which is donor and which is recipient appears to be genetically determined. Conjugation should not be confused with sexual reproduction, as it does not involve the fusion of gametes or formation of a zygote.

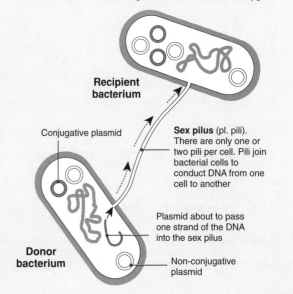

Recipient bacterium

Conjugative plasmid

Sex pilus (pl. pili). There are only one or two pili per cell. Pili join bacterial cells to conduct DNA from one cell to another

Plasmid about to pass one strand of the DNA into the sex pilus

Donor bacterium

Non-conjugative plasmid

Periodicals: Bacteria

Related activities: Prokaryotic Chromosomes
Weblinks: Bacterial Conjugation Animation

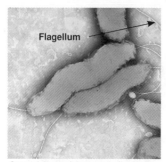

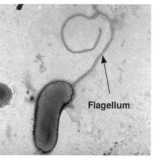

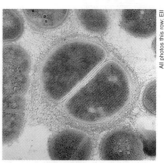

Campylobacter jejuni, a spiral bacterium responsible for foodborne intestinal disease. Note the single flagellum at each end (amphitrichous arrangement).

Helicobacter pylori, a comma-shaped vibrio bacterium that causes stomach ulcers in humans. This bacterium moves by means of multiple polar flagella.

A species of *Spirillum,* a spiral shaped bacterium with a tuft of polar flagella. Most of the species in this genus are harmless aquatic organisms.

Bacteria usually divide by binary fission. During this process, DNA is copied and the cell splits into two cells, as in these gram positive cocci.

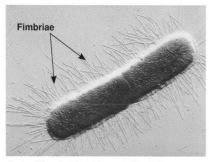

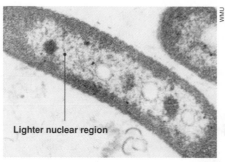

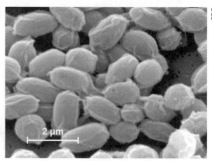

Escherichia coli, a common gut bacterium with **peritrichous** (around the entire cell) **fimbriae**. *E. coli* is a gram negative rod; it does not take up the gram stain but can be counter stained with safranin.

TEM showing *Enterobacter* bacteria, which belong to the family of gut bacteria commonly known as enterics. They are widely distributed in water, sewage, and soil. The family includes motile and non-motile species.

SEM of endospores of *Bacillus anthracis* bacteria, which cause the disease anthrax. These heat-resistant spores remain viable for many years and enable the bacteria to survive in a dormant state.

1. Describe three features distinguishing prokaryotic cells from eukaryotic cells:

   (a) _____

   (b) _____

   (c) _____

2. (a) Describe the function of flagella in bacteria: _____

   _____

   (b) Explain how fimbriae differ structurally and functionally from flagella: _____

   _____

   _____

3. (a) Describe the location and general composition of the bacterial cell wall: _____

   _____

   _____

   (b) Describe how the glycocalyx differs from the cell wall: _____

   _____

4. (a) Describe the main method by which bacteria reproduce: _____

   _____

   (b) Explain how conjugation differs from this usual method: _____

   _____

5. Briefly describe how the artificial manipulation of plasmids has been used for technological applications:

   _____

   _____

   _____

© BIOZONE International 2012
ISBN: 978-1-927173-11-4
Photocopying Prohibited

Animal Cells

Although animal cells and plant cells have many features in common, animal cells do not have a regular shape, and some (such as the phagocytic white blood cells) are quite mobile. The diagram below illustrates the basic ultrastructure of an **intestinal epithelial cell**. It contains organelles common to most relatively unspecialized human cells. The intestine is lined with these columnar epithelial cells. They are taller than they are wide, with the nucleus close to the base and hairlike projections (**microvilli**) on their free surface. Microvilli increase the surface area of the cell, greatly increasing the capacity for absorption.

Structures and Organelles in an Intestinal Epithelial Cell

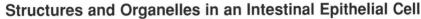

Mitochondrion (pl. mitochondria): 1.5 µm X 2–8 µm. Ovoid organelle bounded by a double membrane. They are the cell's energy transformers, and convert chemical energy into ATP.

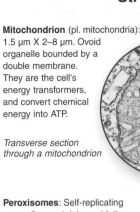

Transverse section through a mitochondrion

Peroxisomes: Self-replicating organelles containing oxidative enzymes, which function to rid the body of toxic substances. They are distinguished from lysosomes by the crystalline core.

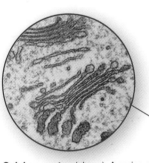

Golgi apparatus (above): A series of flattened, disc-shaped sacs, stacked one on top of the other and connected with the ER. The Golgi stores, modifies, and packages proteins. It 'tags' proteins so that they go to their correct destination.

Cytoplasm: A watery solution containing dissolved substances, enzymes, and the cell organelles and structures.

Ribosomes: These small (20 nm) structures manufacture proteins. Ribosomes are made of ribosomal RNA and protein. They may be free in the cytoplasm or associated with the surface of the endoplasmic reticulum.

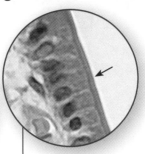

Each epithelial cell has many small projections, called microvilli, visible in this micrograph as a fuzzy brush border (arrowed).

Lysosome: A sac bounded by a single membrane. Lysosomes are pinched off from the Golgi and contain and transport enzymes that break down foreign material. Lysosomes show little internal structure but often contain fragments of degraded material.

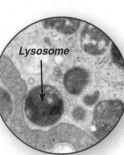
Lysosome

Plasma membrane: 3-10 nm thick phospholipid bilayer with associated proteins and lipids.

Tight junction: impermeable junction binding neighbouring cells together (common in epithelial cells).

Nucleus (below): 5 µm diameter. A large organelle containing most of the cell's DNA. Within the nucleus, the **nucleolus** (*n*) is a dense structure of crystalline protein and nucleic acid involved in ribosome synthesis.

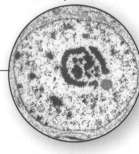

Nuclear pore: A hole in the nuclear membrane. It allows communication between the nucleus and the rest of the cell.

Centrioles: Microtubular structures associated with nuclear division. Under a light microscope, they appear as small, featureless particles, 0.25 µm diameter.

Rough ER: Endoplasmic reticulum with ribosomes attached to its surface. It is where the proteins destined for transport outside of the cell are synthesized.

Rough endoplasmic reticulum showing ribosomes (dark spots)

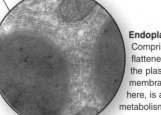

Endoplasmic reticulum (ER): Comprises a network of tubules and flattened sacs. ER is continuous with the plasma membrane and the nuclear membrane. **Smooth ER**, as shown here, is a site for lipid and carbohydrate metabolism, including hormone synthesis.

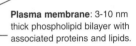

Related activities: Identifying Structures in an Animal Cell
Weblinks: Eukaryotic Cells Interactive Animation, , Review of Cell Structure

RA 2

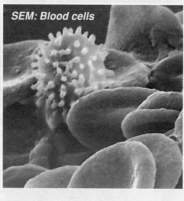

SEM: Blood cells

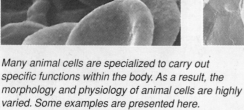

SEM: Skin cells

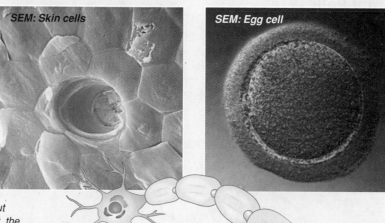

SEM: Egg cell

Many animal cells are specialized to carry out specific functions within the body. As a result, the morphology and physiology of animal cells are highly varied. Some examples are presented here.

Nerve cell

1. Explain what is meant by a generalized cell: _____

_____

_____

2. Discuss how the shape and size of a specialized cell, as well as the number and types of organelles it has, are related to its functional role. Use examples to illustrate your answer:

_____

_____

_____

_____

_____

_____

_____

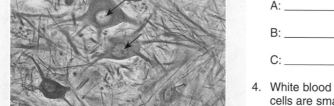

Neurons (nerve cells) in the spinal cord

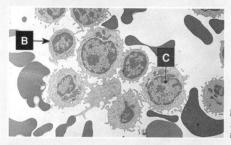

White blood cells and red blood cells (blood smear)

Photos: Ell

3. The two photomicrographs (left) show several types of animal cells. Identify the features indicated by the letters **A-C**:

A: _____

B: _____

C: _____

4. White blood cells are mobile, phagocytic cells, whereas red blood cells are smaller than white blood cells and, in humans, lack a nucleus.

(a) In the photomicrograph (below, left), circle a white blood cell and a red blood cell:

(b) With respect to the features that you can see, explain how you made your decision.

_____

_____

_____

_____

5. Name and describe one structure or organelle present in generalized animal cells but absent from plant cells:

_____

_____

© BIOZONE International 2012
ISBN: 978-1-927173-11-4
Photocopying Prohibited

Plant Cells

Eukaryotic cells have a similar basic structure, although they may vary tremendously in size, shape, and function. Certain features are common to almost all eukaryotic cells, including their three main regions: a **nucleus** (usually located near the center of the cell), surrounded by a watery **cytoplasm**, which is itself enclosed by the **plasma membrane**. Plant cells share many structures and organelles in common with animal cells, but also have several unique features. Plant cells are enclosed in a cellulose cell wall, which gives them a regular and uniform appearance. The cell wall protects the cell, maintains its shape, and prevents excessive water uptake. It provides rigidity to plant structures but permits the free passage of materials into and out of the cell.

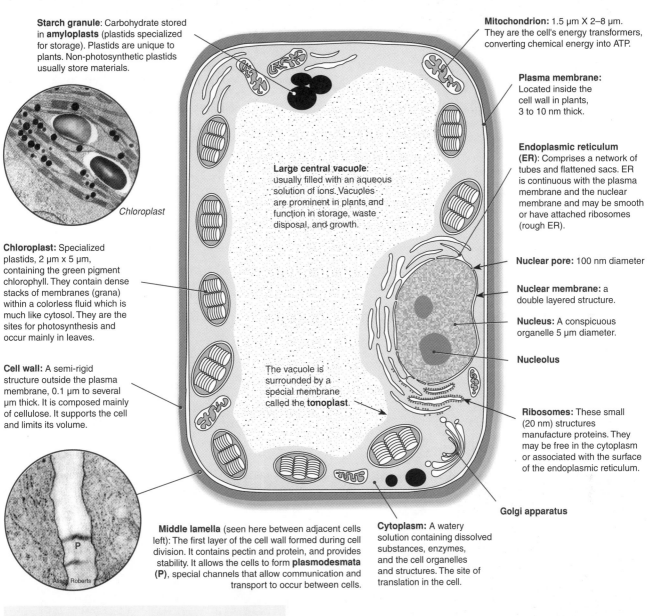

Starch granule: Carbohydrate stored in **amyloplasts** (plastids specialized for storage). Plastids are unique to plants. Non-photosynthetic plastids usually store materials.

Chloroplast

Chloroplast: Specialized plastids, 2 μm x 5 μm, containing the green pigment chlorophyll. They contain dense stacks of membranes (grana) within a colorless fluid which is much like cytosol. They are the sites for photosynthesis and occur mainly in leaves.

Cell wall: A semi-rigid structure outside the plasma membrane, 0.1 μm to several μm thick. It is composed mainly of cellulose. It supports the cell and limits its volume.

Mitochondrion: 1.5 μm X 2–8 μm. They are the cell's energy transformers, converting chemical energy into ATP.

Plasma membrane: Located inside the cell wall in plants, 3 to 10 nm thick.

Endoplasmic reticulum (ER): Comprises a network of tubes and flattened sacs. ER is continuous with the plasma membrane and the nuclear membrane and may be smooth or have attached ribosomes (rough ER).

Large central vacuole: usually filled with an aqueous solution of ions. Vacuoles are prominent in plants and function in storage, waste disposal, and growth.

Nuclear pore: 100 nm diameter

Nuclear membrane: a double layered structure.

Nucleus: A conspicuous organelle 5 μm diameter.

Nucleolus

The vacuole is surrounded by a special membrane called the **tonoplast**.

Ribosomes: These small (20 nm) structures manufacture proteins. They may be free in the cytoplasm or associated with the surface of the endoplasmic reticulum.

Golgi apparatus

Middle lamella (seen here between adjacent cells left): The first layer of the cell wall formed during cell division. It contains pectin and protein, and provides stability. It allows the cells to form **plasmodesmata (P)**, special channels that allow communication and transport to occur between cells.

Cytoplasm: A watery solution containing dissolved substances, enzymes, and the cell organelles and structures. The site of translation in the cell.

Alison Roberts

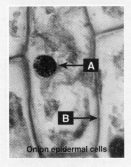
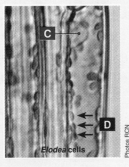

Onion epidermal cells · *Elodea* cells · Photos RCN

1. The photographs (left) show plant cells as seen by a light microscope. Identify the basic features labelled **A-D**:

A: _____

B: _____

C: _____

D: _____

2. Describe three structures/organelles present in generalized plant cells but absent from animal cells:

(a) _____

(b) _____

(c) _____

© BIOZONE International 2012
ISBN: 978-1-927173-11-4
Photocopying Prohibited

Related activities: *Identifying Structures in a Plant Cell*
Weblinks: *Eukaryotic Cells Interactive Animation, Review of Cell Structure*

2

The Cell's Cytoskeleton

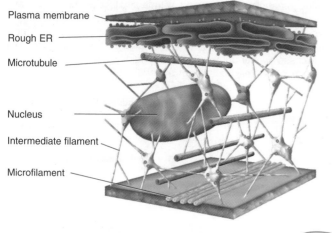

Plasma membrane
Rough ER
Microtubule
Nucleus
Intermediate filament
Microfilament

The cell's cytoplasm is not a fluid filled space; it contains a complex network of fibers called the **cytoskeleton**. The cytoskeleton provides tension and so provides structural support to maintain the cell's shape. The cytoskeleton is made up of three proteinaceous elements: microfilaments, intermediate filaments, and microtubules. Each has a distinct size structure and protein composition, and a specific role in cytoskeletal function. Cilia and flagella are made up of microtubules and for this reason they are considered to be part of the cytoskeleton.

The elements of the cytoskeleton are dynamic; they move and change to alter the cell's shape, move materials within the cell, and move the cell itself. Movement of materials is achieved through the action of motor proteins, which transport material by 'walking' along cytoskeletal 'tracks', hydrolyzing ATP at each step.

Microfilaments

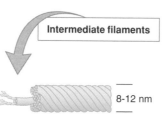

7 nm

Actin subunit

Intermediate filaments

8-12 nm

Microtubules

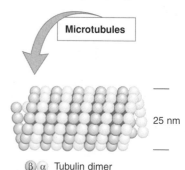

25 nm

β α Tubulin dimer

	Microfilaments	Intermediate filaments	Microtubules
Protein subunits	Actin	Fibrous proteins, e.g. keratin	α and β tubulin dimers
Structure	Two intertwined strands	Fibers wound into thicker cables	Hollow tubes
Functions	• Maintain cell shape • Motility (pseudopodia) • Contraction (muscle) • Cytokinesis of cell division	• Maintain cell shape • Anchor nucleus and organelles	• Maintain cell shape • Motility (cilia and flagella) • Move chromosomes (spindle) • Move organelles

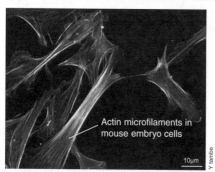

Actin microfilaments in mouse embryo cells

10μm

Y tambe

Intermediate filaments surrounding nucleus

NIH

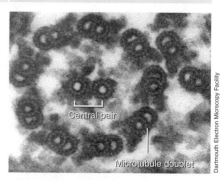

Central pair

Microtubule doublet

Dartmouth Electron Microscopy Facility

Microfilaments are long polymers of the protein actin. Microfilaments can grow and shrink as actin subunits are added or taken away from either end. Networks of microfilaments form a matrix that helps to define the cell's shape. Actin microfilaments are also involved in cell division (during cytokinesis) and in muscle contraction.

Intermediate filaments can be composed of a number of different fibrous proteins and are defined by their size rather than composition. The protein subunits are wound into cables around 10 nm in diameter. Intermediate filaments form a dense network within and projecting from the nucleus, helping to anchor it in place.

Microtubules are the largest cytoskeletal components and grow or shrink in length as tubulin subunits are added or subtracted from one end. The are involved in movement of material within the cell and in moving the cell itself. This EM shows a cilia from *Chlamydomonas*, with the 9+2 arrangement of microtubular doublets.

1. Describe the role that all components of the cytoskeleton have in common: _____

2. Explain the importance of the cytoskeleton being a dynamic structure: _____
   _____

3. Explain how the presence of a cytoskeleton could aid in directing the movement of materials within the cell:
   _____
   _____

A 2

Related activities: Cell Structures and Organelles

© BIOZONE International 2012
ISBN: 978-1-927173-11-4
Photocopying Prohibited

Cell Structures and Organelles

The tables below and on the two following pages provide a format to summarize information about the structures and organelles of typical eukaryotic cells. Complete the tables using the list provided and by referring to other pages in this topic. The first cell component has been completed for you as a guide and the log scale of measurements (top of next page) illustrates the relative sizes of some cellular structures. **List of structures and organelles**: *cell wall, mitochondrion, chloroplast, cell junctions, centrioles, ribosome, endoplasmic reticulum, Golgi apparatus, nucleus, flagellum, cytoskeleton,* and *vacuoles.*

Cell Component	Details	Present in		Visible under light microscope
		Plant cells	Animal cells	
(a) Double layer of phospholipids (called the lipid bilayer). Proteins	Name: Plasma (cell surface) membrane Location: Surrounding the cell Function: Gives the cell shape and protection. It also regulates the movement of substances into and out of the cell.	YES	YES	YES *(but not at the level of detail shown in diagram)*
(b) Large subunit, Small subunit	Name: Location: Function:			
(c) Outer membrane, Inner membrane, Matrix, Cristae	Name: Location: Function:			
(d) Secretory vesicles budding off, Cisternae, Transfer vesicles from the smooth endoplasmic reticulum	Name: Location: Function:			
(e) Ribosomes, transport pathway, Rough, Smooth, Vesicles budding off, Flattened membrane sacs	Name: Location: Function:			
(f) Grana comprise stacks of thylakoids, Stroma, Lamellae	Name: Location: Function:			

© BIOZONE International 2012
ISBN: 978-1-927173-11-4
Photocopying Prohibited

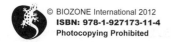

Periodicals: Cellular factories

Related activities: Animal Cells, Plant Cells
Weblinks: Cell Size and Scale

RA 2

The internal structure of Cells

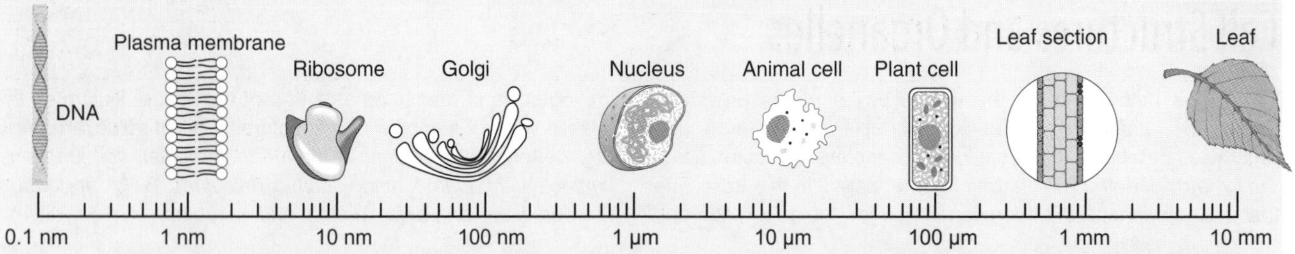

DNA

Plasma membrane

Ribosome

Golgi

Nucleus

Animal cell

Plant cell

Leaf section

Leaf

| 0.1 nm | 1 nm | 10 nm | 100 nm | 1 µm | 10 µm | 100 µm | 1 mm | 10 mm |

Cell Component	Details	Present in		Visible under light microscope
		Plant cells	Animal cells	
(g) 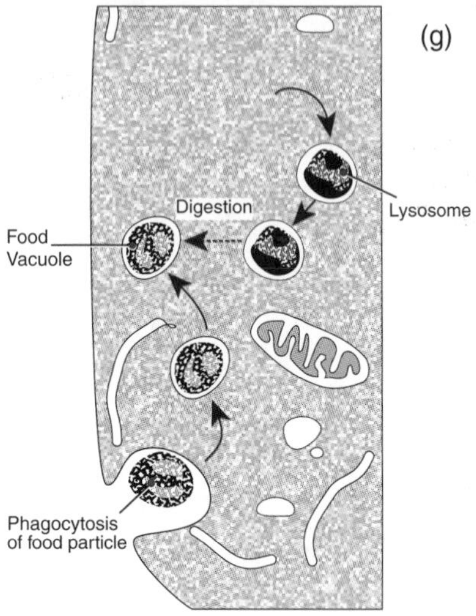 Food Vacuole — Digestion — Lysosome — Phagocytosis of food particle	Name: Lysosome and food vacuole Location: Function:			
(h) 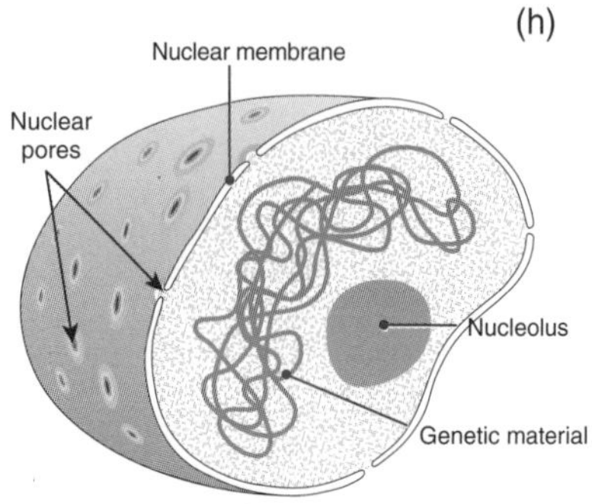 Nuclear membrane — Nuclear pores — Nucleolus — Genetic material	Name: Location: Function:			
(i) 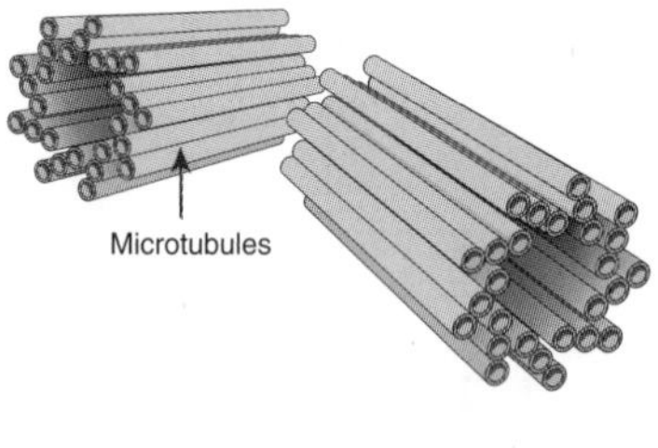 Microtubules	Name: Location: Function:			
(j) 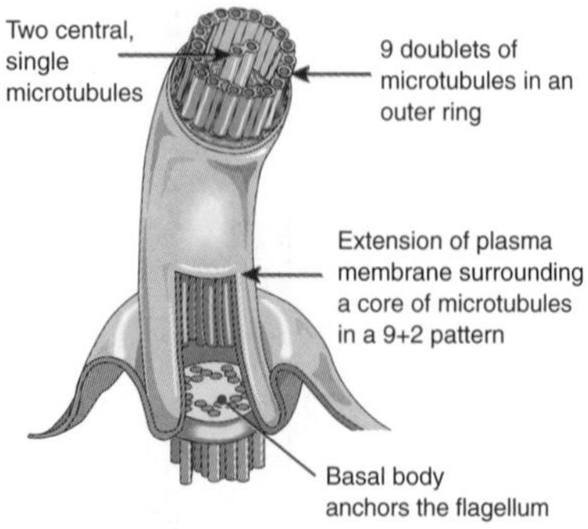 Two central, single microtubules — 9 doublets of microtubules in an outer ring — Extension of plasma membrane surrounding a core of microtubules in a 9+2 pattern — Basal body anchors the flagellum	Name: Location: Function:			

Cell Component	Details	Present in		Visible under light microscope
		Plant cells	Animal cells	
(k) 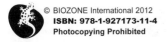 Plasma membrane / Organelle / Microtubule / Intermediate filament / Microfilament	Name: Location: Function:			
(l) Middle lamella / Pectins / Hemicelluloses / Cellulose fibres	Name: Cellulose cell wall Location: Function:			
(m) Tight junction / Desmosome / Gap junction / Extracellular matrix	Name: Cell junctions Location: Function:			

The internal structure of Cells

Cell Processes

All of the organelles and other structures in the cell have functions. The cell can be compared to a factory with an assembly line. Organelles in the cell provide the equivalent of the power supply, assembly line, packaging department, repair and maintenance, transport system, and the control center.

The sum total of all the processes occurring in a cell is known as **metabolism**. Some of these processes store energy in molecules (**anabolism**) while others release that stored energy (**catabolism**). Below is a summary of the major processes that take place in a cell.

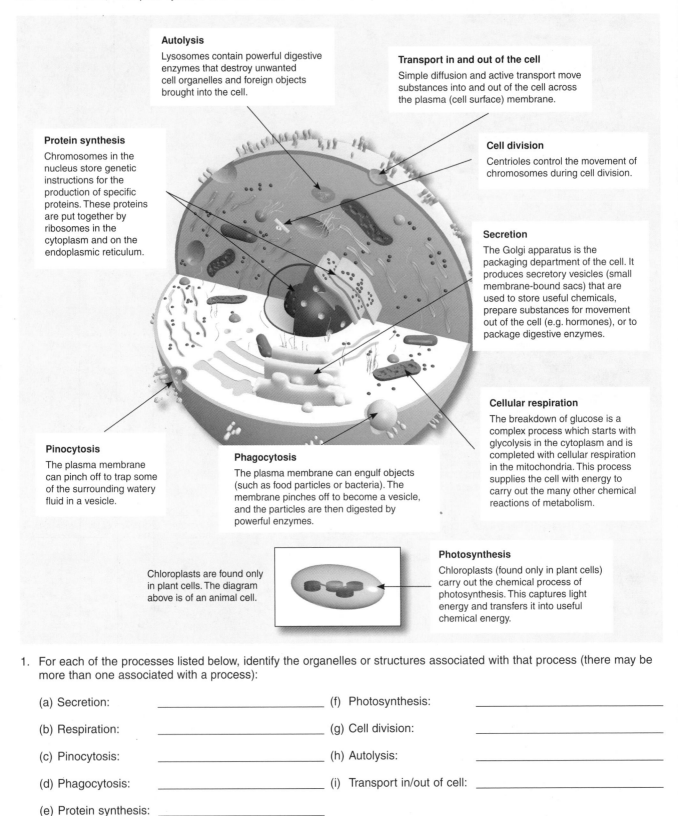

Autolysis
Lysosomes contain powerful digestive enzymes that destroy unwanted cell organelles and foreign objects brought into the cell.

Transport in and out of the cell
Simple diffusion and active transport move substances into and out of the cell across the plasma (cell surface) membrane.

Protein synthesis
Chromosomes in the nucleus store genetic instructions for the production of specific proteins. These proteins are put together by ribosomes in the cytoplasm and on the endoplasmic reticulum.

Cell division
Centrioles control the movement of chromosomes during cell division.

Secretion
The Golgi apparatus is the packaging department of the cell. It produces secretory vesicles (small membrane-bound sacs) that are used to store useful chemicals, prepare substances for movement out of the cell (e.g. hormones), or to package digestive enzymes.

Cellular respiration
The breakdown of glucose is a complex process which starts with glycolysis in the cytoplasm and is completed with cellular respiration in the mitochondria. This process supplies the cell with energy to carry out the many other chemical reactions of metabolism.

Pinocytosis
The plasma membrane can pinch off to trap some of the surrounding watery fluid in a vesicle.

Phagocytosis
The plasma membrane can engulf objects (such as food particles or bacteria). The membrane pinches off to become a vesicle, and the particles are then digested by powerful enzymes.

Chloroplasts are found only in plant cells. The diagram above is of an animal cell.

Photosynthesis
Chloroplasts (found only in plant cells) carry out the chemical process of photosynthesis. This captures light energy and transfers it into useful chemical energy.

1. For each of the processes listed below, identify the organelles or structures associated with that process (there may be more than one associated with a process):

   (a) Secretion: _____

   (b) Respiration: _____

   (c) Pinocytosis: _____

   (d) Phagocytosis: _____

   (e) Protein synthesis: _____

   (f) Photosynthesis: _____

   (g) Cell division: _____

   (h) Autolysis: _____

   (i) Transport in/out of cell: _____

2. Explain what is meant by **metabolism** and describe an example of a metabolic process: _____

   _____

   _____

Related activities: Animal Cells, Plant Cells

Golgi and the Endoplasmic Reticulum

Cells produce a range of organic polymers made up of repeating units of smaller molecules. The synthesis, packaging and movement of these **macromolecules** inside the cell involves a number of membrane bound organelles, as indicated below. These organelles provide compartments where the enzyme systems involved can be isolated.

The **Golgi** comprises stacks of flattened membranes in the shape of curved sacs. At its *cis-* face, the Golgi receives transport vesicles from the ER. Transported substances are modified, stored and shipped from the *trans-* face to the surface of the cell or other destinations.

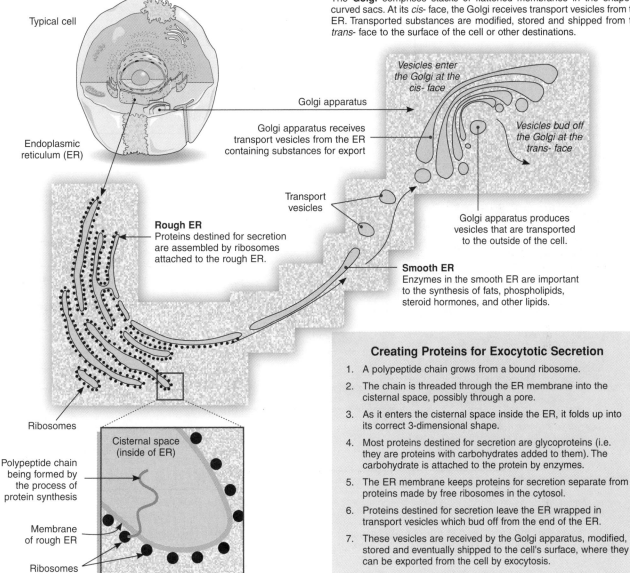

Typical cell

Endoplasmic reticulum (ER)

Golgi apparatus

Golgi apparatus receives transport vesicles from the ER containing substances for export

Vesicles enter the Golgi at the *cis-* face

Vesicles bud off the Golgi at the *trans-* face

Transport vesicles

Golgi apparatus produces vesicles that are transported to the outside of the cell.

Rough ER
Proteins destined for secretion are assembled by ribosomes attached to the rough ER.

Smooth ER
Enzymes in the smooth ER are important to the synthesis of fats, phospholipids, steroid hormones, and other lipids.

Ribosomes

Polypeptide chain being formed by the process of protein synthesis

Cisternal space (inside of ER)

Membrane of rough ER

Ribosomes

Creating Proteins for Exocytotic Secretion

1. A polypeptide chain grows from a bound ribosome.
2. The chain is threaded through the ER membrane into the cisternal space, possibly through a pore.
3. As it enters the cisternal space inside the ER, it folds up into its correct 3-dimensional shape.
4. Most proteins destined for secretion are glycoproteins (i.e. they are proteins with carbohydrates added to them). The carbohydrate is attached to the protein by enzymes.
5. The ER membrane keeps proteins for secretion separate from proteins made by free ribosomes in the cytosol.
6. Proteins destined for secretion leave the ER wrapped in transport vesicles which bud off from the end of the ER.
7. These vesicles are received by the Golgi apparatus, modified, stored and eventually shipped to the cell's surface, where they can be exported from the cell by exocytosis.

1. Using examples, explain what is meant by a macromolecule: _____

_____

2. Why are polypeptides requiring transport synthesized by membrane-bound (rather than free) ribosomes?

_____

_____

3. Why are most proteins destined for secretion from the cell glycoproteins? _____

_____

4. Briefly describe the roles of the following organelles in the production of macromolecules:

(a) Rough ER: _____

(b) Smooth ER: _____

(c) Golgi apparatus: _____

(d) Transport vesicles: _____

Identifying Structures in an Animal Cell

Our current knowledge of cell ultrastructure has been made possible by the advent of electron microscopy. Transmission electron microscopy is the most frequently used technique for viewing cellular organelles. When viewing TEMs, the cellular organelles may appear to be quite different depending on whether they are in transverse or longitudinal section.

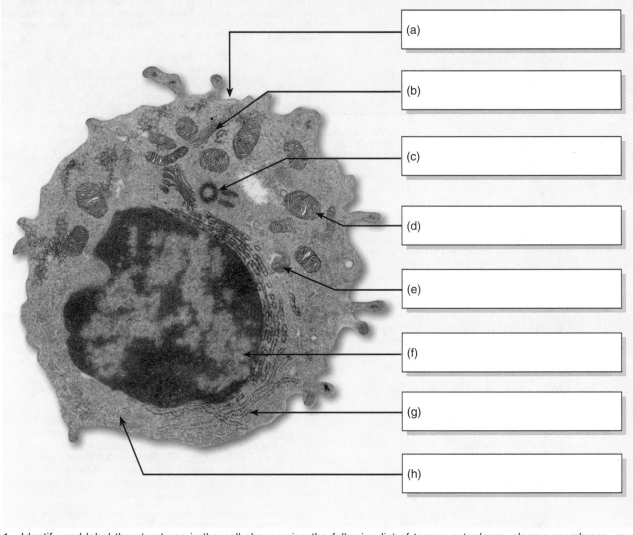

1. Identify and label the structures in the cell above using the following list of terms: *cytoplasm, plasma membrane, rough endoplasmic reticulum, mitochondrion, nucleus, centriole, Golgi apparatus, lysosome*

2. Which of the organelles in the EM above are shown in both transverse and longitudinal section?

_____

3. Why do plants lack the mobile phagocytic cells typical of animal cells? _____

_____

_____

4. The animal cell pictured above is a lymphocyte. Describe the features that suggest to you that:

   (a) It has a role in producing and secreting proteins: _____

   _____

   (b) It is metabolically very active: _____

   _____

5. What features of the lymphocyte cell above identify it as eukaryotic? _____

_____

_____

Identifying Structures in a Plant Cell

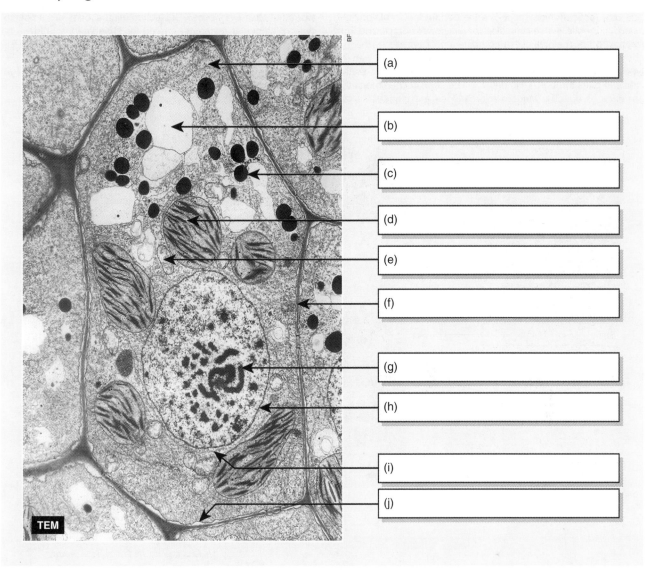

(a)

(b)

(c)

(d)

(e)

(f)

(g)

(h)

(i)

(j)

BF

TEM

1. Study the diagrams on the other pages in this chapter to familiarise yourself with the structures found in eukaryotic cells. Identify and label the ten structures in the cell above using the following list of terms: *nuclear membrane, cytoplasm, endoplasmic reticulum, mitochondrion, starch granules, chromosome, vacuole, plasma membrane, cell wall, chloroplast*

2. State how many cells, or parts of cells, are visible in the electron micrograph above: _____

3. Describe the features that identify this cell as a plant cell: _____

_____

_____

4. (a) Explain where cytoplasm is found in the cell: _____

_____

(b) Describe what cytoplasm is made up of: _____

_____

5. Describe two structures, pictured in the cell above, that are associated with storage:

(a) _____

_____

(b) _____

_____

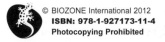 © BIOZONE International 2012
ISBN: 978-1-927173-11-4
Photocopying Prohibited

Optical Microscopes

The light (optical) microscope is an important tool in biology and using it correctly is an essential skill. High power **compound light microscopes** (below) use a combination of lenses to magnify objects up to several hundred times. A specimen viewed with this type of microscope must be thin and mostly transparent so that light can pass through it. No detail will be seen in specimens that are thick or opaque. Modern microscopes are binocular (have two adjustable eyepieces). Dissecting microscopes are a special type of binocular microscope used for observations at low total magnification (X4 to X50), where a large working distance between the objectives and stage is required. A dissecting microscope has two separate lens systems, one for each eye. Such microscopes produce a 3-D view of the specimen and are sometimes called stereo microscopes for this reason.

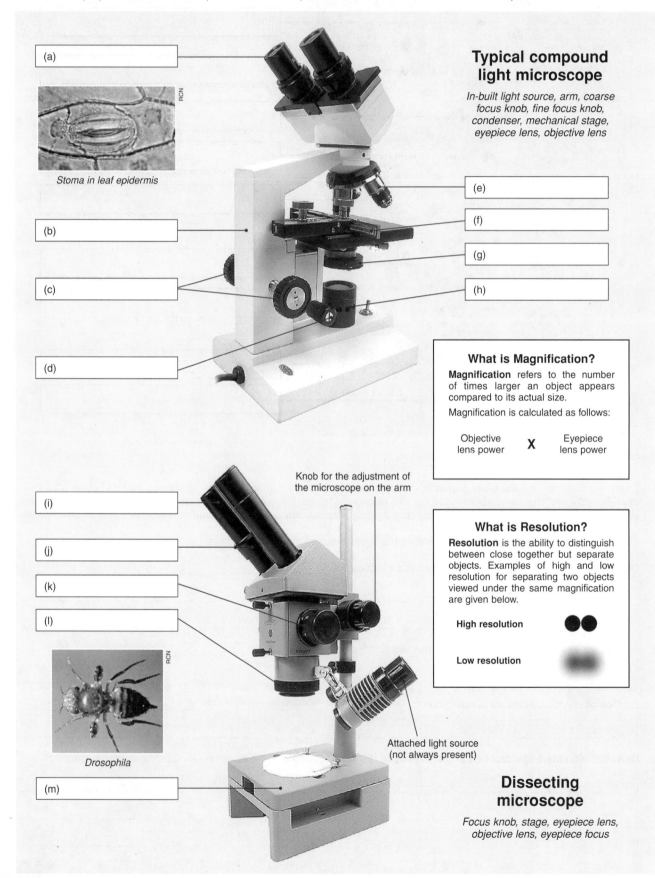

(a)

Stoma in leaf epidermis

(b)

(c)

(d)

Typical compound light microscope

In-built light source, arm, coarse focus knob, fine focus knob, condenser, mechanical stage, eyepiece lens, objective lens

(e)

(f)

(g)

(h)

What is Magnification?

Magnification refers to the number of times larger an object appears compared to its actual size.

Magnification is calculated as follows:

$$\text{Objective lens power} \quad \times \quad \text{Eyepiece lens power}$$

Knob for the adjustment of the microscope on the arm

(i)

(j)

(k)

(l)

What is Resolution?

Resolution is the ability to distinguish between close together but separate objects. Examples of high and low resolution for separating two objects viewed under the same magnification are given below.

High resolution

Low resolution

Drosophila

(m)

Attached light source (not always present)

Dissecting microscope

Focus knob, stage, eyepiece lens, objective lens, eyepiece focus

Related activities: *Plant Cells, Animal Cells*
Web links: *Light Microscopy Basics*

Periodicals:
Light microscopy

© BIOZONE International 2012
ISBN: 978-1-927173-11-4
Photocopying Prohibited

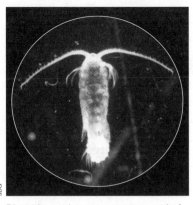 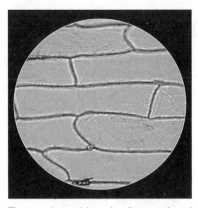 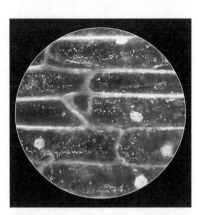

Dissecting microscopes are used for identifying and sorting organisms, observing microbial cultures, and dissections.

These onion epidermal cells are viewed with standard **bright field** lighting. Very little detail can be seen (only cell walls) and the cell nuclei are barely visible.

Dark field illumination is excellent for viewing specimens that are almost transparent. The nuclei of these onion epidermal cells are clearly visible.

1. Label the two photographs on the previous page, the compound light microscope (a) to (h) and the dissecting microscope (i) to (m). Use words from the lists supplied for each image.

2. Determine the magnification of a microscope using:

   (a) 15 X eyepiece and 40 X objective lens: _____   (b) 10 X eyepiece and 60 X objective lens: _____

3. Describe the main difference between a compound light microscope and a dissecting microscope: _____

   _____

   _____

4. What type of microscope would you use to:

   (a) Count stream invertebrates in a sample: _____   (b) Observe cells in mitosis: _____

5. (a) Distinguish between **magnification** and **resolution** (resolving power):_____

   _____

   (b) Explain the benefits of a higher resolution: _____

   _____

   _____

6. Below is a list of ten key steps taken to set up a microscope and optimally view a sample. The steps have been mixed up. Put them in their **correct order** by numbering each step:

   ☐ Focus and center the specimen using the high objective lens. Adjust focus using the fine focus knob only.

   ☐ Adjust the illumination to an appropriate level by adjusting the iris diaphragm and the condenser. The light should appear on the slide directly below the objective lens, and give an even amount of illumination.

   ☐ Rotate the objective lenses until the shortest lens is in place (pointing down towards the stage).
   This is the lowest / highest power objective lens (delete one).

   ☐ Place the slide on the microscope stage. Secure with the sample clips.

   ☐ Fine tune the illumination so you can view maximum detail on your sample.

   ☐ Focus and center the specimen using the medium objective lens. Focus firstly with the coarse focus knob, then with the fine focus knob (if needed).

   ☐ Turn on the light source.

   ☐ Focus and center the specimen using the low objective lens. Focus firstly with the coarse focus knob, then with the fine focus knob.

   ☐ Focus the eyepieces to adjust your view.

   ☐ Adjust the distance between the eyepieces so that they are comfortable for your eyes.

© BIOZONE International 2012
ISBN: 978-1-927173-11-4
Photocopying Prohibited

Interpreting Electron Micrographs

The photographs below were taken using a **transmission electron microscope** (TEM). They show the ultrastructure of some organelles. Remember that these photos are showing only **parts of cells**, **not whole cells**. Some of the photographs show more than one type of organelle. The questions refer to the main organelle indicated in the photo.

1. (a) State which kind of cell this is: _____

   (b) Identify the structure labelled A: _____

   _____

   (c) Describe the function of this structure: _____

   _____

   (d) Identify the structure labelled B: _____

   _____

2. (a) Name this organelle (arrowed): _____

   (b) State which kind of cell(s) this organelle would be found in:

   _____

   (c) Describe the function of this organelle: _____

   _____

   _____

3. (a) Name the large, circular organelle: _____

   (b) State which kind of cell(s) this organelle would be found in:

   _____

   (c) Describe the function of this organelle: _____

   _____

   _____

   (d) Label **two** regions that can be seen **inside** this organelle.

4. (a) Name and label the ribbon-like organelle in this photograph (arrowed):

   _____

   (b) State which kind of cell(s) this organelle is found in:

   _____

   (c) Describe the function of these organelles: _____

   _____

   (d) Name the dark 'blobs' attached to the organelle you have labelled:

   _____

5. (a) Name this large circular structure (arrowed): _____

   (b) State which kind of cell(s) this structure would be found in: _____

   _____

   (c) Describe the function of this structure: _____

   _____

   (d) Label three features relating to this structure in the photograph.

© BIOZONE International 2012
ISBN: 978-1-927173-11-4
Photocopying Prohibited

RA 2

Related activities: Animal Cells, Plant Cells
Web links: Eukaryotic Cells Interactive Animation

Microscopy Techniques

Specimens are often prepared in some way before viewing in order to highlight features and reveal details. A **wet mount** is a temporary preparation in which a specimen and a drop of fluid are trapped under a thin coverslip. Wet mounts are used to view live microscopic organisms but can also be used to view suspensions such as blood. A wet mount improves a sample's appearance and enhances visible detail. **Stains** and dyes can also be used to highlight specific components or structures. Most stains are used on dead specimens. these are called **non-viable stains**. Stains applied to living material are called **vital stains**.

Making a Temporary Wet Mount

1 **Sectioning:** Very thin sections of fresh material are cut with a razorblade.

2 **Mounting:** The thin section(s) are placed in the center of a clean glass microscope slide and covered with a drop of mounting liquid (e.g. water, glycerol, or stain). A coverslip is placed on top to exclude air (below).

3 If too much liquid is added the coverslip will sit too high, and it will be difficult to focus on the sample. If not enough liquid is used, it may quickly evaporate and the sample may dry up before it can be fully examined.

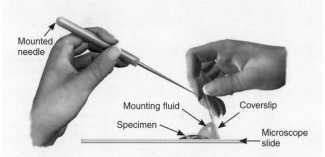

A mounted needle is used to support the coverslip and lower it gently over the specimen. This avoids including air in the mount.

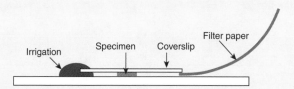

If a specimen is already mounted, a drop of stain can be placed at one end of the coverslip and drawn through using filter paper (above). Water can be drawn through in the same way to remove excess stain.

Staining Techniques

Some commonly used stains		
Stain	**Final color**	**Used for**
Iodine solution	blue-black	Starch
Crystal violet	purple	Gram staining
Aniline sulfate	yellow	lignin
Methylene blue	blue	Nuclei

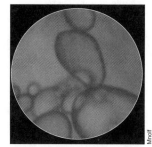

Iodine solution stains starch containing organelles, such as **potato amyloplasts**, blue-black.

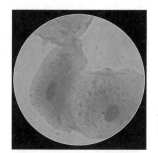

Methylene blue is a commonly used temporary stain for animal cells. It makes **nuclei** more visible.

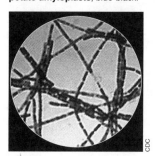

The gram stain contains **crystal violet**, which binds strongly to the peptidoglycan component in the cell walls of **gram positive bacteria**, turning them purple.

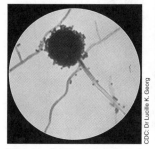

Vital (or viable) **stains** do not immediately harm living cells. **Trypan blue** distinguishes living and dead cells, and is used to study fungal hyphae (above).

1. Why are many microscope samples wet mounted prior to viewing under a microscope? _____

_____

2. What is the main purpose of using a stain? _____

_____

_____

3. What is the difference between a **vital** and **non-viable** stain? _____

_____

4. Identify a stain that would be appropriate for improving identification of the following:

(a) Fungal hyphae: _____   (d) Lignin in a plant root section: _____

(b) Starch in potato cells: _____   (e) Nuclei in cheek cells: _____

(c) Cell wall of bacteria: _____

Related activities: Optical Microscopes

RA 2

Calculating Linear Magnification

Microscopes produce an enlarged (magnified) image of an object allowing it to be observed in greater detail than is possible with the naked eye. **Magnification** refers to the number of times larger an object appears compared to its actual size. The degree of magnification possible depends upon the type of microscopy used. **Linear magnification** is calculated by taking a ratio of the image height to the object's actual height. If this ratio is greater than one, the image is enlarged, if it is less than one, it is reduced. To calculate magnification, all measurements should be converted to the same units. Most often, you will be asked to calculate an object's actual size, in which case you will be told the size of the object, as viewed through the microscope, and given the magnification.

Calculating Linear Magnification: A Worked Example

1 Measure the body length of the bed bug image (right). Your measurement should be 40 mm (*not* including the body hairs and antennae).

2 Measure the length of the scale line marked 1.0 mm. You will find it is 10 mm long. The magnification of the scale line can be calculated using equation 1 (below right).

The magnification of the scale line is **10** (10 mm / 1 mm)

NB: The magnification of the bed bug image will also be 10x because the scale line and image are magnified to the same degree.

3 Calculate the actual (real) size of the bed bug using equation 2 (right):

The actual size of the bed bug is **4 mm** (40 mm / 10 x magnification)

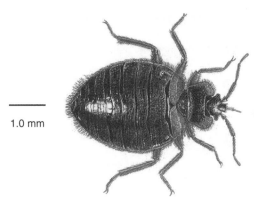

1.0 mm

Microscopy Equations

1. Magnification $=\dfrac{\text{size of the image}}{\text{actual size of object}}$

2. Actual object size $=\dfrac{\text{size of the image}}{\text{magnification}}$

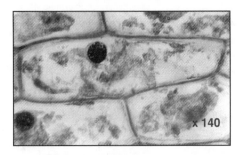

x 140

1. The bright field microscopy image on the left is of onion epidermal cells. The measured length of the onion cell in the center of the photograph is 52 000 µm (52 mm). The image has been magnified 140 x. Calculate the actual size of the cell:

_____

_____

_____

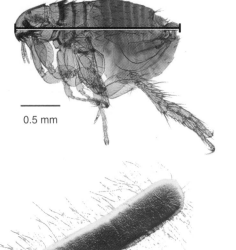

0.5 mm

2. The image of the flea (left) has been captured using light microscopy.

(a) Calculate the magnification using the scale line on the image:

_____

_____

_____

(b) The body length of the flea is indicated by a line. Measure along the line and calculate the actual length of the flea:

_____

_____

_____

3. The image size of the *E.coli* cell (left) is 43 mm, and its actual size is 2 µm. Using this information, calculate the magnification of the image:

_____

_____

_____

© BIOZONE International 2012
ISBN: 978-1-927173-11-4
Photocopying Prohibited

DA 1   *Related activities: Cell Sizes*

Periodicals:
Size does matter

Making Biological Drawings

Many observational studies made using microscopes will require you to make accurate representations of what you see. Although some observations will be made using relatively low power (X40) microscopy, some histological preparations require higher magnifications to identify the finer structure of the tissue. Tissue sections will usually be provided as longitudinal (LS) or transverse sections (TS). When you have access to both TS and LS images from the same specimen it is also possible to extrapolate the three dimensional shape of the structure under view. Observational drawing from a microscope is a skill that must be developed. It requires relaxed viewing (right) in which the image is viewed with one eye, while the other eye attends to the drawing being made. Attention should be given to the symmetry and proportions of the structure, accurate labelling, statement of magnification and sectioning, and stain used, if this is appropriate. In this activity, you will practice the skills required to translate what is viewed into a good biological drawing.

1. **Materials**: Use clear pencil lines on good quality paper. You will need a sharp HB pencil and a good quality eraser.

2. **Size and positioning**: Center your diagram on the page, not in a corner. This will leave room for labels. The drawing should be large enough to easily represent all the details you see without crowding. Show only as much as is necessary for an understanding of the structure.

3. **Accuracy**: Your drawing should be a complete, accurate representation of what you have observed, and should communicate your understanding of the material to anyone who looks at it. A biological drawing is distinct from a diagram, which is idealized and may contain more structure than can be seen in one section. Proportions should be accurate. If necessary, measure the lengths of various parts with a ruler. If viewing through a microscope, estimate them as a proportion of the field of view, then translate these proportions onto the page. When drawing shapes that indicate a discrete outline, make sure the line is complete.

4. **Technique**: Use only simple, narrow lines. Represent depth by stippling and use it only when it is essential to your drawing. Look at the specimen while you are drawing it.

6. **Labels**: All parts of your drawing must be labelled accurately. Labelling lines should be drawn with a ruler and should not cross. Where possible, keep label lines vertical or horizontal.

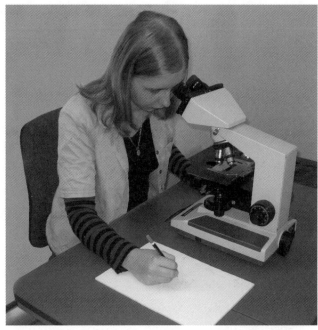

Above: Use relaxed viewing when drawing at the microscope. Use one eye (the left for right handers) to view and the right eye to view and direct your drawing.

Label the drawing with an explanatory **title**, identifying the subject, **magnification** or a **scale** to indicate size, names of structures, and any movements you see in living specimens.

Remember that drawings are intended as a record and as a means of encouraging close observation; artistic ability is not necessary. Before you turn in a drawing, ask yourself if you know what every line represents. If you do not, look more closely at the material. ***Draw what you see, not what you think you see!***

Examples of acceptable biological drawings: The diagrams below show two acceptable biological drawings. The left shows a whole organism and its size is indicated by a scale. The right shows plant tissue. It is not necessary to show many cells even though your view through the microscope may show them. You need enough to show their structure and how they are arranged. Scale is indicated by stating how many times larger it has been drawn. Do not confuse this with what magnification it was viewed at under the microscope. **T.S.** indicates a *transverse section*.

Cyclopoid copepod

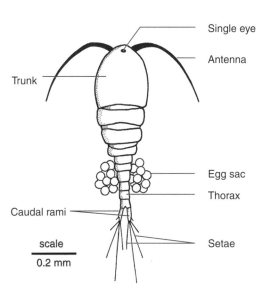

Collenchyma T.S.
from *Helianthus* stem Magnification x 450

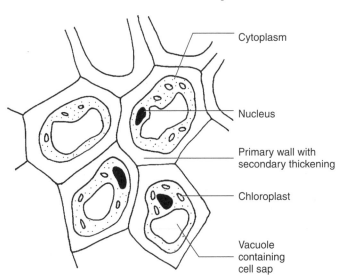

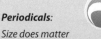

Periodicals:
Size does matter

Related activities: Optical Microscopes
Web links: Scientific Drawings for Biological Courses

A 2

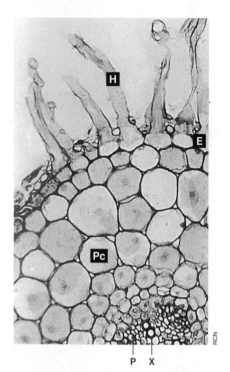

Specimen used for drawing

The photograph above is a light microscope view of a stained transverse section (cross section) of a root from a *Ranunculus* (buttercup) plant. It shows the arrangement of the different tissues in the root. The vascular bundle is at the center of the root, with the larger, central xylem vessels (**X**) and smaller phloem vessels (**P**) grouped around them. The root hair cells (**H**) are arranged on the external surface and form part of the epidermal layer (**E**). Parenchyma cells (**Pc**) make up the bulk of the root's mass. The distance from point **X** to point **E** on the photograph (above) is about 0.15 mm (150 µm).

An Unacceptable Biological Drawing

The diagram below is an example of how *not* to produce a biological drawing; it is based on the photograph to the left. There are many aspects of the drawing that are unacceptable. The exercise below asks you to identify the errors in this student's attempt.

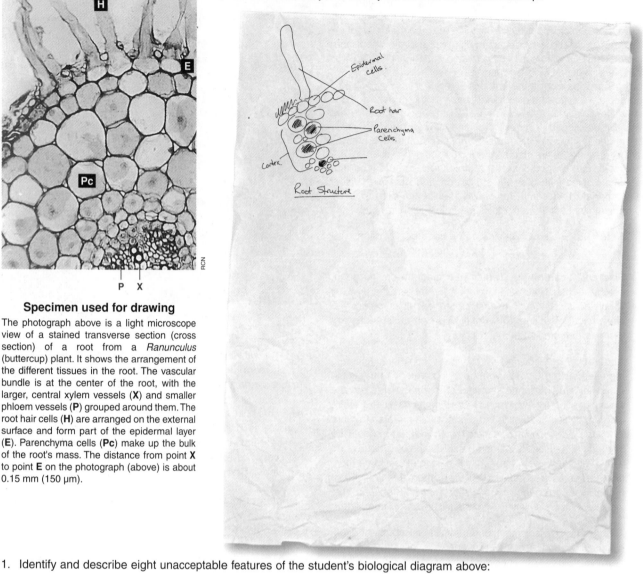

1. Identify and describe eight unacceptable features of the student's biological diagram above:

   (a) _____

   (b) _____

   (c) _____

   (d) _____

   (e) _____

   (f) _____

   (g) _____

   (h) _____

2. In the remaining space next to the 'poor example' (above) or on a blank piece of paper, attempt your own version of a biological drawing for the same material, based on the photograph above. Make a point of correcting all of the errors that you have identified in the sample student's attempt.

3. Why is an accurate biological drawing more valuable to a scientific investigation than an 'artistic' one?

   _____

   _____

   _____

© BIOZONE International 2012
ISBN: 978-1-927173-11-4
Photocopying Prohibited

KEY TERMS: Crossword

Complete the crossword below, which will test your understanding of key terms in this chapter and their meanings.

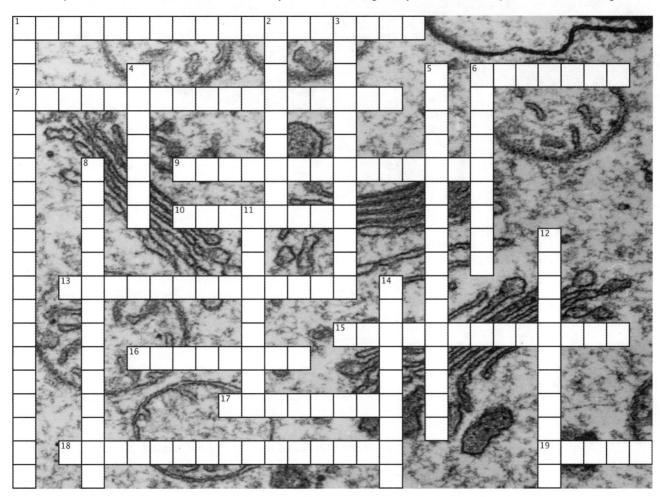

Clues Across

1. An advanced microscope that uses electron beams to produce high resolution images (2 words: 8, 10).

6. Membrane bound area within a eukaryotic cell where the chromosomes are found.

7. A type of microscope in which lenses use light to magnify objects (2 words: 7, 10)

9. Lipid bilayer membrane surrounding the cell. Proteins are embedded in it and are responsible for the passage of material into and out of the cell (2 words: 6, 8)

10. Membrane bound vacuolar organelle that contains enzymes that form part of the intracellular digestive system.

13. How many times larger an image is than the original object.

15. Organelle responsible for producing the cell's energy. It appears oval in shape with an outer double membrane and a convoluted interior membrane. It has its own circular DNA.

16. Long, filamentous appendages in bacteria used for locomotion.

17. A structure, present in plants and bacteria, which is found outside the plasma membrane and gives rigidity to the cell (2 words: 4, 4)

18. Cells with a membrane bound nucleus and organelles.(2 words:10,5)

19. A chemical that binds to parts of the cell and allows those parts to be seen more easily under a microscope.

Clues Down

1. An organelle comprising a convoluted membranous stack and divided into rough and smooth regions. It plays a part in protein and membrane synthesis. (2 words: 11, 9)

2. Small structures comprising RNA and protein that are found in all cells. They function in translation of the genetic code (mRNA into proteins).

3. A network of actin filaments and microtubules within the cytosol that provide structure and assist the movement of materials within the cell.

4. A membrane-bound cavity in the cytoplasm of eukaryotic cells, usually filled with an aqueous solution of ions. They are much larger in plant cells than in animal cells.

5. Cells that lack a membrane-bound nucleus or organelles. (2 words:10, 5)

6. An organelle found within the nucleus that contains ribosomal RNA and is associated with the part of the chromosome that codes for rRNA.

8. Organelle that resembles a series of flattened stacks. It modifies and packages proteins and also performs a secretory function by budding off vesicles (2 words: 5, 9)

11. A structural and functional part of the cell usually bound within its own membrane. Examples include the mitochondria and lysosomes.

12. An organelle found in photosynthetic organisms such as plants, which contains chlorophyll and in which the reactions of photosynthesis take place.

14. The watery contents of the cell within the plasma membrane, but excluding the contents of the nucleus.

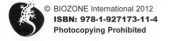

Cell Membranes and Transport

Enduring Understanding

2.B
4.A

Key concepts

▶ Cellular metabolism depends on the transport of substances across cellular membranes.

▶ Cell size is limited by surface area to volume ratio.

▶ New cells arise through cell division.

▶ Cellular diversity arises through specialization from stem cell progenitors.

▶ Regulation of cell division is important in development.

▶ Emergent properties are a feature of increasing complexity in biological systems.

Key terms

active transport
amphipathic
aquaporin
carrier protein
cell wall
channel protein
concentration gradient
diffusion
endocytosis
exocytosis
facilitated diffusion
fluid mosaic model
glycolipid
glycoprotein
hypertonic
hypotonic
ion pump
isotonic
osmosis
passive transport
phagocytosis
phospholipid
pinocytosis
plasma membrane
plasmolysis
selectively permeable
surface area: volume ratio
transmembrane protein
turgor

Essential Knowledge

☐ 1. Use the **KEY TERMS** to compile a glossary for this topic.

The Structure of Cell Membranes (2.B.1, 4.A.1) pages 37-38, 41,59-60, 63-64

☐ 2. Describe the role of the **plasma membrane** in separating the internal environment of the cell from the external environment.

☐ 3. Describe the **fluid mosaic model** of the plasma membrane, including the significance of the **amphipathic** character of the phospholipids that make up the structural framework of the membrane and the role of **transmembrane proteins**, **glycoproteins**, and **glycolipids**.

☐ 4. Explain how the properties of the embedded proteins contribute to the selectively permeable nature of the membrane. Include reference to **aquaporins**, and embedded **channel proteins** and **carrier proteins**.

☐ 5. Describe and evaluate the experimental evidence for the current model of membrane structure.

☐ 6. Distinguish between the plasma membrane and the **cell wall** of plants, bacteria, algae, fungi, and some Archaea. Recognize that the cell wall lies outside the plasma membrane and provides structural boundary to the cell and a permeability barrier for some substances.

Transport Across Membranes (2.B.2)   pages 61-62, 64-70

☐ 7. Describe and explain **diffusion**, **facilitated diffusion**, and **osmosis**, identifying them as **passive transport** processes. Explain the factors affecting diffusion rates across membranes: **membrane thickness**, **surface area**, and **concentration gradient**.

☐ 8. Explain the role of passive transport in the import of resources and export of waste products. Using an example, e.g. the transport of glucose (a polar molecule) into red blood cells, explain how membrane proteins are involved in the facilitated diffusion of charged and polar molecules across a membrane.

☐ 9. Explain why cell size is limited by the rate of diffusion. Recall the significance of **surface area to volume ratio** to cells and relate this to organism size.

☐ 10. Explain **turgor** and **plasmolysis** in plant cells. With respect to solutions of differing solute concentration, explain **hypotonic**, **isotonic**, and **hypertonic**.

☐ 11. Distinguish between passive transport and **active transport**, identifying the involvement of membrane proteins and energy in active transport processes.

☐ 12. Using examples, describe and explain active transport processes in cells, including **ion pumps**, **endocytosis**, and **exocytosis**.

Periodicals:
Listings for this chapter are on page 383

Weblinks:
www.thebiozone.com/
weblink/AP1-3114.html

BIOZONE APP:
Student Review Series
Cell Membranes & Transport

The Structure of Membranes

All cells have a plasma membrane that forms the outer limit of the cell. Bacteria, fungi, and plant cells have a cell wall outside this, but it is quite distinct and outside the plasma membrane. Membranes are also found inside eukaryotic cells as part of membranous **organelles**. Present day knowledge of membrane structure has been built up as a result of many observations and experiments. The original model of membrane structure, proposed by Davson and Danielli, was the unit membrane (a lipid bilayer coated with protein). This model was later modified after the discovery that the protein molecules were embedded *within* the bilayer rather than coating the outside. The now-accepted model of membrane structure is the **fluid mosaic model** described below.

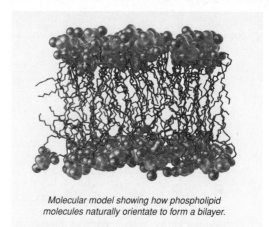

Molecular model showing how phospholipid molecules naturally orientate to form a bilayer.

The Fluid Mosaic Model of Membrane Structure

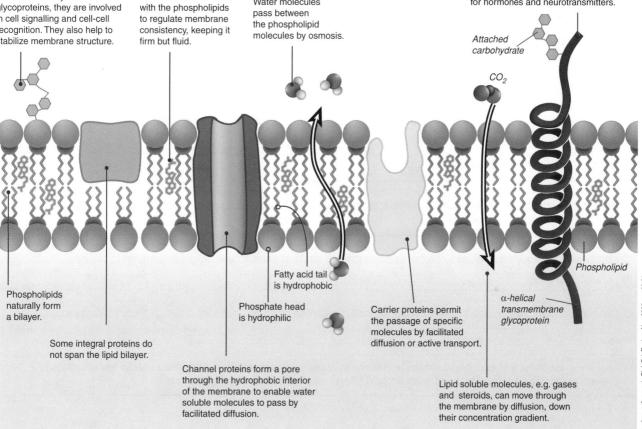

Glycolipids in membranes are phospholipids with attached carbohydrate. Like glycoproteins, they are involved in cell signalling and cell-cell recognition. They also help to stabilize membrane structure.

Cholesterol is a packing molecule and interacts with the phospholipids to regulate membrane consistency, keeping it firm but fluid.

Water molecules pass between the phospholipid molecules by osmosis.

Glycoproteins are proteins with attached carbohydrate. They are important in membrane stability, in cell-cell recognition, and in cell signalling, acting as receptors for hormones and neurotransmitters.

Attached carbohydrate

CO_2

Phospholipids naturally form a bilayer.

Some integral proteins do not span the lipid bilayer.

Channel proteins form a pore through the hydrophobic interior of the membrane to enable water soluble molecules to pass by facilitated diffusion.

Fatty acid tail is hydrophobic

Phosphate head is hydrophilic

Carrier proteins permit the passage of specific molecules by facilitated diffusion or active transport.

Lipid soluble molecules, e.g. gases and steroids, can move through the membrane by diffusion, down their concentration gradient.

α-helical transmembrane glycoprotein

Phospholipid

Based on a diagram in Biol. Sci. Review, Nov. 2009, pp. 20-21

1. Identify the component(s) of the plasma membrane involved in:

   (a) Facilitated diffusion: _____   (c) Cell signaling: _____

   (b) Active transport: _____   (d) Regulating membrane fluidity: _____

2. How do the properties of phospholipids contribute to their role in forming the structural framework of membranes?

   _____

   _____

3. (a) Describe the modern fluid mosaic model of membrane structure: _____

   _____

   _____

   _____

Periodicals:
Border control
The fluid mosaic model

Related activities: *Lipids, The Role of Membranes in Cells*
Web links: *Membrane Structure Tutorial*

RA 2

(b) Explain how the fluid mosaic model accounts for the observed properties of cellular membranes:

_____

_____

_____

_____

_____

_____

4. Explain the importance of each of the following to cellular function:

(a) High membrane surface area:_____

_____

(b) Channel proteins and carrier proteins in the plasma membrane: _____

_____

5. Non-polar (lipid-soluble) molecules diffuse more rapidly through membranes than polar (lipid-insoluble) molecules:

(a) Explain the reason for this: _____

_____

(b) Discuss the implications of this to the transport of substances into the cell through the plasma membrane:

_____

_____

6. Describe the purpose of cholesterol in plasma membranes: _____

_____

7. List three substances that need to be transported **into** all kinds of animal cells, in order for them to survive:

(a) _____ (b) _____ (c) _____

8. List two substances that need to be transported **out** of all kinds of animal cells, in order for them to survive:

(a) _____ (b) _____

9. Use the symbol for a phospholipid molecule (below) to draw a **simple labelled diagram** to show the structure of a plasma membrane (include features such as lipid bilayer and various kinds of proteins):

Symbol for phospholipid

The Role of Membranes in Cells

Many of the important structures and organelles in cells are composed of, or are enclosed by, membranes. These include: the endoplasmic reticulum, mitochondria, nucleus, Golgi apparatus, chloroplasts, lysosomes, vesicles and the cell plasma membrane itself. All membranes within eukaryotic cells share the same basic structure as the plasma membrane that encloses the entire cell. They perform a number of critical functions in the cell: compartmentalizing regions of different function within the cell, controlling the entry and exit of substances, and fulfilling a role in recognition and communication between cells. Some of these roles are described below and electron micrographs of the organelles involved are on the following page.

Isolation of enzymes
Membrane-bound lysosomes contain enzymes for the destruction of wastes and foreign material. Peroxisomes are the site for destruction of the toxic and reactive molecule, hydrogen peroxide (formed as a result of some cellular reactions).

Role in lipid synthesis
The smooth ER is the site of lipid and steroid synthesis.

Containment of DNA
The nucleus is surrounded by a nuclear envelope of two membranes, forming a separate compartment for the cell's genetic material.

Role in protein and membrane synthesis
Some protein synthesis occurs on free ribosomes, but much occurs on membrane-bound ribosomes on the rough endoplasmic reticulum. Here, the protein is synthesized directly into the space within the ER membranes. The rough ER is also involved in membrane synthesis, growing in place by adding proteins and phospholipids.

Cell communication and recognition
The proteins embedded in the membrane act as receptor molecules for hormones and neurotransmitters. Glycoproteins and glycolipids stabilize the plasma membrane and act as cell identity markers, helping cells to organize themselves into tissues, and enabling foreign cells to be recognized.

Packaging and secretion
The Golgi apparatus is a specialized membrane-bound organelle which produces lysosomes and compartmentalizes the modification, packaging and secretion of substances such as proteins and hormones.

Transport processes
Channel and carrier proteins are involved in selective transport across the plasma membrane. The level of cholesterol in the membrane influences permeability and transport functions.

Entry and export of substances
The plasma membrane may take up fluid or solid material and form membrane-bound vesicles (or larger vacuoles) within the cell. Membrane-bound transport vesicles move substances to the inner surface of the cell where they can be exported from the cell by exocytosis.

Energy transfer
The reactions of cellular respiration (and photosynthesis in plants) take place in the membrane-bound energy transfer systems occurring in mitochondria and chloroplasts respectively. See the example explained below.

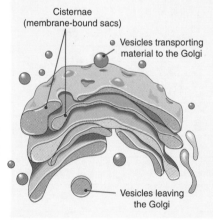

Cisternae (membrane-bound sacs)
Vesicles transporting material to the Golgi
Vesicles leaving the Golgi

Compartmentation within Membranes

Membranes play an important role in separating regions within the cell (and within organelles) where particular reactions occur. Specific enzymes are therefore often located in particular organelles. The reaction rate is controlled by controlling the rate at which substrates enter the organelle and therefore the availability of the raw materials required for the reactions.

The Golgi (diagram left and TEM right) modifies, sorts, and packages macromolecules for cell secretion. Enzymes within the cisternae modify proteins by adding carbohydrates and phosphates. To do this, the Golgi imports the substances it needs from the cytosol.

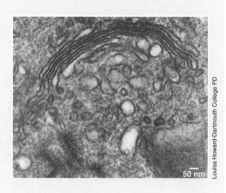

50 nm

Cell Membranes and Transport

1. Discuss the importance of membrane systems and organelles in providing compartments within the cell:

Related activities: Cell Structures and Organelles, Golgi and the Endoplasmic Reticulum
Web links: Cell Membranes

A 2

Functional Roles of Membranes in Cells

The **nuclear membrane**, which surrounds the nucleus, regulates the passage of genetic information to the cytoplasm and may also protect the DNA from damage.

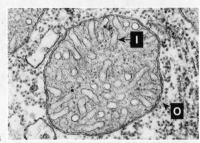

Mitochondria have an outer membrane (**O**) which controls the entry and exit of materials involved in aerobic respiration. Inner membranes (**I**) are the site of enzyme activity.

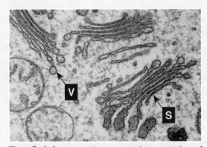

The **Golgi apparatus** comprises stacks of membrane-bound sacs (**S**). It is involved in packaging materials for transport or export from the cell as secretory vesicles (**V**).

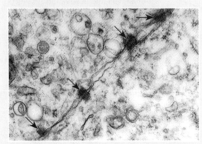

The **plasma membrane** surrounds the cell. In this photo, intercellular junctions called **desmosomes**, which connect neighboring cells, are indicated with arrows.

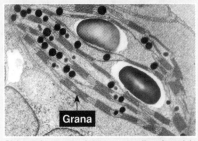

Chloroplasts are large organelles found in plant cells. The stacked membrane systems of chloroplasts (grana) trap light energy which is then used to fix carbon into 6-C sugars.

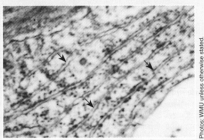

This EM shows stacks of rough endoplasmic reticulum (arrows). The membranes are studded with ribosomes, which synthesize proteins into the intermembrane space.

2. Match each of the following organelles with the correct description of its functional role in the cell:

chloroplast, rough endoplasmic reticulum, lysosome, smooth endoplasmic reticulum, mitochondrion, Golgi apparatus

(a) Active in synthesis, sorting, and secretion of cell products: _____

(b) Digestive organelle where macromolecules are hydrolyzed: _____

(c) Organelle where most cellular respiration occurs and most ATP is generated: _____

(d) Active in membrane synthesis and synthesis of secretory proteins: _____

(e) Active in lipid and hormone synthesis and secretion: _____

(f) Photosynthetic organelle converts light energy to chemical energy stored in sugar molecules: _____

3. Explain how the membrane surface area is increased within cells and organelles: _____

_____

4. Discuss the various functional roles of membranes in cells: _____

_____

_____

_____

_____

5. (a) Name a cellular organelle that possesses a membrane: _____

(b) Describe the membrane's purpose in this organelle: _____

_____

_____

_____

_____

How Do We Know? Membrane Structure

Cellular membranes play many extremely important roles in cells, and understanding their structure is central to understanding cellular function. Moreover, understanding the structure and function of membrane proteins is essential to understanding cellular transport processes, and cell recognition and signaling. Cellular membranes are far too small to be seen clearly using light microscopy, and certainly any detail is impossible to resolve. Since early last century, scientists have known that membranes comprised a lipid bilayer with associated proteins. But how did they elucidate just how these molecules were organized?

The answers were provided with electron microscopy, and one technique in particular – **freeze fracture**. As the name implies, freeze fracture, at its very simplest level, is the freezing of a cell and then cleaving it so that it fractures in a certain way. Scientists can then use electron microscopy to see the indentations and outlines of the structures remaining after cleavage. Membranes are composed of two layers held together by weak intermolecular bonds, so they cleave into two halves when fractured. This provides views of the inner surfaces of the membrane.

The procedure involves several steps:

▶ The tissue is prefixed using a cross linking agent. This alters the strength of the internal and external parts of the membrane.
▶ The cell is fixed to immobilise any mobile macromolecules.
▶ The specimen is passed through a sequential series of glycerol solutions of increasing strength. This protects the cells from bursting when placed into the cryomaterial.
▶ The specimen is frozen using liquid propane cooled by liquid nitrogen. The specimens are mounted on gold supports and cooled briefly before transfer to the freeze-etch machine.
▶ Specimen is cleaved in a helium-vented vacuum at -150°C. A razor blade cooled to -170°C acts as both a cold trap for water and the cleaving instrument.
▶ At this stage the specimen may be evaporated a little to produce some relief in the surface of the fracture (known as etching) so that a 3-dimensional effect occurs.
▶ For viewing under EM, a replica of the specimen is made and coated in gold or platinum to ~3 nm thick. This produces a shadow effect that allows structures to be seen clearly. A layer of carbon around 30 nm thick is used to stabilize the specimen.
▶ The samples are then raised to room temperature and placed into distilled water or digestive enzymes, which allows the replica to separate from the sample. The replica is then rinsed several times in distilled water before it is ready for viewing.

The freeze fracture technique provided the necessary supporting evidence for the current fluid mosaic model of membrane structure. When cleaved, proteins in the membrane left impressions that showed they were embedded into the membrane and not a continuous layer on the outside as earlier models proposed.

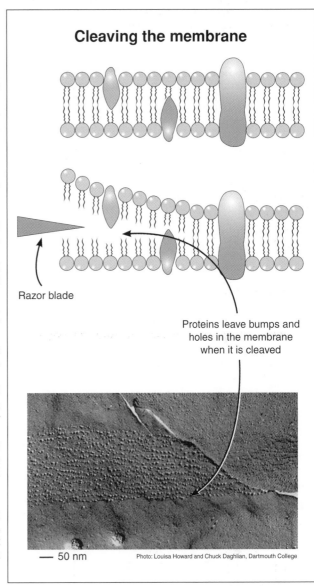

Cleaving the membrane

Razor blade

Proteins leave bumps and holes in the membrane when it is cleaved

— 50 nm   Photo: Louisa Howard and Chuck Daghlian, Dartmouth College

1. Describe the principle of freeze fracture and explain why it is such a useful technique for studying membrane structure:

_____

_____

_____

2. Explain how this freeze-fracture studies provided evidence for our current model of membrane structure: _____

_____

_____

3. An earlier model of membrane structure was the unit membrane; a phospholipid bilayer with a protein coat. Explain how the freeze-fracture studies showed this model to be flawed:

_____

_____

_____

Related activities: The Structure of Membranes   **RA 2**

Cell Membranes and Transport

Passive Transport Processes

The molecules that make up substances are constantly moving about in a random way. This random motion causes molecules to disperse from areas of high to low concentration. This movement is called **diffusion**. The molecules move down a **concentration gradient**. Diffusion and **osmosis** (diffusion of water molecules across a selectively permeable membrane) are **passive** processes, and use no energy. Diffusion occurs freely across membranes, as long as the membrane is permeable to that molecule (selectively permeable membranes allow the passage of some molecules but not others). Each type of molecule diffuses down its own concentration gradient. Diffusion of molecules in one direction does not hinder the movement of other molecules. Diffusion is important in allowing exchanges with the environment and in the regulation of cell water content.

Diffusion is the movement of particles from regions of high to low concentration (down a **concentration gradient**), with the end result being that the molecules become evenly distributed. In biological systems, diffusion often occurs across selectively permeable membranes.

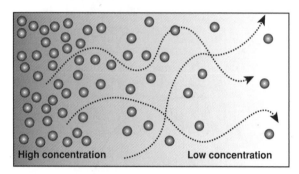

High concentration ...... **Low concentration**

Concentration gradient

If molecules are free to move, they move from high to low concentration until they are evenly dispersed.

Factors affecting rates of diffusion

Concentration gradient:	Diffusion rates will be higher when there is a greater difference in concentration between two regions.
The distance involved:	Diffusion over shorter distances occurs at a greater rate than diffusion over larger distances.
The area involved:	The larger the area across which diffusion occurs, the greater the rate of diffusion.
Barriers to diffusion:	Thicker barriers slow diffusion rate. Pores in a barrier enhance diffusion.

Fick's law	$\dfrac{\text{Surface area of membrane} \quad \times \quad \text{Difference in concentration across the membrane}}{\text{Length of the diffusion path (thickness of the membrane)}}$

These factors are expressed in Fick's law, which governs the rate of diffusion of substances within a system. Temperature also affects diffusion rates; at higher temperatures molecules have more energy and move more rapidly.

Diffusion Through Membranes

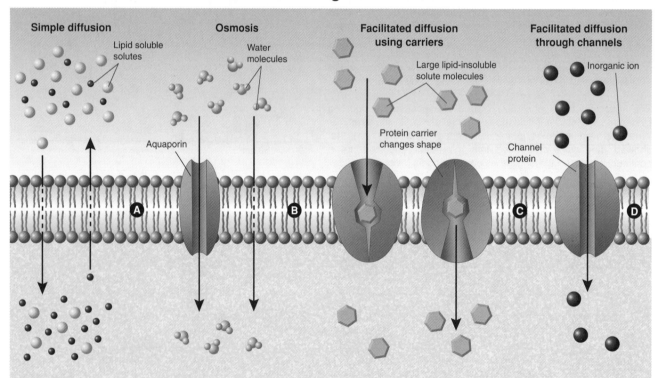

Simple diffusion — Lipid soluble solutes

Osmosis — Water molecules — Aquaporin

Facilitated diffusion using carriers — Large lipid-insoluble solute molecules — Protein carrier changes shape

Facilitated diffusion through channels — Inorganic ion — Channel protein

A: Some molecules (e.g. gases and lipid soluble molecules) diffuse directly across the plasma membrane. Two-way diffusion is common in biological systems, EXAMPLE: At the alveolar surface of the lung, oxygen diffuses into the blood and CO_2 diffuses out.

B: Osmosis is the diffusion of water across a selectively permeable membrane (in this case, the plasma membrane). Some water can diffuse directly through the lipid bilayer, but diffusion rate is increased by protein channels in the membrane called **aquaporins**.

C: A lipid-insoluble molecule is aided across the membrane by **carrier mediated facilitated diffusion**. This involves a trans-membrane carrier protein specific to the molecule being transported EXAMPLE: Glucose transport into red blood cells.

D: Small polar molecules and ions diffuse rapidly across the membrane by **channel-mediated facilitated diffusion**. Special channel proteins create hydrophilic pores that allow some solutes, usually inorganic ions, to pass through. EXAMPLE: Na^+ entering nerve cells.

Related activities: Active and Passive Transport Summary
Weblinks: Cellular Transport

Periodicals:
Getting in and out

Osmotic Gradients and Water Movement

Osmosis is the diffusion of water molecules, across a selectively permeable membrane, from higher to lower concentration of water molecules (sometimes described as from lower to higher solute concentration). Water always diffuses in this direction. The cytoplasm contains dissolved substances (**solutes**). When cells are placed in a solution of different concentration, there is an **osmotic gradient** between the external environment and the inside of the cell. In plant cells, the rigid cell wall is also important. When a plant cell takes up water, it swells until the cell contents exert a pressure on the cell wall. The cell wall is rigid and the pressure from the cytoplasm is called the wall pressure or **turgor pressure**. Turgor is important in plant support.

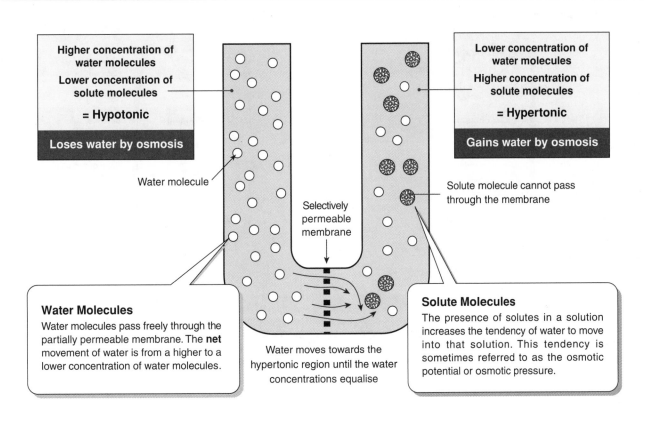

Higher concentration of water molecules

Lower concentration of solute molecules

= Hypotonic

Loses water by osmosis

Water molecule

Lower concentration of water molecules

Higher concentration of solute molecules

= Hypertonic

Gains water by osmosis

Solute molecule cannot pass through the membrane

Selectively permeable membrane

Water Molecules

Water molecules pass freely through the partially permeable membrane. The **net** movement of water is from a higher to a lower concentration of water molecules.

Water moves towards the hypertonic region until the water concentrations equalise

Solute Molecules

The presence of solutes in a solution increases the tendency of water to move into that solution. This tendency is sometimes referred to as the osmotic potential or osmotic pressure.

1. Describe two properties of an exchange surface that would facilitate rapid diffusion rates:

   (a) _____ (b) _____

2. Describe two biologically important features of diffusion:

   (a) _____

   (b) _____

3. Describe how facilitated diffusion is achieved for:

   (a) Small polar molecules and ions: _____

   _____

   (b) Glucose: _____

   _____

4. How are concentration gradients maintained across membranes? _____

   _____

5. Describe the role of aquaporins in the rapid movement of water through some cells: _____

   _____

6. (a) What happens if a cell takes up sucrose by active transport? _____

   _____

   (b) Describe a situation where this occurs in plants: _____

Cell Membranes and Transport

Water Relations in Plant Cells

The plasma membrane of cells is a selectively permeable membrane and osmosis is the main way by which water enters and leaves the cell. When the external water concentration is the same as that of the cell there is no net movement of water. Two systems (cell and environment) with the same water concentration are termed isotonic. The diagram below illustrates two different situations: when the external water concentration is higher than the cell (**hypotonic**) and when it is lower than the cell (**hypertonic**).

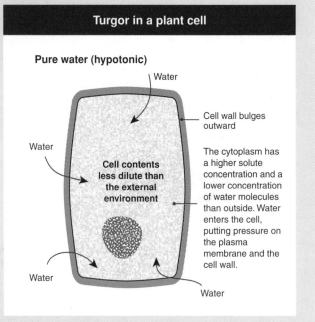

Plasmolysis in a plant cell	Turgor in a plant cell

Hypertonic salt solution

Water

Cell wall is freely permeable to water molecules.

Water

Water concentration in the cell is higher than outside.

Cell contents more dilute than the external environment

Cytoplasm

Plasma membrane

Water

Water

Pure water (hypotonic)

Water

Cell wall bulges outward

Water

The cytoplasm has a higher solute concentration and a lower concentration of water molecules than outside. Water enters the cell, putting pressure on the plasma membrane and the cell wall.

Cell contents less dilute than the external environment

Water

Water

In a **hypertonic** solution, the external water concentration is lower than the water concentration of the cell. Water leaves the cell and, because the cell wall is rigid, the cell membrane shrinks away from the cell wall. This process is termed **plasmolysis** and the cell becomes **flaccid** (turgor pressure = 0). Complete plasmolysis is irreversible; the cell cannot recover by taking up water.

In a **hypotonic** solution, the external water concentration is higher than the cell cytoplasm. Water enters the cell, causing it to swell tight. A wall (turgor) pressure is generated when enough water has been taken up to cause the cell contents to press against the cell wall. Turgor pressure rises until it offsets further net influx of water into the cell (the cell is turgid). The rigid cell wall prevents cell rupture.

7. Describe what would happen to an animal cell (e.g. a red blood cell) if it was placed into:

(a) Pure water: _____

(b) A hypertonic solution: _____

(c) A hypotonic solution: _____

8. *Paramecium* is a freshwater protozoan. Describe the problem it has in controlling the amount of water inside the cell:

_____

_____

9. Fluid replacements are usually provided for heavily perspiring athletes after endurance events.

(a) Identify the preferable tonicity of these replacement drinks (isotonic, hypertonic, or hypotonic): _____

(b) Give a reason for your answer: _____

_____

10. The malarial parasite lives in human blood. Relative to the tonicity of the blood, the parasite's cell contents would be hypertonic / isotonic / hypotonic (circle the correct answer).

11. (a) Explain the role of cell wall pressure in generating cell turgor in plants: _____

_____

(b) Discuss the role of cell turgor in plants: _____

_____

_____

_____

© BIOZONE International 2012
ISBN: 978-1-927173-11-4
Photocopying Prohibited

Ion Pumps

Diffusion alone cannot supply the cell's entire requirements for molecules (and ions). Some molecules (e.g. glucose) are required by the cell in higher concentrations than occur outside the cell. Others (e.g. sodium) must be removed from the cell in order to maintain fluid balance. These molecules must be moved across the plasma membrane by active transport mechanisms. **Active transport** requires the expenditure of energy because the molecules (or ions) must be moved **against** their concentration gradient. The work of active transport is performed by specific carrier proteins in the membrane. These transport proteins

harness the energy of ATP to pump molecules from a low to a high concentration. When ATP transfers a phosphate group to the carrier protein, the protein changes its shape in such a way as to move the bound molecule across the membrane. Three types of membrane pump are illustrated below. The sodium-potassium pump (below, center) is found in almost all animal cells and is common in plant cells also. The concentration gradient created by ion pumps such as this and the proton pump (left) is often coupled to the transport of molecules such as glucose (e.g. in the intestine) as shown below right.

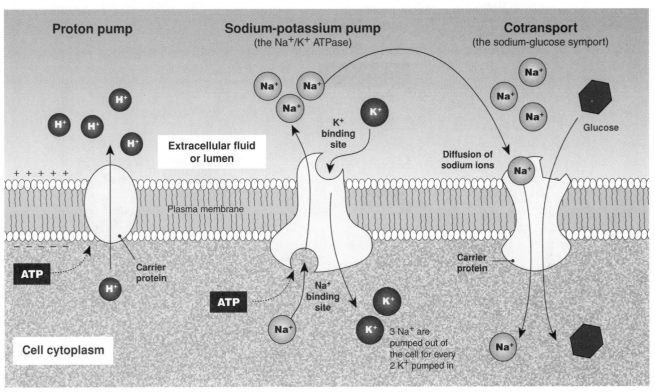

Proton pumps

ATP driven proton pumps use energy to remove hydrogen ions (H⁺) from inside the cell to the outside. This creates a large difference in the proton concentration either side of the membrane, with the inside of the plasma membrane being negatively charged. This potential difference can be coupled to the transport of other molecules.

Sodium-potassium pump

The sodium-potassium pump is a specific protein in the membrane that uses energy in the form of ATP to exchange sodium ions (Na⁺) for potassium ions (K⁺) across the membrane. The unequal balance of Na⁺ and K⁺ across the membrane creates large concentration gradients that can be used to drive transport of other substances (e.g. cotransport of glucose).

Cotransport (coupled transport)

A gradient in sodium ions drives the active transport of **glucose** in intestinal epithelial cells. The specific transport protein couples the return of Na⁺ down its concentration gradient to the transport of glucose into the intestinal epithelial cell. A low intracellular concentration of Na⁺ (and therefore the concentration gradient) is maintained by a sodium-potassium pump.

1. Why is ATP required for membrane pump systems to operate? _____

_____

2. (a) Explain what is meant by cotransport: _____

_____

(b) How is cotransport used to move glucose into the intestinal epithelial cells? _____

_____

_____

(c) What happens to the glucose that is transported into the intestinal epithelial cells? _____

_____

3. Describe two consequences of the extracellular accumulation of sodium ions: _____

_____

_____

© BIOZONE International 2012
ISBN: 978-1-927173-11-4
Photocopying Prohibited

Periodicals:
How biological membranes
achieve selective transport

Related activities: Active and Passive Transport Summary
Weblinks: Cellular Transport, Symport

Disturbances to Ion Transport

Regulation of ion transport across a plasma membrane is important for maintaining the correct balance of ions inside and outside the cell. Maintaining this balance is in turn an important way of regulating the amount of water in a cell, as water moves by osmosis from areas of low solute concentration to high solute concentration. Disturbances to the correct flow of ions can affect this balance. In the example of cholera infection below, transport channels for chloride ions out of the cell are permanently opened, causing an incorrect balance of ions and resulting in a massive loss of electrolytes and water from the cells.

Mechanism of Cholera Induced Dehydration

Mechanism of Oral Rehydration Salts

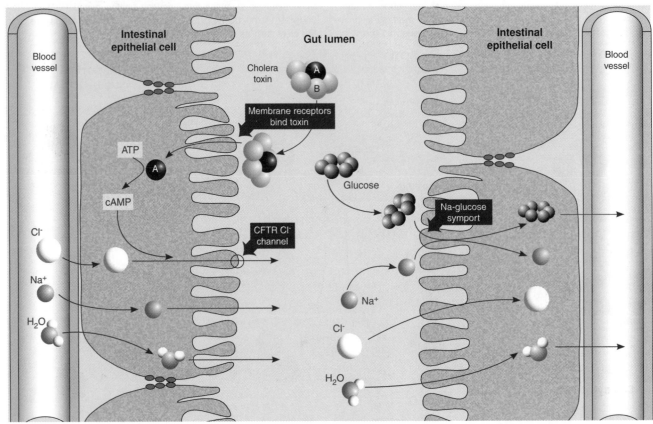

Cholera toxin contains two subunits (A and B), which bind to receptors on the cell membrane. Subunit A enters the cell and activates pathways that cause the production of cAMP from ATP. cAMP opens the Cystic Fibrosis Transmembrane conductance Regulator (CFTR) channel, causing the loss of chloride ions (Cl⁻) into the gut lumen. This results in a negative charge within the lumen. Sodium follows down its **electrochemical gradient** (the electrical potential and a difference in the chemical concentration across a membrane) while water follows down its osmotic gradient. These are replaced from the blood, causing a drop in blood volume.

Glucose and sodium enter the gut lumen in high concentrations and are transported into the cell by the sodium-glucose symport (ion pump). The influx of sodium causes a positive charge within the cell and results in Cl⁻ moving down its electrochemical gradient and back into the cell. Water follows down its osmotic gradient. Water can then reenter the blood to restore blood volume (rehydration). This effect can occur rapidly, but oral rehydration must continue until the patient's immune system is able to eliminate the cholera bacteria.

1. How does the cholera toxin cause a loss of water into the gut lumen? _____

_____

_____

_____

2. Oral rehydration salts are used to treat cholera and contain high concentrations of glucose and sodium. Explain how these are able to replace the water and electrolytes lost from a cell:

_____

_____

_____

3. Why was it important to understand ion transport in the intestine when devising a treatment for cholera?

_____

_____

_____

Exocytosis and Endocytosis

Most cells carry out **cytosis**: a form of **active transport** involving the infolding or outfolding of the plasma membrane. The ability of cells to do this is a function of the flexibility of the plasma membrane. Cytosis results in bulk transport into or out of the cell and is achieved through the localized activity of microfilaments and microtubules in the cell cytoskeleton. **Endocytosis** involves

material being engulfed. It typically occurs in protozoans and certain white blood cells of the mammalian defence system (phagocytes). **Exocytosis** involves the release of material from vesicles or vacuoles that have fused with the plasma membrane. Exocytosis is typical of cells that export material (secretory cells).

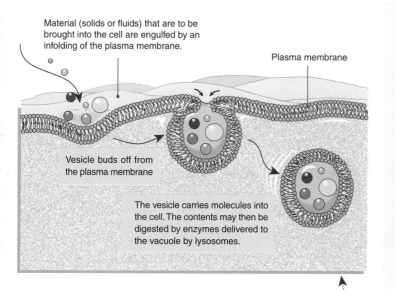

Material (solids or fluids) that are to be brought into the cell are engulfed by an infolding of the plasma membrane.

Plasma membrane

Vesicle buds off from the plasma membrane

The vesicle carries molecules into the cell. The contents may then be digested by enzymes delivered to the vacuole by lysosomes.

Both endocytosis and exocytosis require energy in the form of ATP

Endocytosis

Endocytosis (left) occurs by invagination (infolding) of the plasma membrane, which then forms vesicles or vacuoles that become detached and enter the cytoplasm. There are two main types of endocytosis:

Phagocytosis: 'cell-eating'
Phagocytosis involves the cell engulfing **solid material** to form large vesicles or vacuoles (e.g. food vacuoles). Examples: Feeding in *Amoeba*, phagocytosis of foreign material and cell debris by neutrophils and macrophages. Some endocytosis is **receptor mediated** and is triggered when receptor proteins on the extracellular surface of the plasma membrane bind to specific substances. Examples include the uptake of lipoproteins by mammalian cells.

Pinocytosis: 'cell-drinking'
Pinocytosis involves the non-specific uptake of **liquids** or fine suspensions into the cell to form small pinocytic vesicles. Pinocytosis is used primarily for absorbing extracellular fluid. Examples: Uptake in many protozoa, some cells of the liver, and some plant cells.

Areas of enlargement

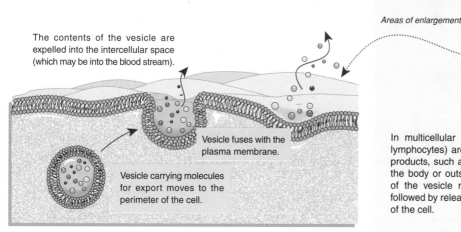

The contents of the vesicle are expelled into the intercellular space (which may be into the blood stream).

Vesicle fuses with the plasma membrane.

Vesicle carrying molecules for export moves to the perimeter of the cell.

Exocytosis

In multicellular organisms, several types of cells (e.g. lymphocytes) are specialized to manufacture and export products, such as proteins, from the cell to elsewhere in the body or outside it. Exocytosis (left) occurs by fusion of the vesicle membrane and the plasma membrane, followed by release of the vesicle's contents to the outside of the cell.

1. Distinguish between **phagocytosis** and **pinocytosis**: _____

_____

2. Describe an example of phagocytosis and identify the cell type involved: _____

_____

3. Describe an example of exocytosis and identify the cell type involved: _____

_____

4. Why is cytosis affected by changes in oxygen level, whereas diffusion is not? _____

_____

5. How does each of the following substances enter a living macrophage (for help, see *Passive Transport Processes*):

(a) Oxygen: _____ (c) Water: _____

(b) Cellular debris: _____ (d) Glucose: _____

Periodicals:
What is endocytosis?

Related activities: Active and Passive Transport Summary
Weblinks: Cellular Transport

Active and Passive Transport Summary

Cells have a need to move materials both into and out of the cell. Raw materials and other molecules necessary for metabolism must be accumulated from outside the cell. Some of these substances are scarce outside of the cell and some effort is required to accumulate them. Waste products and molecules for use in other parts of the body must be 'exported' out of the cell.

Some materials (e.g. gases and water) move into and out of the cell by **passive transport** processes, without the expenditure of energy on the part of the cell. Other molecules (e.g. sucrose) are moved into and out of the cell using **active transport**. Active transport processes involve the expenditure of energy in the form of ATP, and therefore use oxygen.

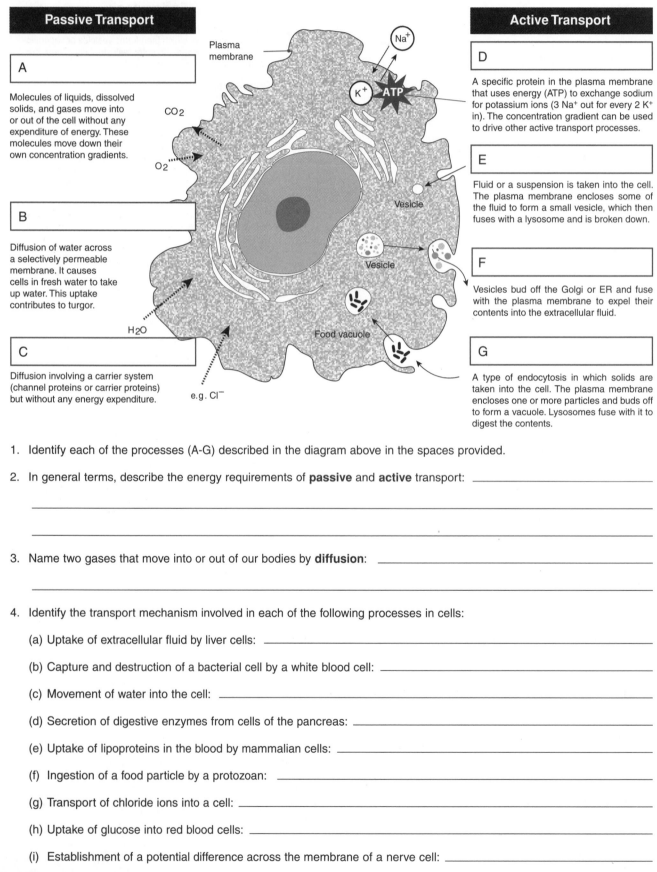

Passive Transport

A

Molecules of liquids, dissolved solids, and gases move into or out of the cell without any expenditure of energy. These molecules move down their own concentration gradients.

B

Diffusion of water across a selectively permeable membrane. It causes cells in fresh water to take up water. This uptake contributes to turgor.

C

Diffusion involving a carrier system (channel proteins or carrier proteins) but without any energy expenditure.

Plasma membrane

CO_2

O_2

H_2O

e.g. Cl^-

Na^+

K^+ ATP

Vesicle

Vesicle

Food vacuole

Active Transport

D

A specific protein in the plasma membrane that uses energy (ATP) to exchange sodium for potassium ions (3 Na^+ out for every 2 K^+ in). The concentration gradient can be used to drive other active transport processes.

E

Fluid or a suspension is taken into the cell. The plasma membrane encloses some of the fluid to form a small vesicle, which then fuses with a lysosome and is broken down.

F

Vesicles bud off the Golgi or ER and fuse with the plasma membrane to expel their contents into the extracellular fluid.

G

A type of endocytosis in which solids are taken into the cell. The plasma membrane encloses one or more particles and buds off to form a vacuole. Lysosomes fuse with it to digest the contents.

1. Identify each of the processes (A-G) described in the diagram above in the spaces provided.

2. In general terms, describe the energy requirements of **passive** and **active** transport: _____

_____

_____

3. Name two gases that move into or out of our bodies by **diffusion**: _____

_____

4. Identify the transport mechanism involved in each of the following processes in cells:

   (a) Uptake of extracellular fluid by liver cells: _____

   (b) Capture and destruction of a bacterial cell by a white blood cell: _____

   (c) Movement of water into the cell: _____

   (d) Secretion of digestive enzymes from cells of the pancreas: _____

   (e) Uptake of lipoproteins in the blood by mammalian cells: _____

   (f) Ingestion of a food particle by a protozoan: _____

   (g) Transport of chloride ions into a cell: _____

   (h) Uptake of glucose into red blood cells: _____

   (i) Establishment of a potential difference across the membrane of a nerve cell: _____

Related activities: Passive Transport Processes, Ion Pumps, Exocytosis and Endocytosis
Weblinks: Cellular Transport

KEY TERMS: Mix and Match

INSTRUCTIONS: Test your vocabulary by matching each term to its definition, as identified by its preceding letter code.

active transport

amphipathic

aquaporin

carrier protein

cell wall

channel protein

concentration gradient

diffusion

endocytosis

exocytosis

facilitated diffusion

fluid mosaic model

glycolipids

glycoproteins

hypertonic

hypotonic

ion pump

isotonic

osmosis

passive transport

phagocytosis

pinocytosis

plasma membrane

plasmolysis

selectively permeable

surface area: volume ratio

transmembrane protein

turgor

A Passive movement of water molecules across a selectively permeable membrane down a concentration gradient.

B The model for membrane structure which proposes a double phospholipid bilayer in which proteins and cholesterol are embedded.

C A type of passive transport, facilitated by transport proteins.

D Protein that spans the plasma membrane.

E The process in plant cells where the plasma membrane pulls away from the cell wall as a result of the loss of water through osmosis.

F The energy-requiring movement of substances across a biological membrane against a concentration gradient.

G A solution with lower solute concentration relative to another solution (across a membrane).

H Active transport in which molecules are engulfed by the plasma membrane, forming a phagosome or food vacuole within the cell.

I The gradual difference in the concentration of solutes as a function of distance through the solution.

J The force exerted outward on a plant cell wall by the water contained in the cell.

K Lipids with attached carbohydrates which serve as markers for cellular recognition.

L Solutions of equal solute concentration are termed this.

M This relationship determines capacity for effective diffusion in a cell.

N The uptake of liquids or fine suspensions by endocytosis.

O The passive movement of molecules from high to low concentration.

P The movement of substances across a biological membrane without energy expenditure.

Q A solution with higher solute concentration relative to another solution (across a membrane).

R A structure, present in plants and bacteria, which is found outside the plasma membrane and gives rigidity to the cell.

S A selectively-permeable phospholipid bilayer forming the boundary of all cells.

T A transmembrane protein that moves ions across a plasma membrane against their concentration gradient.

U A specific form of endocytosis involving the engulfment of solid particles by the plasma membrane.

V Protein that provides a channel through the plasma membrane for small polar molecules and ions.

W Protein channel that increases the diffusion rate of water through the plasma membrane.

X Protein in the plasma membrane that facilitates the diffusion of a specific lipid insoluble molecule.

Y Active transport process by which membrane-bound secretory vesicles fuse with the plasma membrane and release the vesicle contents into the external environment.

Z A membrane that acts selectively to allow some substances, but not others, to pass.

AA Membrane-bound proteins with attached carbohydrates, involved in cell to cell interactions.

BB Possessing both hydrophilic and hydrophobic (lipophilic) properties.

Cell Membranes and Transport

Cellular Communication

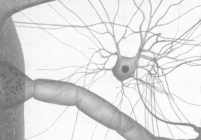

Key concepts

▶ Communication occurs via transduction of signals received from cells, organisms, or the environment.

▶ Cell signaling pathways share common features in all organisms.

▶ Cells communicate by direct contact and over short and long distances via signaling molecules.

▶ Second messengers are important in the functioning of many signal cascades.

▶ Changes in a signal transduction pathway can alter the cellular response.

Key terms

autocrine signaling
autoinducer
communication
cyclic AMP
endocrine cells
endocrine signaling
G-protein linked receptor
hormone
ligand
morphogen
neurotransmitters
paracrine signaling
phosphorylation cascade
protein kinase
quorum sensing
receptor
second messenger
signal cascade
signal transduction
signal transduction pathway

Essential Knowledge

☐ 1. Use the **KEY TERMS** to compile a glossary for this topic.

Common Features of Cell Communication (3.D.1)   pages 73-75, 174, 189

☐ 2. Explain the basis of communication, i.e. the transduction of signals from cells, organisms, or the environment. Understand and explain why there is strong selective pressure for the correct and appropriate **signal transduction**.

☐ 3. Using an example such as **quorum sensing** in bacteria, describe and explain the **signal transduction pathways** for response in single-celled organisms.

☐ 4. Using examples, explain how signal transduction pathways coordinate the activities of cells in multicellular organisms. Examples could include:
   – temperature-dependent sex determination in some vertebrates
   – signal transduction pathways for DNA repair
   – epinephrine-induced breakdown of glycogen to glucose
   – interactions between immune cells in the immune response

☐ 5. Describe features of cell signaling pathways. Understand and describe how the basic chemical processes for cell communication are indicative of common ancestry (e.g. the hedgehog signaling pathway in bilateral animals).

Signal Transduction Pathways (3.D.2-3.D.3)   pages 74-82

☐ 6. Using an appropriate example, e.g. contact between cells of the immune system, describe how cells communicate by cell-to-cell contact (**paracrine signaling**).

☐ 7. Describe and explain how cells communicate over short distances using local regulators such as **neurotransmitters**, **autoinducers** for bacterial luminescence, and **morphogens** such as hedgehog (in *Drosophila*), sonic hedgehog (in mammals), and epidermal growth factor (humans).

☐ 8. Using appropriate examples of **hormones**, describe and explain how signals from **endocrine cells** can be transported in the blood or hemolymph to influence the activity of distant target cells.

☐ 9. Using an example, explain the process of **signal transduction**, including with recognition of the **ligand** by a **receptor** molecule, initiation of transduction, relay of the signal, and the cellular response.

☐ 10. Explain the role of **second messengers**, such as **cyclic AMP**, in the functioning of signal cascades.

☐ 11. Describe the role of protein modification and **phosphorylation cascades** in signal transduction pathways.

Changes in Signal Transduction Pathways (3.D.4)   pages 83-84

☐ 12. Describe and explain how changes in a signal transduction pathway can alter cellular response. Use examples to show how blocked or defective signal transduction pathways can be detrimental or preventative.

Periodicals:
Listings for this
chapter are on page 382

Weblinks:
www.thebiozone.com/
weblink/AP1-3114.html

BIOZONE APP:
Student Review Series
Cell Structure

Communication between Unicellular Organisms

Unicellular organisms gather information about their environment and respond to it appropriately by using signaling pathways triggered by chemicals in the environment. These chemicals can relay information about how close a food source is, how many of the same bacteria are nearby (**quorum sensing**), or the location of a favorable environment. Chemical substances produced by unicellular organisms can also be used to communicate between individual cells and influence their cellular processes.

Bioluminescence in Bacteria

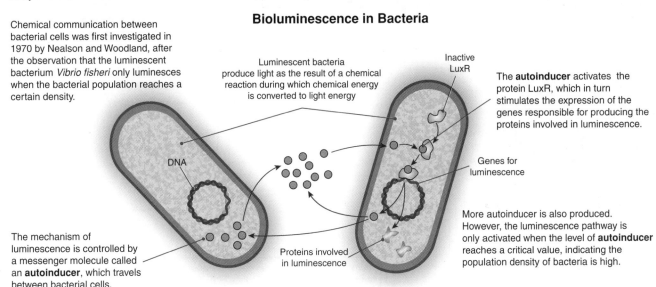

Chemical communication between bacterial cells was first investigated in 1970 by Nealson and Woodland, after the observation that the luminescent bacterium *Vibrio fisheri* only luminesces when the bacterial population reaches a certain density.

Luminescent bacteria produce light as the result of a chemical reaction during which chemical energy is converted to light energy

DNA

The mechanism of luminescence is controlled by a messenger molecule called an **autoinducer**, which travels between bacterial cells.

Proteins involved in luminescence

Inactive LuxR

The **autoinducer** activates the protein LuxR, which in turn stimulates the expression of the genes responsible for producing the proteins involved in luminescence.

Genes for luminescence

More autoinducer is also produced. However, the luminescence pathway is only activated when the level of **autoinducer** reaches a critical value, indicating the population density of bacteria is high.

Control of Chemotaxis in Bacteria

Chemotaxis is the movement or orientation of an organism along a chemical concentration gradient either toward or away from the chemical stimulus. Chemotaxis in bacteria is controlled by **signal transduction**, which affects the direction of flagella rotation. Clockwise rotation of the flagellum causes tumbling and is the native condition (occurs without receptor stimulation) in bacteria. Tumbling occurs when the molecule CheA activates the molecule CheY, which attaches to the flagellum motor and causes its clockwise rotation.

When an **attractant molecule** (a chemical the bacterium wants to move towards) is encountered, the action of CheA is inhibited so that it can't activate CheY. Flagellum rotation is anticlockwise, causing the bacterium to swim straight (a run). After a second, the bacterium becomes insensitive to the attractant molecule and resumes tumbling. Encountering another attractant molecule causes it to start another run. In this way bacteria can move towards an attractant source, as a greater concentration of attractant molecules causes more runs and less tumbling. The protein CheZ is also able to directly inhibit CheY and stop it binding to the motor complex.

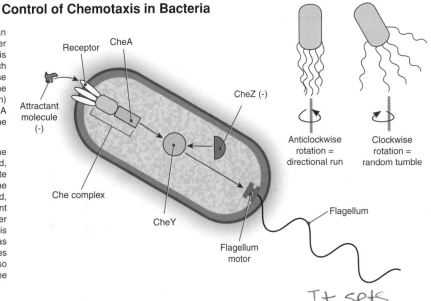

Receptor CheA

Attractant molecule (-)

Che complex

CheY

CheZ (-)

Flagellum motor

Flagellum

Anticlockwise rotation = directional run

Clockwise rotation = random tumble

1. (a) Explain how the autoinducer molecule in luminescent bacteria signals when to luminesce: _It sets_ _up a signal transduction pathway._

_____

(b) Explain how this enables the bacterium to detect the population density: _They start producing_ _more luciferase_

(c) How might this information be of survival advantage to the bacterium? _____

_____

2. Explain how the Che molecules allow a bacterium to move towards or away from a chemical source: _____

_____

_____

Periodicals:
How and why bacteria communicate

Related activities: Types of Cell Signaling
Weblinks: Bacterial Chemotaxis

Types of Cell Signaling

Cells use **signals** (chemical messengers) to communicate and to gather information about, and respond to, changes in their cellular environment. The signaling and response process is called a **signal transduction pathway**, and often involves a number of enzymes and molecules in a **signal cascade**. A signal cascade results in a response in the target cell. Cell signaling pathways are categorized primarily on the distance over which the signal molecule travels to reach its target cell, and generally fall into three categories. The **endocrine** pathway involves the transport of **hormones** over large distances through the blood or hemolymph. During **paracrine** signaling, the signal travels an intermediate distance to act upon neighboring cells. **Autocrine** signaling involves a cell producing and reacting to its own signal. These three pathways are illustrated below.

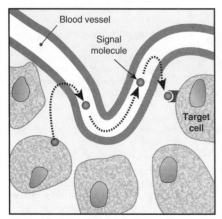

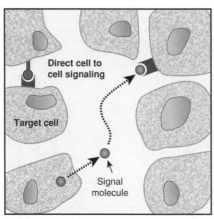

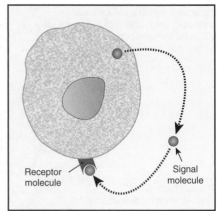

Endocrine signaling: Hormone signals are released by ductless endocrine glands and carried long distances through the body by the circulatory system to the target cells. Examples include sex hormones, growth factors, and neurohormones such as dopamine.

Paracrine signaling: Signals released from a cell act upon target cells within the immediate vicinity. The chemical messenger can be transferred through the extracellular fluid (e.g. at synapses) or directly between cells, which is important during embryonic development.

Autocrine signaling: Cells produce and react to their own signals. In vertebrates, when a foreign antibody enters the body, some T-cells (lymphocytes) produce a growth factor to stimulate their own production. The increased number of T-cells helps to fight the infection.

Signaling and T-Cell Activation

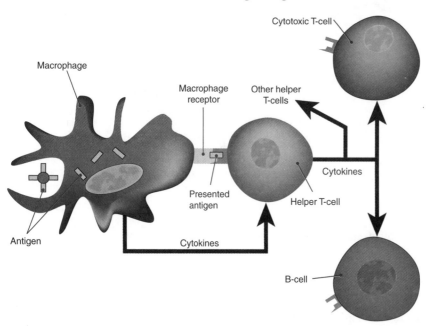

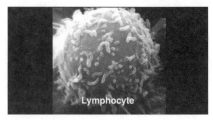

Helper T-cells are activated by direct cell-to-cell signaling and by paracrine signaling using **cytokines** from macrophages.

Macrophages ingest antigens, process them, and present them on the cell surface where they are recognized by helper T-cells. The helper T-cell binds to the antigen and to the macrophage receptor, which leads to activation of the helper T-cell.

The macrophage also produces and releases cytokines, which enhance T-cell activation. The activated T-cell then releases more cytokines which causes the proliferation of other helper T-cells (positive feedback) and helps to activate cytotoxic T-cells and antibody-producing B-cells.

1. Briefly describe the three types of cell signaling:

   (a) _Endocrine - hormone signals_

   (b) _Paracrine - target cells within the immediate vicinity_

   (c) _Autocrine - cells produce and react to their own signal_

2. Identify the components that all three cell signaling types have in common: _____

   _____

3. Activation of helper T-cells involves which signaling pathway(s)? _____

Related activities: Hormonal Regulation
Weblinks: Hormones, Receptors and Target cells

Periodicals:
New jobs for ancient chaperones

© BIOZONE International 2012
ISBN: 978-1-927173-11-4
Photocopying Prohibited

Signals and Signal Cascades

Signals that activate processes within the cell can be external **signal molecules** or external conditions that cause internal changes. These signals often cause a cascade of reactions that result in a cellular outcome. For example, the molecule epinephrine causes the breakdown of glycogen in mammals, and ionizing radiation activates the protein kinase ATM (Ataxia Telangiectasia Mutated) which initiates the repair of DNA. ATM belongs to a family of proteins that have been highly conserved throughout eukaryote evolution and is a key component in the repair of damaged DNA. Damage to a DNA molecule from ionizing radiation, such as UV light, induces a signal cascade that results in the repair of the DNA, arrest of cell cycle, or **apoptosis**.

Initiating DNA Repair

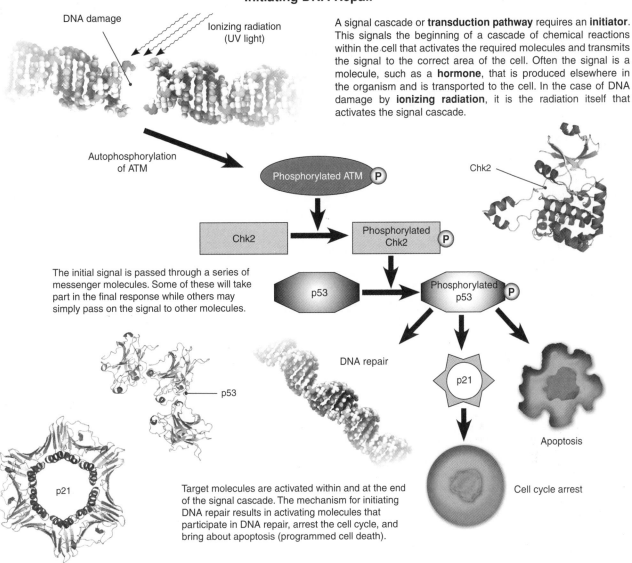

A signal cascade or **transduction pathway** requires an **initiator**. This signals the beginning of a cascade of chemical reactions within the cell that activates the required molecules and transmits the signal to the correct area of the cell. Often the signal is a molecule, such as a **hormone**, that is produced elsewhere in the organism and is transported to the cell. In the case of DNA damage by **ionizing radiation**, it is the radiation itself that activates the signal cascade.

The initial signal is passed through a series of messenger molecules. Some of these will take part in the final response while others may simply pass on the signal to other molecules.

Target molecules are activated within and at the end of the signal cascade. The mechanism for initiating DNA repair results in activating molecules that participate in DNA repair, arrest the cell cycle, and bring about apoptosis (programmed cell death).

1. (a) Describe how DNA repair is signaled in cells: _____

_____

_____

(b) Explain why a signal cascade can be useful in certain cell processes: _____

_____

_____

2. Use the example of DNA damage to explain why there is a strong selection pressure for the correct functioning of signal transduction pathways:

_____

_____

_____

Neurotransmitters

Neurotransmitters are chemicals that allow the transmission of signals between neurons. They are found in the axon endings of neurons and are released into the space between one neuron and the next (the **synaptic cleft**) after a depolarization or hyperpolarization of the nerve ending. Neurotransmitters can be classified into amino acids, peptides, or monoamines. The many neurotransmitters produce various responses depending on their location in the body. They can be excitatory (likely to cause an action potential in the receiving neuron) or inhibitory (causing hyperpolarization) depending on the receptor they activate.

Neurotransmitters Carry Signals Between Neurons

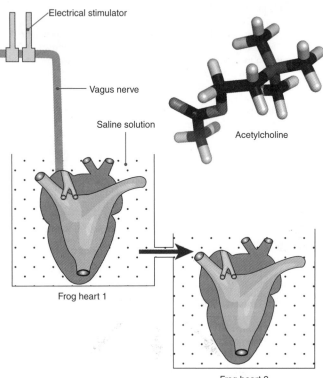

Acetylcholine

Electrical stimulator

Vagus nerve

Saline solution

Frog heart 1

Frog heart 2

Chemical signaling between neurons was first demonstrated in 1921 by Otto Loewi. In his experiment, the still beating hearts of two frogs were placed in connected flasks filled with saline solution. The vagus nerve (parasympathetic) of the first heart was still attached and was stimulated by electricity to reduce its rate of beating. After a delay, the rate of beating in the second heart also slowed. Increasing the beating rate in the first heart caused an increase in the beating rate in the second heart, showing electrical stimulus of the first heart caused it to release a chemical into the saline solution that then affected the heartbeat of the second heart. The chemical was found to be **acetylcholine**.

Neurotransmitters

Name	Postsynaptic effect	
Acetylcholine	Excitatory/inhibitory	Responsible for the stimulation of muscles. Found in sensory neurons and the autonomic nervous system.
Norepinephrine	Excitatory	Brings the nervous system into high alert. Increases heart rate and blood pressure.
Dopamine	Excitatory/inhibitory	Associated with reward mechanisms in the brain. Produces the "feel good" feeling.
Gamma amino butyric acid (GABA)	Inhibitory	Inhibits excitatory neurotransmitters that can cause anxiety.
Glutamate	Excitatory	Found in the central nervous system and concentrated in the brain.
Serotonin	Inhibitory	Serotonin is strongly involved in regulation of mood and perception.
Endorphin	Excitatory	Involved in pain reduction and pleasure.

1. Describe the purpose of a neurotransmitter: _____

_____

2 (a) Explain why stimulating the first frog heart with electricity caused it to change its beating rate:

_____

_____

_____

(b) Explain why the second heart in the experiment reduced its beating rate after a delay:

_____

_____

3. Why can some neurotransmitters be both excitatory and inhibitory?_____

_____

_____

_____

© BIOZONE International 2012
ISBN: 978-1-927173-11-4
Photocopying Prohibited

Neurohormones

A **neurohormone** is any hormone produced and released by specialized **neurosecretory cells**, which function as both nerve cells and endocrine cells. The **hypothalamus** is responsible for synthesizing and secreting several neurohormones. It has an important role in linking the nervous system to the endocrine system via the pituitary, with which it has a close structural and functional relationship. The **posterior pituitary** is **neural** in origin and is essentially an extension of the hypothalamus. Its neurosecretory cells release oxytocin and ADH directly into the blood in response to nerve impulses. The **anterior pituitary** is connected to the hypothalamus by blood vessels and receives neurohormones from the hypothalamus via a capillary network. These **hypothalamic releasing hormones** regulate the secretion of the anterior pituitary's hormones.

The Role of the Hypothalamus

Neurosecretory Cells

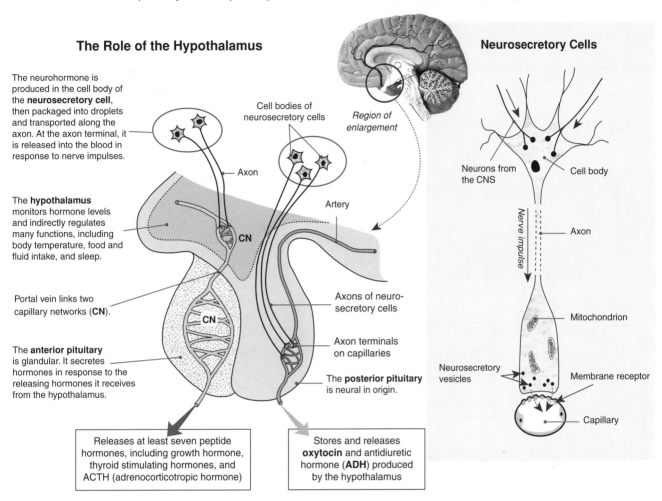

The neurohormone is produced in the cell body of the **neurosecretory cell**, then packaged into droplets and transported along the axon. At the axon terminal, it is released into the blood in response to nerve impulses.

The **hypothalamus** monitors hormone levels and indirectly regulates many functions, including body temperature, food and fluid intake, and sleep.

Portal vein links two capillary networks (**CN**).

The **anterior pituitary** is glandular. It secretes hormones in response to the releasing hormones it receives from the hypothalamus.

Cell bodies of neurosecretory cells

Region of enlargement

Axon

Artery

CN

CN

Axons of neuro-secretory cells

Axon terminals on capillaries

The **posterior pituitary** is neural in origin.

Neurons from the CNS

Cell body

Nerve impulse

Axon

Mitochondrion

Neurosecretory vesicles

Membrane receptor

Capillary

Releases at least seven peptide hormones, including growth hormone, thyroid stimulating hormones, and ACTH (adrenocorticotropic hormone)

Stores and releases **oxytocin** and antidiuretic hormone (**ADH**) produced by the hypothalamus

1. (a) Explain how the anterior and posterior pituitary differ with respect to their relationship to the hypothalamus:

_____

_____

(b) Explain how these differences relate to the nature of the hormonal secretions for each region: _____

_____

_____

2. Describe the role of the neurohormones released by the hypothalamus: _____

_____

3. Explain why the adrenal and thyroid glands atrophy if the pituitary gland ceases to function: _____

_____

_____

4. Although the anterior pituitary is often called the master gland, the hypothalamus could also claim that title. Explain:

_____

_____

Related activities: Hormones and their Effects
Web links: Control of Endocrine Activity

RA 2

Hormonal Regulation

The endocrine system regulates the body's metabolic processes by releasing chemical messengers (**hormones**) into the blood, which transports them to **target cells**. Hormones are potent chemical regulators: they are produced in minute quantities yet can have a large effect on metabolism. The endocrine system comprises endocrine cells (organized into endocrine glands), and the hormones they produce. Unlike exocrine glands (e.g. sweat and salivary glands), endocrine glands are ductless glands, secreting hormones directly into the bloodstream rather than through a duct or tube. Some organs (e.g. the pancreas) have both endocrine and exocrine regions, but these are structurally and functionally distinct. The basis of hormonal control and the role of negative feedback mechanisms in regulating hormone levels are described below.

The Mechanism of Hormone Action

Endocrine cells produce hormones and secrete them into the bloodstream where they are distributed throughout the body. Although hormones are broadcast throughout the body, they affect only specific **target cells**. These target cells have receptors on the plasma membrane which recognize and bind the hormone (see inset below). The binding of hormone and receptor triggers the response in the target cell. Cells are unresponsive to a hormone if they do not have the appropriate receptors.

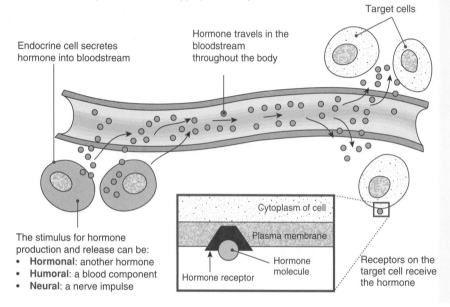

Target cells

Hormone travels in the bloodstream throughout the body

Endocrine cell secretes hormone into bloodstream

The stimulus for hormone production and release can be:
- **Hormonal**: another hormone
- **Humoral**: a blood component
- **Neural**: a nerve impulse

Cytoplasm of cell

Plasma membrane

Hormone molecule

Hormone receptor

Receptors on the target cell receive the hormone

Antagonistic Hormones

Insulin secretion

Raises blood glucose level

Lowers blood glucose level

Glucagon secretion

The effects of one hormone are often counteracted by an opposing hormone. Feedback mechanisms adjust the balance of the two hormones to maintain a physiological function. Example: insulin acts to decrease blood glucose and glucagon acts to raise it.

1. (a) Explain what is meant by **antagonistic hormones** and describe an example of how two such hormones operate:

_____

_____

Example: _____

_____

(b) Explain the role of feedback mechanisms in adjusting hormone levels (explain using an example if this is helpful):

_____

_____

2. How do hormones bring about a response in target cells even though all cells may come into contact with the hormone:

_____

_____

3. Explain how hormonal control differs from nervous system control with respect to the following:

(a) The speed of hormonal responses is slower: _____

_____

(b) Hormonal responses are generally longer lasting: _____

_____

© BIOZONE International 2012
ISBN: 978-1-927173-11-4
Photocopying Prohibited

Related activities: Types of Cell Signaling
Web links: Control of Endocrine Activity

Hormones and Their Effects

The **hypothalamus** is located at the base of the brain, just above the pituitary gland. Information comes to the hypothalamus through sensory pathways from the sense organs. On the basis of this information, the hypothalamus controls and integrates many of the body's activities, including the reflex activity of the **autonomic nervous system**. The pituitary gland comprises two regions: the **posterior pituitary** and the **anterior pituitary**. Interactions between the hypothalamus, the pituitary gland, and the adrenal glands constitute what is called the hypothalamic-pituitary-adrenal (HPA) axis. This axis is a multi-step biochemical pathway in which each step in the pathway passes on information to the next but also receives feedback for its own regulation. It constitutes a major part of the neuroendocrine system, controlling reactions to both short and long term stress and regulating many physiological processes, including digestion, immune response, emotion, sexuality, and energy storage and expenditure.

ANTERIOR PITUITARY

The anterior pituitary releases at least seven **peptide hormones** (below) into the blood from simple secretory cells. The release of these hormones is regulated by releasing and inhibiting hormones from the hypothalamus.

POSTERIOR PITUITARY

The posterior pituitary develops as an extension of the hypothalamus. The release of its two hormones, oxytocin and antidiuretic hormone, occurs directly as a result of nervous input to the hypothalamus.

Hypothalamus

Hormones from the anterior pituitary

Hormones from the posterior pituitary

Thyroid Stimulating Hormone
Increases synthesis and secretion of thyroid hormones (T_3 and thyroxine) from the thyroid gland.

Growth Hormone
Stimulates protein synthesis and growth in most tissues; one of the main regulators of metabolism.

Prolactin
Stimulates growth of the mammary glands and synthesis of milk protein.

Oxytocin
Acts on the mammary glands to stimulate milk ejection (let down). Oxytocin also acts on the uterus to stimulate contraction during labor.

Adrenocorticotropic Hormone
Usually abbreviated to ACTH, this hormone increases synthesis and secretion of hormones (aldosterone and cortisol) from the **adrenal cortex**.

Antidiuretic Hormone
Acts on the kidney to increase water reabsorption from the filtrate (urine).

Follicle Stimulating Hormone
In females: Stimulates maturation of the ovarian follicles.
In males: Increases the production of sperm in the seminiferous tubules.

Luteinizing Hormone
In females: Stimulates secretion of ovarian hormones, ovulation, and formation of the corpus luteum.
In males: Stimulates synthesis and secretion of testosterone.

Melanophore Stimulating Hormone
Increases melanin synthesis and dispersal in the skin's melanocytes.

Related activities: Neurohormones
Weblinks: Hypothalamic, Pituitary, Endocrine Axis , Drag and Drop Hormone Match

Effects of Growth Hormone

Growth hormone (GH) is released in response to GHRH (growth hormone releasing hormone) from the hypothalamus. GH acts both directly and indirectly to affect metabolic activities associated with growth.

GH directly stimulates metabolism of fat, but its major role is to stimulate the liver and other tissues to secrete IGF-1 (Insulin-like Growth Factor 1) and through this stimulate bone and muscle growth. GH secretion is regulated via negative feedback:

High levels of IGF-1

► *suppress secretion of GHRH*

► *stimulate release of somato-statin from the hypothalamus. Somatostatin suppresses GH secretion (not shown).*

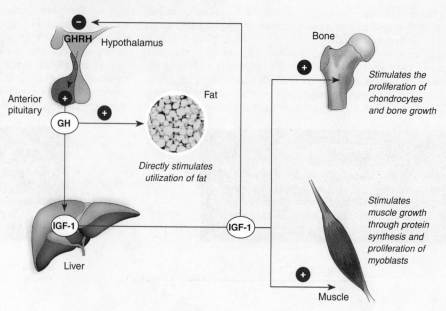

Hypothalamus

Anterior pituitary

GH

Fat

Directly stimulates utilization of fat

Liver

IGF-1

IGF-1

Bone

Stimulates the proliferation of chondrocytes and bone growth

Stimulates muscle growth through protein synthesis and proliferation of myoblasts

Muscle

1. (a) Describe the metabolic effects of growth hormone: _____

    _____

 (b) Predict the effect of chronic **GH deficiency** in infancy: _____

    _____

 (c) Predict the effect of chronic **GH hypersecretion** in infancy: _____

    _____

 (d) Describe the two main mechanisms through which the secretion of growth hormone is regulated:

    _____

    _____

2. "The pituitary releases a number of hormones that regulate the secretion of hormones from other glands". Discuss this statement with reference to growth hormone (GH) and **thyroid stimulating hormone** (TSH):

    _____

    _____

    _____

    _____

3. Using the example of TSH and its target tissue (the thyroid), explain how the release of anterior pituitary hormones is regulated. Include reference to the role of negative feedback mechanisms in this process:

    _____

    _____

    _____

    _____

    _____

4. Iodine is needed to produce thyroid hormones. Explain why the thyroid enlarges in response to an iodine deficiency:

    _____

    _____

Signal Transduction

Signal transduction describes the process by which an extracellular signal brings about an intracellular response. A **signal cascade** is initiated by a signaling molecule, such as a hormone, binding to a receptor of a target cell. The signal cascade **amplifies** the original signal and results in a specific cellular response (e.g. enzyme activation). Water soluble hormones operate by interacting with transmembrane receptors and activating a **second messenger system** (e.g. cyclic AMP, Ca^{2+}, or inositol triphosphate), which

links the hormone to the cellular response. Signal transduction pathways involving phosphorylation cascades, such as the one illustrated below for epinephrine, involve the activation of **protein kinases**. Steroid hormones, being lipid-soluble, are able to enter the cell freely to interact directly with intracellular receptors and a membrane receptor is not involved. In both cases, the response of the target cell is recognized by the hormone-producing cell through a feedback signal and the hormone is degraded.

Structure of a Transmembrane Receptor

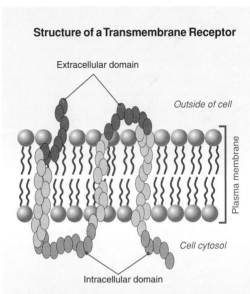

Extracellular domain

Outside of cell

Plasma membrane

Cell cytosol

Intracellular domain

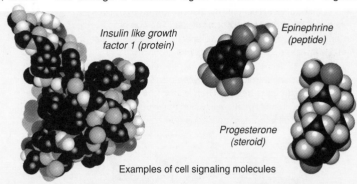

Insulin like growth factor 1 (protein)

Epinephrine (peptide)

Progesterone (steroid)

Examples of cell signaling molecules

The binding sites of cell receptors are very specific: they only bind certain **ligands** (signal molecules). This stops them from reacting to every signal bombarding the cell. Receptors fall into two main categories:

▶ **Cytoplasmic receptors**: Cytoplasmic receptors, located within the cell cytoplasm, bind ligands which are able to cross the plasma membrane unaided.

▶ **Transmembrane receptors**: These span the cell membrane and bind ligands which cannot cross the plasma membrane on their own. They have an extracellular domain outside the cell, and an intracellular domain within the cell cytosol (left).

G-protein linked receptors (below), ion channels, and protein kinases are examples of transmembrane receptors.

Second Messenger Activation of a Signal Cascade

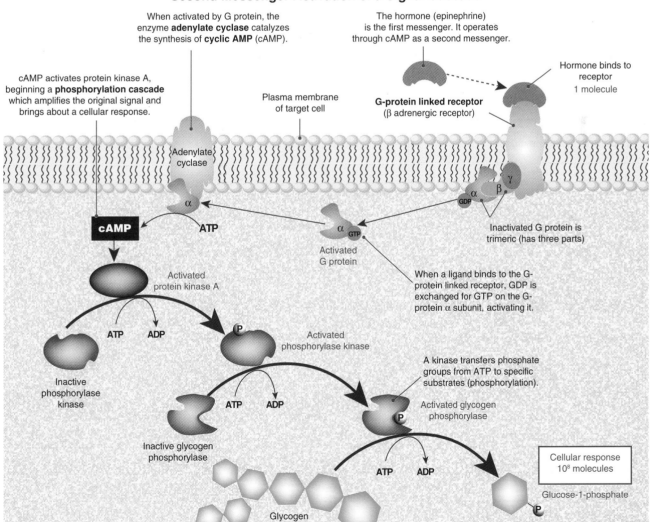

When activated by G protein, the enzyme **adenylate cyclase** catalyzes the synthesis of **cyclic AMP** (cAMP).

The hormone (epinephrine) is the first messenger. It operates through cAMP as a second messenger.

cAMP activates protein kinase A, beginning a **phosphorylation cascade** which amplifies the original signal and brings about a cellular response.

Plasma membrane of target cell

Hormone binds to receptor
1 molecule

G-protein linked receptor
(β adrenergic receptor)

Adenylate cyclase

α
GDP
β
γ

Inactivated G protein is trimeric (has three parts)

cAMP

ATP

α
GTP

Activated G protein

When a ligand binds to the G-protein linked receptor, GDP is exchanged for GTP on the G-protein α subunit, activating it.

Activated protein kinase A

ATP   **ADP**

P

Activated phosphorylase kinase

A kinase transfers phosphate groups from ATP to specific substrates (phosphorylation).

Inactive phosphorylase kinase

Inactive glycogen phosphorylase

ATP   **ADP**

P

Activated glycogen phosphorylase

Glycogen

ATP   **ADP**

Cellular response
10^8 molecules

Glucose-1-phosphate
P

Related activities: Signals and Signal Cascades   *Weblinks*: Second Messengers, G Protein Signal Transduction (IP₃), G Protein Coupled Receptors

Gene Activation by Steroid Hormones

Steroid hormones are steroids (fatty molecules) that act as hormones. They have wide-ranging metabolic effects and are involved in regulating metabolic activity, inflammation, immune function, salt and water balance, development of sexual characteristics, and response to stress and injury.

Steroid hormones are lipid soluble and diffuse freely from the blood through the plasma membrane and into the cytoplasm of target cells. In the cytoplasm, the steroid binds to a specific steroid receptor (**transcription factor**) to form a functional **hormone-receptor complex**. The complex is able to enter the nucleus and bind to specific DNA sequences to induce transcription of its target genes.

The best-studied steroid-binding receptors are called nuclear receptors (examples below). Their ability to interact directly with DNA and control gene expression makes them important in embryonic development and adult homeostasis.

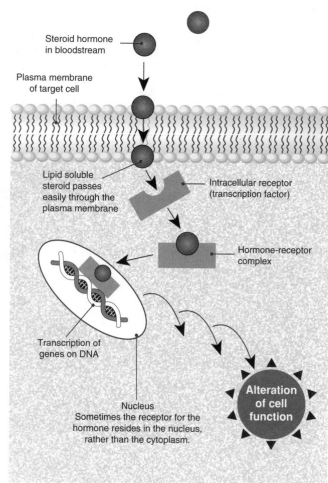

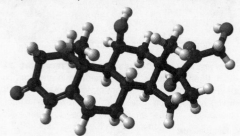

Cortisol binds to a glucocorticoid receptor. Cortisol is a steroid hormone (a glucocorticoid) produced in the adrenal cortex (cortical region of the adrenal glands). Its release is triggered by ACTH from the anterior pituitary gland. Cortisol is involved in glucose metabolism and in mediating the response to stress.

1. Explain why a second messenger is needed to convey a signal inside a cell from a water soluble first messenger:

   _____

   _____

   _____

2. Explain why each molecule in the cAMP signal cascade is phosphorylated: _____

   _____

   _____

3. Describe and explain the role of G proteins in coupling the receptor to the cellular response: _____

   _____

   _____

4. Describe the basis of signal amplification and explain its significance: _____

   _____

   _____

   _____

5. (a) How does the action of steroid hormones differ from that of water soluble signaling molecules?

   _____

   (b) Suggest why steroid hormones are important in developmental processes: _____

   _____

   _____

Related activities: *Cell Signaling, First Messenger Systems*
Weblinks: *Signal Transduction, First and Second Messenger System*

Effect of Blocking Signals

Almost as soon as the importance of signal transduction pathways in the body's myriad functions and responses was realized, their importance in the development of disease became obvious. If any of these pathways were to fail there would be significant deleterious effects. Similarly any over activity in these pathways could be equally deleterious. Many modern medical drugs focus on restoring these signaling pathways when they fail or blocking the effects of overactive pathways. The negative effects of many toxic substances are caused by their ability to block signal transduction pathways.

Signals, Responses, and Inhibitors

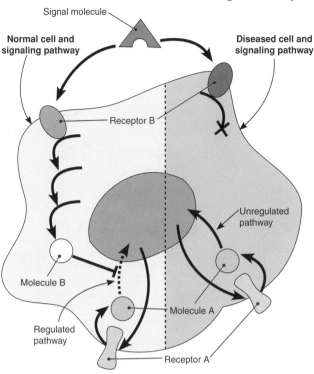

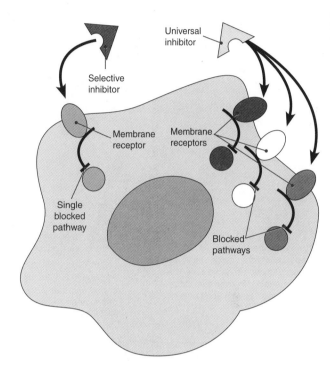

When choosing a drug to correct a transduction pathway, it is important to understand how the pathway operates. The signaling pathways of diseased cells (e.g. tumor cells) are often overactive (e.g. overexpressing receptors for growth factors). In the example above, receptor A signals the production of more receptor A via the messenger molecule A. Receptor B initiates a signal cascade activating molecule B, which inactivates molecule A and so regulates receptor A's production. The diseased cell fails to make molecule B and so cannot regulate receptor A's production.

Drug selection also involves the question of how many receptors are affected by a particular drug. Universal signal transduction inhibitors affect diseased cells by blocking many signal pathways, but also affect healthy cells, and so can cause damage outside the target area. Selective signal transduction inhibitors target a single overactive pathway, thus causing no damage outside the target area. However the specificity of these drugs may mean many different types are required to treat a disease effectively.

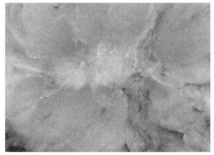

Breast cancer (like all cancers) is the result of cells dividing out of control. In breast cancer, the HER2 receptor protein has been identified as a major cause.

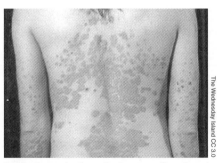

Psoriasis results from the hyper-proliferation of keratinocytes (epidermal cells) which is driven by EGFR (epidermal growth factor receptor) causing scaly patches or plaques on the skin.

Finger prick test for diabetics

Type II (adult onset) diabetes is caused by the failure of cells to respond to signaling by the hormone insulin, which results in the failure of cells to take up glucose.

1. (a) Explain the link between signal transduction and disease: _____

_____

_____

(b) Explain why understanding the link is important: _____

_____

_____

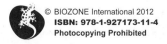

Related activities: Signals and Signal Cascades
Weblinks: Insecticides

RA 2

Pesticides and Signals

Many pesticides work by affecting the signaling of nerves cells by either blocking uptake of signaling molecules or facilitating the uptake of far greater amounts than normal.

Cholinesterase inhibitors work by blocking the enzyme responsible for breaking down acetylcholine after it has transmitted a message across a synapse. The result is the continual firing of the receiving neuron causing over stimulation and death.

Neonicotoid insecticides mimic the action of acetylcholine in synapses. This again causes the over stimulation of the nerve cells and results in death.

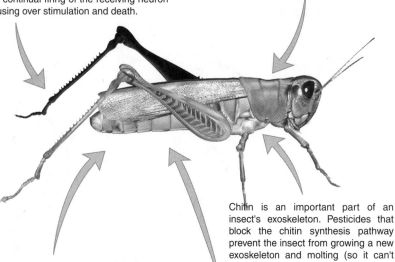

Chitin is an important part of an insect's exoskeleton. Pesticides that block the chitin synthesis pathway prevent the insect from growing a new exoskeleton and molting (so it can't grow), eventually killing the insect.

Organochlorines inhibit the GABA receptor, which itself has an inhibitory function in nerve cells. The result is again the over stimulation of the nervous system and death.

Pyrethrins act on the sodium channels in neurons that pump sodium across the membrane, preventing them from closing and causing over stimulation and death.

Action of Antihistamine

Hayfever and similar allergic reactions occur when mast cells produce the chemical histamine as part of an immune response to foreign particles such as pollen. Histamine triggers the inflammatory response and results in swelling, red eyes, and a runny nose.

Antihistamines block the histamine receptor of cells and so block the signal for the inflammatory response.

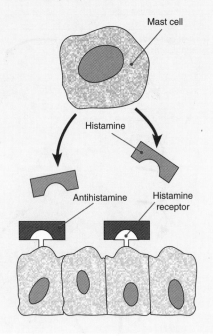

Mast cell

Histamine

Antihistamine

Histamine receptor

2. (a) Describe the action of a universal signal transduction inhibitor: _____

_____

_____

(b) Describe the action of a selective transduction inhibitor: _____

_____

_____

(c) Explain the role of these drugs in controlling a disease: _____

_____

_____

3. (a) Describe the general function of insecticides that act on the nervous system: _____

_____

_____

(b) Explain why pesticides based on cholinesterase inhibitors are also toxic to humans while pesticides based on chitin synthesis inhibitors are not:

_____

_____

_____

4. Explain how antihistamines prevent hayfever: _____

_____

_____

KEY TERMS: Word Find

Use the clues below to find the relevant key terms in the WORD FIND grid

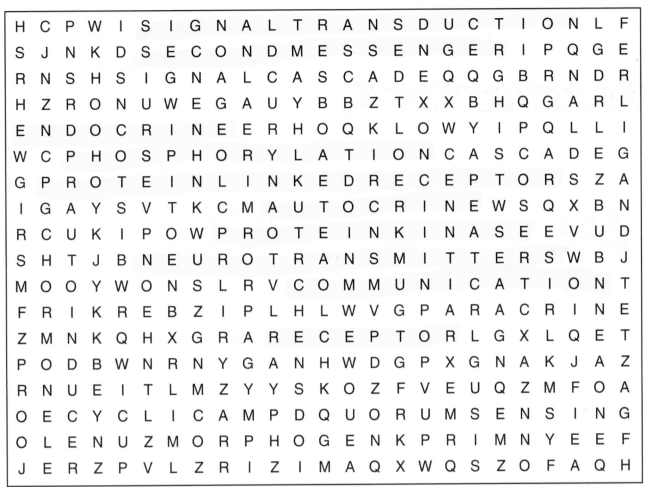

H	C	P	W	I	S	I	G	N	A	L	T	R	A	N	S	D	U	C	T	I	O	N	L	F
S	J	N	K	D	S	E	C	O	N	D	M	E	S	S	E	N	G	E	R	I	P	Q	G	E
R	N	S	H	S	I	G	N	A	L	C	A	S	C	A	D	E	Q	Q	G	B	R	N	D	R
H	Z	R	O	N	U	W	E	G	A	U	Y	B	B	Z	T	X	X	B	H	Q	G	A	R	L
E	N	D	O	C	R	I	N	E	E	R	H	O	Q	K	L	O	W	Y	I	P	Q	L	L	I
W	C	P	H	O	S	P	H	O	R	Y	L	A	T	I	O	N	C	A	S	C	A	D	E	G
G	P	R	O	T	E	I	N	L	I	N	K	E	D	R	E	C	E	P	T	O	R	S	Z	A
I	G	A	Y	S	V	T	K	C	M	A	U	T	O	C	R	I	N	E	W	S	Q	X	B	N
R	C	U	K	I	P	O	W	P	R	O	T	E	I	N	K	I	N	A	S	E	E	V	U	D
S	H	T	J	B	N	E	U	R	O	T	R	A	N	S	M	I	T	T	E	R	S	W	B	J
M	O	O	Y	W	O	N	S	L	R	V	C	O	M	M	U	N	I	C	A	T	I	O	N	T
F	R	I	K	R	E	B	Z	I	P	L	H	L	W	V	G	P	A	R	A	C	R	I	N	E
Z	M	N	K	Q	H	X	G	R	A	R	E	C	E	P	T	O	R	L	G	X	L	Q	E	T
P	O	D	B	W	N	R	N	Y	G	A	N	H	W	D	G	P	X	G	N	A	K	J	A	Z
R	N	U	E	I	T	L	M	Z	Y	Y	S	K	O	Z	F	V	E	U	Q	Z	M	F	O	A
O	E	C	Y	C	L	I	C	A	M	P	D	Q	U	O	R	U	M	S	E	N	S	I	N	G
O	L	E	N	U	Z	M	O	R	P	H	O	G	E	N	K	P	R	I	M	N	Y	E	E	F
J	E	R	Z	P	V	L	Z	R	I	Z	I	M	A	Q	X	W	Q	S	Z	O	F	A	Q	H

Signaling that involves a cell producing and reacting to its own signal.

A messenger molecule that travels between bacteria participating in quorum sensing.

The exchange of information between cells, normally by chemical messengers.

An important second messenger derived from ATP.

Signaling that involves hormones released by ductless endocrine glands and carried through the body by the circulatory system to target cells.

A transmembrane receptor that senses molecules outside the cell and activates signal transduction pathways inside the cell by activating an associated G-protein.

A chemical messenger that induces a specific physiological response.

A molecule that binds to another.

A compound that influences the development or organization of cells or tissue by forming a concentration gradient.

Chemical messengers that are released by axon terminals in response to an action potential.

Signaling involving chemical messaging whose effect is only exerted at short range on neighboring cells.

A sequence of chemical reactions in which one enzyme activates another by adding a phosphate group. The chain reaction eventually phosphorylates and activates one or many target proteins.

An enzyme which modifies another protein by adding a phosphate group (phosphorylation).

A stimulus and response system used by bacteria to coordinate and regulate gene expression in response to population density.

A protein attached to a membrane or surface of a cell that is able to bind a signaling molecule, usually causing a change in the activity of the cell.

A molecule that relays signals from receptors on the cell surface to a target molecule inside the cell.

A series of enzyme driven reactions in which the product of one reaction (usually an activated enzyme) activates the next, usually in response to a signal molecule binding to a receptor.

A mechanism that converts a mechanical or chemical stimulus to a cell into a specific cellular response.

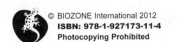

Sources of Variation

Sources of genetic variation	• Meiosis and fertilization • Gene and chromosome mutation • Aneuploidy • Viral replication and variation

Chromosomes and Cell Division

Mitosis and the cell cycle	• The cell cycle and its regulation • Mitosis and its role in organisms
Meiosis and crossing over	• The role of meiosis in organisms • Stages in meiotic division • Meiosis and variation
Connect 3.A.2 with 3.C.2	*Biological systems have multiple mechanisms to increase genetic variation.*

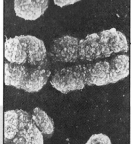
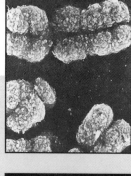

Organisms have mechanisms to generate genetic variation in the offspring. Variable phenotypes are the raw material on which evolution operates.

The inheritance of traits may be predictable, but may also be the product of multiple genes and/or processes.

Mendelian genetics / Non-Mendelian inheritance

Mendelian genetics	• Mendel's laws • Dominance: Basic genetic crosses • Lethal alleles
Non-Mendelian inheritance	• Linkage and recombination • Sex linkage and sex limited traits • Non-nuclear inheritance • Multiple genes • Pedigrees and genetic counseling • Genes and environment
Sex determination	• Sex determination • The role of environment
Connect 3.A.3 with 3.C.1	*Non-disjunction during meiosis can result in heritable chromosomal aberrations such as trisomy.*

The Chromosomal Basis of Inheritance

DNA and RNA

Nucleic acids	• Structure and role of DNA and RNA • Organization of genetic information • DNA replication
Gene expression	• Flow of genetic information • Transcription and translation • Prokaryotes vs eukaryotes • Proteins and phenotype
Genetic engineering	• What is genetic modification? • Genetic engineering techniques • Applications of genetic engineering
Connect 3.A.1 with 3.C	*DNA can be altered through mutation. Heritable mutations can affect offspring.*

DNA (and sometimes RNA) is the primary source of heritable information. Phenotypes are determined by protein activity.

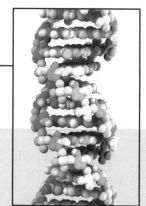

Genetics

Important in this section ...

- *Understand the structure and role of DNA.*
- *Understand how meiosis and sexual reproduction produce variation and how this is inherited.*
- *Develop an understanding of the techniques and applications of nucleic acid technology.*

Expression of genetic information involves cellular and molecular mechanisms.

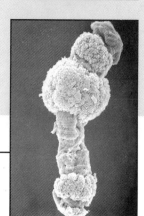

Development	• Cellular differentiation • The expression of specific genes
Regulation of gene expression	• Homeotic genes • The timing of development • Effect of mutation on development • Genetic transplantation experiments • Gene regulation by microRNAs • Apoptosis in normal development
Control of transcription	• Transcriptional control in prokaryotes • Control of transcription in eukaryotes
Connect 2.E.1 with 1.A.1	*The mechanisms regulating development are subject to strong selection pressures.*

Regulation of Gene Expression

DNA and RNA

Key concepts

▶ DNA controls the behavior of cells. In most cases, information flows from DNA to RNA to protein.

▶ DNA is a self-replicating molecule constructed according to strict base-pairing rules.

▶ The genetic code, through transcription and translation, contains the information to construct proteins.

▶ The heritable information in organisms can be manipulated using a few basic techniques.

▶ Genetic manipulation of organisms can have wide applications in industry, agriculture, and medicine.

Key terms

amino acids
anticodon
base-pairing rule
coding strand
codon
DNA
DNA ligase
DNA polymerase
DNA replication
exons
gel electrophoresis
gene expression
genetic code
genetic engineering
helicase
hydrogen bonding
introns
lagging strand
leading strand
nucleic acids
nucleotides
Okazaki fragments
polymerase chain reaction
protein
replication fork
restriction enzyme
reverse transcriptase
RNA (mRNA, rRNA, tRNA)
RNAi
start codon / stop codon
template strand
transcription
transgenic organism
translation

Essential Knowledge

☐ 1. Use the **KEY TERMS** to compile a glossary for this topic.

The Blueprint of Life *(3.A.1: a-b, 4.A.1)* pages 88-111

☐ 2. Recall the structure of **nucleotides** and **nucleic acids**. Describe the Watson-Crick double-helix model of **DNA** structure, including reference to the **anti-parallel** nature of DNA, the **base-pairing rule**, and **hydrogen bonding**.

☐ 3. Describe the structure and function of **mRNA, tRNA, rRNA**. Explain the role of **RNAi**. Contrast the structure and function of RNA and DNA.

☐ 4. Describe the **semi-conservative replication** of DNA. Demonstrate understanding of the **base-pairing rule** for creating a complementary strand from a single strand of DNA.

☐ 5. Recognize retroviruses as a special case of alternate information flow (RNA to DNA) and describe the role of the enzyme **reverse transcriptase** in this.

Gene Expression *(3.A.1: c-d, 4.A.1)* pages 112-121

☐ 6. Describe and explain the main features of the **genetic code**.

☐ 7. Identify the two stages of **gene expression** as transcription and translation. Explain how these processes differ in prokaryotes and eukaryotes.

☐ 8. Describe **transcription**, including the significance of the **coding** (sense) **strand** and **template** (antisense) **strand**. Describe post-transcriptional modifications of the mRNA transcript in eukaryotic cells. Explain the significance of **introns** with respect to the production of a functional mRNA molecule.

☐ 9. Recall the structure of **amino acids** and how they form the primary structure of proteins (polypeptides). Describe and explain **translation**, including the role of **tRNA molecules, ribosomes, start codons**, and **stop codons**.

☐ 10. Explain how the activities of proteins determine phenotype.

Genetic Engineering *(3.A.1: e)* pages 122-139

☐ 11. Explain how the heritable information of DNA (and sometimes RNA) can be manipulated with **genetic engineering** techniques. Your should understand:
(a) How **restriction enzymes** are used to manipulate and analyze DNA.
(b) The role of **gel electrophoresis** in identifying DNA fragments.
(c) The role of **polymerase chain reaction (PCR)** in **DNA amplification**

☐ 12. Describe some outcomes of DNA manipulation, including any of the following:
(a) Production of a genetically engineered food, such as golden rice.
(b) Production of **transgenic** animals, e.g. for expression of a specific trait.
(c) Production of pharmaceuticals, e.g. human insulin.

Periodicals:
Listings for this
chapter are on page 382

Weblinks:
www.thebiozone.com/
weblink/AP1-3114.html

BIOZONE APP:
Student Review Series
The Genetic Code

The Genome

Genome research has become an important field of genetics. A **genome** is the entire haploid complement of genetic material of a cell or organism. Each species has a unique genome, although there is a small amount of genetic variation between individuals within a species. For example, in humans the average genetic difference is one in every 500-1000 bases. Every cell in an individual has a complete copy of the genome. The base sequence shown below is the total DNA sequence for the genome of a virus. There are nine genes in the sequence, coding for nine different proteins. At least 2000 times this amount of DNA would be found in a single bacterial cell. Half a million times the quantity of DNA would be found in the genome of a single human cell. The first gene has been highlighted blue, while the start and stop codes are in blue rectangles.

Genome for the φX174 bacterial virus

Start — The blue area represents the nucleotide sequence for a single gene

(Full DNA base sequence of the φX174 genome shown as dense rows of nucleotides; the first gene is highlighted, with the start code ATG and stop code AATGA marked.)

φX174 bacteriophage

This virus consists of DNA packaged within a protein coat made up of a 20-sided polyhedron. Spikes made of protein at each of the 12 corners are used to attach itself to a bacterial cell.

The entire DNA sequence for the virus is made up of just 9 genes

1. Explain what is meant by the **genome** of an organism: _the entire haploid complement of genetic material of a cell or organism_

2. Determine the number of bases, kilobases, and megabases in this genome (100 bases in each row, except the last):

   1 kb = 1 kilobase = 1000 bases   **1 Mb** = 1 megabase = 1 000 000 bases

   (a) Bases: _____   (b) Kilobases: _5.6_   (c) Megabases: _.56_

3. Determine how many bases are present in the gene shown above (in the blue area): _1.7 Kb_

4. State whether the genome of the virus above is **small, average** or **large** in size compared to those viruses listed in the table on the page *DNA Molecules*:

   average

Prokaryotic Chromosome Structure

DNA is a universal carrier of genetic information but the way in which it is packaged in the cells of prokaryotes and eukaryotes is fundamentally different. Unlike eukaryotic chromosomes, the prokaryotic chromosome is not enclosed in a nuclear membrane and is not associated with protein. It is a single circular (rather than linear) molecule of double stranded DNA and is located in a nuclear region called the **nucleoid**, which is in direct contact with the cytoplasm. The nucleoid can be variously shaped, and in actively growing cells it may occupy as much as 20% of the cell's volume. The chromosome is attached to the plasma membrane, and proteins in the plasma membrane are responsible for DNA replication and segregation of the new chromosome to a daughter

cell in cell division. As well as the bacterial chromosome, bacteria often contain small circular, double-stranded DNA molecules called **plasmids**. Plasmids are not connected to the main bacterial chromosome and they replicate independently of it. They usually contain 5-100 genes that are not crucial to cell survival under normal conditions and they may be gained or lost without harming the cell. However, in certain environments, they may confer a selective advantage as they may carry genes for properties such as antibiotic resistance, heavy metal tolerance, and synthesis of certain enzymes. The horizontal gene transfer of plasmid DNA by bacterial **conjugation** is a major factor in the spread of drug resistance and in rapid bacterial evolution.

Organization of the Prokaryotic Chromosome

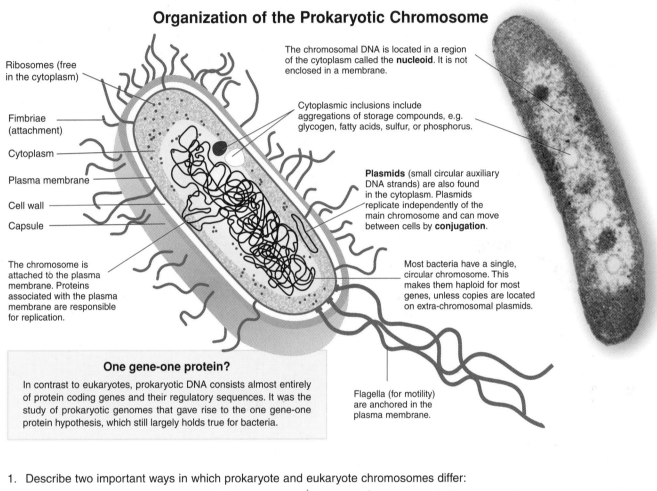

One gene-one protein?

In contrast to eukaryotes, prokaryotic DNA consists almost entirely of protein coding genes and their regulatory sequences. It was the study of prokaryotic genomes that gave rise to the one gene-one protein hypothesis, which still largely holds true for bacteria.

1. Describe two important ways in which prokaryote and eukaryote chromosomes differ:

   (a) _In prokaryotyps the chromosomes are not enclosed in a nuclear membrane_

   (b) _Eukaryote chromosome are located in the the nucleoid_

2. Explain the consequences to protein synthesis of the prokaryotic chromosome being free in the cytoplasm:

   Proteins associated with the plasma membrane are resposible for replication of the chromosome

3. Most of the bacterial genome comprises protein coding genes and their regulatory sequences. Explain the consequence of this to the relative sizes of bacterial and eukaryotic chromosomes:

   Prokaryotic genomes gave rise to one gene-one protein hypothesis

Related activities: Plasmid DNA, Eukaryotic Chromosome Structure
Weblinks: Structure and Function of Bacterial Cells

RA 2

Plasmid DNA

How genetic information is stored depends on the organism. Prokaryotes store most of their genetic information in one large chromosome. However, a small percentage can be found as independently replicating, circular, extra-chromosomal piece of DNA known as **plasmids**. Plasmids may carry important genes, such as those for the production of toxins that eliminate prokaryote competitors. Plasmids are less common in eukaryotes but some species, such as the yeast *Saccharomyces,* do have them. The genetic material from viruses may form plasmid-like structures called episomes once they have infected a cell.

Features of a plasmid

- Small in size (usually not bigger than 1000-2000 kb)
- Forms circular loops
- Extra-chromosomal
- Self replicating

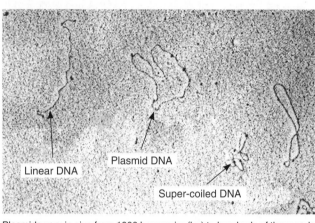

Linear DNA

Plasmid DNA

Super-coiled DNA

Wiki CC3.0

Plasmids vary in size from 1000 base pairs (bp) to hundreds of thousands of base pairs. In bacteria, they play an important role in providing extra genetic material that confers properties such as antibiotic resistance.

Plasmids can be transferred between bacterial cells by a process of plasmid transfer called **conjugation**. Conjugation enables bacteria to obtain genetic material from other individuals by **horizontal gene transfer** (transfer of genetic material to individuals other than offspring) and it is an important mechanism for the spread of antibiotic resistance.

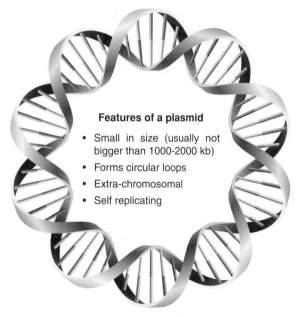

Auxin production: interferes with auxin regulated plant development processes

Cytokinin production (controls cell division in plants)

Opine synthesis

Transfer DNA region (transferred to host cell) (20,000 bp)

Border of transfer region (25 bp)

Border of transfer region (25 bp)

Catabolism of **opines** (used as a nitrogen and energy source by *Agrobacterium*)

Virulence region A - H

Origin of replication

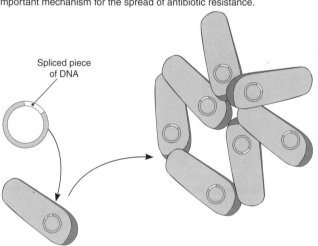

Spliced piece of DNA

The bacterium *Agrobacterium tumefaciens* often contains the T*i* (tumor inducing) plasmid. This plasmid is able to transfer genetic material into plant cells and causes crown gall disease. Several regions on the plasmid (identified above) help it to infect plants. The plasmid is just over 200,000 bp long and contains 196 genes. The mapping of its genes has made it of great importance in the creation of transgenic plants.

Plasmids have provided a tool with which to introduce novel genetic material into an organism. A new gene can be spliced into a plasmid and the plasmid inserted into a recipient organism (e.g. a bacterium). The bacteria will then produce the product encoded by the gene. This methodology has enabled the industrial-scale production of valuable gene products, such as human insulin, from genetically engineered organisms.

1. What is a plasmid? _An extra chromosomal piece of DNA_

2. Explain how a plasmid can convey a survival advantage to bacteria under certain conditions: _They can produce toxins that eliminate prokaryote competion_

3. (a) Why are plasmids (such as the T*i* plasmid) useful to genetic engineers? _The plasmid is just over 200,00 bp long and contains 196 genes_

(b) Into which region of the T*i* plasmid would you insert a gene in order for it to be transferred into a host plant cell? _the cytoplasm_

Related activities: DNA Molecules, Ligation, In Vivo Gene Cloning

Eukaryotic Chromosome Structure

The chromosomes of eukaryote cells (such as those from plants and animals) are complex in their structure compared to those of prokaryotes. The illustration below shows a chromosome during the early stage of meiosis. Here it exists as a chromosome consisting of two chromatids. A non-dividing cell would have chromosomes with the 'equivalent' of a single chromatid only. The chromosome consists of a protein coated DNA strand which coils in three ways during the time when the cell prepares to divide (below).

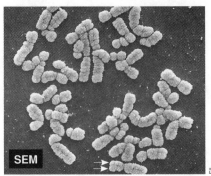

SEM

A cluster of human chromosomes seen during metaphase of cell division. Individual chromatids (arrowed) are difficult to discern on these double chromatid chromosomes.

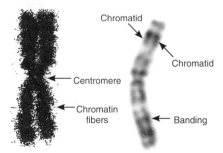

Chromatid

Chromatid

Centromere

Chromatin fibers

Banding

Chromosome TEM   Human chromosome 3

A human chromosome from a dividing white blood cell (above left). Note the compact organization of the chromatin in the two chromatids. The LM photograph (above right) shows the banding visible on human chromosome 3.

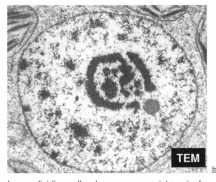

TEM

In non-dividing cells, chromosomes exist as single-armed structures. They are not visible as coiled structures, but are 'unwound' to make the genes accessible for transcription (above).

Looped domains

The evidence for the existence of looped domains comes from the study of giant lampbrush chromosomes in amphibian oocytes (above). Under electron microscopy, the lateral loops of the DNA-protein complex appear brushlike.

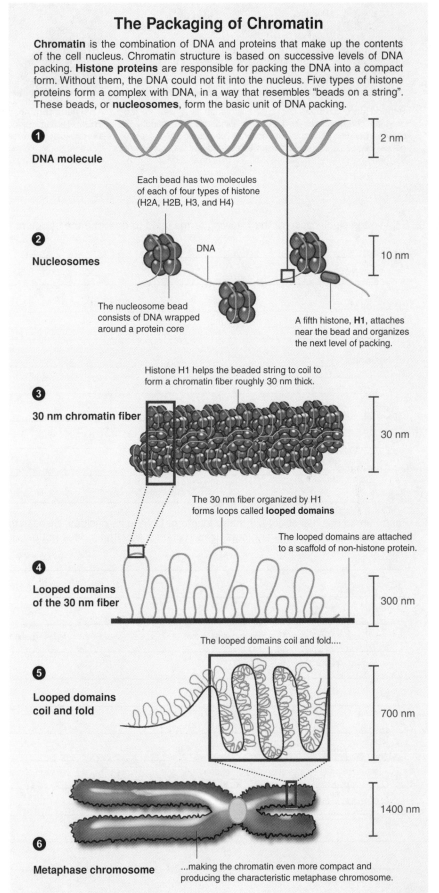

The Packaging of Chromatin

Chromatin is the combination of DNA and proteins that make up the contents of the cell nucleus. Chromatin structure is based on successive levels of DNA packing. **Histone proteins** are responsible for packing the DNA into a compact form. Without them, the DNA could not fit into the nucleus. Five types of histone proteins form a complex with DNA, in a way that resembles "beads on a string". These beads, or **nucleosomes**, form the basic unit of DNA packing.

① **DNA molecule** — 2 nm

② **Nucleosomes** — 10 nm

Each bead has two molecules of each of four types of histone (H2A, H2B, H3, and H4)

DNA

The nucleosome bead consists of DNA wrapped around a protein core

A fifth histone, **H1**, attaches near the bead and organizes the next level of packing.

Histone H1 helps the beaded string to coil to form a chromatin fiber roughly 30 nm thick.

③ **30 nm chromatin fiber** — 30 nm

The 30 nm fiber organized by H1 forms loops called **looped domains**

The looped domains are attached to a scaffold of non-histone protein.

④ **Looped domains of the 30 nm fiber** — 300 nm

The looped domains coil and fold....

⑤ **Looped domains coil and fold** — 700 nm

— 1400 nm

⑥ **Metaphase chromosome**

...making the chromatin even more compact and producing the characteristic metaphase chromosome.

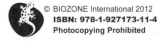

Periodicals: Control center

Related activities: DNA Molecules
Weblinks: Chromosome Structure, DNA Packing

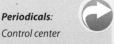

RA 2

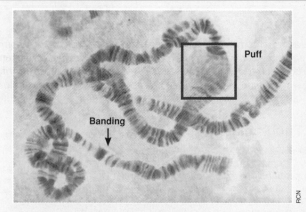

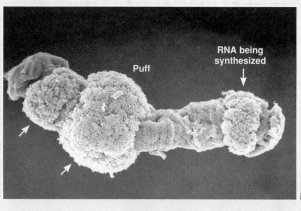

Banded chromosome: This light micrograph is a view of the polytene chromosomes in a salivary gland cell of a sandfly. It shows a banding pattern that is thought to correspond to groups of genes. Regions of chromosome **puffing** are thought to occur where the genes are being transcribed into mRNA (see SEM on right).

A **polytene chromosome** viewed with a scanning electron microscope (SEM). The arrows indicate localized regions of the chromosome that are uncoiling to expose their genes (puffing) to allow transcription of those regions. Polytene chromosomes are a type of chromosome consisting of a large bundle of chromatids bound tightly together.

1. Explain the significance of the following terms used to describe the structure of chromosomes:

(a) DNA: _____

_____

(b) Chromatin: _____

_____

(c) Histone: _____

_____

(d) Centromere: _____

_____

(e) Chromatid: _____

_____

2. Each human cell has about a 1 metre length of DNA in its nucleus. Discuss the mechanisms by which this DNA is packaged into the nucleus and organized in such a way that it does not get ripped apart during cell division:

_____

_____

_____

_____

_____

_____

_____

_____

_____

_____

_____

_____

_____

_____

_____

_____

© BIOZONE International 2012
ISBN: 978-1-927173-11-4
Photocopying Prohibited

Karyotypes

The diagram below shows the **karyotype** of a normal human. Karyotypes are prepared from the nuclei of cultured white blood cells that are 'frozen' at the metaphase stage of mitosis (see the photo circled on the next page). A photograph of the chromosomes is then cut up and the chromosomes are rearranged on a grid so that the homologous pairs are placed together. Homologous pairs are identified by their general shape, length, and the pattern of banding produced by a special staining technique. Karyotypes for a human male and female are shown below. The **male karyotype** has 44 autosomes, a single X chromosome, and a Y chromosome (written as 44 + XY), whereas the **female karyotype** shows two X chromosomes (written as 44 + XX).

Typical Layout of a Human Karyotype

1   2   3   4   5

6   7   8   9   10   11   12

Variable region →

13   14   15   16   17   18

19   20   21   22   Y   X

A scanning electron micrograph (SEM) of human chromosomes clearly showing their double chromatids.

This SEM shows the human X and Y chromosomes. Although these two are the sex chromosomes, they are not homologous.

Karyotypes for different species

The term **karyotype** refers to the chromosome complement of a cell or a whole organism. In particular, it shows the number, size, and shape of the chromosomes as seen during metaphase of mitosis. The diagram on the left depicts the human karyotype. Chromosome numbers vary considerably among organisms and may differ markedly between closely related species:

Organism	Chromosome number (2N)
Vertebrates	
human	46
chimpanzee	48
gorilla	48
horse	64
cattle	60
dog	78
cat	38
rabbit	44
rat	42
turkey	82
goldfish	94
Invertebrates	
fruit fly, *Drosophila*	8
housefly	12
honey bee	32 or 16
Hydra	32
Plants	
cabbage	18
broad bean	12
potato	48
orange	18, 27 or 36
barley	14
garden pea	14
Ponderosa pine	24

NOTE: The number of chromosomes is not a measure of the quantity of genetic information.

DNA and RNA

1. (a) What is a **karyotype**? _____

_____

(b) What information can it provide? _____

_____

2. Distinguish between **autosomes** and **sex chromosomes**: _____

_____

_____

_____

Related activities: Human Karyotype Exercise

Weblinks: Making a Karyotype

RA 2

Preparing a Karyotype

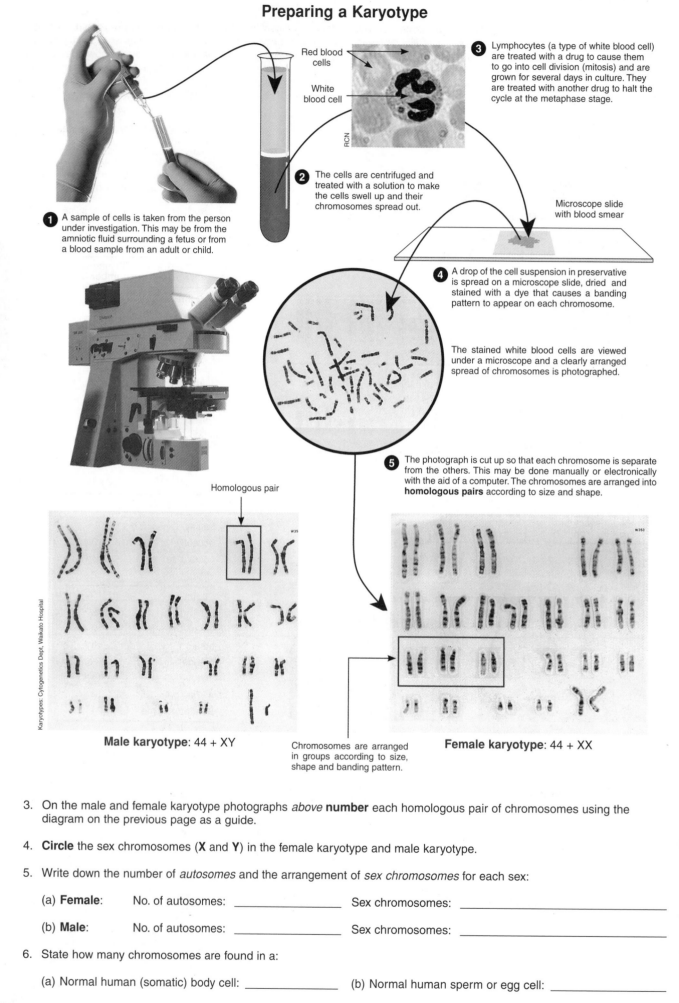

Red blood cells

White blood cell

RCN

3 Lymphocytes (a type of white blood cell) are treated with a drug to cause them to go into cell division (mitosis) and are grown for several days in culture. They are treated with another drug to halt the cycle at the metaphase stage.

2 The cells are centrifuged and treated with a solution to make the cells swell up and their chromosomes spread out.

Microscope slide with blood smear

1 A sample of cells is taken from the person under investigation. This may be from the amniotic fluid surrounding a fetus or from a blood sample from an adult or child.

4 A drop of the cell suspension in preservative is spread on a microscope slide, dried and stained with a dye that causes a banding pattern to appear on each chromosome.

The stained white blood cells are viewed under a microscope and a clearly arranged spread of chromosomes is photographed.

5 The photograph is cut up so that each chromosome is separate from the others. This may be done manually or electronically with the aid of a computer. The chromosomes are arranged into **homologous pairs** according to size and shape.

Homologous pair

W35

Karyotypes: Cytogenetics Dept, Waikato Hospital

Male karyotype: 44 + XY

Chromosomes are arranged in groups according to size, shape and banding pattern.

W352

Female karyotype: 44 + XX

3. On the male and female karyotype photographs *above* **number** each homologous pair of chromosomes using the diagram on the previous page as a guide.

4. **Circle** the sex chromosomes (**X** and **Y**) in the female karyotype and male karyotype.

5. Write down the number of *autosomes* and the arrangement of *sex chromosomes* for each sex:

(a) **Female**: No. of autosomes: _____ Sex chromosomes: _____

(b) **Male**: No. of autosomes: _____ Sex chromosomes: _____

6. State how many chromosomes are found in a:

(a) Normal human (somatic) body cell: _____ (b) Normal human sperm or egg cell: _____

Human Karyotype Exercise

Each chromosome has distinctive features that make it distinguishable from others. Chromosomes are stained in a special technique that gives them a banded appearance in which the banding pattern represents regions of the chromosome containing up to many hundreds of genes. Cut out the chromosomes below and arrange them on the *Record Sheet* in order to determine the sex and chromosome condition of the individual whose karyotype is shown. The karyotypes presented on the previous pages and the hints on how to recognize chromosome pairs can be used to help you complete this activity.

Distinguishing Characteristics of Chromosomes

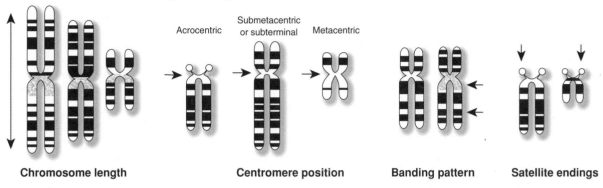

Chromosome length   Acrocentric   Submetacentric or subterminal   Metacentric   Centromere position   Banding pattern   Satellite endings

DNA and RNA

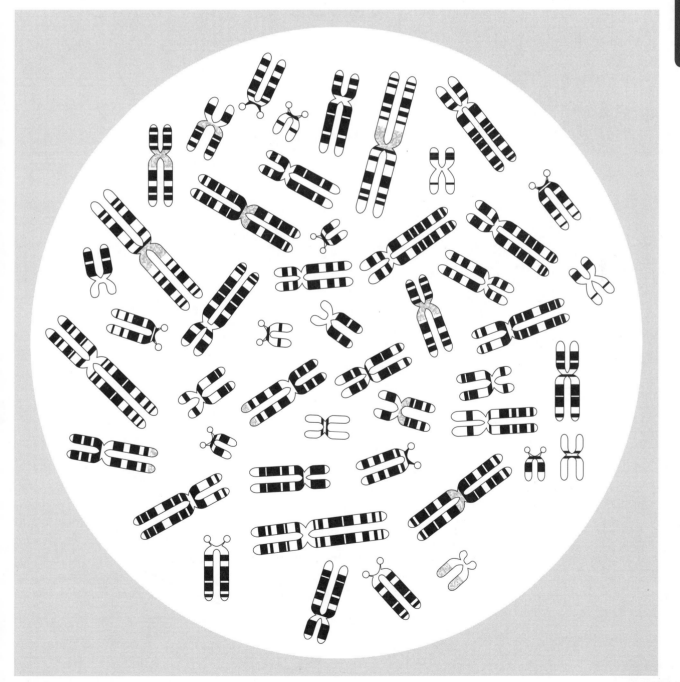

Related activities: Karyotypes

This page is left blank deliberately

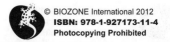

1. Cut out the chromosomes on page 95 and arrange them on the record sheet below in their homologous pairs.

2. (a) Determine the sex of this individual: **male** or **female** (circle one)

(b) State whether the individual's *chromosome arrangement* is: **normal** or **abnormal** (circle one)

(c) If the arrangement is *abnormal*, state in what way and name the syndrome displayed: _____

_____

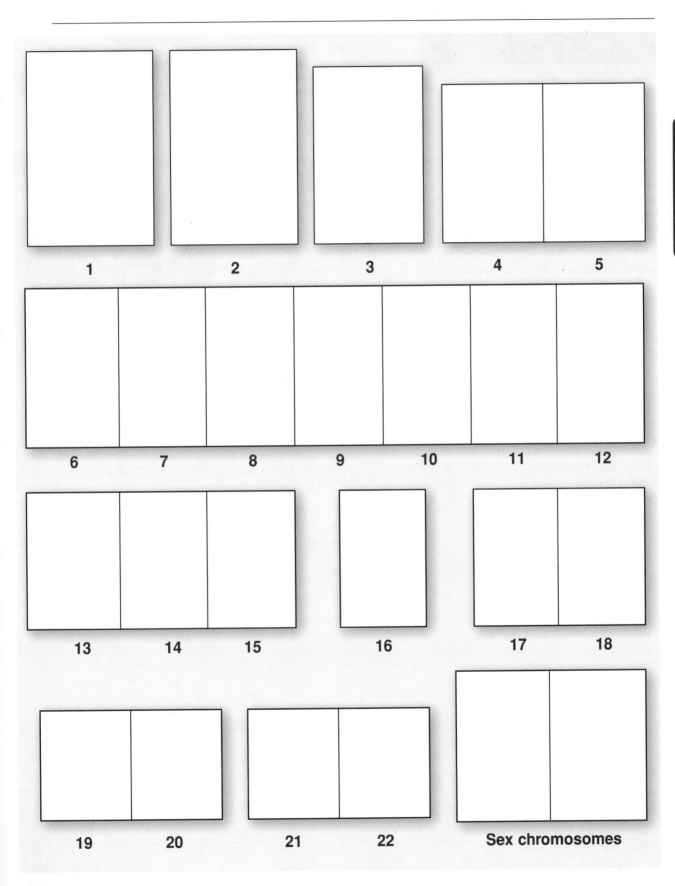

Does DNA Really Carry the Code?

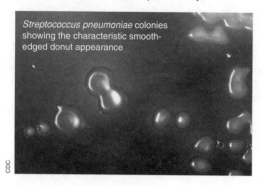

Streptococcus pneumoniae colonies showing the characteristic smooth-edged donut appearance

Scientists had known about DNA since the end of the 19th century, but its role in storing information remained unknown until the 1940s, and its structure remained a mystery for another decade after that. In 1928, experiments by British scientist Fredrick Griffith gave the first indications that DNA was responsible for passing on information. Griffith had been working with two strains of the bacteria Streptococcus pneumoniae. Only one strain (the pathogenic strain) caused pneumonia and it was easily identified because it formed colonies with smooth edges. The other, benign strain formed colonies with rough edges. When mice were injected with the pathogenic strain they developed pneumonia and died. The mice injected with the benign strain did not. Mice injected with the heat-killed pathogenic strain did not develop pneumonia either. This showed that the disease was not caused by a chemical associated with the bacteria, or a response by the body to the bacteria, it was the bacterial cells themselves. In a second experiment, Griffith mixed the benign strain with the heat-killed pathogenic strain and injected it into healthy mice. To his surprise, the mice developed pneumonia. When bacteria from the mice were recovered and cultured they produced colonies identical to the pathogenic strain. Somehow the harmless bacteria had acquired information from the dead pathogenic strain. Griffith called this process **transformation**.

In 1944, American scientists, led by Oswald Avery, continued with Griffith's experiments. They made an extract from the heat-killed pathogenic strain and treated it with chemicals to destroy any lipids, carbohydrates, or proteins. This was mixed with the benign strain and transformation still occurred. This established that no proteins, lipids, or carbohydrates were responsible for the transformation. When another identical extract was treated with chemicals that break down DNA, the transformation did not take place - the benign strain failed to acquire the information required to cause pneumonia. From this it was deduced that DNA was the unit that was carrying the information from one bacteria to another.

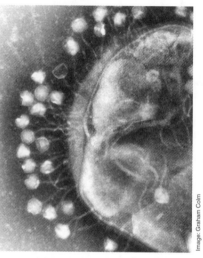

Another experiment in 1952 by Alfred Hershey, confirmed what the other two experiments had shown. Hershey worked with viruses, which were known to have DNA and to transfer information to their host. However, there was debate over whether the information was transferred by the DNA or by the protein coat of the virus. Hershey used radioactive sulfur and radioactive phosphorus to mark different parts of the virus. The sulfur was incorporated into the protein coat while the phosphorus was incorporated into the viral DNA. The viruses were then mixed with bacteria and the infected bacteria analysed. The bacteria were found to contain radioactive phosphorus but not radioactive sulfur, showing that the virus had indeed passed information to its host by injecting its own DNA.

1. How did Griffith confirm that it was the bacteria causing the pneumonia and not something else?

The bacteria formed colonies with smooth edges while others formed colonies with rough edges

2. Why were sulfur and phosphorus used in Hershey's experiment?

The sulfur was incorporated into the protein coat while the phosphorus was incorporated in the the viral DNA.

3. Why is it important to conduct two different experiments (e.g. Avery's and Hershey's) when investigating a hypothesis?

To see if the hypothesis is correct even under different circumstances.

Related activities: DNA Molecules

DNA Molecules

Even the smallest DNA molecules are extremely long. The DNA from the small *Polyoma* virus, for example, is 1.7 μm long; about three times longer than the longest proteins. The DNA comprising a bacterial chromosome is 1000 times longer than the cell into which it has to fit. The amount of DNA present in the nucleus of the cells of eukaryotic organisms varies widely from one species to another. In vertebrate sex cells, the quantity of DNA ranges from 40,000 kb to 80,000 000 kb, with humans about in the middle of the range. The traditional focus of DNA research has been on those DNA sequences that code for proteins, yet protein-coding DNA accounts for less than 2% of the DNA in human chromosomes. The rest of the DNA, once dismissed as non-coding 'evolutionary junk', is now recognized as giving rise to functional RNA molecules, many of which have already been identified as having important regulatory functions. While there is no clear correspondence between the complexity of an organism and the number of protein-coding genes in its genome, this is not the case for non-protein-coding DNA. The genomes of more complex organisms contain much more of this so-called "non-coding" DNA. These RNA-only 'hidden' genes tend to be short and difficult to identify, but the sequences are highly conserved and clearly have a role in inheritance, development, and health.

Total length of DNA in viruses, bacteria, and eukayotes

Taxon	Organism	Base pairs (in 1000s, or kb)	Length
Viruses	Polyoma or SV40	5.1	1.7 μm
	Lambda phage	48.6	17 μm
	T2 phage	166	56 μm
	Vaccinia	190	65 μm
Bacteria	Mycoplasma	760	260 μm
	E. coli (from human gut)	4600	1.56 mm
Eukaryotes	Yeast	13 500	4.6 mm
	Drosophila (fruit fly)	165 000	5.6 cm
	Human	2 900 000	99 cm

Giant lampbrush chromosomes

Lampbrush chromosomes are large chromosomes found in amphibian eggs, with lateral loops of DNA that produce a brushlike appearance under the microscope. The two scanning electron micrographs (below and right) show minute strands of DNA giving a fuzzy appearance in the high power view.

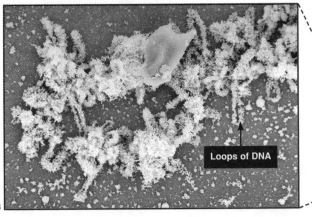

Loops of DNA

Kilobase (kb)

A kilobase (kb) is 1000 base pairs of a double-stranded nucleic acid molecule (or 1000 bases of a single-stranded molecule). One kb of double stranded DNA has a length of approximately 0.34 μm (1 μm = 1/1000 mm).

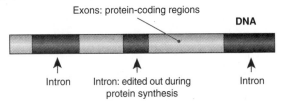

Exons: protein-coding regions

DNA

Intron   Intron: edited out during protein synthesis   Intron

Most protein-coding genes in eukaryotic DNA are not continuous and may be interrupted by 'intrusions' of other pieces of DNA. Protein-coding regions (**exons**) are interrupted by non-protein-coding regions called **introns**. Introns range in frequency from 1 to over 30 in a single 'gene' and also in size (100 to more than 10,000 bases). Introns are edited out of the protein-coding sequence during protein synthesis, but probably, after processing, go on to serve a regulatory function.

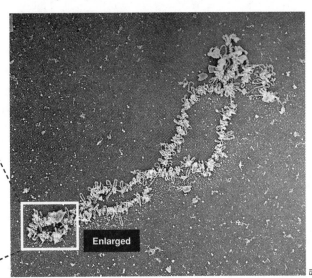

Enlarged

1. Consult the table above and make the following comparisons. Determine how much more DNA is present in:

   (a) The bacterium *E. coli* compared to the Lambda Phage virus: _____ 4,551.4 _____

   (b) Human cells compared to the bacteria *E. coli*: _____ 2,895,400 _____

2. What proportion of DNA in a eukaryotic cell is used to code for proteins or structural RNA? ___ 10% ___

3. Describe two reasons why geneticists have reevaluated their traditional view that one gene codes for one polypeptide:

   (a) _____ Chromatin organization _____

   (b) _____ DNA replication _____

Periodicals:
DNA: 50 years of the double helix

Related activities: The Simplest Case: Genes to Proteins

D 1

RNA Molecules

RNA plays vital roles in transcribing and translating DNA, forming messenger RNA (mRNA), transfer RNA (tRNA), and ribosomal RNA (rRNA). RNA is also involved in processes such as modifying mRNA after transcription and regulating translation.

Regulation of translation is achieved by destroying specific mRNA targets using short RNA lengths, which may be exogenous (short interfering RNAs) or endogenous (microRNAs). Mechanisms of **RNA interference** (RNAi) are illustrated below.

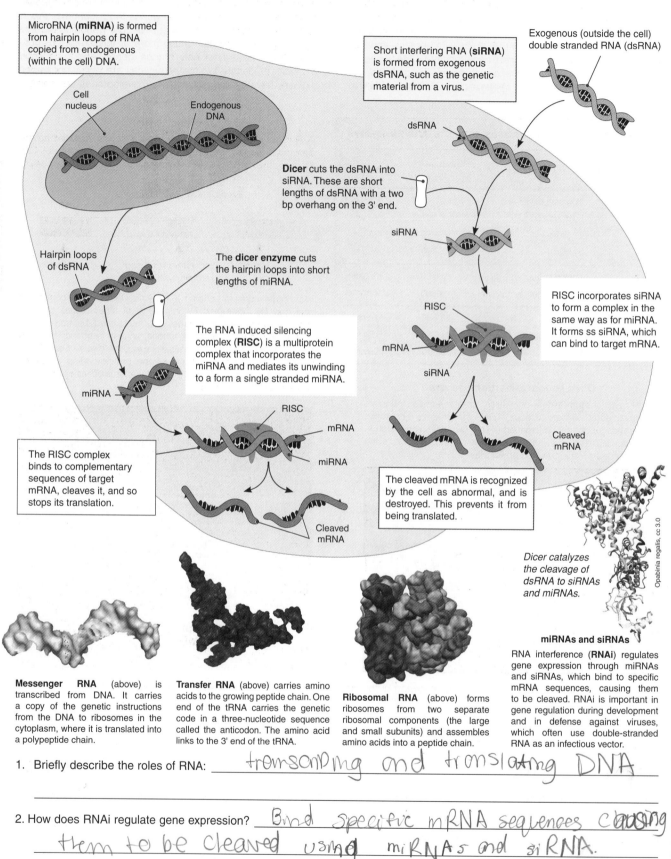

MicroRNA (**miRNA**) is formed from hairpin loops of RNA copied from endogenous (within the cell) DNA.

Short interfering RNA (**siRNA**) is formed from exogenous dsRNA, such as the genetic material from a virus.

Exogenous (outside the cell) double stranded RNA (dsRNA)

Cell nucleus

Endogenous DNA

dsRNA

Dicer cuts the dsRNA into siRNA. These are short lengths of dsRNA with a two bp overhang on the 3' end.

siRNA

Hairpin loops of dsRNA

The **dicer enzyme** cuts the hairpin loops into short lengths of miRNA.

The RNA induced silencing complex (**RISC**) is a multiprotein complex that incorporates the miRNA and mediates its unwinding to a form a single stranded miRNA.

RISC

mRNA

siRNA

RISC incorporates siRNA to form a complex in the same way as for miRNA. It forms ss siRNA, which can bind to target mRNA.

miRNA

RISC

mRNA

miRNA

The RISC complex binds to complementary sequences of target mRNA, cleaves it, and so stops its translation.

The cleaved mRNA is recognized by the cell as abnormal, and is destroyed. This prevents it from being translated.

Cleaved mRNA

Cleaved mRNA

Dicer catalyzes the cleavage of dsRNA to siRNAs and miRNAs.

Opabinia regalis, cc 3.0

Messenger RNA (above) is transcribed from DNA. It carries a copy of the genetic instructions from the DNA to ribosomes in the cytoplasm, where it is translated into a polypeptide chain.

Transfer RNA (above) carries amino acids to the growing peptide chain. One end of the tRNA carries the genetic code in a three-nucleotide sequence called the anticodon. The amino acid links to the 3' end of the tRNA.

Ribosomal RNA (above) forms ribosomes from two separate ribosomal components (the large and small subunits) and assembles amino acids into a peptide chain.

miRNAs and siRNAs

RNA interference (**RNAi**) regulates gene expression through miRNAs and siRNAs, which bind to specific mRNA sequences, causing them to be cleaved. RNAi is important in gene regulation during development and in defense against viruses, which often use double-stranded RNA as an infectious vector.

1. Briefly describe the roles of RNA: _transcribing and translating DNA_

2. How does RNAi regulate gene expression? _Bind specific mRNA sequences causing them to be cleaved using miRNAs and siRNA._

Related activities: Transcription in Eukaryotes, Translation
Weblinks: siRNA Activity

Creating a DNA Model

Although DNA molecules can be enormous in terms of their molecular size, they are made up of simple repeating units called **nucleotides**. A number of factors control the way in which these nucleotide building blocks are linked together. These factors cause the nucleotides to join together in a predictable way. This is referred to as the **base pairing rule** and can be used to construct a complementary DNA strand from a template strand, as illustrated in the exercise below:

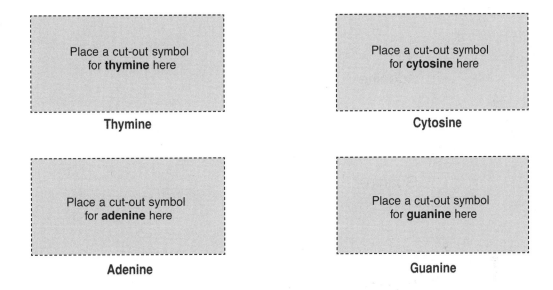

DNA Base Pairing Rule

Adenine	is always attracted to	**Thymine**	A ⟷ T
Thymine	is always attracted to	**Adenine**	T ⟷ A
Cytosine	is always attracted to	**Guanine**	C ⟷ G
Guanine	is always attracted to	**Cytosine**	G ⟷ C

1. Cut out page 103 and separate each of the 24 nucleotides by cutting along the columns and rows (see arrows indicating two such cutting points). Although drawn as geometric shapes, these symbols represent chemical structures.

2. Place one of each of the four kinds of nucleotide on their correct spaces below:

> Place a cut-out symbol
> for **thymine** here

Thymine

> Place a cut-out symbol
> for **cytosine** here

Cytosine

> Place a cut-out symbol
> for **adenine** here

Adenine

> Place a cut-out symbol
> for **guanine** here

Guanine

3. Identify and **label** each of the following features on the *adenine* nucleotide immediately above:
phosphate, **sugar**, **base**, **hydrogen bonds**

4. Create one strand of the DNA molecule by placing the 9 correct 'cut out' nucleotides in the labelled spaces on the following page (DNA molecule). Make sure these are the right way up (with the **P** on the left) and are aligned with the left hand edge of each box. Begin with thymine and end with guanine.

5. Create the complementary strand of DNA by using the base pairing rule above. Note that the nucleotides have to be arranged upside down.

6. Under normal circumstances, it is not possible for adenine to pair up with guanine or cytosine, nor for any other mismatches to occur. Describe the two factors that prevent a mismatch from occurring:

   (a) Factor 1: _The number of hydrogen bond attractven points_

   _____

   (b) Factor 2: _The size (length of the base_

   _____

7. Once you have checked that the arrangement is correct, you may glue, paste or tape these nucleotides in place.

> **NOTE:** There may be some value in keeping these pieces loose in order to practice the base pairing rule. For this purpose, *removable tape* would be best.

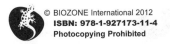
© BIOZONE International 2012
ISBN: 978-1-927173-11-4
Photocopying Prohibited

Related activities: Nucleotides and Nucleic Acids, DNA Molecules

PA 2

DNA and RNA

DNA Molecule

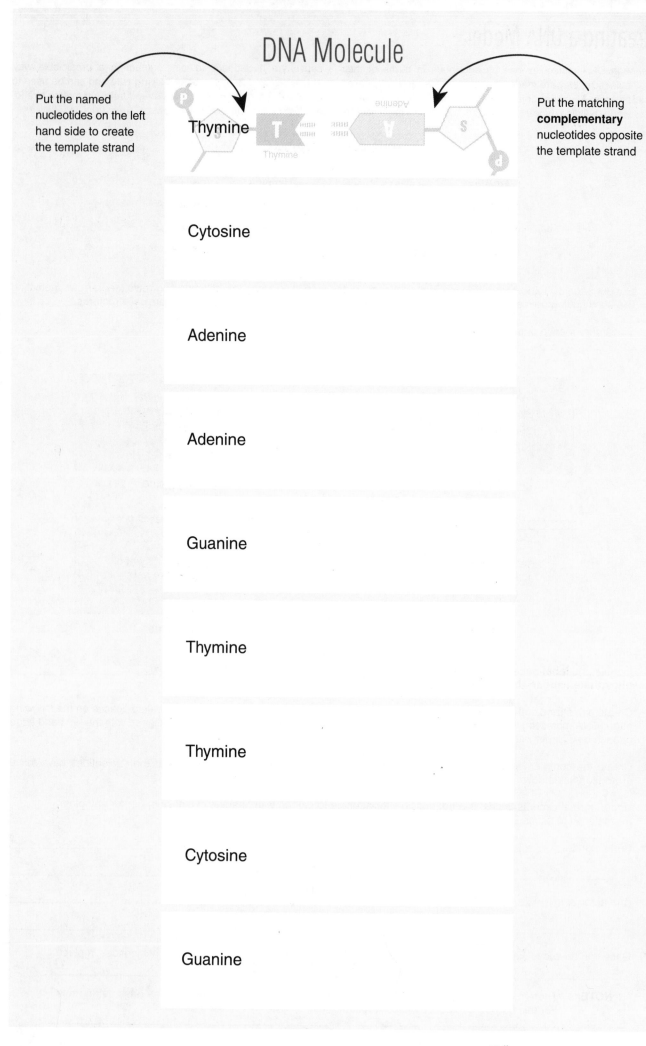

Put the named nucleotides on the left hand side to create the template strand

Thymine

Put the matching **complementary** nucleotides opposite the template strand

Cytosine

Adenine

Adenine

Guanine

Thymine

Thymine

Cytosine

Guanine

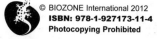

DNA Replication

Cells carry out the process of **DNA replication** (DNA duplication) prior to cell division (mitosis and meiosis). This process ensures that each resulting cell is able to receive a complete set of genes from the original cell. After the DNA has replicated, each chromosome is made up of two chromatids, which are joined at the centromere. Each chromatid contains half original (parent) DNA and half new (daughter) DNA. The two chromatids will become separated during cell division to form two separate chromosomes. During DNA replication, new nucleotides become added at a region called the **replication fork**. The position of the replication fork moves along the chromosome as the replication progresses. This whole process occurs simultaneously for each chromosome of a cell and the entire process is tightly controlled by enzymes.

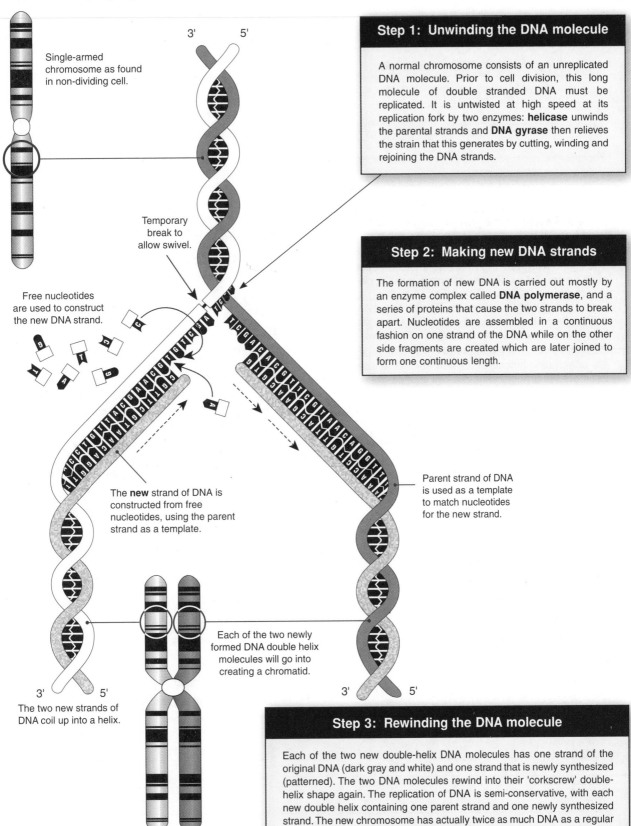

Single-armed chromosome as found in non-dividing cell.

Temporary break to allow swivel.

Free nucleotides are used to construct the new DNA strand.

The new strand of DNA is constructed from free nucleotides, using the parent strand as a template.

Parent strand of DNA is used as a template to match nucleotides for the new strand.

Each of the two newly formed DNA double helix molecules will go into creating a chromatid.

The two new strands of DNA coil up into a helix.

Replicated chromosome ready for cell division.

Step 1: Unwinding the DNA molecule

A normal chromosome consists of an unreplicated DNA molecule. Prior to cell division, this long molecule of double stranded DNA must be replicated. It is untwisted at high speed at its replication fork by two enzymes: **helicase** unwinds the parental strands and **DNA gyrase** then relieves the strain that this generates by cutting, winding and rejoining the DNA strands.

Step 2: Making new DNA strands

The formation of new DNA is carried out mostly by an enzyme complex called **DNA polymerase**, and a series of proteins that cause the two strands to break apart. Nucleotides are assembled in a continuous fashion on one strand of the DNA while on the other side fragments are created which are later joined to form one continuous length.

Step 3: Rewinding the DNA molecule

Each of the two new double-helix DNA molecules has one strand of the original DNA (dark gray and white) and one strand that is newly synthesized (patterned). The two DNA molecules rewind into their 'corkscrew' double-helix shape again. The replication of DNA is semi-conservative, with each new double helix containing one parent strand and one newly synthesized strand. The new chromosome has actually twice as much DNA as a regular (non-replicated) chromosome. The two chromatids will become separated in the cell division process to form two separate chromosomes.

DNA and RNA

© BIOZONE International 2012
ISBN: 978-1-927173-11-4
Photocopying Prohibited

Periodicals: DNA polymerase

Related activities: Mitosis and the Cell Cycle, The Genetic Code
Weblinks: DNA Replication

A 2

1. State the purpose of DNA replication: _each cell is able to recieve a comple set of ge_

2. Summarize the three main steps involved in DNA replication:

   (a) _unwind the DNA molecule_

   (b) _Making DNA strand_

   (c) _Rewing DNA Strands_

3. For a cell with 22 chromosomes, state how many chromatids would exist following DNA replication: _____

4. Discuss the importance of enzymes in DNA replication: _helicase unwinds the parental strand and DNA gyrase relieves the strain cutting, unwinding than rejoining the DNA molecule_

5. DNA replication occurs during the S (synthesis) phase of the **cell cycle**. This is part of a larger phase called interphase. It is the phase in which the cell is not dividing (in mitosis).

   The light micrograph (right) shows a section of cells in an onion root tip. These cells have a cell cycle of approximately 24 hours. The cells can be seen to be in various stages of the cell cycle. By counting the number of cells in the various stages it is possible to calculate how long the cell spends in each stage of the cycle.

   Count and record the number of cells in the image which are undergoing mitosis and those that are in interphase. Estimate the amount of time a cell spends in each phase.

Onion Root Tip Cells

Stage	No. of cells	% of total cells	Estimated time in stage
Interphase			
Mitosis			
Total		100	

6. Match the statements in the table below to form complete sentences, then put the sentences in order to make a coherent paragraph about DNA replication and its role:

   The enzymes also proofread the DNA during replication...   ...is required before mitosis or meiosis can occur.

   DNA replication is the process by which the DNA molecule...   ...by enzymes.

   Replication is tightly controlled...   ...to correct any mistakes.

   After replication, the chromosome...   ...and half new DNA.

   DNA replication...   ...during mitosis.

   The chromatids separate...   ...is copied to produce two identical DNA strands.

   A chromatid contains half original ...   ...is made up of two chromatids.

   Write the complete paragraph here: _DNA replication is required before mitosis or meiosis can occur. DNA replicats is the process by which the DNA molecule is copped too produe two identical DNA strenst. Replication is tightly controlled by enzymes. The enzymes also proodread the DNA during replicate to correct any mistakes. The chromatids seperase during mitosis._

Enzyme Control of DNA Replication

The sequence of enzyme controlled events in DNA replication is shown below (1-5). Although shown as separate, many of the enzymes are found clustered together as enzyme complexes. These enzymes are also able to 'proof-read' the new DNA strand as it is made and correct mistakes. The polymerase enzyme can only work in one direction, so that one new strand is constructed as a continuous length (the leading strand) while the other new strand is made in short segments to be later joined together (the lagging strand). **NOTE** that the nucleotides are present as deoxynucleoside triphosphates. When hydrolyzed, these provide the energy for incorporating the nucleotide into the strand.

DNA replication occurs during interphase of the cell cycle at an astounding rate. As many as 4000 nucleotides per second are replicated. This explains how under ideal conditions, bacterial cells with as many as 4 million nucleotides, can complete a cell cycle in about 20 minutes.

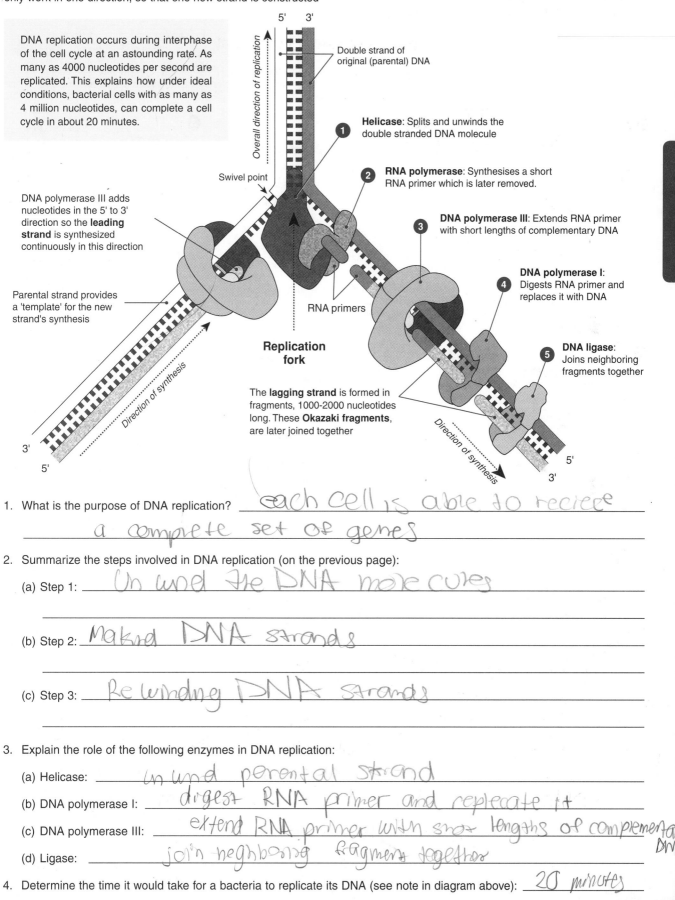

Double strand of original (parental) DNA

Helicase: Splits and unwinds the double stranded DNA molecule

RNA polymerase: Synthesises a short RNA primer which is later removed.

DNA polymerase III: Extends RNA primer with short lengths of complementary DNA

DNA polymerase I: Digests RNA primer and replaces it with DNA

DNA ligase: Joins neighboring fragments together

DNA polymerase III adds nucleotides in the 5' to 3' direction so the **leading strand** is synthesized continuously in this direction

Parental strand provides a 'template' for the new strand's synthesis

Swivel point

Overall direction of replication

Direction of synthesis

RNA primers

Replication fork

The **lagging strand** is formed in fragments, 1000-2000 nucleotides long. These **Okazaki fragments**, are later joined together

Direction of synthesis

1. What is the purpose of DNA replication? _each cell is able to reciece a complete set of genes_

2. Summarize the steps involved in DNA replication (on the previous page):

   (a) Step 1: _Un wind the DNA molecules_

   (b) Step 2: _Makind DNA strands_

   (c) Step 3: _Re winding DNA strands_

3. Explain the role of the following enzymes in DNA replication:

   (a) Helicase: _un wind perental strand_

   (b) DNA polymerase I: _digest RNA primer and replecate it_

   (c) DNA polymerase III: _extend RNA primer with shor lengths of complementar DNA_

   (d) Ligase: _join neghboring fragment together_

4. Determine the time it would take for a bacteria to replicate its DNA (see note in diagram above): _20 minutes_

Related activities: DNA Replication
Weblinks: DNA Replication (Advanced)

Meselson and Stahl's Experiment

When Watson and Crick identified the structure of DNA in 1953 it became apparent that its structure could help to explain how DNA was replicated. Three models were proposed. Watson and Crick proposed the **semi-conservative model** in which each DNA strand served as a template, forming a new DNA molecule that was half old and half new DNA. The **conservative model** proposed that the original DNA served as a complete template so that the resulting DNA comprised two completely new strands. The **dispersive model** proposed that the two new DNA molecules had part new and part old DNA interspersed (mixed) throughout them. **Meselson and Stahl** devised a simple experiment using *E. coli* grown in differing isotopes of nitrogen, to determine which theory was correct.

Meselson and Stahl's Experiment

E. coli were grown for several generations in a medium containing a **heavy nitrogen isotope** (^{15}N). Once all the bacterial DNA contained ^{15}N, they were transferred to a medium containing a **light nitrogen isotope** (^{14}N). After the transfer, newly synthesized DNA would contain ^{14}N and old DNA would contain ^{15}N.

1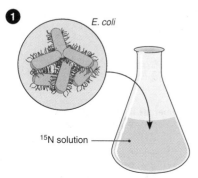

E. coli

^{15}N solution

E. coli were grown in a nutrient solution containing ^{15}N. After 14 generations, all the bacterial DNA contained ^{15}N. A sample is removed. This is **generation 0**.

2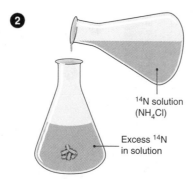

^{14}N solution (NH$_4$Cl)

Excess ^{14}N in solution

Generation 0 is added to a solution containing excess ^{14}N (as NH$_4$Cl). During replication, new DNA will incorporate ^{14}N and be 'lighter' than the original DNA (which contains only ^{15}N).

3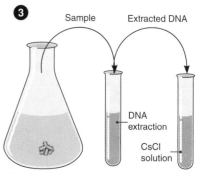

Sample   Extracted DNA

DNA extraction

CsCl solution

Every generation (~ 20 minutes), a sample is taken and treated to release the DNA. The DNA is placed in a CsCl solution which provides a density gradient for separation of the DNA.

4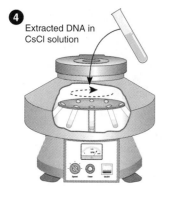

Extracted DNA in CsCl solution

Samples are spun in a high speed ultracentrifuge at 140,000 *g* for 20 hours. Heavier ^{15}N DNA moves closer to the bottom of the test tube than light ^{14}N DNA or intermediate ^{14}N/ ^{15}N DNA.

5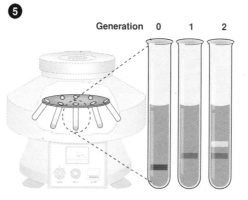

Generation 0   1   2

All the DNA in the generation 0 sample moved to the bottom of the test tube. All the DNA in the generation 1 sample moved to an intermediate position. At generation 2 half the DNA was at the intermediate position and half was near the top of the test tube. In subsequent generations, more DNA was near the top and less was in the intermediate position.

Models for DNA Replication

Conservative   Semi-conservative   Dispersive

1. Describe each of the DNA replication models:

   (a) Conservative: _orginal DNA served as a complete template so that the resulting DNA comprised two completely new stra_

   (b) Semi-converative: _DNA strand served as a template forming. a new DNA molecule that was half old and new DNA_

   (c) Dispersive: _two new DNA molecules had part new and old DNA interspersed mixed throughout them_

2. Explain why the *E. coli* were grown in an ^{15}N solution before being transferred to an ^{14}N solution: _____

   _____

   _____

Related activities: Modeling DNA Replication
Weblinks: Meselson and Stahl Animation

Modeling DNA Replication

There were three possible ways in which DNA could replicate. Meselson and Stahl's experiment was able to determine which method was used by starting with parent DNA that was heavier than would normally be expected. They were then able to analyze the relative weight of the replicated DNA to work out the correct replication method.

Instructions:

1. Cut out the DNA shapes provided on this page.

2. Intertwine the first pair (labelled 0) of heavy ^{15}N (black) DNA. This forms Generation 0 (parental DNA).

3. Use the descriptions of the three possible models for DNA replication on page 108 to model DNA replication in semi-conservative, conservative, and dispersive DNA replication.

4. For each replication method, record in the spaces provided on page 111 the percentage of **heavy** ^{15}N-^{15}N (black-black), **intermediate** ^{15}N-^{14}N (black-gray), **light** ^{14}N-^{14}N (gray-gray), or other DNA molecules formed.

5. For the dispersive model you will need to cut the DNA along the dotted lines and then stick them back together in the dispersed sequence with tape. Construct the dispersive model **LAST**.

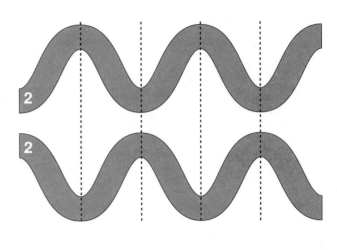

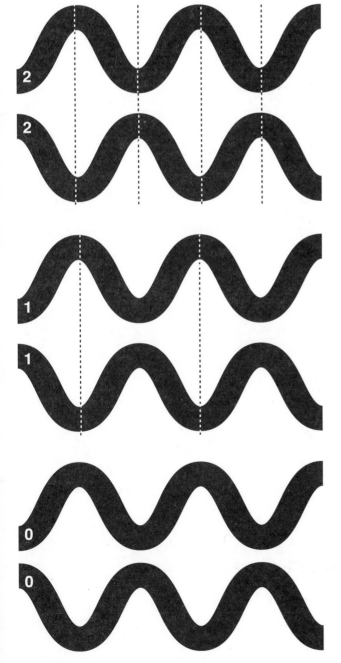

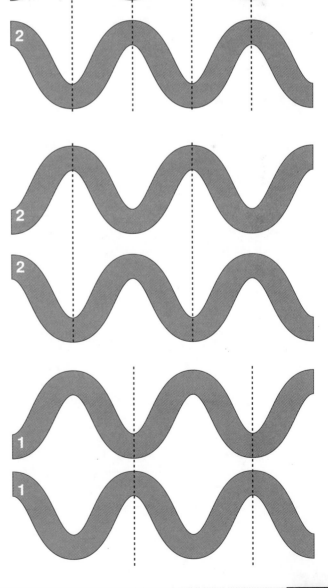

Related activities: Meselson and Stahl's Experiment
Weblinks: Meselson and Stahl Animation

RP

DNA and RNA

This page is left blank deliberately

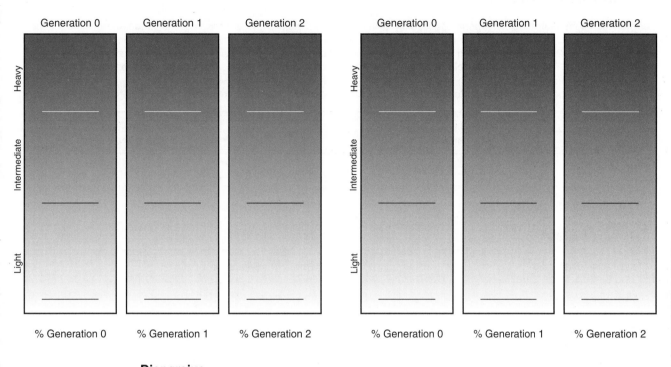

Conservative

| Generation 0 | Generation 1 | Generation 2 |

Heavy / Intermediate / Light

% Generation 0   % Generation 1   % Generation 2

Semiconservative

| Generation 0 | Generation 1 | Generation 2 |

Heavy / Intermediate / Light

% Generation 0   % Generation 1   % Generation 2

Dispersive

| Generation 0 | Generation 1 | Generation 2 |

Heavy / Intermediate / Light

% Generation 0   % Generation 1   % Generation 2

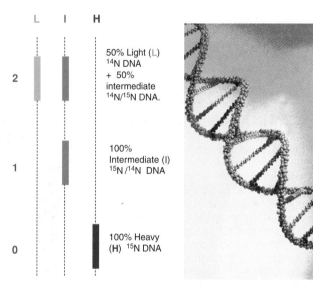

L   I   H

2 — 50% Light (L) ^{14}N DNA + 50% intermediate $^{14}N/^{15}N$ DNA.

1 — 100% Intermediate (I) $^{15}N/^{14}N$ DNA

0 — 100% Heavy (H) ^{15}N DNA

The results from Meselson's and Stahl's are shown graphically above. All of the generation 1 DNA contained one light strand (^{14}N) and one heavy strand (^{15}N) to produce an intermediate density. At generation 2, 50% of the DNA was light and 50% was intermediate DNA.

1. (a) Compare your modeling results to the results gained by Meselson and Stahl to decide which of the three DNA replication models is supported by the data:

_____

_____

(b) Was Watson and Crick's proposal correct? _____

2. Identify the replication model that fits the following data:

(a) 100% of generation 0 is "heavy DNA", 50% of generation 1 is "heavy" and 50% is "light", and 25% of generation 2 is "heavy" and 75% is "light":

_____

(b) 100% of generation 0 is "heavy DNA", 100% of generation 1 is "intermediate DNA", and 100% generation 2 lies between the "intermediate" and "light" DNA regions:

_____

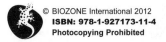

The Genetic Code

The genetic information that codes for the assembly of amino acids is stored as three-letter codes, called **codons**. Each codon represents one of 20 amino acids used in the construction of polypeptide chains. The **mRNA amino acid table** (below) can be used to identify the amino acid encoded by each of the mRNA codons. Note that the code is **degenerate** in that for each amino

acid, there may be more than one codon (there is redundancy in the code). Most of this degeneracy involves the third nucleotide of a codon. The genetic code is almost **universal**; all living organisms on Earth, from viruses and bacteria, to plants and humans, share the same code (the rare exceptions represent mutations that have occurred over the long history of evolution).

Amino acid		Codons that code for this amino acid	No.	Amino acid		Codons that code for this amino acid	No.
Ala	Alanine	GCU, GCC, GCA, GCG	4	Leu	Leucine	UAA,UUG,CUU,CUC, CUA, CUG	6
Arg	Arginine	CGU, CGC, CGA, CGG AGA, AGG	6	Lys	Lysine	AAA, AAG	2
Asn	Asparagine	AAU, AAC	2	Met	Methionine	AUG	1
Asp	Aspartic acid	GAU, GAC	2	Phe	Phenylalanine	UUU, UUC	2
Cys	Cysteine	UGU, UGC	2	Pro	Proline	CCU, CCC, CCA, CCG	4
Gln	Glutamine	CAA, CAG	2	Ser	Serine	UCU, UCC, UCA, UCG AGU, AGC	6
Glu	Glutamic acid	GAA, GAG	2	Thr	Threonine	ACU, ACC, ACA, ACG	4
Gly	Glycine	GGU, GGC, GGA, GGG	4	Trp	Tryptophan	UGG	1
His	Histidine	CAU, CAC	2	Tyr	Tyrosine	UAU, UAC	2
Ile	Isoleucine	AUU, AUC, AUA	3	Val	Valine	GUU, GUC, GUA, GUG	4

mRNA-Amino Acid Table

How to read the table: The table on the right is used to 'decode' the genetic code as a sequence of amino acids in a polypeptide chain, from a given mRNA sequence. To work out which amino acid is coded for by a codon (triplet of bases) look for the first letter of the codon in the row label on the left hand side. Then look for the column that intersects the same row from above that matches the second base. Finally, locate the third base in the codon by looking along the row from the right hand end that matches your codon.

Example: Determine **CAG**

C on the left row,
A on the top column,
G on the right row
CAG is Gln **(glutamine)**

Read second letter here / Read first letter here / **Second Letter** / Read third letter here

First Letter		U	C	A	G		Third Letter
U		UUU Phe UUC Phe UUA Leu UUG Leu	UCU Ser UCC Ser UCA Ser UCG Ser	UAU Tyr UAC Tyr UAA STOP UAG STOP	UGU Cys UGC Cys UGA STOP UGG Trp		U C A G
C		CUU Leu CUC Leu CUA Leu CUG Leu	CCU Pro CCC Pro CCA Pro CCG Pro	CAU His CAC His CAA Gln CAG Gln	CGU Arg CGC Arg CGA Arg CGG Arg		U C A G
A		AUU Ile AUC Ile AUA Ile AUG Met	ACU Thr ACC Thr ACA Thr ACG Thr	AAU Asn AAC Asn AAA Lys AAG Lys	AGU Ser AGC Ser AGA Arg AGG Arg		U C A G
G		GUU Val GUC Val GUA Val GUG Val	GCU Ala GCC Ala GCA Ala GCG Ala	GAU Asp GAC Asp GAA Glu GAG Glu	GGU Gly GGC Gly GGA Gly GGG Gly		U C A G

1. Use the **mRNA-amino acid table** (above) to list in the table above all the **codons** that code for each of the amino acids and the number of different codons that can code for each amino acid (the first amino acid has been done for you).

2. (a) How many amino acids could be coded for if a codon consisted of just two bases? _16_

   (b) Why is this number of bases inadequate to code for the 20 amino acids required to make proteins?
   There are not enough combinations with the 4 base alphabet to code for the 20 amino acids

3. Describe the consequence of the degeneracy of the genetic code to the likely effect of a change to one base in a triplet:
   More than one triplet may encode some amino acids

© BIOZONE International 2012
ISBN: 978-1-927173-11-4
Photocopying Prohibited

The Simplest Case: Genes to Proteins

The traditionally held view of genes was as sections of DNA coding only for protein. This view has been revised in recent years with the discovery that much of the nonprotein-coding DNA encodes functional RNAs; it is not all non-coding "junk" DNA as was previously assumed. In fact, our concept of what constitutes a gene is changing rapidly and now encompasses all those segments of DNA that are transcribed (to RNA). This activity considers only the simplest scenario: one in which the gene codes for a functional protein. **Nucleotides**, the basic unit of genetic information, are read in groups of three (**triplets**). Some triplets have a special controlling function in the making of a polypeptide chain. The equivalent of the triplet on the mRNA molecule is the **codon**. Three codons can signify termination of the amino acid chain (UAG, UAA and UGA in the mRNA code). The codon AUG is found at the beginning of every gene (on mRNA) and marks the starting point for reading the gene. The genes required to form a functional end-product (in this case, a functional protein) are collectively called a **transcription unit**.

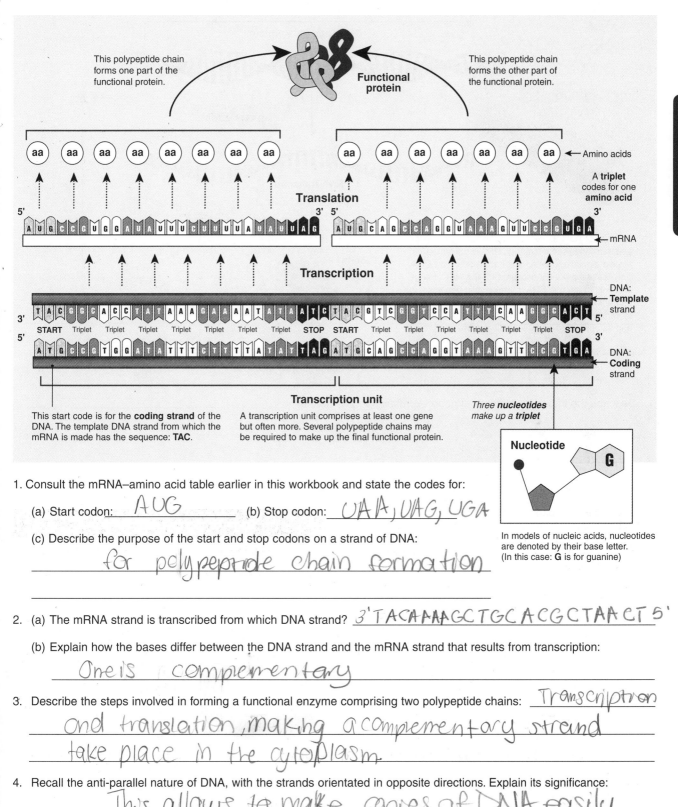

This polypeptide chain forms one part of the functional protein.

This polypeptide chain forms the other part of the functional protein.

Functional protein

aa ← Amino acids

A **triplet** codes for one **amino acid**

Translation

mRNA

Transcription

DNA: **Template** strand

START Triplet Triplet Triplet Triplet Triplet Triplet STOP START Triplet Triplet Triplet Triplet Triplet Triplet STOP

DNA: **Coding** strand

Transcription unit

This start code is for the **coding strand** of the DNA. The template DNA strand from which the mRNA is made has the sequence: **TAC**.

A transcription unit comprises at least one gene but often more. Several polypeptide chains may be required to make up the final functional protein.

Three **nucleotides** make up a **triplet**

Nucleotide

G

In models of nucleic acids, nucleotides are denoted by their base letter. (In this case: **G** is for guanine)

1. Consult the mRNA–amino acid table earlier in this workbook and state the codes for:

   (a) Start codon: _AUG_   (b) Stop codon: _UAA, UAG, UGA_

   (c) Describe the purpose of the start and stop codons on a strand of DNA:

   for polypeptide chain formation

2. (a) The mRNA strand is transcribed from which DNA strand? _3'TACAAAGCTGCACGCTAACT 5'_

   (b) Explain how the bases differ between the DNA strand and the mRNA strand that results from transcription:

   One is complementary

3. Describe the steps involved in forming a functional enzyme comprising two polypeptide chains: _Transcription and translation, making a complementary strand take place in the cytoplasm._

4. Recall the anti-parallel nature of DNA, with the strands orientated in opposite directions. Explain its significance:

   This allows to make copies of DNA easily

Periodicals:
What is a gene?

Related activities: Gene Expression: The Genetic Code
Weblinks: Review of Protein Synthesis

DNA and RNA

Gene Expression: Prokaryotes vs Eukaryotes

The process of transferring the information encoded in a gene to its functional gene product is called **gene expression**. The central dogma of molecular biology for the past 50 years or so has stated that genetic information, encoded in DNA, is transcribed as molecules of RNA, which are then translated into the amino acid sequences that make up proteins. The established opinion was often stated as "one gene-one protein" and proteins were assumed to be the main regulatory agents for the cell (including its gene expression). The one gene-one protein model is supported by studies of prokaryotic genomes, where the DNA consists almost entirely of protein-coding genes and their regulatory sequences.

Genes and Gene Expression in Prokaryotes

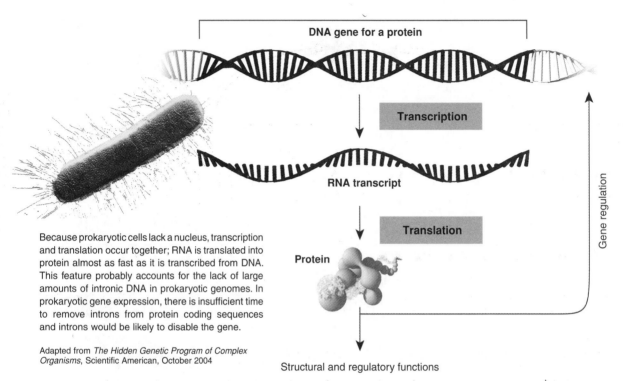

Because prokaryotic cells lack a nucleus, transcription and translation occur together; RNA is translated into protein almost as fast as it is transcribed from DNA. This feature probably accounts for the lack of large amounts of intronic DNA in prokaryotic genomes. In prokaryotic gene expression, there is insufficient time to remove introns from protein coding sequences and introns would be likely to disable the gene.

Adapted from *The Hidden Genetic Program of Complex Organisms*, Scientific American, October 2004

1. Describe the important features of gene expression in prokaryotes: <u>Genes in the DNA are</u> <u>transcribed into mRNA. This cenallow mRNA to reverse</u> <u>transcription. DNA can be inserted into the host's chromosomes</u>

2. The traditional (old) view of gene expression in eukaryotes (table right) was based on a modification of the one gene-one protein model. This model does not adequately explain gene expression in eukaryotes, but it is probably still appropriate for prokaryotes. Suggest why:
<u>Lengths of non-coding</u>
<u>DNA in eukaryote DNA that</u>
<u>can be edited out</u>
<u>during protein</u>
<u>synthesis</u>

Gene Expression in Eukaryotes	
The Old View	**The New View**
Introns are spliced out of a primary RNA transcript	Introns are spliced out of a primary RNA transcript
All the exon RNA is translated into proteins.	Not all exon RNA is translated into proteins. Non-protein-coding exonic DNA may have its own function or may contribute to microRNAs
Introns are junk DNA with no function; they are degraded and recycled.	Introns are processed into microRNAs which are involved in regulating development.

3. How is gene expression in prokaryotes fundamentally different from gene expression in eukaryotes?
<u>Prokaryotes lack a nucleus so transcription</u>
<u>and translation occurs together</u>

Related activities: The Simplest Case: Genes to Proteins
Weblinks: Transcription in Prokaryotes

Periodicals:
Gene structure and expression

© BIOZONE International 2012
ISBN: 978-1-927173-11-4
Photocopying Prohibited

Transcription in Eukaryotes

Transcription is the process by which the code contained in the DNA molecule is transcribed (rewritten) into a **mRNA** molecule. Transcription is under the control of the cell's metabolic processes which must activate a gene before this process can begin. The enzyme that directly controls the process is RNA polymerase, which makes a strand of mRNA using the single strand of DNA (the **template strand**) as a template (hence the term). The enzyme transcribes only a gene length of DNA at a time and therefore recognizes start and stop signals (codes) at the beginning and end of the gene. Only RNA polymerase is involved in mRNA synthesis; it causes the unwinding of the DNA as well. It is common to find several RNA polymerase enzyme molecules on the same gene at any one time, allowing a high rate of mRNA synthesis to occur.

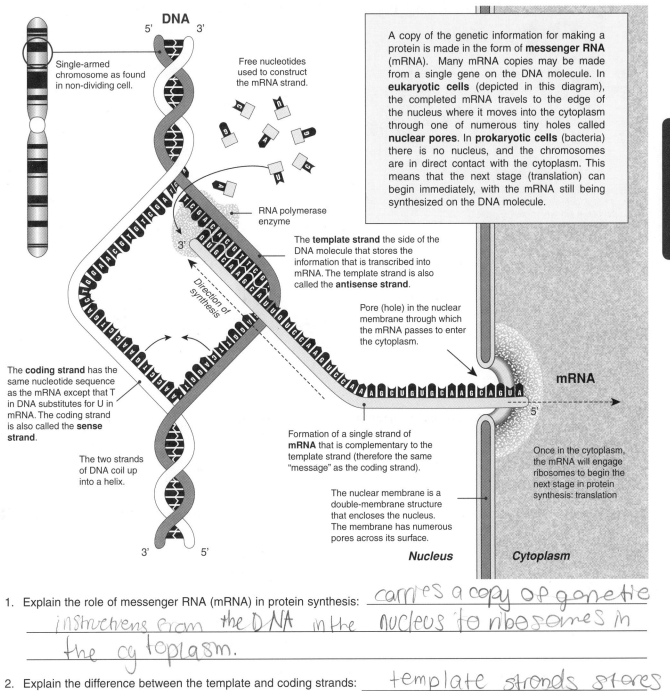

A copy of the genetic information for making a protein is made in the form of **messenger RNA** (mRNA). Many mRNA copies may be made from a single gene on the DNA molecule. In **eukaryotic cells** (depicted in this diagram), the completed mRNA travels to the edge of the nucleus where it moves into the cytoplasm through one of numerous tiny holes called **nuclear pores**. In **prokaryotic cells** (bacteria) there is no nucleus, and the chromosomes are in direct contact with the cytoplasm. This means that the next stage (translation) can begin immediately, with the mRNA still being synthesized on the DNA molecule.

DNA

Single-armed chromosome as found in non-dividing cell.

Free nucleotides used to construct the mRNA strand.

RNA polymerase enzyme

The **template strand** the side of the DNA molecule that stores the information that is transcribed into mRNA. The template strand is also called the **antisense strand**.

Pore (hole) in the nuclear membrane through which the mRNA passes to enter the cytoplasm.

The **coding strand** has the same nucleotide sequence as the mRNA except that T in DNA substitutes for U in mRNA. The coding strand is also called the **sense strand**.

The two strands of DNA coil up into a helix.

Formation of a single strand of **mRNA** that is complementary to the template strand (therefore the same "message" as the coding strand).

The nuclear membrane is a double-membrane structure that encloses the nucleus. The membrane has numerous pores across its surface.

mRNA

Once in the cytoplasm, the mRNA will engage ribosomes to begin the next stage in protein synthesis: translation

Nucleus

Cytoplasm

DNA and RNA

1. Explain the role of messenger RNA (mRNA) in protein synthesis: _carries a copy of genetic instructions from the DNA in the nucleus to ribosomes in the cytoplasm._

2. Explain the difference between the template and coding strands: _template strands stores the information that is transcribed into mRNA_

3. For the following triplets on the DNA, determine the **codon** sequence for the mRNA that would be synthesized:

(a) Triplets on the DNA: T A C   T A G   C C G   C G A   T T T

Codons on the mRNA: AUG AUC GGU AAA

(b) Triplets on the DNA: T A C   A A G   C C T   A T A   A A A

Codons on the mRNA: AUG UUC GGA UAU UUU

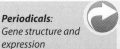

Periodicals: Gene structure and expression

Related activities: Gene Expression: Prokaryotes vs Eukaryotes
Weblinks: Animation of Transcription

Post Transcriptional Modification

Human DNA contains only 25,000 genes, but produces 90,000 different proteins. Each gene must therefore produce more than one protein. This is achieved by both **post transcriptional** and **post translational modification**. Primary mRNA molecules contain exons and introns. Usually **introns** are removed after transcription and the **exons** are spliced together, this is post transcriptional modification. However, the number of exons joined together and the way they are spliced together is not always the same. This creates variations of the polypeptide chain that results. These mechanisms allow for the production of the diverse range of proteins.

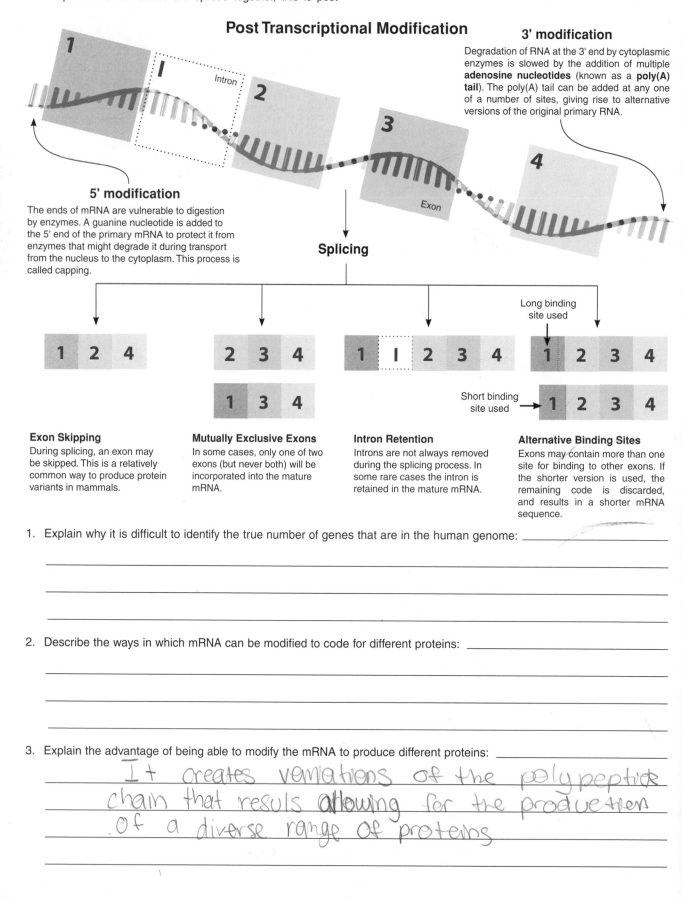

Post Transcriptional Modification

3' modification

Degradation of RNA at the 3' end by cytoplasmic enzymes is slowed by the addition of multiple **adenosine nucleotides** (known as a **poly(A) tail**). The poly(A) tail can be added at any one of a number of sites, giving rise to alternative versions of the original primary RNA.

5' modification

The ends of mRNA are vulnerable to digestion by enzymes. A guanine nucleotide is added to the 5' end of the primary mRNA to protect it from enzymes that might degrade it during transport from the nucleus to the cytoplasm. This process is called capping.

Splicing

Long binding site used

Short binding site used

Exon Skipping
During splicing, an exon may be skipped. This is a relatively common way to produce protein variants in mammals.

Mutually Exclusive Exons
In some cases, only one of two exons (but never both) will be incorporated into the mature mRNA.

Intron Retention
Introns are not always removed during the splicing process. In some rare cases the intron is retained in the mature mRNA.

Alternative Binding Sites
Exons may contain more than one site for binding to other exons. If the shorter version is used, the remaining code is discarded, and results in a shorter mRNA sequence.

1. Explain why it is difficult to identify the true number of genes that are in the human genome: _____

_____

_____

_____

2. Describe the ways in which mRNA can be modified to code for different proteins: _____

_____

_____

_____

3. Explain the advantage of being able to modify the mRNA to produce different proteins: _____

It creates variations of the polypeptide chain that resuls allowing for the production of a diverse range of proteins

_____

_____

Related activities: *Transcription in Eukaryotes*

Periodicals:
The alternative genome
The hidden genetic program

© BIOZONE International 2012
ISBN: 978-1-927173-11-4
Photocopying Prohibited

Translation

The diagram at the bottom of the page shows the translation phase of protein synthesis. The scene shows how a single mRNA molecule can be 'serviced' by many ribosomes at the same time. The ribosome on the right is in a more advanced stage of constructing a polypeptide chain because it has 'translated' more of the mRNA than the ribosome on the left. The anticodon at the base of each tRNA must make a perfect complementary match with the codon on the mRNA before the amino acid is released. Once released, the amino acid is added to the growing polypeptide chain by enzymes.

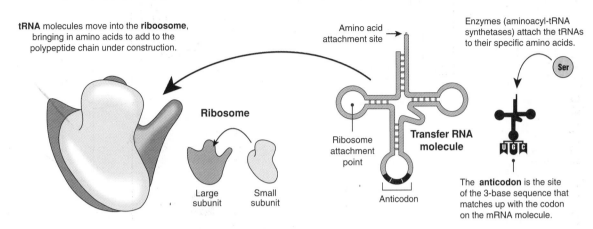

tRNA molecules move into the **riboosome**, bringing in amino acids to add to the polypeptide chain under construction.

Ribosome

Large subunit

Small subunit

Amino acid attachment site

Enzymes (aminoacyl-tRNA synthetases) attach the tRNAs to their specific amino acids.

Ser

Ribosome attachment point

Transfer RNA molecule

UGC

Anticodon

The **anticodon** is the site of the 3-base sequence that matches up with the codon on the mRNA molecule.

Ribosomes are made up of a complex of ribosomal RNA (rRNA) and proteins. They exist as two separate sub-units (above) until they are attracted to a binding site on the mRNA molecule, when they join together. Ribosomes have binding sites that attract transfer RNA (**tRNA**) molecules loaded with amino acids. The tRNA molecules are about 80 nucleotides in length and are made under the direction of genes in the chromosomes. There is a different tRNA molecule for each of the different possible anticodons (see the diagram below) and, because of the degeneracy of the genetic code, there may be up to six different tRNAs carrying the same amino acid.

DNA and RNA

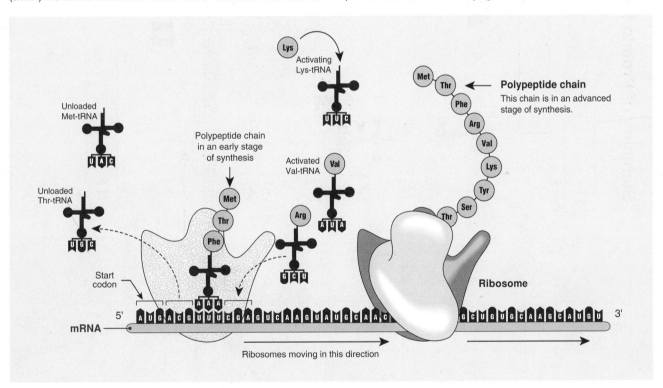

Lys

Activating Lys-tRNA

UUC

Unloaded Met-tRNA

UAC

Polypeptide chain in an early stage of synthesis

Met

Thr

Phe

Start codon

Unloaded Thr-tRNA

UGC

Activated Val-tRNA

Val

Arg

AUA

Arg

GCU

Met

Thr

Phe

Arg

Val

Lys

Tyr

Ser

Thr

Polypeptide chain
This chain is in an advanced stage of synthesis.

Ribosome

mRNA

5' AUGACGUUUCGAGUCAAGUAUGCAAC   GCUGUGCAAGCAUGU 3'

Ribosomes moving in this direction

1. For the following codons on the mRNA, determine the **anticodons** for each tRNA that would deliver the amino acids:

Codons on the mRNA: U A C   U A G   C C G   C G A   U U U

Anticodons on the tRNAs: AUG AUC GGC GCU AAA

2. There are many different types of tRNA molecules, each with a different anticodon (HINT: see the mRNA table).

(a) How many different tRNA types are there, each with a unique anticodon? 64

(b) Explain your answer: There are 64 possible codons for mRNA but three are terminater codons

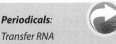
Periodicals:
Transfer RNA

Related activities: The Genetic Code
Weblinks: Polyribosomes

RA 2

Protein Synthesis: Translating the Code

Summary of Protein Synthesis in a Eukaryotic Cell

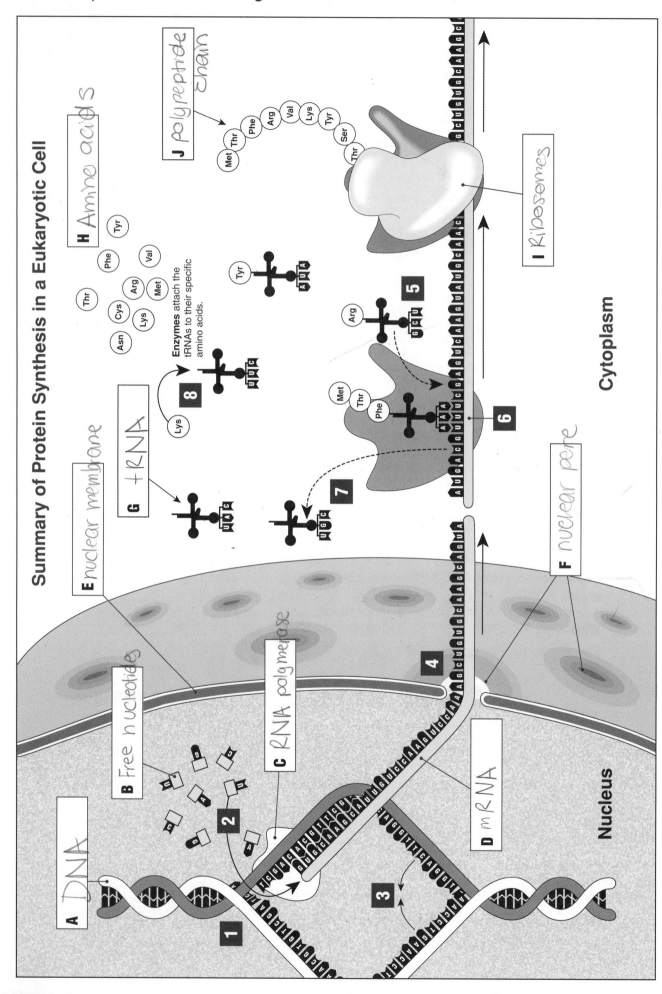

J Polypeptide chain

H Amino acids

Enzymes attach the tRNAs to their specific amino acids.

G tRNA

E nuclear membrane

B Free nucleotides

C RNA polymerase

A DNA

D mRNA

F nuclear pore

I Ribosomes

Cytoplasm

Nucleus

Related activities: *The Simplest Case: Genes to Proteins*
Weblinks: *Review of Protein Synthesis*

Periodicals:
Gene structure and expression

The diagram opposite shows an overview of the process of protein synthesis. Each of the structures involved is labeled with a letter (**A-J**), while the major steps in the process are identified with numbers (**1-8**).

1. Using the word list provided below, identify each of the structures marked with a letter. Write the name of that structure in the spaces provided on the diagram.

DNA, nuclear pore, free nucleotides, tRNA, RNA polymerase enzyme,

amino acids, mRNA, ribosome, nuclear membrane, polypeptide chain

2. Match each of the processes (identified on the diagram with numbers 1-8) to the correct summary of the process provided below. Write the process number next to the appropriate sentence.

8   tRNA molecule is recharged with another amino acid of the same type, ready to take part in protein synthesis

5   tRNA molecule brings in the correct amino acid to the ribosome

1   Unwinding the DNA molecule

3   DNA rewinds into double helix structure

6   Anti-codon on the tRNA matches with the correct codon on the mRNA and drops off the amino acid

7   tRNA leaves the ribosome

4   mRNA moves through nuclear pore in nuclear membrane to the cytoplasm

2   mRNA synthesis: nucleotides added to the growing strand of messenger RNA molecule

3. Explain the purpose of protein synthesis: _The make building materials,_ _enzymes or Other regulatory chemicals_

4. Name the three different types of RNA involved in protein synthesis: _tRNA, mRNA_ _and RNA polymerase_

5. Outline three structural or functional differences between RNA and DNA:

(a) _DNA is a linear double helix_

(b) _DNA bases are AGCT   RNA's are AGUC_

(c) _DNA is a deoxyribose_

6. How are nucleic acids attached to tRNA? _through enzymes_

7. (a) Name the general process taking place in the **nucleus**: _DNA replication_

(b) Name the general process taking place in the **cytoplasm**: _____

8. Consult the *mRNA-amino acid table* earlier in this workbook. Explain the result of a point mutation involving a change to the third base in a nucleotide as follows:

(a) UUU changes to UUC: _more than one codon for the same_ _amino acid_

(b) UUU changes to UUA: _a change of a single amino acid_ _in the polypeptide chain_

(c) Which of these mutations is likely to result in a change to the protein produced? _UUU to UUA_

Analyzing a DNA Sample

The nucleotide (base sequence) of a section of DNA can be determined using DNA sequencing techniques The base sequence determines the amino acid sequence of the resultant protein therefore the DNA tells us what type of protein that gene encodes. This exercise reviews the areas of DNA replication, transcription, and translation using an analysis of a gel electrophoresis column. **Attempt it after you have completed the rest of this topic**. Remember that the gel pattern represents the sequence in the synthesized strand.

1. Determine the amino acid sequence of a protein from the nucleotide sequence of its DNA, with the following steps:
   (a) Determine the sequence of **synthesized DNA** in the gel
   (b) Convert it to the complementary sequence of the **sample DNA**
   (c) Complete the **mRNA** sequence
   (d) Determine the **amino acid** sequence by using the 'mRNA amino acid table' in this workbook.

   NOTE: The nucleotides in the gel are read from bottom to top and the sequence is written in the spaces provided from left to right (the first 4 have been done for you).

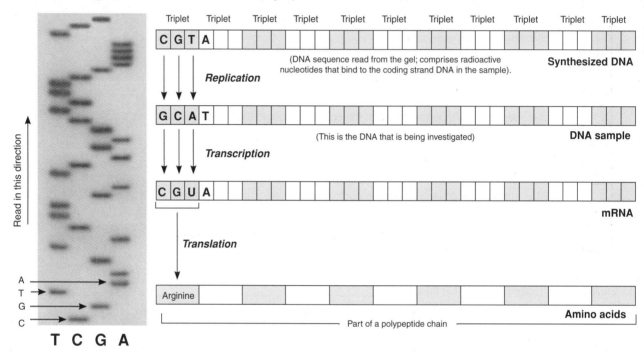

2. For each single strand DNA sequence below, write the base sequence for the **complementary DNA** strand:

   (a) DNA: T A C   T A G   C C G   C G A   T T T   A C A   A T T

       DNA: _____

   (b) DNA: T A C   G C C   T T A   A A G   G G C   C G A   A T C

       DNA: _____

   (c) Identify the cell process that this exercise represents: _____

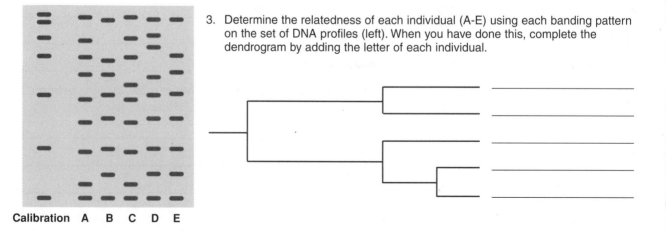

3. Determine the relatedness of each individual (A-E) using each banding pattern on the set of DNA profiles (left). When you have done this, complete the dendrogram by adding the letter of each individual.

Calibration   A   B   C   D   E

Related activities: The Genetic Code, Gel Electrophoresis

Proteins and Phenotype

Proteins play important roles in the development of phenotypes. An example of the effect they have can be seen in a comparison of normal red blood cells and sickle cells. The mutation of a single nucleotide in the gene that codes for the β chain of the hemoglobin complex causes sickle cell disease, which has radical effects on the functioning of the body. Under low oxygen conditions, the mutations causes hemoglobin molecules (HbS) that tend to stick together, forming chains which distort the cell, resulting in a sickle shape. The cell is less flexible and carries oxygen less efficiently than normal red blood cells.

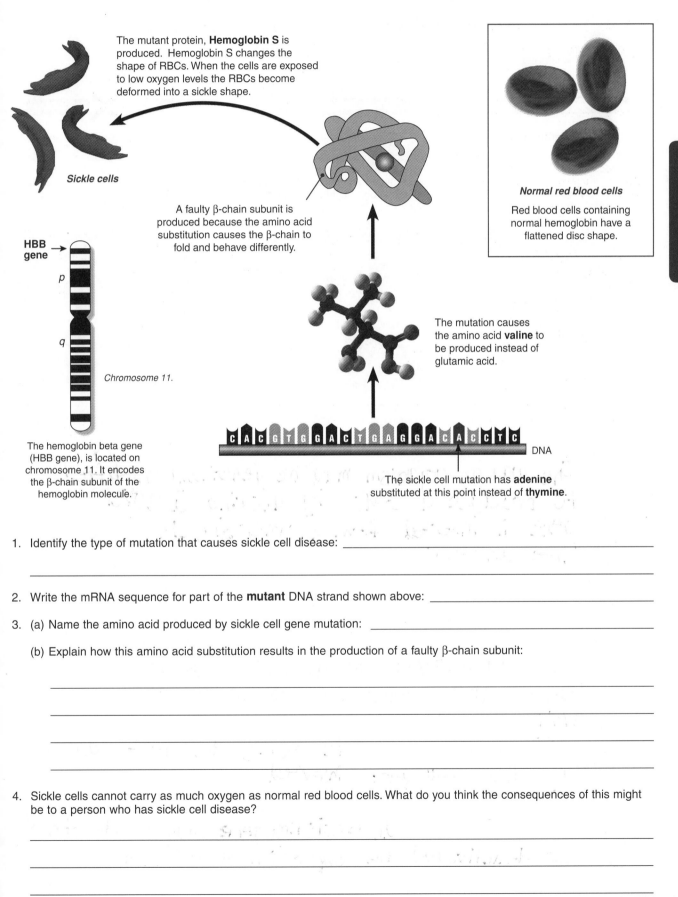

The mutant protein, **Hemoglobin S** is produced. Hemoglobin S changes the shape of RBCs. When the cells are exposed to low oxygen levels the RBCs become deformed into a sickle shape.

Sickle cells

A faulty β-chain subunit is produced because the amino acid substitution causes the β-chain to fold and behave differently.

HBB gene

p

q

Chromosome 11.

The hemoglobin beta gene (HBB gene), is located on chromosome 11. It encodes the β-chain subunit of the hemoglobin molecule.

Normal red blood cells

Red blood cells containing normal hemoglobin have a flattened disc shape.

The mutation causes the amino acid **valine** to be produced instead of glutamic acid.

C A C G T G G A C T G A G G A C A C C T C   DNA

The sickle cell mutation has **adenine** substituted at this point instead of **thymine**.

1. Identify the type of mutation that causes sickle cell disease: _____

_____

2. Write the mRNA sequence for part of the **mutant** DNA strand shown above: _____

3. (a) Name the amino acid produced by sickle cell gene mutation: _____

   (b) Explain how this amino acid substitution results in the production of a faulty β-chain subunit:

   _____

   _____

   _____

   _____

4. Sickle cells cannot carry as much oxygen as normal red blood cells. What do you think the consequences of this might be to a person who has sickle cell disease?

   _____

   _____

   _____

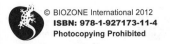

Periodicals:
Genetics of sickle cell anaemia

Related activities: Proteins

A 2

What is Genetic Engineering?

The genetic modification of organisms is a vast industry, and the applications of the technology are exciting and far reaching. It brings new hope for medical cures, promises to increase yields in agriculture, and has the potential to help solve the world's pollution and resource crises. Organisms with artificially altered DNA are referred to as **genetically modified organisms** or

GMOs. They may be modified in one of three ways (outlined below). Some of the current and proposed applications of gene technology raise complex ethical and safety issues. The benefits of their use must be carefully weighed against the risks to human health, as well as the health and well-being of other organisms and the environment as a whole.

Producing Genetically Modified Organisms (GMOs)

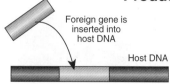
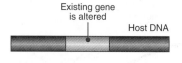

Add a foreign gene

A novel (foreign) gene is inserted from another species. This will enable the GMO to express the trait coded by the new gene. Organisms genetically altered in this way are referred to as **transgenic**.

Alter an existing gene

An existing gene may be altered to make it express at a higher level (e.g. produce more growth hormone) or in a different way (in tissue that would not normally express it). This method is also used for gene therapy.

Delete or 'turn off' a gene

An existing gene may be deleted or deactivated (switched off) to prevent the expression of a trait (e.g. the deactivation of the ripening gene in tomatoes produced the Flavr-Savr tomato).

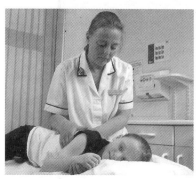

Human insulin, used to treat diabetic patients, is now produced using transgenic bacteria.

Gene therapy could be used treat genetic disorders, such as cystic fibrosis.

Manipulating gene action is one way in which to control processes such as ripening in fruit.

1. Using examples, discuss the ways in which an organism may be genetically modified (to produce a GMO):

One way an organism may be genetically modified to produce a GMO by allowing a novel gene is inserted from another specie into another

2. Explain how human needs or desires have provided a stimulus for the development of the following biotechnologies:

(a) Gene therapy: *Alter an exsisting gene to make it express at a higher level or in a different a way*

(b) The production and use of transgenic organisms: *to express the trait coded by the new gene inserted*

(c) Plant micropropagation (tissue culture): *an exsisting gene may be deleted or deactivated the expression of a trait*

Related activities: Applications of GMOs
Weblinks: Biotechnology Tmeline

Restriction Enzymes

One of the essential tools of genetic engineering is a group of special **restriction enzymes** (also known as restriction endonucleases). These have the ability to cut DNA molecules at very precise sequences of 4 to 8 base pairs called **recognition sites**. These enzymes are the "molecular scalpels" that allow genetic engineers to cut up DNA in a controlled way. Although first isolated in 1970, these enzymes were discovered earlier in many bacteria. The purified forms of these bacterial restriction enzymes are used today as tools to cut DNA (see table on the next page for examples). Enzymes are named according to the bacterial species from which they were first isolated. By using a 'tool kit' of over 400 restriction enzymes recognizing about 100 recognition sites, genetic engineers can isolate, sequence, and manipulate individual genes derived from any type of organism. The sites at which the fragments of DNA are cut may result in overhanging "sticky ends" or non-overhanging "blunt ends". Pieces may later be joined together using an enzyme called **DNA ligase** in a process called **ligation**.

DNA and RNA

Sticky End Restriction Enzymes

1 A **restriction enzyme** cuts the double-stranded DNA molecule at its specific **recognition site** (see the table on the following page for a representative list of restriction enzymes and their recognition sites).

2 The cuts produce a DNA fragment with two **sticky ends** (ends with exposed nucleotide bases at each end). The piece it is removed from is also left with sticky ends.

Restriction enzymes may cut DNA leaving an overhang or sticky end, without its complementary sequence opposite. DNA cut in such a way is able to be joined to other exposed end fragments of DNA with matching sticky ends. Such joins are specific to their recognition sites.

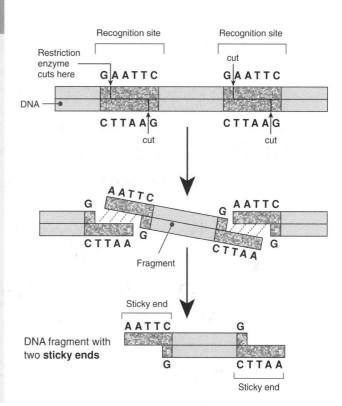

DNA fragment with two **sticky ends**

Blunt End Restriction Enzymes

1 A **restriction enzyme** cuts the double-stranded DNA molecule at its specific recognition site (see the table opposite for a representative list of restriction enzymes and their recognition sites).

2 The cuts produce a DNA fragment with two **blunt ends** (ends with no exposed nucleotide bases at each end). The piece it is removed from is also left with blunt ends.

It is possible to use restriction enzymes that cut leaving no overhang. DNA cut in such a way is able to be joined to any other blunt end fragment, but tends to be nonspecific because there are no sticky ends as recognition sites.

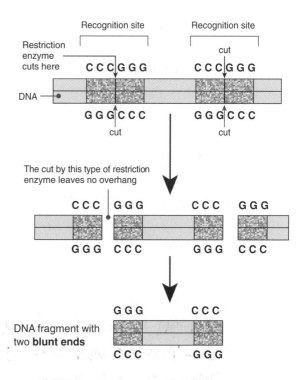

DNA fragment with two **blunt ends**

Related activities: Ligation
Weblinks: DNA Interactive: Cut and Paste

A 3

Origin of Restriction Enzymes

Restriction enzymes have been isolated from many bacteria. It was observed that certain *bacteriophages* (viruses that infect bacteria) could not infect bacteria other than their usual hosts. The reason was found to be that other potential hosts could destroy almost all of the phage DNA using *restriction enzymes* present naturally in their cells; a defence mechanism against the entry of foreign DNA. Restriction enzymes are named according to the species they were first isolated from, followed by a number to distinguish different enzymes isolated from the same organism.

Recognition sites for selected restriction enzymes

Enzyme	Source	Recognition sites
EcoRI	*Escherichia coli* RY13	G A A T T C
BamHI	*Bacillus amyloliquefaciens* H	G G A T C C
HaeIII	*Haemophilus aegyptius*	G G C C
HindIII	*Haemophilus influenzae* Rd	A A G C T T
HpaI	*Haemophilus parainfluenzae*	G T T A A C
HpaII	*Haemophilus parainfluenzae*	C C G G
MboI	*Moraxella bovis*	G A T C
NotI	*Norcardia otitidis-caviarum*	G C G G C C G C
TaqI	*Thermus aquaticus*	T C G A

1. Explain the following terms, identifying their role in recombinant DNA technology:

   (a) Restriction enzyme: _cuts the double-stranded DNA molecule at its specific regonition site_

   (b) Recognition site: _DNA with two sticky end_

   (c) Sticky end: _ends with exposed nucleotide based at each end_

   (d) Blunt end: _ends w/ no exposed nucleotide bases at each end_

2. The action of a specific sticky end restriction enzyme is illustrated on the previous page (top). Use the table above to:

   (a) Name the **restriction enzyme** used: _HpaI_

   (b) Name the organism from which it was first isolated: _haemophilus parainfluenzae_

   (c) State the **base sequence** for this restriction enzyme's recognition site: _GTTAAC_

3. A genetic engineer wants to use the restriction enzyme **Bam**HI to cut the DNA sequence below:

   (a) Consult the table above and state the recognition site for this enzyme: _GGATCC_

   (b) Circle every **recognition site** on the DNA sequence below that could be cut by the enzyme **Bam**HI:

```
        10             20             30             40             50             60
|AATGGGTACG|CACAGTGGAT|CCACGTAGTA|TGCGATGCGT|AGTGTTTATG|GAGAGAAGAA|
        70             80             90            100            110            120
|AACGCGTCGC|CTTTTATCGA|TGCTGTACGG|ATGCGGAAGT|GGCGATGAGG|ATCCATGCAA|
       130            140            150            160            170            180
|TCGCGGCCGA|TCGXGTAATA|TATCGTGGCT|GCGTTTATTA|TCGTGACTAG|TAGCAGTATG|
       190            200            210            220            230            240
|CGATGTGACT|GATGCTATGC|TGACTATGCT|ATGTTTTTAT|GCTGGATCCA|GCGTAAGCAT|
       250            260            270            280            290            300
|TTCGCTGCGT|GGATCCCATA|TCCTTATATG|CATATATTCT|TATACGGATC|GCGCACGTTT|
```

   (c) State how many fragments of DNA were created by this action: _____

4. When restriction enzymes were first isolated in 1970, there were not many applications for them. Now, they are an important tool in genetic engineering. Describe the human needs and demands that have driven the development and use of restriction enzymes in genetic engineering:

   The human needs and demands that have driven the development and use of restriction enzymes in genetic engineering is the need to.

Ligation

DNA fragments produced using restriction enzymes may be reassembled by a process called **ligation**. Pieces are joined together using an enzyme called **DNA ligase**. DNA of different origins produced in this way is called **recombinant DNA** (because it is DNA that has been recombined from different sources). The combined techniques of using restriction enzymes and ligation are the basic tools of genetic engineering (also known as recombinant DNA technology).

Creating a Recombinant DNA Plasmid

1. If two pieces of DNA are cut by the same restriction enzyme, they will produce fragments with matching **sticky ends** (ends with exposed nucleotide bases at each end).

2. When two such matching sticky ends come together, they can join by base-pairing. This process is called **annealing.** This can allow DNA fragments from a different source, perhaps a **plasmid**, to be joined to the DNA fragment.

3. The joined fragments will usually form either a linear molecule or a circular one, as shown here for a **plasmid**. However, other combinations of fragments can occur.

4. The fragments of DNA are joined together by the enzyme **DNA ligase**, producing a molecule of **recombinant DNA**.

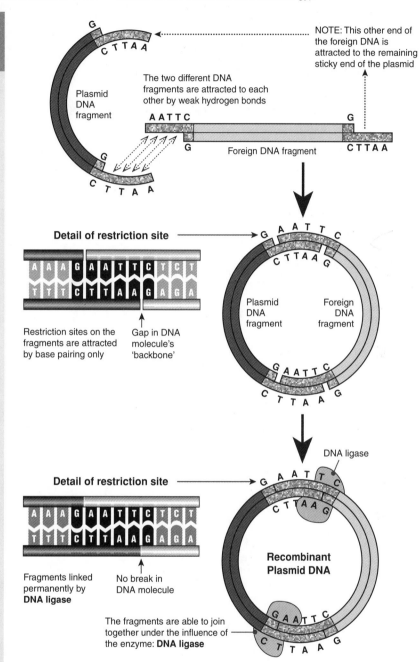

NOTE: This other end of the foreign DNA is attracted to the remaining sticky end of the plasmid

The two different DNA fragments are attracted to each other by weak hydrogen bonds

Plasmid DNA fragment

Foreign DNA fragment

Detail of restriction site

Restriction sites on the fragments are attracted by base pairing only

Gap in DNA molecule's 'backbone'

Plasmid DNA fragment

Foreign DNA fragment

Detail of restriction site

Fragments linked permanently by **DNA ligase**

No break in DNA molecule

DNA ligase

Recombinant Plasmid DNA

The fragments are able to join together under the influence of the enzyme: **DNA ligase**

DNA and RNA

---

1. Explain in your own words the two main steps in the process of joining two DNA fragments together:

   (a) Annealing: *bring the matching sticky ends together*

   (b) DNA ligase: *an enzyme that join the DNA fragments together*

2. Refer to the activity "DNA Replication", in the topic *The Role of DNA*, and state the usual role of DNA ligase in a cell:
   link the fragments together

3. Explain why **ligation** can be considered the *reverse* of the **restriction enzyme** process: *it*
   re assembles the DNA fragment together

Related activities: Restriction Enzymes, DNA Replication

Gel Electrophoresis

Gel electrophoresis is a method that separates large molecules (including nucleic acids or proteins) on the basis of size, electric charge, and other physical properties. Such molecules possess a slight electric charge (see DNA below). To prepare DNA for gel electrophoresis the DNA is often cut up into smaller pieces. This is done by mixing DNA with restriction enzymes in controlled conditions for about an hour. Called **restriction digestion**, it produces a range of DNA fragments of different lengths. During electrophoresis, molecules are forced to move through the pores of a **gel** (a jelly-like material), when the electrical current is applied. Active electrodes at each end of the gel provide the driving force. The electrical current from one electrode repels the molecules while the other electrode simultaneously attracts the molecules. The frictional force of the gel resists the flow of the molecules, separating them by size. Their rate of migration through the gel depends on the strength of the electric field, size and shape of the molecules, and on the ionic strength and temperature of the buffer in which the molecules are moving. After staining, the separated molecules in each lane can be seen as a series of bands spread from one end of the gel to the other.

Analyzing DNA using Gel Electrophoresis

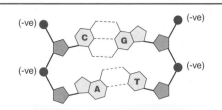

DNA is negatively charged because the phosphates (black) that form part of the backbone of a DNA molecule have a negative charge.

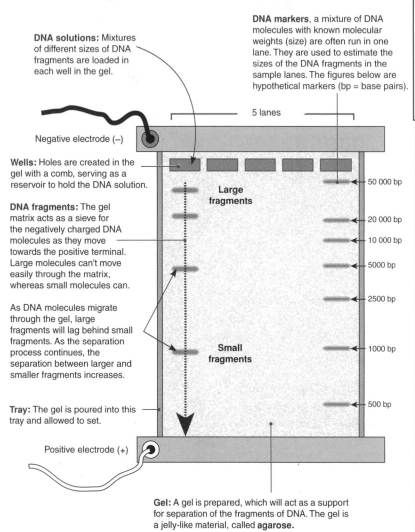

DNA solutions: Mixtures of different sizes of DNA fragments are loaded in each well in the gel.

DNA markers, a mixture of DNA molecules with known molecular weights (size) are often run in one lane. They are used to estimate the sizes of the DNA fragments in the sample lanes. The figures below are hypothetical markers (bp = base pairs).

5 lanes

Negative electrode (−)

Wells: Holes are created in the gel with a comb, serving as a reservoir to hold the DNA solution.

DNA fragments: The gel matrix acts as a sieve for the negatively charged DNA molecules as they move towards the positive terminal. Large molecules can't move easily through the matrix, whereas small molecules can.

As DNA molecules migrate through the gel, large fragments will lag behind small fragments. As the separation process continues, the separation between larger and smaller fragments increases.

Tray: The gel is poured into this tray and allowed to set.

Positive electrode (+)

Gel: A gel is prepared, which will act as a support for separation of the fragments of DNA. The gel is a jelly-like material, called **agarose.**

Large fragments

Small fragments

50 000 bp
20 000 bp
10 000 bp
5000 bp
2500 bp
1000 bp
500 bp

Steps in the process of gel electrophoresis of DNA

1. A tray is prepared to hold the gel matrix.

2. A gel comb is used to create holes in the gel. The gel comb is placed in the tray.

3. Agarose gel powder is mixed with a buffer solution (this carries the DNA in a stable form). The solution is heated until dissolved and poured into the tray and allowed to cool.

4. The gel tray is placed in an electrophoresis chamber and the chamber is filled with buffer, covering the gel. This allows the electric current from electrodes at either end of the gel to flow through the gel.

5. DNA samples are mixed with a "loading dye" to make the DNA sample visible. The dye also contains glycerol or sucrose to make the DNA sample heavy so that it will sink to the bottom of the well.

6. A safety cover is placed over the gel, electrodes are attached to a power supply and turned on.

7. When the dye marker has moved through the gel, the current is turned off and the gel is removed from the tray.

8. DNA molecules are made visible by staining the gel with **methylene blue** or ethidium bromide which binds to DNA and will fluoresce in UV light.

1. Explain the purpose of gel electrophoresis: _to seperate large molecules on the basis of size, electric charge and other physical properties_

2. Describe the two forces that control the speed at which fragments pass through the gel:

   (a) _electric charge_

   (b) _size_

3. Explain why the smallest fragments travel through the gel the fastest: _There is a less frictional force of the gel resistence_

© BIOZONE International 2012
ISBN: 978-1-927173-11-4
Photocopying Prohibited

Related activities: Nucleotides and Nucleic Acids
Weblinks: DNA Extraction, Gel Electrophoresis

12

Polymerase Chain Reaction

Many procedures in DNA technology (such as DNA sequencing and DNA profiling) require substantial amounts of DNA to work with. Some samples, such as those from a crime scene or fragments of DNA from a long extinct organism, may be difficult to get in any quantity. The diagram below describes the laboratory process called **polymerase chain reaction** (PCR).

Using this technique, vast quantities of DNA identical to the original samples can be created. This process is often termed **DNA amplification**. Although only one cycle of replication is shown below, the following cycles replicate DNA at an exponential rate. PCR can be used to make billions of copies of DNA in only a few hours.

A Single Cycle of the Polymerase Chain Reaction

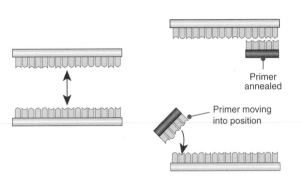

DNA polymerase: A thermally stable form of the enzyme is used (e.g. *Taq polymerase*). This is extracted from thermophilic bacteria.

Primer annealed

Primer moving into position

Nucleotides

Direction of synthesis

1 A DNA sample (called **target DNA**) is obtained. It is denatured (DNA strands are separated) by heating at 98°C for 5 minutes.

2 The sample is cooled to 60°C. Primers are annealed (bonded) to each DNA strand. In PCR, the primers are short strands of DNA; they provide the starting sequence for DNA extension.

3 DNA polymerase binds to the primers and, using the free nucleotides, synthesizes complementary strands of DNA.

4 After one cycle, there are now two copies of the original DNA.

Repeat for about 25 cycles

Repeat cycle of heating and cooling until enough copies of the target DNA have been produced

Loading tray
Prepared samples in tiny PCR tubes are placed in the loading tray and the lid is closed.

Temperature control
Inside the machine are heating and refrigeration mechanisms to rapidly change the temperature.

Dispensing pipette
Pipettes with disposable tips are used to dispense DNA samples into the PCR tubes.

Thermal Cycler

Amplification of DNA can be carried out with simple-to-use machines called thermal cyclers. Once a DNA sample has been prepared, in just a few hours the amount of DNA can be increased billions of times. Thermal cyclers are in common use in the biology departments of universities, as well as other kinds of research and analytical laboratories. The one pictured on the left is typical of this modern piece of equipment.

DNA quantitation
The amount of DNA in a sample can be determined by placing a known volume in this quantitation machine. For many genetic engineering processes, a minimum amount of DNA is required.

Controls
The control panel allows a number of different PCR programmes to be stored in the machine's memory. Carrying out a PCR run usually just involves starting one of the stored programmes.

1. Explain the purpose of PCR: _The purpose of PCR is to make vast quantities of identical DNA to the original sample created._

Periodicals:
DNA polymerase

Related activities: In Vivo Gene Cloning

RDA 3

DNA and RNA

2. Describe how the **polymerase chain reaction** works: _____

_____

_____

_____

_____

_____

_____

_____

3. Describe three situations where only very small DNA samples may be available for sampling and PCR could be used:

(a) _____

_____

(b) _____

_____

(c) _____

_____

4. After only two cycles of replication, four copies of the double-stranded DNA exist. Calculate how much a DNA sample will have increased after:

(a) 10 cycles: _____   (b) 25 cycles: _____

5. The risk of contamination in the preparation for PCR is considerable.

(a) Describe the effect of having a single molecule of unwanted DNA in the sample prior to PCR:

_____

_____

_____

(b) Describe two possible sources of DNA contamination in preparing a PCR sample:

Source 1: _____

Source 2: _____

(c) Describe two precautions that could be taken to reduce the risk of DNA contamination:

Precaution 1: _____

_____

Precaution 2: _____

_____

6. Describe two other genetic engineering/genetic manipulation procedures that require PCR amplification of DNA:

(a) _____

_____

_____

(b) _____

_____

_____

Amazing Organisms, Amazing Enzymes

Before the 1980s scientists knew of only a few organisms that could survive in extreme conditions. Indeed, many scientists believed that life in highly saline or high temperature and pressure environments was impossible. That view changed with the discovery of bacteria inhabiting the deep sea hydrothermal vents. They tolerate temperatures over 110°C and pressures of over 200 atmospheres. Bacteria were also found in volcanic hot pools on land, some surviving at temperatures in excess of 80°C. Most enzymes are denatured at temperatures above 40°C, but these **thermophilic** bacteria have enzymes that are fully functional at high temperatures. This discovery led to the development of one of the most important techniques in biotechnology, the **polymerase chain reaction** (PCR).

PCR is a technique, first described in the 1970s, that allows scientists to copy and multiply a piece of DNA millions of times. The DNA is heated to 98°C so that it separates into single strands and polymerase enzyme is added to synthesize new DNA strands from supplied free nucleotides. This earlier technique was labor intensive and expensive because the polymerase denatured at the high temperatures and had to be replaced every cycle. In 1985, a thermophilic polymerase (*Taq* **polymerase**) was isolated from the bacterium *Thermophilus aquaticus,* which inhabited the hot springs of Yellowstone National Park. Isolating this enzyme enabled automation of the PCR process, because the polymerase was stable throughout multiple cycles of synthesis. This led to an rapid growth in biotechnology, and gene technology in particular, because DNA samples could be easily copied for sequencing.

Searching for novel compounds in organisms from extreme environments is important in the development of new biotechnologies. Organisms must have compounds that can work in their specific environment, and the identification and extraction of these may allow them to be adapted for human use. For example, the Antarctic sea sponge *Kirkpatrickia variolosa* produces an alkaloid excreted as a toxic defence to prevent other organisms growing nearby. Tests indicate that this same chemical may have biological activity against cancer cells. Compounds from other sponge species are currently being assessed to treat a range of diseases including cancer, AIDS, tuberculosis and other bacterial infections, and cystic fibrosis.

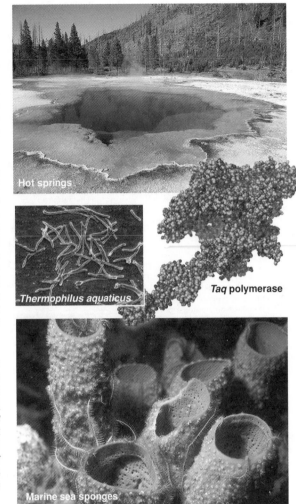

Hot springs

Thermophilus aquaticus

Taq polymerase

Marine sea sponges

DNA and RNA

1. Why was PCR not a viable technique until the mid 1980s? _____

_____

_____

_____

2. Explain why *Taq* polymerase was so important in the development of PCR: _____

_____

_____

_____

_____

3. Explain how investigating the lifestyles of other organisms can lead to advances in unrelated areas of science:

_____

_____

_____

_____

_____

_____

_____

© BIOZONE International 2012
ISBN: 978-1-927173-11-4
Photocopying Prohibited

Related activities: Polymerase Chain Reaction

A 2

Preparing a Gene for Cloning

Gene cloning is the process of making large quantities of a piece of DNA once it has been isolated. Its purpose is to yield large quantities of an individual gene or its protein product when the gene is expressed. Many copies of a gene can be made using PCR, which amplifies the DNA. Genes can also be cloned when they are part of an organism, in a technique called *in vivo* cloning. A gene of interest (e.g. a human gene) is inserted into the DNA of a vector, resulting in a **recombinant DNA molecule** called a **molecular clone**. This technique uses the self-replicating properties of the vector to make copies of

the gene. The genes of interest are rarely ready for cloning in their native form because they include pieces of non-protein coding DNA, called **introns**, which must be removed. Molecular biologists have a handy tool in the form of an enzyme, called **reverse transcriptase**, which makes this possible. Reverse transcriptase is a common name for an enzyme that functions as a RNA-dependent DNA polymerase and it is used to copy RNA into DNA. This task is integral to gene cloning because it produces a reconstructed gene that is ready for amplification.

Preparing a Gene for Cloning

1. Explain the role of restriction enzymes in preparing a clone: _____

2. (a) Explain why introns are removed before cloning a gene: _____

   (b) Describe the role of reverse transcriptase in this process: _____

3. Describe the normal role of reverse transcriptase: _____

Gene Cloning

It is possible to use the internal replication machinery of a cell to clone a gene, or even many genes, at once. By using cells to copy desired genes, it is also possible to produce any protein product the genes may code for. Recombinant DNA techniques (restriction digestion and ligation) are used to insert a gene of interest into the DNA of a vector (e.g. plasmid or viral DNA). This produces a **recombinant DNA molecule** that can used to transmit the gene of interest to another organism. To be useful, all vectors must be able to replicate inside their host organism, they must have one or more sites at which a restriction enzyme

can cut, and they must have some kind of **genetic marker** that allows them to be easily identified. Viruses, and organisms such as bacteria and yeasts have DNA that behaves in this way. Bacterial plasmids are commonly used because they are easy to manipulate, their restriction sites are well known, and they are readily taken up by cells in culture. Once the recombinant plasmid vector (containing the desired gene) has been taken up by bacterial cells, and those cells are identified, the gene can be replicated many times as the bacteria grow and divide.

Cloning a Human Gene

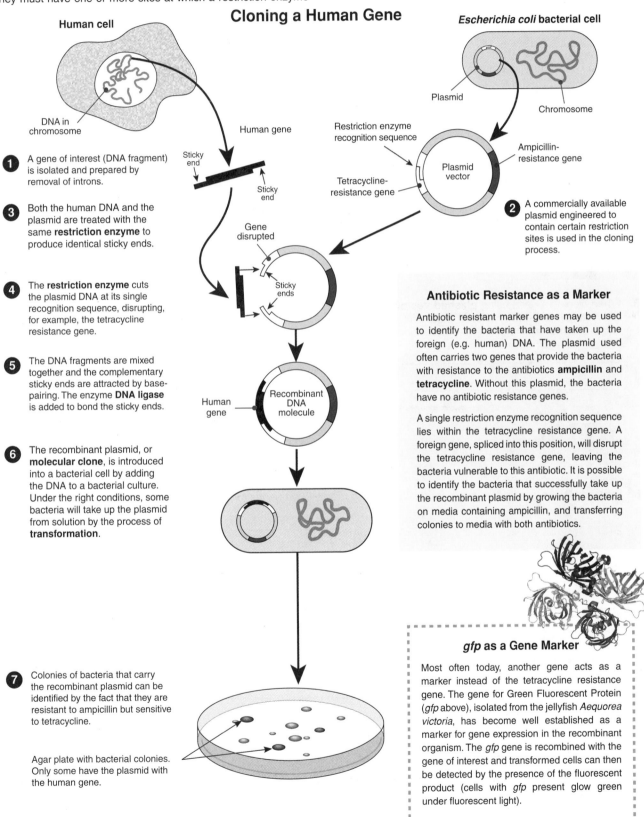

Human cell

Escherichia coli bacterial cell

DNA in chromosome

Plasmid

Chromosome

1 A gene of interest (DNA fragment) is isolated and prepared by removal of introns.

Human gene

Sticky end

Sticky end

Restriction enzyme recognition sequence

Ampicillin-resistance gene

Plasmid vector

Tetracycline-resistance gene

2 A commercially available plasmid engineered to contain certain restriction sites is used in the cloning process.

3 Both the human DNA and the plasmid are treated with the same **restriction enzyme** to produce identical sticky ends.

Gene disrupted

Sticky ends

4 The **restriction enzyme** cuts the plasmid DNA at its single recognition sequence, disrupting, for example, the tetracycline resistance gene.

5 The DNA fragments are mixed together and the complementary sticky ends are attracted by base-pairing. The enzyme **DNA ligase** is added to bond the sticky ends.

Human gene

Recombinant DNA molecule

6 The recombinant plasmid, or **molecular clone**, is introduced into a bacterial cell by adding the DNA to a bacterial culture. Under the right conditions, some bacteria will take up the plasmid from solution by the process of **transformation**.

7 Colonies of bacteria that carry the recombinant plasmid can be identified by the fact that they are resistant to ampicillin but sensitive to tetracycline.

Agar plate with bacterial colonies. Only some have the plasmid with the human gene.

Antibiotic Resistance as a Marker

Antibiotic resistant marker genes may be used to identify the bacteria that have taken up the foreign (e.g. human) DNA. The plasmid used often carries two genes that provide the bacteria with resistance to the antibiotics **ampicillin** and **tetracycline**. Without this plasmid, the bacteria have no antibiotic resistance genes.

A single restriction enzyme recognition sequence lies within the tetracycline resistance gene. A foreign gene, spliced into this position, will disrupt the tetracycline resistance gene, leaving the bacteria vulnerable to this antibiotic. It is possible to identify the bacteria that successfully take up the recombinant plasmid by growing the bacteria on media containing ampicillin, and transferring colonies to media with both antibiotics.

gfp as a Gene Marker

Most often today, another gene acts as a marker instead of the tetracycline resistance gene. The gene for Green Fluorescent Protein (*gfp* above), isolated from the jellyfish *Aequorea victoria*, has become well established as a marker for gene expression in the recombinant organism. The *gfp* gene is recombined with the gene of interest and transformed cells can then be detected by the presence of the fluorescent product (cells with *gfp* present glow green under fluorescent light).

© BIOZONE International 2012
ISBN: 978-1-927173-11-4
Photocopying Prohibited

DNA and RNA

Related activities: Preparing a Gene for Cloning, Replication in Bacteriophages
Weblinks: Gene Cloning

RA 3

1. Explain why it might be desirable to use *in vivo* methods to clone genes rather than PCR: _____

_____

_____

_____

2. Explain when it may not be desirable to use bacteria to clone genes: _____

_____

_____

3. Explain how a human gene is removed from a chromosome and placed into a plasmid. _____

_____

_____

_____

_____

4. A bacterial plasmid replicates at the same rate as the bacteria. If a bacteria containing a recombinant plasmid replicates and divides once every thirty minutes, calculate the number of plasmid copies there will be after twenty four hours:

_____

_____

5. When cloning a gene using **plasmid vectors**, the bacterial colonies containing the recombinant plasmids are mixed up with colonies that have none. All the colonies look identical, but some have taken up the plasmids with the human gene, and some have not. Explain how the colonies with the recombinant plasmids are identified:

_____

_____

_____

_____

_____

6. Explain why the *gfp* marker is a more desirable gene marker than genes for antibiotic resistance:

_____

_____

_____

_____

7. Bacteriophages are viruses that infect bacteria:

    (a) What feature of bacteriophages make them useful for genetic engineering? _____

    _____

    (b) How could a bacteriophage be used to clone a gene? _____

    _____

    _____

    _____

    _____

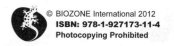

Using Recombinant Bacteria

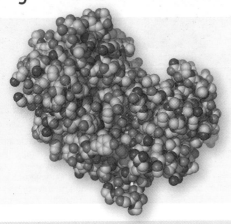

The Issue

► **Chymosin** (also known as **rennin**) is an enzyme that digests milk proteins. It is the active ingredient in rennet, a substance used by cheesemakers to clot milk into curds.

► Traditionally rennin is extracted from "chyme", i.e. the stomach secretions of suckling calves (hence its name of chymosin).

► By the 1960s, a shortage of chymosin was limiting the volume of cheese produced.

► Enzymes from fungi were used as an alternative but were unsuitable because they caused variations in the cheese flavor.

Concept 1
Enzymes are proteins made up of amino acids. The amino acid sequence of chymosin can be determined and the mRNA coding sequence for its translation identified.

Concept 2
Reverse transcriptase can be used to synthesize a DNA strand from the mRNA. This process produces DNA without the introns, which cannot be processed by bacteria.

Concept 3
DNA can be cut at specific sites using **restriction enzymes** and rejoined using **DNA ligase.** New genes can be inserted into self-replicating bacterial **plasmids**.

Concept 4
Under certain conditions, bacteria are able to lose or take up plasmids from their environment. Bacteria are readily grown in vat cultures at little expense.

Concept 5
The protein in made by the bacteria in large quantities.

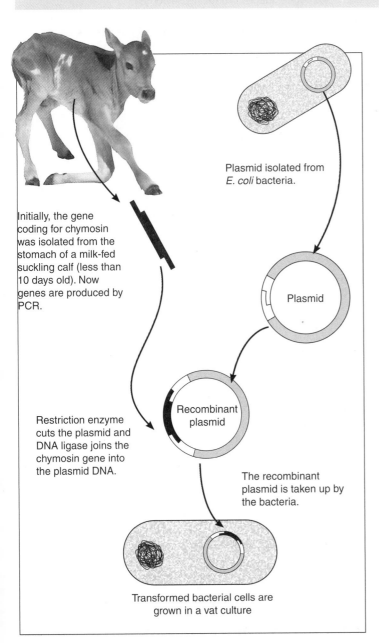

Plasmid isolated from *E. coli* bacteria.

Initially, the gene coding for chymosin was isolated from the stomach of a milk-fed suckling calf (less than 10 days old). Now genes are produced by PCR.

Plasmid

Restriction enzyme cuts the plasmid and DNA ligase joins the chymosin gene into the plasmid DNA.

Recombinant plasmid

The recombinant plasmid is taken up by the bacteria.

Transformed bacterial cells are grown in a vat culture

Techniques

The amino acid sequence of chymosin is first determined and the RNA codons for each amino acid identified.

mRNA matching the identified sequence is isolated from the stomach of young calves. **Reverse transcriptase** is used to transcribe mRNA into DNA. The DNA sequence can also be made synthetically once the sequence is determined.

The DNA is amplified using PCR.

Plasmids from *E. coli* bacteria are isolated and cut using **restriction enzymes.** The DNA sequence for chymosin is inserted using **DNA ligase**.

Plasmids are returned to *E. coli* by placing the bacteria under conditions that induce them to take up plasmids.

Outcomes

The transformed bacteria are grown in vat culture. Chymosin is produced by *E. coli* in packets within the cell that are separated during the processing and refining stage.

Recombinant chymosin entered the marketplace in 1990. It established a significant market share because cheesemakers found it to be cost effective, of high quality, and in consistent supply. Most cheese is now produced using recombinant chymosin such as CHY-MAX.

Further Applications

A large amount of processing is required to extract chymosin from *E.coli*. There are now a number of alternative bacteria and fungi that have been engineered to produce the enzyme. Most chymosin is now produced using the fungi *Aspergillus niger* and *Kluyveromyces lactis*. Both are produced in a similar way as that described for *E. coli*.

Periodicals:
Tailor-made proteins

Related activities: Restriction Enzymes, Ligation, Production of Insulin, In Vivo Gene Cloning

Enzymes from GMOs are widely used in the baking industry. Maltogenic alpha amylase from *Bacillus subtilis* bacteria is used as an anti-staling agent to prolong shelf life. Hemicellulases from *B. subtilis* and xylanase from the fungus *Aspergillus oryzae* are used for improvement of dough, crumb structure, and volume during the baking process.

Lipase from *Aspergillus oryzae* is used in processing of palm oil to produce low cost cocoa butter substitutes (above), which have a similar 'mouth feel' to cocoa butter.

Acetolactate decarboxylase from *B. subtilis* is one of several enzymes used in the brewing industry. It reduces maturation time of the beer by by-passing a rate-limiting step.

1. Describe the main use of chymosin: _____
   _____

2. What was the traditional source of chymosin? _____
   _____

3. Summarize the key concepts that led to the development of the technique for producing chymosin:

   (a) Concept 1: _____

   (b) Concept 2: _____

   (c) Concept 3: _____

   (d) Concept 4: _____

4. Discuss how the gene for chymosin was isolated and how the technique could be applied to isolating other genes:
   _____
   _____
   _____
   _____
   _____
   _____
   _____
   _____

5. Describe three advantages of using chymosin produced by GE bacteria over chymosin from traditional sources:

   (a) _____
   _____

   (b) _____
   _____

   (c) _____

6. Explain why the fungus *Aspergillus niger* is now more commonly used to produce chymosin instead of *E. coli*:
   _____
   _____
   _____

© BIOZONE International 2012
ISBN: 978-1-927173-11-4
Photocopying Prohibited

Golden Rice

The Issue

▶ **Beta-carotene** (β-carotene) is a precursor to **vitamin A** which is involved in many functions including vision, immunity, fetal development, and skin health.

▶ Vitamin A deficiency is common in developing countries where up to 500,000 children suffer from night blindness, and death rates due to infections are high due to a lowered immune response.

▶ Providing enough food containing useful quantities of β-carotene is difficult and expensive in many countries.

Concept 1	Concept 2	Concept 3	Concept 4
Rice is a staple food in many developing countries. It is grown in large quantities and is available to most of the population, but it lacks many of the essential nutrients required by the human body for healthy development. It is low in β-carotene.	Rice plants produce β-carotene but not in the edible rice **endosperm**. Engineering a new biosynthetic pathway would allow β-carotene to be produced in the endosperm. Genes expressing enzymes for carotene synthesis can be inserted into the rice genome.	The enzyme **carotene desaturase (CRT1)** in the soil bacterium *Erwinia uredovora*, catalyses multiple steps in carotenoid biosynthesis. **Phytoene synthase (PSY)** overexpresses a colorless carotene in the daffodil plant *Narcissus pseudonarcissus*.	DNA can be inserted into an organism's genome using a suitable **vector**. *Agrobacterium tumefaciens* is a tumor-forming bacterial plant pathogen that is commonly used to insert novel DNA into plants.

The Development of Golden Rice

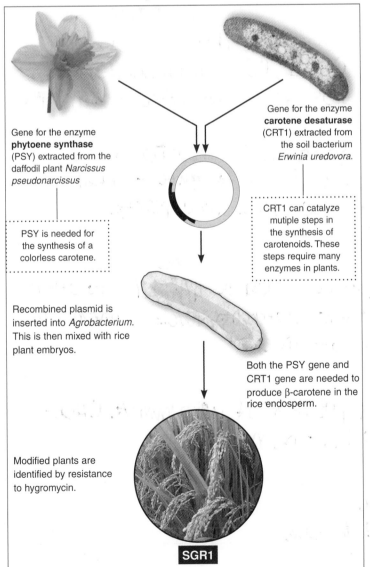

Gene for the enzyme **phytoene synthase** (PSY) extracted from the daffodil plant *Narcissus pseudonarcissus*

Gene for the enzyme **carotene desaturase** (CRT1) extracted from the soil bacterium *Erwinia uredovora*.

PSY is needed for the synthesis of a colorless carotene.

CRT1 can catalyze mutiple steps in the synthesis of carotenoids. These steps require many enzymes in plants.

Recombined plasmid is inserted into *Agrobacterium*. This is then mixed with rice plant embryos.

Both the PSY gene and CRT1 gene are needed to produce β-carotene in the rice endosperm.

Modified plants are identified by resistance to hygromycin.

SGR1

Techniques

The **PSY** gene from daffodils and the **CRT1** gene from *Erwinia uredovora* are sequenced.

DNA sequences are synthesized into packages containing the CRT1 or PSY gene, terminator sequences, and **endosperm specific promoters** (these ensure expression of the gene only in the edible portion of the rice).

The *Ti* **plasmid** from *Agrobacterium* is modified using restriction enzymes and DNA ligase to delete the tumor-forming gene and insert the synthesized DNA packages. A gene for resistance to the antibiotic **hygromycin** is also inserted so that transformed plants can be identified later. The parts of the *Ti* plasmid required for plant transformation are retained.

Modified *Ti* plasmid is inserted into the bacterium.

Agrobacterium is incubated with rice plant embryo. Transformed embryos are identified by their resistance to hygromycin.

Outcomes

The rice produced had endosperm with a distinctive yellow color. Under greenhouse conditions golden rice (**SGR1**) contained 1.6 µg per g of carotenoids. Levels up to five times higher were produced in the field, probably due to improved growing conditions.

Further Applications

Further research on the action of the PSY gene identified more efficient methods for the production of β-carotene. The second generation of golden rice now contains up to 37 µg per g of carotenoids. Golden rice was the first instance where a complete biosynthetic pathway was engineered. The procedures could be applied to other food plants to increase their nutrient levels.

Periodicals:
Rice, risk, and regulations,
The engineering of crop plants

Related activities: *Restriction Enzymes, Ligation,*
In Vivo Gene Cloning

All photos USDA

The ability of *Agrobacterium* to transfer genes to plants is exploited for crop improvement. The tumor-inducing *Ti* plasmid is modified to delete the tumor-forming gene and insert a gene coding for a desirable trait. The parts of the *Ti* plasmid required for plant transformation are retained.

Soybeans are one of the many food crops that have been genetically modified for broad spectrum herbicide resistance. The first GM soybeans were planted in the US in 1996. By 2007, nearly 60% of the global soybean crop was genetically modified; the highest of any other crop plant.

GM cotton was produced by inserting the gene for the BT toxin into its genome. The bacterium *Bacillus thuringiensis* naturally produces BT toxin, which is harmful to a range of insects, including the larvae that eat cotton. The BT gene causes cotton to produce this insecticide in its tissues.

1. Describe the basic methodology used to create golden rice: The basic methodology used to create golden rice using both the PSY and CRT1 genes to produce β-carotene in the rice endosperm

2. Explain how scientists ensured β-carotene was produced in the endosperm: The DNA sequences are synthesyzed into packages containing CRT1, PSY, terminater sequence and endosperm specific premoter

3. What property of *Agrobacterium tumefaciens* makes it an ideal vector for introducing new genes into plants? The temer-ferming gene and inserta genic coding for a desirable trait to make it ideal veeter for introducing new genes into plants.

4. (a) How could this new variety of rice reduce disease in developing countries? The PSY gene and β-carotene could be appwed to other feed to increase their nutrents levels which can create a new variety of rice.

   (b) Absorption of vitamin A requires sufficient dietary fat. Explain how this could be problematic for the targeted use of golden rice in developing countries: β-carotene is a precuser of vitamin A. Without this golden rice couldn't be made

5. As well as increasing nutrient content as in golden rice, other traits of crop plants are also desirable. For each of the following traits, suggest features that could be desirable in terms of increasing yield:

   (a) Grain size or number: would increase

   (b) Maturation rate: the rate would decrease

   (c) Pest resistance: would increase

13

Production of Insulin

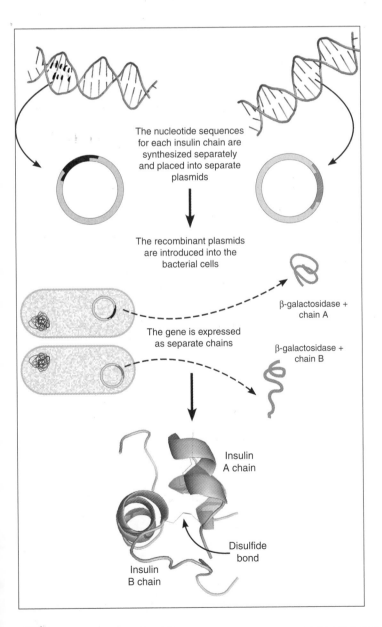

Insulin
B chain

Insulin
A chain

The Issue

▶ **Type 1 diabetes mellitus** is a metabolic disease caused by a lack of **insulin**. Around 25 people in every 100,000 suffer from type 1 diabetes.

▶ It is treatable only with injections of insulin.

▶ In the past, insulin was taken from the pancreases of cows and pigs and purified for human use. The method was expensive and some patients had severe allergic reactions to the foreign insulin or its contaminants.

Concept 1
DNA can be cut at specific sites using **restriction enzymes** and joined together using **DNA ligase**. New genes can be inserted into self-replicating bacterial **plasmids** at the point where the cuts are made.

Concept 2
Plasmids are small, circular pieces of DNA found in some bacteria. They usually carry genes useful to the bacterium. *E. coli* plasmids can carry promoters required for the transcription of genes.

Concept 3
Under certain conditions, Bacteria are able to lose or pick up plasmids from their environment. Bacteria can be readily grown in vat cultures at little expense.

Concept 4
The DNA sequences coding for the production of the two polypeptide chains (A and B) that form human insulin can be isolated from the human genome.

The nucleotide sequences for each insulin chain are synthesized separately and placed into separate plasmids

The recombinant plasmids are introduced into the bacterial cells

β-galactosidase + chain A

The gene is expressed as separate chains

β-galactosidase + chain B

Insulin A chain

Disulfide bond

Insulin B chain

Techniques

The **gene** is **chemically synthesized** as two nucleotide sequences, one for the **insulin A chain** and one for the **insulin B chain**. The two sequences are small enough to be inserted into a plasmid.

Plasmids are extracted from *Escherichia coli*. The gene for the bacterial enzyme **β-galactosidase** is located on the plasmid. To make the bacteria produce insulin, the insulin gene must be linked to the **β-galactosidase** gene, which carries a promoter for transcription.

Restriction enzymes are used to cut plasmids at the appropriate site and the A and B insulin sequences are inserted. The sequences are joined with the plasmid DNA using **DNA ligase**.

The **recombinant plasmids** are inserted back into the bacteria by placing them together in a culture that favors plasmid uptake by bacteria.

The bacteria are then grown and multiplied in vats under carefully controlled growth conditions.

Outcomes

The product consists partly of β-galactosidase, joined with either the A or B chain of insulin. The chains are extracted, purified, and mixed together. The A and B insulin chains connect via **disulfide cross linkages** to form the functional insulin protein. The insulin can then be made ready for injection in various formulations.

Further Applications

The techniques involved in producing human insulin from genetically modified bacteria can be applied to a range of human proteins and hormones. Proteins currently being produced include human growth hormone, interferon, and factor VIII.

Related activities: Restriction Enzymes, Ligation

A 2

38

Insulin production in *Saccharomyces*

Yeast cells are **eukaryotic** and hence are much larger than bacterial cells. This enables them to accommodate much larger plasmids and proteins within them.

The gene for human insulin is inserted into a plasmid. The yeast plasmid is larger than that of *E.coli*, so the entire gene can be inserted in one piece rather than as two separate pieces.

Cleavage site

The **proinsulin** protein that is produced folds into a specific shape and is cleaved by the yeast's own cellular enzymes, producing the completed insulin chain.

By producing insulin this way, the secondary step of combining the separate protein chains is eliminated, making the refining process much simpler.

Cleavage site

1. Describe the three major problems associated with the traditional method of obtaining insulin to treat diabetes:

   (a) _____

   (b) _____

   (c) _____

2. Explain the reasoning behind using *E. coli* to produce insulin and the benefits that GM technology has brought to diabetics:

   _____

   _____

   _____

   _____

3. Explain why, when using *E. coli*, the insulin gene is synthesized as two separate A and B chain nucleotide sequences:

   _____

   _____

4. Why are the synthetic nucleotide sequences ('genes') 'tied' to the β-galactosidase gene?_____

   _____

   _____

5. Yeast (*Saccharomyces cerevisiae*) is also used in the production of human insulin. Discuss the differences in the production of insulin using yeast and *E. coli* with respect to:

   (a) Insertion of the gene into the plasmid: _____

   _____

   _____

   (b) Secretion and purification of the protein product: _____

   _____

   _____

   _____

© BIOZONE International 2012
ISBN: 978-1-927173-11-4
Photocopying Prohibited

Food for the Masses

It is estimated that by 2050 the world population will reach between 9 and 10 billion people. Currently **1 billion people** (one sixth of the world's population) are **undernourished**. If trends continue, 1.5 billion people will be living under the threat of starvation by 2050, and by 2100 (if global warming is taken into account) nearly half the world's population could be threatened with food shortages. The solution to the problem of food production is complicated. Most of the Earth's arable land has already been developed and currently uses 37% of the Earth's land area, leaving little room to grow more crops or farm more animals. Development of new

fast growing and high yield crops appears to be a major part of the solution, but many crops can only be grown under a narrow range of conditions or are susceptible to disease. Moreover, the farming and irrigation of some areas is difficult, costly, and damaging to the environment because vast amounts of water are diverted from their natural courses. **Genetic modification** of plants may help to solve some of these looming problems by producing plants that will require less intensive culture or that will grow in areas previously considered not arable.

DNA and RNA

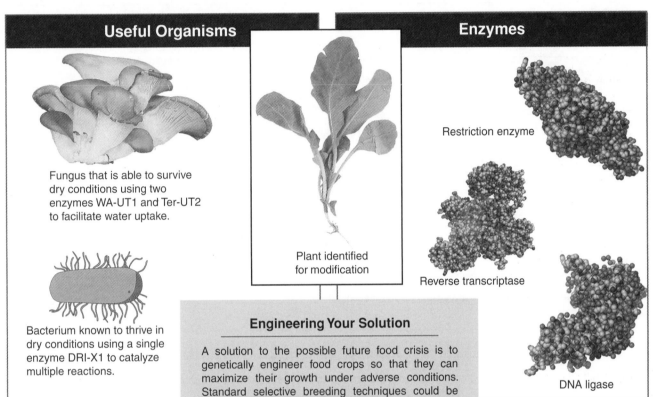

Useful Organisms

Fungus that is able to survive dry conditions using two enzymes WA-UT1 and Ter-UT2 to facilitate water uptake.

Bacterium known to thrive in dry conditions using a single enzyme DRI-X1 to catalyze multiple reactions.

Plant identified for modification

Enzymes

Restriction enzyme

Reverse transcriptase

DNA ligase

Engineering Your Solution

A solution to the possible future food crisis is to genetically engineer food crops so that they can maximize their growth under adverse conditions. Standard selective breeding techniques could be used to do this, but in some plants this may not be possible or feasible and it may require more time than is available. A selection of genetic tools and organisms with useful characteristics are described. **Your task** is to use the items shown to devise a technique to successfully create a plant that could be successfully farmed in semi-desert environments such as sub-Saharan Africa. The following page will take you through the procedure. Not all the items will need to be used.

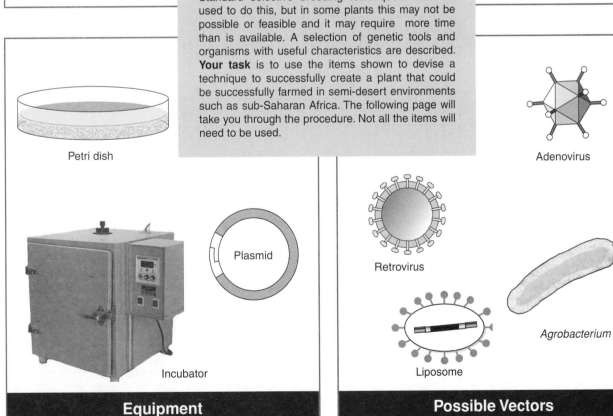

Petri dish

Plasmid

Incubator

Retrovirus

Liposome

Adenovirus

Agrobacterium

Equipment

Possible Vectors

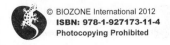

Related activities: Restriction Enzymes, Ligation, Golden Rice, Production of Insulin

RA 2

1. Identify the organism you would chose as a 'donor' of drought survival genes and explain your choice:

_____

2. Describe a process to identify and isolate the required gene(s) and identify the tools to be used: _____

_____

_____

_____

_____

3. Identify a vector for the transfer of the isolated gene(s) into the crop plant and explain your decision: _____

_____

4. Explain how the isolated gene(s) would be integrated into the vector's genome: _____

_____

_____

_____

_____

_____

5. (a) Explain how the vector will transform the identified plant: _____

_____

_____

   (b) Identify the stage of development at which the plant would most easily be transformed. Explain your choice:

_____

_____

_____

6. Explain how the transformed plants could be identified: _____

_____

_____

_____

_____

_____

7. Explain how a large number of plants can be grown from the few samples that have taken up the new DNA:

_____

_____

_____

_____

_____

_____

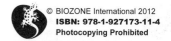 © BIOZONE International 2012
ISBN: 978-1-927173-11-4
Photocopying Prohibited

Applications of GMOs

Techniques for genetic manipulation are now widely applied throughout modern biotechnology: in food and enzyme technology, in industry and medicine, and in agriculture and horticulture. Microorganisms are among the most widely used GMOs, with applications ranging from pharmaceutical production and vaccine development to environmental clean-up. Crop plants are also popular candidates for genetic modification although their use, as with much of genetic engineering of higher organisms, is controversial and sometimes problematic.

Application of GMOs

Extending shelf life

Some fresh produce (e.g. tomatoes) have been engineered to have an extended keeping quality. In the case of tomatoes, the gene for ripening has been switched off, delaying the process of softening in the fruit.

Pest or herbicide resistance

Plants can be engineered to produce their own insecticide and become pest resistant. Genetically engineered herbicide resistance is also common. In this case, chemical weed killers can be used freely without crop damage.

Crop improvement

Gene technology is now an integral part of the development of new crop varieties. Crops can be engineered to produce higher protein levels or to grow in inhospitable conditions (e.g. salty or arid conditions).

Environmental clean-up

Some bacteria have been engineered to thrive on waste products, such as liquefied newspaper pulp or oil. As well as degrading pollutants and wastes, the bacteria may be harvested as a commercial protein source.

Biofactories

Transgenic bacteria are widely used to produce desirable products: often hormones or proteins. Large quantities of a product can be produced using bioreactors (above). Examples: insulin production by recombinant yeast, production of bovine growth hormone.

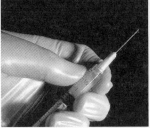

Vaccine development

The potential exists for multipurpose vaccines to be made using gene technology. Genes coding for vaccine components (e.g. viral protein coat) are inserted into an unrelated live vaccine (e.g. polio vaccine), and deliver proteins to stimulate an immune response.

Livestock improvement using transgenic animals

Transgenic sheep have been used to enhance wool production in flocks (above, left). The keratin protein of wool is largely made of a single amino acid, cysteine. Injecting developing sheep with the genes for the enzymes that generate cysteine produces woollier transgenic sheep. In some cases, transgenic animals have been used as biofactories. Transgenic sheep carrying the human gene for a protein, α-1-antitrypsin produce the protein in their milk. The antitrypsin is extracted from the milk and used to treat hereditary emphysema.

1. In a short account discuss one of the applications of GMOs described above:

Related activities: Production of Insulin, Golden Rice, Using Recombinant Bacteria

RA 3

KEY TERMS: Mix and Match

INSTRUCTIONS: Test your vocabulary by matching each term to its definition, as identified by its preceding letter code.

amino acid

anticodon

base-pairing rule

coding strand

codon

DNA

DNA ligase

DNA polymerase

DNA replication

exons

gel electrophoresis

gene expression

genetic code

genetic engineering

helicase

hydrogen bonding

introns

lagging strand

leading strand

nucleic acids

nucleotides

Okazaki fragments

protein

replication fork

RNA (mRNA, rRNA, tRNA)

RNAi

stop codon

template strand

transcription

translation

A Single stranded nucleic acid that consists of nucleotides that contain ribose sugar.

B Organic macromolecules composed of linear chains of amino acids joined together by peptide bonds and then organized, e.g. through folding, into a functional structure.

C The process by which genetic information is used to produce a functional gene product.

D A set of rules by which information encoded in DNA or mRNA is translated into proteins.

E One of the three codons signaling the end of a protein-coding sequence.

F The region of a transfer RNA with a sequence of three bases that are complementary to a codon in the messenger RNA.

G The strand of the DNA double helix that is orientated in a 3' to 5' manner and which is replicated in fragments.

H A structure, created by DNA helicase, that forms during DNA replication.

I The semi-conservative process by which two identical DNA molecules are produced from a single double-stranded DNA molecule.

J The process of creating an equivalent RNA copy of a sequence of DNA.

K An enzyme that separates two annealed DNA strands using energy from ATP hydrolysis.

L Form of intermolecular bonding between hydrogen and an electronegative atom such as oxygen.

M The sequence of DNA that is read during the synthesis of mRNA.

N Universally found macromolecules composed of chains of nucleotides. These molecules carry genetic information within cells.

O An enzyme that catalyzes the incorporation of deoxyribonucleotides into a DNA strand.

P Organic compound consisting of a carboxyl, an amine and an R group (where R may be one of 20 different atomic groupings). Polymerized by peptide bonds to form proteins.

Q Macromolecule consisting of many millions of units containing a phosphate group, sugar and a base (A,T, C or G). Stores the genetic information of the cell.

R The structural units of nucleic acids, DNA and RNA.

S Technique used for the separation of nucleic acids or protein molecules using an electric field applied to a (buffered) gel matrix.

T The techniques of altering the genetic makeup of cells or organisms by the selective removal, insertion, or modification of DNA.

U A process in which small fragments of exogenous or endogenous RNA are used to regulate gene expression by the targeted destruction of mRNA.

V A sequence of three adjacent nucleotides constituting the code for an amino acid.

W The stage of gene expression in which mRNA is decoded to produce a polypeptide.

X An enzyme that links together two DNA strands that have double-strand break.

Y Nucleic acid sequences that are represented in the mature form of an RNA molecule.

Z Relatively short pieces of DNA created on the lagging strand during DNA replication.

AA DNA regions within a gene that are not translated into protein.

BB The rule governing the pairing of complementary bases in DNA.

CC The DNA strand with the same base sequence as the RNA transcript produced (although with thymine replaced by uracil in mRNA).

DD The strand of the DNA double helix that is oriented in a 5' to 3' manner and is replicated in one continuous piece.

© BIOZONE International 2012
ISBN: 978-1-927173-11-4
Photocopying Prohibited

Chromosomes and Cell Division

Key concepts

▶ In the cell cycle, interphase alternates with cell division. New cells arise through cell division.

▶ Internal controls regulate the cell cycle.

▶ Cellular diversity arises through specialization from stem cell progenitors.

▶ Meiosis is a reduction division and is essential for sexual reproduction.

▶ Gametic meiosis produces haploid gametes.

Key terms

anaphase
bivalent
cell cycle
cellular differentiation
cell division
centrioles
chromatid
chromosome
crossing over
crossover frequency
cyclin-dependent kinases
cytokinesis
differentiation
diploid (2N)
DNA replication
fertilization
gamete
haploid (1N)
homologous chromosomes
independent assortment
interphase
maternal chromosome
meiosis
metaphase
mitosis
nuclear division
paternal chromosome
potency
prophase
recombination
sex chromosome
somatic cell
specialized cell
sporic meiosis
synapsis
telophase

Periodicals:
Listings for this
chapter are on page 382

Weblinks:
www.thebiozone.com/
weblink/AP1-3114.html

BIOZONE APP:
Student Review Series
Processes in the Nucleus

Essential Knowledge

☐ 1. Use the **KEY TERMS** to compile a glossary for this topic.

The Cell Cycle (3.A.2: a) pages 145-148, 193-194, 201

☐ 2. Describe the **cell cycle** in eukaryotes, recognizing **interphase** and **mitosis**. Describe the events in the three stages of interphase: G_1, S, and G_2.

☐ 3. Explain how the cell cycle is regulated through a series of internal controls or checkpoints. Explain how the cell cycle is regulated by external signals or growth factors and describe the role of **cyclin-dependent kinases** in this.

☐ 4. Recognize that most of the cell cycle is spent in interphase, which alternates with mitosis and cytokinesis (**cell division**).

☐ 5. Recognize that **differentiation** of cells into **specialized cell** types involves controlled modifications in gene expression, so that daughter cells follow a certain developmental path. Once specialized, cells may cease to divide.

Mitosis (3.A.2: b) pages 144-148

☐ 6. Describe the role of mitosis in growth and repair, and asexual reproduction.

☐ 7. Describe mitosis as a continuous process, with distinct structural stages. Recognize and summarize the events occurring in each of the following stages in mitosis: **prophase**, **metaphase**, **anaphase**, and **telophase**.

☐ 8. State the cellular outcome of mitosis followed by **cytokinesis**. Contrast cytokinesis in plant and animal cells.

Meiosis (3.A.2: c) pages 144, 149-155

☐ 9. Know that **meiosis**, like mitosis, involves DNA replication during interphase in the parent cell, but that this is followed by two cycles of nuclear division. Know that meiosis is a **reduction division** and explain what this means.

☐ 10. Summarize the principal events in meiosis and their significance, including:

(a) **Synapsis** and formation of **bivalents**.
(b) **Chiasma** formation and exchange of genetic material between **chromatids** in the first, (reduction) division.
(c) Separation of chromatids (second division) and production of haploid cells.

☐ 11. Describe the behavior of **homologous chromosomes** (and their associated **alleles**) during meiosis and **fertilization**, with reference to:

• The **recombination** of segments of maternal and paternal homologous chromosomes in **crossing over**.
• The **independent assortment** of **maternal** and **paternal chromosomes**.
• The random fusion of gametes during **fertilization**.

Explain how these events create new allele combinations in the gametes.

Cell Division

The life cycle of a diploid sexually reproducing organism, such as a human, with **gametic meiosis** is illustrated below. In this life cycle, **gametogenesis** involves meiotic division to produce male and female gametes for the purpose of sexual reproduction. The life cycle in flowering plants is different in that the gametes are produced through mitosis in haploid gametophytes. The male gametes are produced inside the pollen grain and the female gametes are produced inside the embryo sac of the ovule. The gametophytes develop and grow from haploid spores, which are produced from **sporic meiosis**.

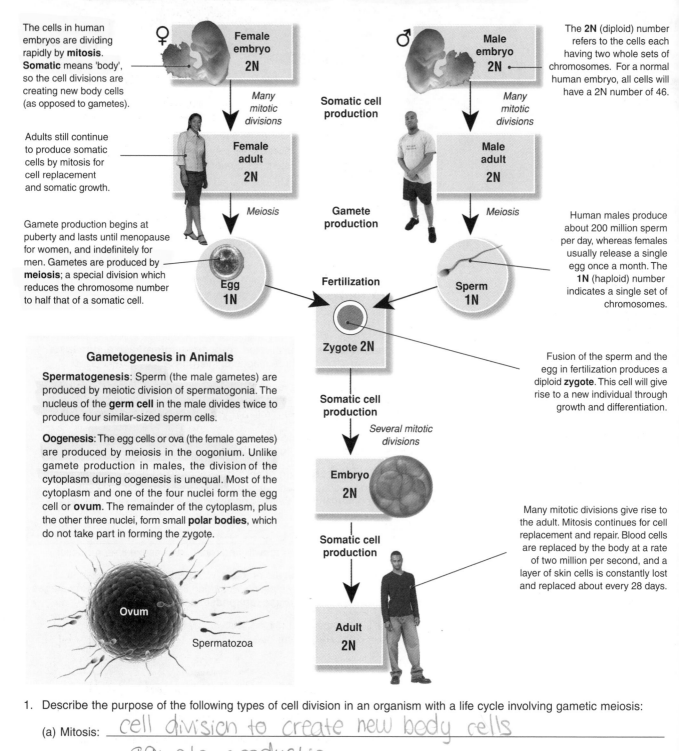

The cells in human embryos are dividing rapidly by **mitosis**. **Somatic** means 'body', so the cell divisions are creating new body cells (as opposed to gametes).

Adults still continue to produce somatic cells by mitosis for cell replacement and somatic growth.

Gamete production begins at puberty and lasts until menopause for women, and indefinitely for men. Gametes are produced by **meiosis**; a special division which reduces the chromosome number to half that of a somatic cell.

The **2N** (diploid) number refers to the cells each having two whole sets of chromosomes. For a normal human embryo, all cells will have a 2N number of 46.

Human males produce about 200 million sperm per day, whereas females usually release a single egg once a month. The **1N** (haploid) number indicates a single set of chromosomes.

Fusion of the sperm and the egg in fertilization produces a diploid **zygote**. This cell will give rise to a new individual through growth and differentiation.

Many mitotic divisions give rise to the adult. Mitosis continues for cell replacement and repair. Blood cells are replaced by the body at a rate of two million per second, and a layer of skin cells is constantly lost and replaced about every 28 days.

Female embryo 2N → *Many mitotic divisions* → **Female adult 2N** → *Meiosis* → **Egg 1N**

Somatic cell production

Gamete production

Male embryo 2N → *Many mitotic divisions* → **Male adult 2N** → *Meiosis* → **Sperm 1N**

Fertilization → **Zygote 2N** → **Somatic cell production** → *Several mitotic divisions* → **Embryo 2N** → **Somatic cell production** → **Adult 2N**

Gametogenesis in Animals

Spermatogenesis: Sperm (the male gametes) are produced by meiotic division of spermatogonia. The nucleus of the **germ cell** in the male divides twice to produce four similar-sized sperm cells.

Oogenesis: The egg cells or ova (the female gametes) are produced by meiosis in the oogonium. Unlike gamete production in males, the division of the cytoplasm during oogenesis is unequal. Most of the cytoplasm and one of the four nuclei form the egg cell or **ovum**. The remainder of the cytoplasm, plus the other three nuclei, form small **polar bodies**, which do not take part in forming the zygote.

Ovum

Spermatozoa

1. Describe the purpose of the following types of cell division in an organism with a life cycle involving gametic meiosis:

    (a) Mitosis: _cell division to create new body cells_

    (b) Meiosis: _gamete production_

2. Describe the basic difference between the cell divisions involved in spermatogenesis and oogenesis:
    The division of the cytoplasm during oogenesis is unequal

3. How does gametogenesis differ between humans and flowering plants? _The gametes are produced through mitosis in haploid gametophytes_

Mitosis and the Cell Cycle

Mitosis is part of the **cell cycle** in which an existing cell (the parent cell) divides into two (the daughter cells). Unlike meiosis, mitosis does not result in a change of chromosome numbers and the daughter cells are genetically identical to the parent cell. Although mitosis is part of a continuous cell cycle, it is divided into stages (below). The example below illustrates the cell cycle in a plant cell. Note that **cytokinesis in plant cells** involves construction of a **cell plate** in the middle of the cell where Golgi vesicles release components for the construction of a new cell wall. In animal cells, cytokinesis involves the formation of a constriction that divides the cell in two. It is usually well underway by the end of telophase and does not involve the formation of a cell plate.

The Cell Cycle and Stages of Mitosis in Plants

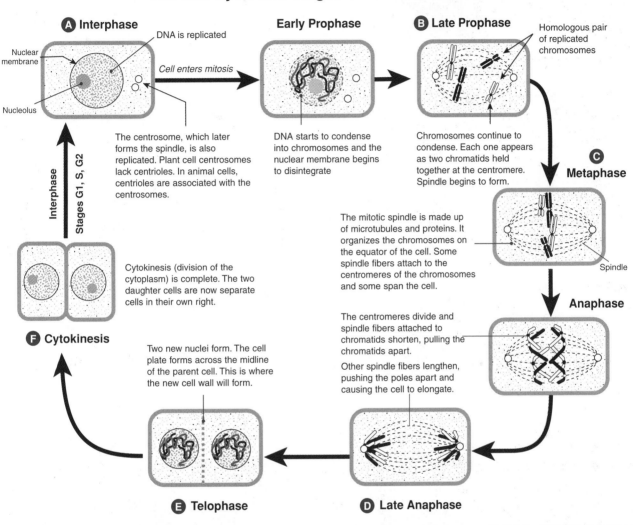

A Interphase

Nuclear membrane

DNA is replicated

Nucleolus

Cell enters mitosis

The centrosome, which later forms the spindle, is also replicated. Plant cell centrosomes lack centrioles. In animal cells, centrioles are associated with the centrosomes.

Interphase — Stages G1, S, G2

Early Prophase

DNA starts to condense into chromosomes and the nuclear membrane begins to disintegrate

B Late Prophase

Homologous pair of replicated chromosomes

Chromosomes continue to condense. Each one appears as two chromatids held together at the centromere. Spindle begins to form.

C Metaphase

The mitotic spindle is made up of microtubules and proteins. It organizes the chromosomes on the equator of the cell. Some spindle fibers attach to the centromeres of the chromosomes and some span the cell.

Spindle

Anaphase

The centromeres divide and spindle fibers attached to chromatids shorten, pulling the chromatids apart.

Other spindle fibers lengthen, pushing the poles apart and causing the cell to elongate.

F Cytokinesis

Cytokinesis (division of the cytoplasm) is complete. The two daughter cells are now separate cells in their own right.

Two new nuclei form. The cell plate forms across the midline of the parent cell. This is where the new cell wall will form.

E Telophase

D Late Anaphase

Chromosomes and Cell Division

The Cell Cycle Overview

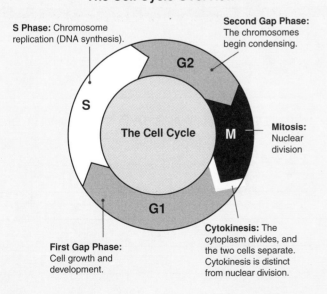

S Phase: Chromosome replication (DNA synthesis).

Second Gap Phase: The chromosomes begin condensing.

G2

S

The Cell Cycle

M — **Mitosis:** Nuclear division

G1

First Gap Phase: Cell growth and development.

Cytokinesis: The cytoplasm divides, and the two cells separate. Cytokinesis is distinct from nuclear division.

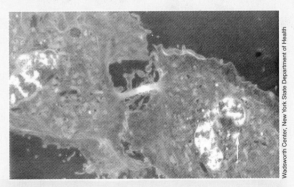

Animal cell **cytokinesis** (above) begins shortly after the sister chromatids have separated in anaphase of mitosis. A contractile ring of microtubular elements assembles in the middle of the cell, next to the plasma membrane, constricting it to form a **cleavage furrow**. In an energy-using process, the cleavage furrow moves inwards, forming a region of abscission where the two cells will separate. In the photograph above, an arrow points to a centrosome, which is still visible near the nucleus.

Wadsworth Center, New York State Department of Health

Periodicals: The cell cycle and mitosis

Related activities: Regulation of the Cell Cycle
Weblinks: Mitosis in an Animal Cell

Mitotic cell division has several purposes (below left). In multicellular organisms, mitosis repairs damaged cells and tissues, and produces the growth in an organism that allows it to reach its adult size. In unicellular organisms, and some small multicellular organisms, cell division allows organisms to reproduce asexually (as in the budding yeast cell cycle below).

The Functions of Mitosis

❶ Growth

In plants, cell division occurs in regions of **meristematic tissue**. In the plant root tip (right), the cells in the root apical meristem are dividing by mitosis to produce new cells. This elongates the root, resulting in **plant growth**.

Root apical meristem

❷ Repair

Photo: AB Sheldon

Some animals, such as this skink (left), detach their limbs as a defense mechanism in a process called autotomy. The limbs can be **regenerated** via the mitotic process, although the tissue composition of the new limb differs slightly from that of the original.

❸ Reproduction

Mitotic division enables some animals to reproduce **asexually**. The cells of this *Hydra* (left) undergo mitosis, forming a 'bud' on the side of the parent organism. Eventually the bud, which is genetically identical to its parent, detaches to continue the life cycle.

Parent

Bud

The Budding Yeast Cell Cycle

Yeasts can reproduce asexually through **budding**. In *Saccharomyces cerevisiae* (baker's yeast), budding involves mitotic division in the parent cell, with the formation of a daughter cell (or bud). As budding begins, a ring of chitin stabilizes the area where the bud will appear and enzymatic activity and turgor pressure act to weaken and extrude the cell wall. New cell wall material is incorporated during this phase. The nucleus of the parent cell also divides in two, to form a daughter nucleus, which migrates into the bud. The daughter cell is genetically identical to its parent cell and continues to grow, eventually separating from the parent cell.

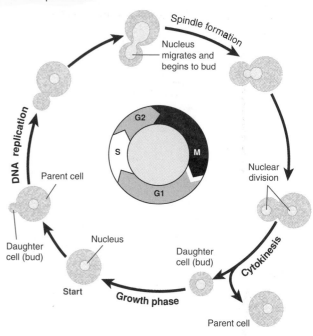

Spindle formation

Nucleus migrates and begins to bud

DNA replication

Parent cell

Nucleus

Daughter cell (bud)

Start

Growth phase

Daughter cell (bud)

Nuclear division

Cytokinesis

Parent cell

G1

G2

S

M

1. The photographs below were taken at various stages of mitosis in a plant cell. They are not in any particular order. Study the diagram on the previous page and determine the stage represented in each photograph (e.g. anaphase).

Photos: RCN

(a) _prophase_ (b) _metaphase_ (c) _anaphase_ (d) _telophase_ (e) _Cytokenisis_

2. State two important changes that chromosomes must undergo before cell division can take place: _____
   _____ They are condensed and copied _____

3. Briefly summarize the stages of the cell cycle by describing what is happening at the points (**A-F**) in the diagram on the previous page:

   A. _____

   B. _____

   C. _____

   D. _____

   E. _____

   F. _____

Regulation of the Cell Cycle

The events of mitosis are virtually the same for all eukaryotes. However, aspects of the cell cycle can vary enormously between species and even between cells of the same organism. For example, the length of the cell cycle varies between cells such as intestinal, liver, and muscle cells. Intestinal cells divide around twice a day, while cells in the liver divide once a year, and those in muscle tissue do not divide at all. If any of these tissues is damaged, however, cell division increases rapidly until the damage is repaired. This variety of length in the cell cycle can be explained by the existence of a regulatory mechanism that is able to slow down or speed up the cell cycle in response to changing conditions.

The Trigger for Mitosis

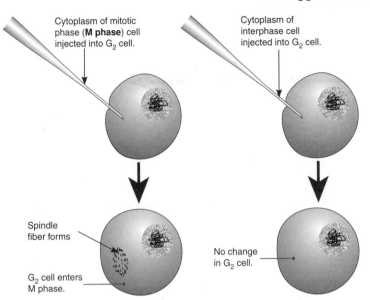

Cytoplasm of mitotic phase (**M phase**) cell injected into G₂ cell.

Cytoplasm of interphase cell injected into G₂ cell.

Spindle fiber forms

G₂ cell enters M phase.

No change in G₂ cell.

Experiments with the eggs of the African clawed frog (*Xenopus laevis*) provided evidence that a substance found in an M-phase cell could induce a G₂ cell to enter M phase. The substance was called **M-phase promoting factor** (**MPF**).

Other studies have shown that MPF is made up of two subunits. The first subunit is a **protein kinase** which activates proteins by transferring a phosphate group from ATP to the protein. The second subunit, called a **cyclin**, activates the first subunit (and thus the first subunit is called a **cyclin-dependent kinase** or **CDK**). CDK is constantly present in the cell, while cyclin is not.

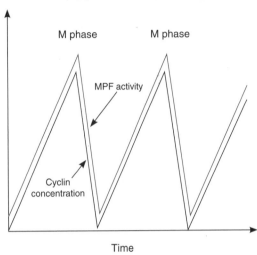

M phase   M phase

MPF activity

Cyclin concentration

Time

Checkpoints

There are three **checkpoints** during the cell cycle. A checkpoint is a critical regulatory point in the cell cycle. At each checkpoint, a set of conditions determines whether or not the cell will continue into the next phase. For example, cell size is important in regulating whether or not the cell can pass through the G₁ checkpoint.

G₂ Checkpoint:
Pass this checkpoint if:
➤ Cell size is large enough.
➤ Replication of chromosomes has been successfully completed.

The Cell Cycle — S, G2, M, G1

G₁ checkpoint
Pass this checkpoint if:
➤ Cell size is large enough.
➤ Sufficient nutrients are available.
➤ Signals from other cells have been received.

Metaphase checkpoint
Pass this checkpoint if:
➤ All chromosomes are attached to the mitotic spindle.

1. (a) Suggest why the cell cycle is shorter in epithelial cells (such as intestinal cells) than in liver or muscle cells:

_____

_____

_____

(b) Describe another situation in which the cell cycle shortens to allow rapid cell division before returning to normal:

_____

_____

_____

Checkpoints, Cyclins, and Cancer

Cell checkpoints give cells a way to ensure that all cellular processes have been completed correctly before entering the next phase. Cancer can result when the pathways regulating the checkpoints fail, so that the cell enters the next phase without reference to the correct completion or functioning of cellular processes. One such checkpoint failure occurs when the G_1 checkpoint fails. Recall that signals from other cells played a part in this checkpoint. Many of these signals are **growth factors** (such as platelet derived growth factor) which stimulate the synthesis of **cyclin**. In cancerous cells, over-production of cyclin leads to uncontrolled growth. The actions of growth factors, cyclin, and CDK are described in the diagram below.

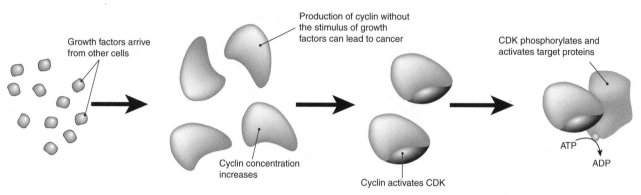

Growth factors arrive from other cells

Production of cyclin without the stimulus of growth factors can lead to cancer

Cyclin concentration increases

Cyclin activates CDK

CDK phosphorylates and activates target proteins

ATP

ADP

2. (a) Explain why the cytoplasm from a M-phase cell could induce a G_2 cell to enter M phase: _____

_____

_____

(b) Explain how the activity of MPF is regulated: _____

_____

_____

_____

3. (a) Explain why the cell must be of a large enough size before passing the G_1 and G_2 checkpoints: _____

_____

_____

_____

(b) i. Which checkpoint ensures that replicated chromosomes will separate correctly? _____

_____

ii. Why is this important? _____

_____

_____

4. (a) Explain why signals from other cells play a part in regulating the cell cycle: _____

_____

_____

_____

(b) How does the over-production of cyclin lead to cancer? _____

_____

_____

_____

_____

© BIOZONE International 2012
ISBN: 978-1-927173-11-4
Photocopying Prohibited

Meiosis

Meiosis is a special type of cell division concerned with producing sex cells (gametes) for the purpose of sexual reproduction. It involves a single chromosomal duplication followed by two successive nuclear divisions, and results in a halving of the diploid chromosome number. Meiosis occurs in the sex organs of plants and animals. If genetic mistakes (**gene** and **chromosome mutations**) occur here, they will be passed on to the offspring (they will be inherited).

Meiosis I

(Reduction division)

The first division separates the homologous chromosomes into two intermediate cells.

Interphase

In a non-dividing cell, the **chromosomes** are not visible as discrete structures because they are uncoiled to make the DNA information available for protein synthesis.

Meiosis is preceded by DNA replication, during which each of the chromosomes replicates. For each chromosome, there are now two genetically identical sister chromatids (as yet unseparated). It is at this stage that gene mutations may occur. These may create new versions of genes (alleles).

Prophase 1

The chromosomes condense. The **homologs**, each consisting of two sister chromatids, pair up in a process called **synapsis** to form **bivalents**. At this stage the arms of the chromatids can become entangled, and segments of chromosome can be exchanged in a process called crossing over.

Metaphase 1

The bivalents line up at the 'equator' (the **metaphase plate**) of the cell in a way that is random. This results in **independent assortment** of maternal and paternal chromosomes.

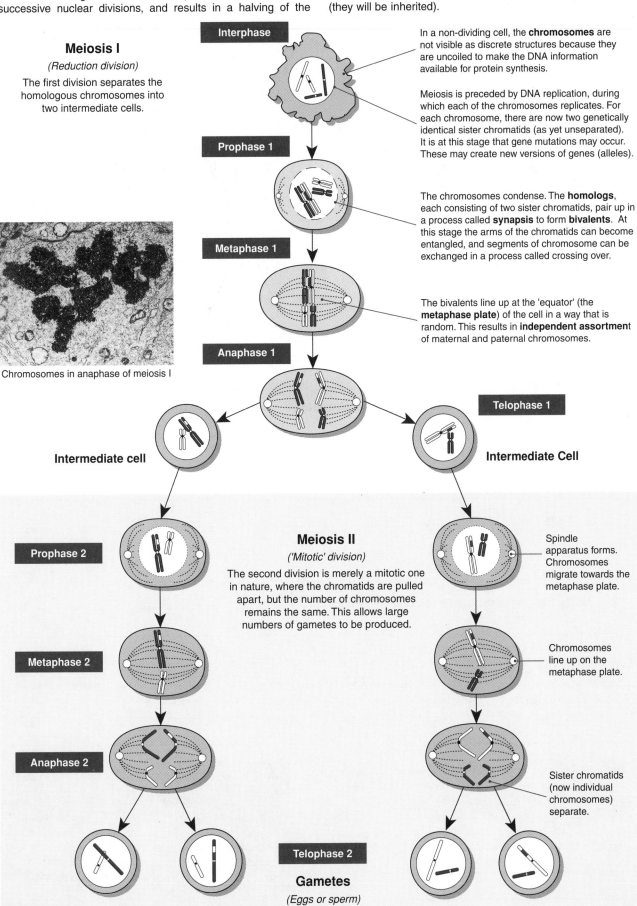

Chromosomes in anaphase of meiosis I

Anaphase 1

Telophase 1

Intermediate cell

Intermediate Cell

Meiosis II

('Mitotic' division)

The second division is merely a mitotic one in nature, where the chromatids are pulled apart, but the number of chromosomes remains the same. This allows large numbers of gametes to be produced.

Prophase 2

Metaphase 2

Anaphase 2

Spindle apparatus forms. Chromosomes migrate towards the metaphase plate.

Chromosomes line up on the metaphase plate.

Sister chromatids (now individual chromosomes) separate.

Telophase 2

Gametes

(Eggs or sperm)

Chromosomes and Cell Division

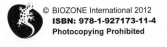
Periodicals:
Mechanisms of meiosis

Related activities: Gene Mutations, Chromosome Mutations
Weblinks: Meiosis Tutorial, Meiosis

RA 1

The meiotic spindle normally distributes chromosomes to daughter cells without error. However, mistakes can occur in which the homologous chromosomes fail to separate properly at anaphase during meiosis I, or sister chromatids fail to separate during meiosis II. In these cases, one gamete receives two of the same type of chromosome and the other gamete receives no copy. This mishap, called non-disjunction, results in abnormal numbers of chromosomes passing to the gametes. If either of the aberrant gametes unites with a normal one at fertilization, the offspring will have an abnormal chromosome number, known as an aneuploidy.

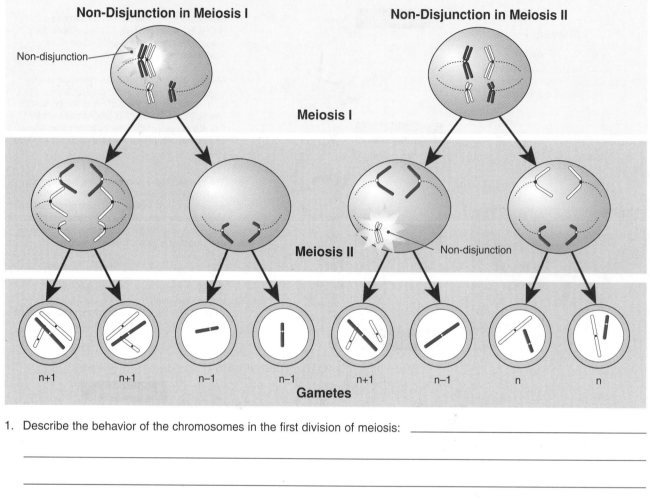

1. Describe the behavior of the chromosomes in the first division of meiosis: _____

_____

_____

_____

_____

_____

2. Describe the behavior of the chromosomes in the second division of meiosis: _____

_____

_____

3. Explain why non-disjunction in meiosis I results in a higher proportion of faulty gametes than non-disjunction in meiosis II:

_____

_____

4. Both these light micrographs (A and B) show chromosomes in metaphase of meiosis. In what way are they different?

_____

_____

_____

_____

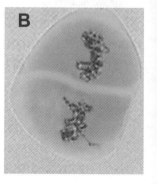

Modeling Meiosis

This practical activity uses ice-block sticks to simulate the production of gametes (sperm and eggs) by **meiosis** and shows you how **crossing over** increases genetic variability. This is demonstrated by studying how two of your own alleles are inherited by the child produced at the completion of the activity. Completing this activity will help you to visualize and understand meiosis, and demonstrate how crossing over contributes to genetic variation in the offspring. It will take 25-45 minutes.

Background

Each of your somatic cells contain 46 chromosomes. You received 23 chromosomes from your mother (**maternal chromosomes**), and 23 chromosomes from your father (**paternal chromosomes**). Therefore, you have 23 homologous (same) pairs. For simplicity, the number of chromosomes studied in this exercise has been reduced to four (two homologous pairs). To study the effect of crossing over on genetic variability, you will look at the inheritance of two of your own traits: the ability to **tongue roll** and **handedness**.

Chromosome #	Phenotype	Genotype
10	Tongue roller	TT, Tt
10	Non-tongue roller	tt
2	Right handed	RR, Rr
2	Left handed	rr

Record your phenotype and genotype for each trait in the table (right). **NOTE:** If you have a dominant trait, you will not know if you are heterozygous or homozygous for that trait, so you can choose either genotype for this activity.

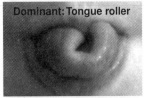

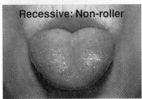

Trait	Phenotype	Genotype
Handedness		
Tongue rolling		

BEFORE YOU START THE SIMULATION: Partner up with a classmate. Your gametes will combine with theirs (fertilization) at the end of the activity to produce a child. Decide who will be the female, and who will be the male. You will need to work with this person again at step 6.

1. Collect four ice-blocks sticks. These represent four chromosomes. Color two sticks blue or mark them with a P. These are the paternal chromosomes. The plain sticks are the maternal chromosomes. Write your initial on each of the four sticks. Label each chromosome with their chromosome number (right).

   Label four sticky dots with the alleles for each of your phenotypic traits, and stick it onto the appropriate chromosome. For example, if you are heterozygous for tongue rolling, the sticky dots with have the alleles **T** and **t**, and they will be placed on chromosome 10. If you are left handed, the alleles will be **r** and **r** and be placed on chromosome 2 (right).

2. Randomly drop the chromosomes onto a table. This represents a cell in either the testes or ovaries. **Duplicate** your chromosomes (to simulate DNA replication) by adding four more identical ice-block sticks to the table (below). This represents **interphase**.

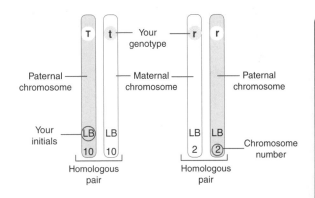

3. Simulate **prophase I** by lining the duplicated chromosome pair with their homologous pair (below). For each chromosome number, you will have four sticks touching side-by-side (A). At this stage **crossing over** occurs. Simulate this by swapping sticky dots from adjoining homologs (B).

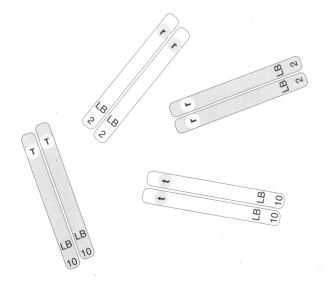

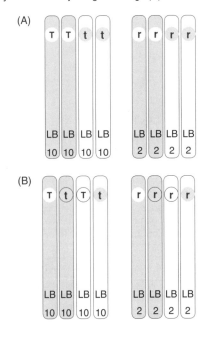

Periodicals: Mechanisms of meiosis

Related activities: Meiosis, Crossing Over
Weblinks: Meiosis Tutorial

Chromosomes and Cell Division

P

4. Randomly align the homologous chromosome pairs to simulate alignment on the metaphase plate (as occurs in **metaphase I**). Simulate **anaphase I** by separating chromosome pairs. For each group of four sticks, two are pulled to each pole.

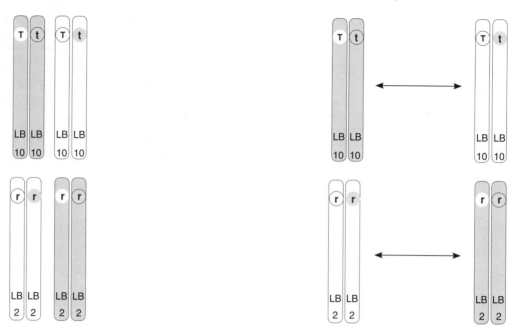

5. **Telophase I:** Two intermediate cells are formed. If you have been random in the previous step, each intermediate cell will contain a mixture of maternal and paternal chromosomes. This is the end of **meiosis 1**.

Now that meiosis 1 is completed, your cells need to undergo **meiosis 2.** Carry out prophase II, metaphase II, anaphase II, and telophase II. Remember, there is no crossing over in meiosis II. At the end of the process each intermediate cell will have produced two haploid gametes (below).

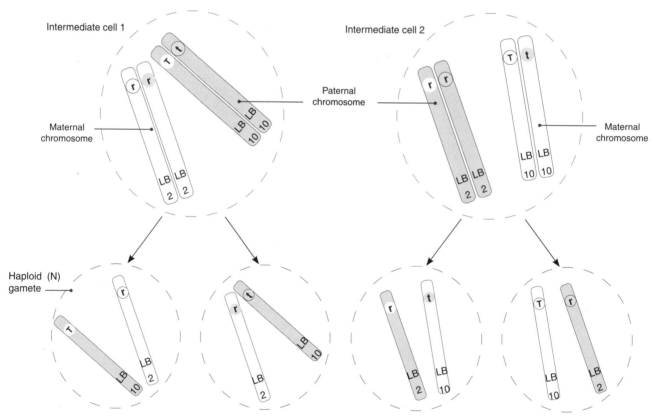

6. Pair up with the partner you chose at the beginning of the exercise to carry out **fertilization**. Randomly select one sperm and one egg cell. The unsuccessful gametes can be removed from the table. Combine the chromosomes of the successful gametes. You have created a child! Fill in the following chart to describe your child's genotype and phenotype for tongue rolling and handedness.

Trait	Phenotype	Genotype
Handedness		
Tongue rolling		

Mitosis vs Meiosis

Cell division is fundamental to all life, as cells arise only by the division of existing cells. All types of cell division begin with replication of the cell's DNA. In eukaryotes, this is followed by division of the nucleus. There are two forms of nuclear division: **mitosis** and **meiosis**, and they have quite different purposes and outcomes. Mitosis is the simpler of the two and produces two identical daughter cells from each parent

cell. Mitosis is responsible for growth and repair processes in multicellular organisms and reproduction in single-celled and asexual eukaryotes. Meiosis involves a **reduction division** in which haploid gametes are produced for the purposes of sexual reproduction. Fusion of haploid gametes in fertilization restores the diploid cell number in the **zygote**. These two fundamentally different types of cell division are compared below.

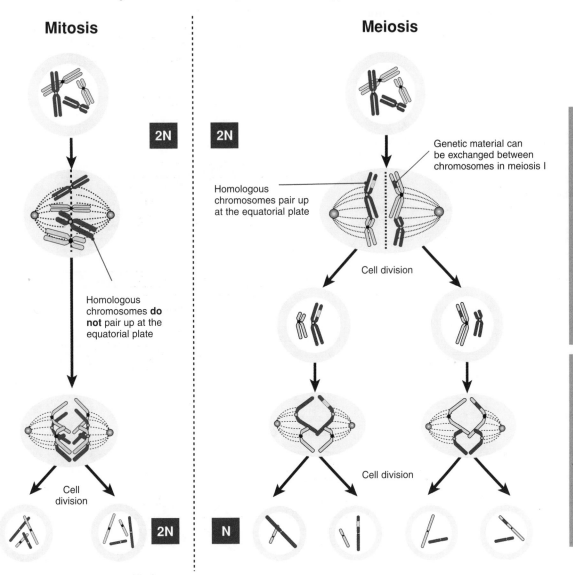

Mitosis

Meiosis

Genetic material can be exchanged between chromosomes in meiosis I

Homologous chromosomes pair up at the equatorial plate

Cell division

Homologous chromosomes **do not** pair up at the equatorial plate

Cell division

Cell division

Cell division

2N / 2N / 2N / N

Meiosis I: Reduction division

Meiosis II: 'Mitotic' division

Chromosomes and Cell Division

1. Explain how mitosis conserves chromosome number while meiosis reduces the number from diploid to haploid:

_____

_____

_____

2. Describe a fundamental difference between the first and second divisions of meiosis: _____

_____

_____

3. Explain how meiosis introduces genetic variability into gametes and offspring (following gamete fusion in fertilization):

_____

_____

_____

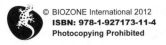

Related activities: Mitosis and the Cell Cycle, Meiosis

A 2

Crossing Over

Crossing over refers to the mutual exchange of pieces of chromosome and involves the swapping of whole groups of genes between the **homologous** chromosomes. This process can occur only during the first division of **meiosis**. Errors in crossing over can result in **chromosome mutations** (see the activity on this topic), which can be very damaging to development. Crossing over can upset expected frequencies of offspring in dihybrid crosses. The frequency of crossing over (COV) between different genes (as followed by inherited, observable traits) can be used to determine the relative positions of genes on a chromosome and provide a **genetic map**. There has been a recent suggestion that crossing over may be necessary to ensure accurate cell division.

Pairing of Homologous Chromosomes
Every somatic cell contains a pair of each type of chromosome, one from each parent. These are called **homologous pairs** or **homologs**. In prophase of meiosis I, the homologs pair up to form **bivalents**. This process is called **synapsis** and it brings the chromatids of the homologs into close contact.

Chiasma Formation and Crossing Over
Synapsis allows the homologous, non-sister chromatids to become entangled and the chromosomes exchange segments. This exchange occurs at regions called **chiasmata** (*sing.* chiasma). In the diagram (centre), a chiasma is forming and the exchange of pieces of chromosome has not yet taken place. Numerous chiasmata may develop between homologs.

Separation
Crossing over produces new allele combinations, a phenomenon known as **recombination**. When the homologs separate in anaphase of meiosis I, each of the chromosomes pictured will have a new mix of alleles that will be passed into the gametes soon to be formed. Recombination is an important source of variation in population gene pools.

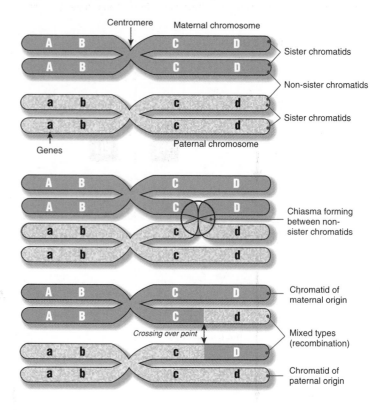

Gamete Formation

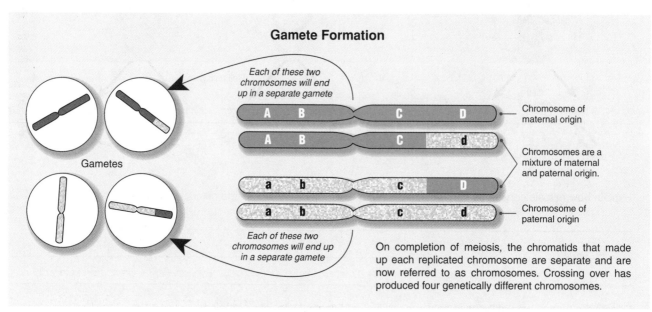

On completion of meiosis, the chromatids that made up each replicated chromosome are separate and are now referred to as chromosomes. Crossing over has produced four genetically different chromosomes.

1. (a) In a general way, describe how crossing over alters the genotype of gametes: _____

    _____

    (b) What is the consequence of this? _____

2. What is the significance of crossing over in the evolution of sexually producing populations? _____

    _____

    _____

© BIOZONE International 2012
ISBN: 978-1-927173-11-4
Photocopying Prohibited

Crossing Over Problems

The diagram below shows a pair of homologous chromosomes about to undergo chiasma formation during the first cell division in the process of meiosis. There are known crossover points along the length of the chromatids (same on all four chromatids shown in the diagram). In the prepared spaces below, draw the gene sequences after crossing over has occurred on three unrelated and separate occasions (it would be useful to use different colored pens to represent the genes from the two different chromosomes). See the diagrams on the previous page as a guide.

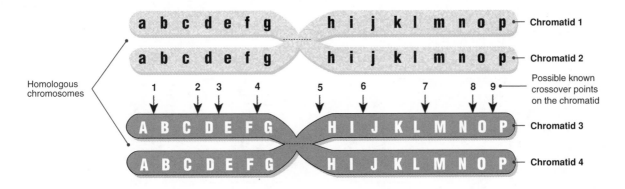

1. Crossing over occurs at a **single** point between the chromosomes above.

   (a) Draw the gene sequences for the four chromatids (on the right), after crossing over has occurred at crossover point: **2**

   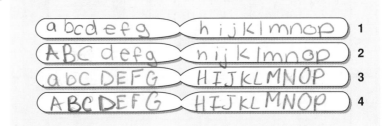

   (b) Which genes have been exchanged with those on its homolog (neighboring chromosome)?

    ABC, abc

2. Crossing over occurs at **two** points between the chromosomes above.

   (a) Draw the gene sequences for the four chromatids (on the right), after crossing over has occurred between crossover points: **6** and **7**.

   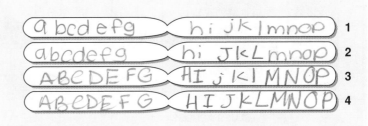

   (b) Which genes have been exchanged with those on its homolog (neighboring chromosome)?

   jkl, JKL

3. Crossing over occurs at **four** points between the chromosomes above.

   (a) Draw the gene sequences for the four chromatids (on the right), after crossing over has occurred between crossover points: **1** and **3**, and **5** and **7**.

   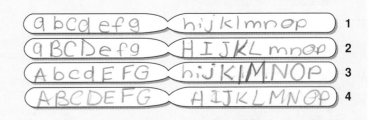

   (b) Which genes have been exchanged with those on its homolog (neighboring chromosome)?

   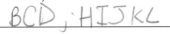 BCD ; HIJKL

4. What would be the genetic consequences if there was no crossing over between chromatids during meiosis?

   _____

   _____

   _____

Related activities: Crossing Over

A 3

Chromosomes and Cell Division

KEY TERMS: Crossword

Complete the crossword below, which will test your understanding of key terms in this chapter and their meanings.

CLUES ACROSS

1. A stage in mitosis when the chromatin condenses and becomes visible as chromosomes.
6. The union of two haploid gametes to reconstitute a diploid cell (the zygote).
10. Meiosis is this type of division.
12. The exchange of alleles between homologous chromosomes during meiosis as a result of crossing over.
15. A chromosome derived from the female parent is called the _ _ _ _ _ _ _ _ chromosome.
17. Paired microtubular structures in animal cells associated with cell division.
18. The stage in cell division when two daughter nuclei appear.
20. The phase of a cell cycle resulting in nuclear division.
21. A pair of associated homologous chromosomes formed after replication. Also called a tetrad.
22. A reduction division in eukaryotic cells in which the number of chromosomes per cell is halved.
23. An exchange of genetic material between homologous chromosomes (2 words: 8, 4).

CLUES DOWN

2. A chromosome derived from the male parent is called the _ _ _ _ _ _ _ _ chromosome.

3. A family of protein kinases with a role in regulating the cell cycle (3 words: 6, 9, 7).
4. Chromosome pairs, one paternal, and maternal, of the same length, centromere position, and staining pattern with genes for the same characteristics at corresponding loci.: 10, 11).
5. The random assortment of maternal and paternal chromosomes during meiosis (2 words: 11, 10).
7. The pairing of homologs during prophase 1 of meiosis.
8. The process by which a less specialized cell becomes a more specialized cell type (2 words: 4, 15).
9. The stage of the cell cycle before cell division begins when the cell grows, the genetic material is duplicated, and the cell prepares to divide.
11. The number of chromosomes in most cells except the gametes.
13. Division of the cytoplasm of a eukaryotic cell to form two daughter cells.
14. The process occurring prior to cell division to produce a copy of all the DNA in the nucleus (2 words: 3, 11).
15. A stage of mitosis in which condensed chromosomes align in the middle of the cell.
16. A stage of mitosis in a eukaryotic cell in which chromosomes separate and chromatids move to opposite poles of the cell.
19. The number of chromosomes in the gamete of an individual.

The Chromosomal Basis of Inheritance

Key concepts

▶ Sexual reproduction introduces variation in the offspring: the raw material for natural selection.

▶ The dominance of alleles can be inferred from the genetic outcomes of crosses.

▶ Lethal alleles and linkage cause departures from expected offspring phenotype ratios.

▶ The inheritance of some traits is dependent on gender.

▶ Gene interactions contribute to genetic variation.

▶ Environmental factors can modify the phenotype encoded by genes.

Key terms

allele
autosome
back cross
cline
codominance (of alleles)
continuous variation
cross
dihybrid cross
discontinuous variation
dominant (of alleles)
F_1 / F_2 generation
genetic counseling
genomic imprinting
genotype
heterozygous
homozygous
incomplete dominance
independent assortment
lethal allele
linkage
monohybrid cross
multiple alleles
non-disjunction
non-nuclear inheritance
pedigree analysis
phenotype
phenotypic plasticity
polygenes (=multiple genes)
Punnett square
pure (true)-breeding
recessive (of alleles)
segregation
selfing
sex chromosome
sex-limited traits
sex linked gene
test cross
trait

Periodicals:
Listings for this chapter are on page 382

Weblinks:
www.thebiozone.com/
weblink/AP1-3114.html

BIOZONE APP:
Student Review Series
Inheritance

Essential Knowledge

□ 1. Use the **KEY TERMS** to compile a glossary for this topic.

The Chromosomal Basis of Inheritance (3.A.3)   pages 145, 158-73, 181-86

□ 2. Recall the role of meiosis and fertilization in generating variation. Understand **segregation** and **independent assortment** of genes on different chromosomes and explain their importance to our understanding of heredity and evolution.

□ 3. Explain how the rules of probability are applied to solving genetic problems.

□ 4. Demonstrate use of the terms commonly used in inheritance studies: **allele, locus, trait, heterozygous, homozygous, genotype, phenotype, cross, test cross, back cross, carrier, offspring, trait, F_1 generation, F_2 generation**.

□ 5. Solve problems involving **monohybrid** and **dihybrid inheritance** of **unlinked, autosomal genes** with a simple **dominant-recessive** inheritance pattern.

□ 6. Describe and explain inheritance involving **codominance, incomplete dominance, multiple alleles**, and **lethal alleles**.

□ 7. Solve problems involving **dihybrid inheritance** of **linked genes**. Understand that the probability of linked genes being inherited together as a unit is a function of the distance between them.

□ 8. Using examples, explain how certain human disorders (e.g. sickle cell disease, Huntington's disease, trisomy 21) can be attributed to the inheritance of single gene traits or to specific chromosomal changes, such as **non-disjunction**.

□ 9. Discuss the ethical, social, and medical issues surrounding human genetic disorders. Describe the use of **pedigree analysis** to illustrate the inheritance of traits in a family tree. Describe the principles and role of **genetic counseling.**

Non-Mendelian Inheritance (3.A.4)   pages 150, 174-180, 187-188, 217-225

□ 10. Recognize that the inheritance of many traits is not explained by simple Mendelian genetics. Using the inheritance of **polygenes (multiple genes)** as an example, explain how non-Mendelian patterns of inheritance can be identified.

□ 11. Distinguish **sex chromosomes** from **autosomes**. Describe examples and solve problems involving different patterns of inheritance involving **sex linked genes** (e.g. red-green color-blindness or hemophilia).

□ 12. Using examples, explain how some traits are **sex-limited**.

□ 13. Explain how some traits are the result of **non-nuclear inheritance**. Describe and explain an example of **genomic imprinting**.

Genes and Environment (4.C.2)   pages 174, 189-190

□ 14. Explain how genotype and environment contribute to phenotypic **variation**. Using examples, explain how environmental factors influence traits. Describe examples of **phenotypic plasticity** and describe its basis.

Alleles

Sexually reproducing organisms in nearly all cases have paired sets of chromosomes, one set coming from each parent. The equivalent chromosomes that form a pair are termed **homologs**. They contain equivalent sets of genes on them. But there is the potential for different versions of a gene to exist in a population and these are termed **alleles**.

Homologous Chromosomes

In sexually reproducing organisms, most cells have a homologous pair of chromosomes (one coming from each parent). This diagram shows the position of three different genes on the same chromosome that control three different traits (A, B and C).

These two different versions of gene A create a condition known as **heterozygous.** Only the dominant allele (A) will determine the phenotype.

When both chromosomes have identical copies of the dominant allele for gene B the organism is said to be **homozygous dominant** for that gene.

When both chromosomes have identical copies of the recessive allele for gene C the organism is said to be **homozygous recessive** for that gene.

Maternal chromosome originating from the egg of this person's mother.

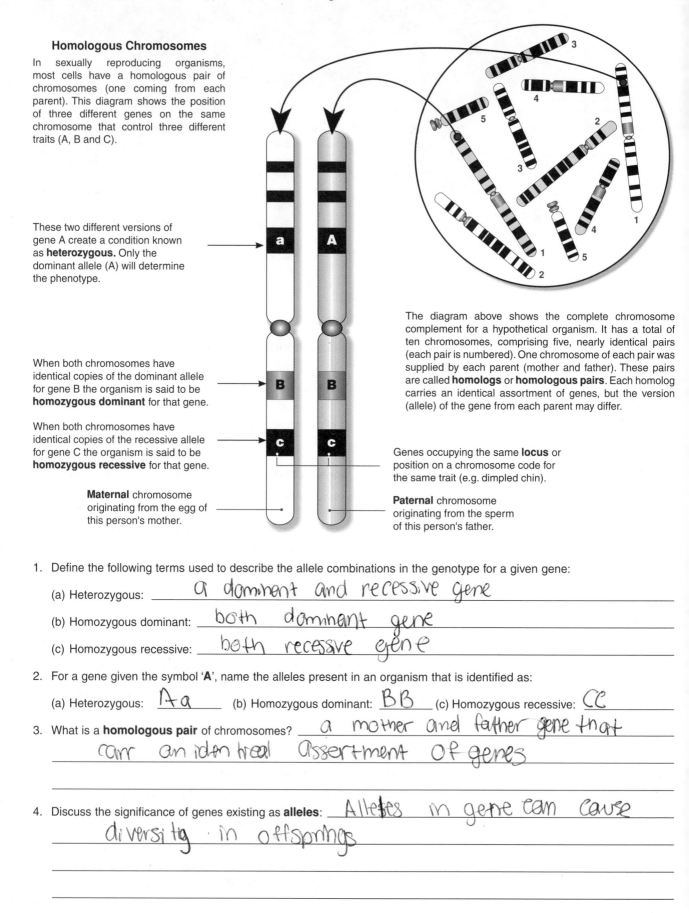

The diagram above shows the complete chromosome complement for a hypothetical organism. It has a total of ten chromosomes, comprising five, nearly identical pairs (each pair is numbered). One chromosome of each pair was supplied by each parent (mother and father). These pairs are called **homologs** or **homologous pairs**. Each homolog carries an identical assortment of genes, but the version (allele) of the gene from each parent may differ.

Genes occupying the same **locus** or position on a chromosome code for the same trait (e.g. dimpled chin).

Paternal chromosome originating from the sperm of this person's father.

1. Define the following terms used to describe the allele combinations in the genotype for a given gene:

   (a) Heterozygous: _a dominent and recessive gene_

   (b) Homozygous dominant: _both dominant gene_

   (c) Homozygous recessive: _both recessve gene_

2. For a gene given the symbol '**A**', name the alleles present in an organism that is identified as:

   (a) Heterozygous: _Aa_   (b) Homozygous dominant: _BB_   (c) Homozygous recessive: _CC_

3. What is a **homologous pair** of chromosomes? _a mother and father gene that_ _carr an iden treal assertment of genes_

4. Discuss the significance of genes existing as **alleles**: _Alleles in gene can cause_ _diversity in offsprings_

Mendel's Pea Plant Experiments

Gregor Mendel (1822-84), pictured right, was an Austrian monk who carried out the pioneering studies of inheritance. Mendel bred pea plants so he could study the inheritance patterns of a number of **traits** (specific characteristics). He showed that characters could be masked in one generation but could reappear in later generations and proposed that inheritance involved the transmission of discrete units of inheritance from one generation to the next. We now call these units of inheritance **genes**.

Mendel examined six phenotypic traits and found that they were inherited in predictable ratios, depending on the phenotypes of the parents. Some of his results from crossing heterozygous plants are tabulated below. The numbers in the results column represent how many offspring had those phenotypic features.

1. Study the **results** for each of the six experiments below. Determine which of the two phenotypes is dominant, and which is recessive. Place your answers in the spaces in the **dominance** column in the table below.

2. Calculate the ratio of dominant phenotypes to recessive phenotypes (to two decimal places). The first one has been done for you (5474 ÷ 1850 = 2.96). Place your answers in the spaces provided in the table below:

Trait	Possible Phenotypes		Results		Dominance	Ratio
Seed shape	Wrinkled	Round	Wrinkled Round **TOTAL**	1850 5474 7324	Dominant: Round Recessive: Wrinkled	2.96: 1
Seed color	Green	Yellow	Green Yellow **TOTAL**	2001 6022 8023	Dominant: Yellow Recessive: green	3.00:1
Pod color	Green	Yellow	Green Yellow **TOTAL**	428 152 580	Dominant: green Recessive: yellow	2.81:1
Flower position	Axial	Terminal	Axial Terminal **TOTAL**	651 207 858	Dominant: axial Recessive: terminal	3.14:1
Pod shape	Constricted	Inflated	Constricted Inflated **TOTAL**	299 882 1181	Dominant: inflated Recessive: constricted	2.94: 1
Stem length	Tall	Dwarf	Tall Dwarf **TOTAL**	787 277 1064	Dominant: tall Recessive: dwarf	2.84:1

3. Mendel's experiments identified that two heterozygous parents should produce offspring in the ratio of three times as many dominant offspring to those showing the recessive phenotype.

(a) Which three of Mendel's experiments provided ratios closest to the theoretical 3:1 ratio?

Pod Shape; Flower position, seed color

(b) Suggest why these results deviated less from the theoretical ratio than the others: Pod color

Periodicals: Mendel's legacy

Related activities: Alleles, Mendel's Laws of Inheritance
Weblinks: Basic Principles of Genetics

DA 2

Mendel's Laws of Inheritance

From his work on the inheritance of phenotypic traits in peas, Mendel formulated a number of ideas about the inheritance of characters. These were later given formal recognition as Mendel's laws of inheritance. These are outlined below.

The Theory of Particulate Inheritance

Characteristics of both parents are passed on to the next generation as discrete entities (genes).

This model explained many observations that could not be explained by the idea of blending inheritance, which was universally accepted prior to this theory. The trait for flower color (right) appears to take on the appearance of only one parent plant in the first generation, but reappears in later generations.

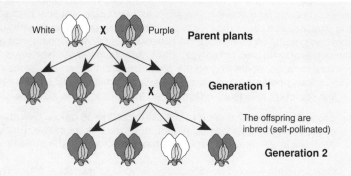

Law of Segregation

*During gametic meiosis, the two members of any pair of alleles **segregate** unchanged and are passed into different gametes, so that each gamete receives only one allele of a pair.*

These gametes are eggs (ova) and sperm cells. The allele in the gamete will be passed on to the offspring.

> **NOTE:** This diagram has been simplified, omitting the stage where the second chromatid is produced for each chromosome.

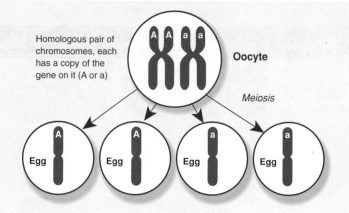

Law of Independent Assortment

Allele pairs separate independently during gamete formation, and traits are passed on to offspring independently of one another (this is only true for unlinked genes).

This diagram shows two genes (A and B) that code for different traits. Each of these genes is represented twice, one copy (allele) on each of two homologous chromosomes. The genes A and B are located on different chromosomes and, because of this, they will be inherited independently of each other i.e. the gametes may contain any combination of the parental alleles.

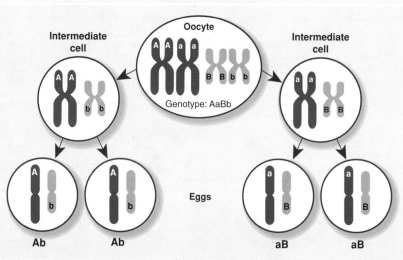

1. State the **property of genetic inheritance** that allows parent pea plants of different flower color to give rise to flowers of a single color in the first generation, with both parental flower colors reappearing in the following generation:

   _____

2. The oocyte is the egg producing cell in the ovary of an animal. In the diagram illustrating the **law of segregation** above:

   (a) State the genotype for the oocyte (adult organism): _____

   (b) State the genotype of each of the **four** gametes: _____

   (c) State how many different kinds of gamete can be produced by this oocyte: _____

3. The diagram illustrating the **law of independent assortment** (above) shows only one possible result of the random sorting of the chromosomes to produce: Ab and aB in the gametes.

   (a) List another possible combination of genes (on the chromosomes) ending up in gametes from the same oocyte:

   _____

   (b) How many different gene combinations are possible for the oocyte? _____4_____

© BIOZONE International 2012
ISBN: 978-1-927173-11-4
Photocopying Prohibited

Related activities: Alleles, Mendel's Pea Plant Experiments

Basic Genetic Crosses

Examine the diagrams below on monohybrid crosses and complete the exercise for dihybrid (two gene) inheritance. A **test cross** is also provided to show how the genotype of a dominant phenotype can be determined. A test cross will yield one of two different results, depending on the genotype of the dominant individual. A **back cross** (not shown) refers to any cross between an offspring and one of its parents (or an individual genetically identical to one of its parents).

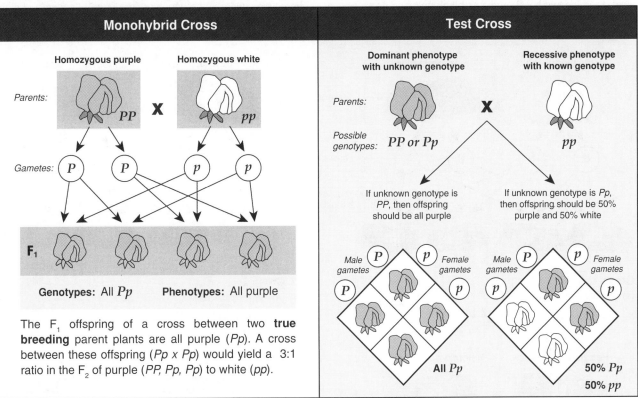

Monohybrid Cross

Homozygous purple (*PP*) X **Homozygous white** (*pp*)

Parents / Gametes: P P p p

F₁

Genotypes: All *Pp*   **Phenotypes:** All purple

The F₁ offspring of a cross between two **true breeding** parent plants are all purple (*Pp*). A cross between these offspring (*Pp x Pp*) would yield a 3:1 ratio in the F₂ of purple (*PP, Pp, Pp*) to white (*pp*).

Test Cross

Dominant phenotype with unknown genotype   X   **Recessive phenotype with known genotype**

Parents:

Possible genotypes: *PP or Pp*   *pp*

If unknown genotype is *PP*, then offspring should be all purple

If unknown genotype is *Pp*, then offspring should be 50% purple and 50% white

Male gametes P   p Female gametes

All *Pp*   **50% *Pp*** / **50% *pp***

Dihybrid Cross

A dihbrid cross studies the inheritance patterns of two genes. In pea seeds, yellow color (*Y*) is dominant to green (*y*) and round shape (*R*) is dominant to wrinkled (r). Each **true breeding** parental plant has matching alleles for each of these characters (*YYRR* or *yyrr*). F₁ offspring will all have the same genotype and phenotype (yellow-round: *YyRr*).

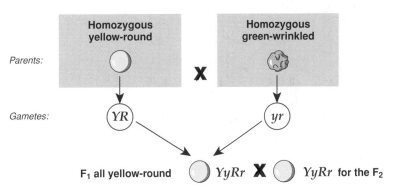

Homozygous yellow-round   X   **Homozygous green-wrinkled**

Parents:

Gametes: *YR*   *yr*

F₁ all yellow-round   *YyRr* **X** *YyRr* **for the F₂**

1. Fill in the Punnett square (below right) to show the genotypes of the F₂ generation.

2. In the boxes below, use fractions to indicate the numbers of each phenotype produced from this cross.

   Yellow-round ☐

   Green-round ☐

   Yellow-wrinkled ☐

   Green-wrinkled ☐

3. Express these numbers as a ratio:

_____

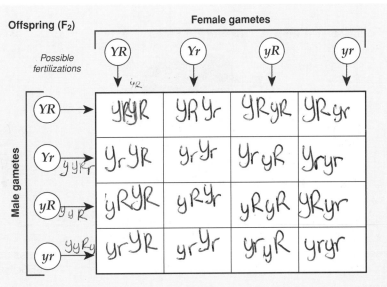

Offspring (F₂)   **Female gametes**

Possible fertilizations: *YR*   *Yr*   *yR*   *yr*

Male gametes

	YR	Yr	yR	yr
YR	YRYR	YRYr	YRyR	YRyr
Yr	YrYR	YrYr	YryR	Yryr
yR	yRYR	yRYr	yRyR	yRyr
yr	yrYR	yrYr	yryR	yryr

Monohybrid Cross

The study of **single-gene inheritance** is achieved by performing **monohybrid crosses**. The six basic types of matings possible among the three genotypes can be observed by studying a pair of alleles that govern coat color in the guinea pig. A dominant allele given the symbol **B** produces **black** hair, and its recessive allele **b** produces white. Each of the parents can produce two types of gamete by the process of **meiosis**. Determine the

genotype and **phenotype frequencies** for the crosses below (enter the frequencies in the spaces provided). For crosses 3 to 6, you must also determine gametes produced by each parent (write these in the circles), and offspring (F₁) genotypes and phenotypes (write in the genotype inside the offspring and state if black or white).

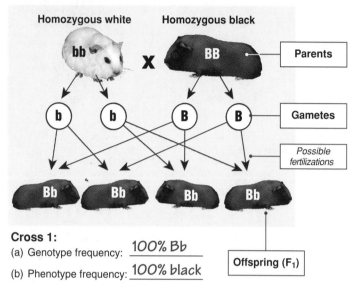

Parents

Gametes

Possible fertilizations

Cross 1:
(a) Genotype frequency: _100% Bb_

(b) Phenotype frequency: _100% black_

Offspring (F₁)

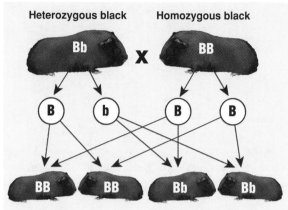

Cross 2:
(a) Genotype frequency: _____

(b) Phenotype frequency: _____

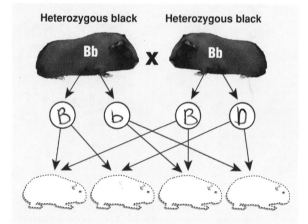

Cross 3:
(a) Genotype frequency: _____

(b) Phenotype frequency: _____

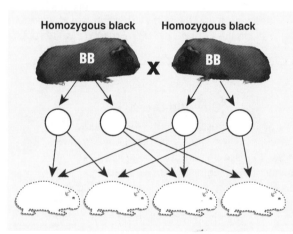

Cross 4:
(a) Genotype frequency: _____

(b) Phenotype frequency: _____

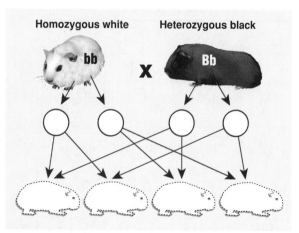

Cross 5:
(a) Genotype frequency: _____

(b) Phenotype frequency: _____

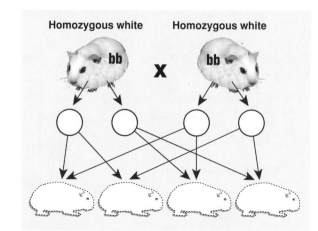

Cross 6:
(a) Genotype frequency: _____

(b) Phenotype frequency: _____

© BIOZONE International 2012
ISBN: 978-1-927173-11-4

Related activities: Basic Genetic Crosses

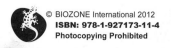

Codominance

Codominance refers to an inheritance pattern in which both alleles in a heterozygote contribute to the phenotype. Both alleles are **independently** and **equally expressed**. One example includes the human blood group AB which is the result of two alleles: A and B, both being equally expressed. Other examples include certain coat colors in horses and cattle. Reddish coat color is equally dominant with white. Animals that have both alleles have coats that are roan-colored (coats with a mix of red and white hairs). The red hairs and white hairs are expressed equally and independently (not blended to produce pink).

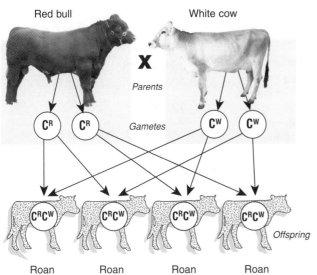

A roan shorthorn heifer

In the shorthorn cattle breed, coat color is inherited. White shorthorn parents always produce calves with white coats. Red parents always produce red calves. However, when a red parent mates with a white one, the calves have a coat color that is different from either parent; a mixture of red and white hairs, called roan. Use the example (left) to help you to solve the problems below.

1. Explain how codominance of alleles can result in offspring with a phenotype that is different from either parent:

Both alleles in a heterozygote contribute to the phenotype

2. A white bull is mated with a roan cow (right):

   (a) Fill in the spaces to show the genotypes and phenotypes for parents and calves:

   (b) What is the phenotype ratio for this cross?

   1:1 red: white

   (c) How could a cattle farmer control the breeding so that the herd ultimately consisted of only red cattle?

   Get rid of white cows and mate only red cows eventually only having red

3. A farmer has only roan cattle on his farm. He suspects that one of the neighbors' bulls may have jumped the fence to mate with his cows earlier in the year because half the calves born were red and half were roan. One neighbor has a red bull, the other has a roan.

   (a) Fill in the spaces (right) to show the genotype and phenotype for parents and calves.

   (b) Which bull serviced the cows? **red** or ~~roan~~ (delete one)

4. Describe the classical phenotypic ratio for a codominant gene resulting from the cross of two heterozygous parents (e.g. a cross between two roan cattle):

   _1:1_

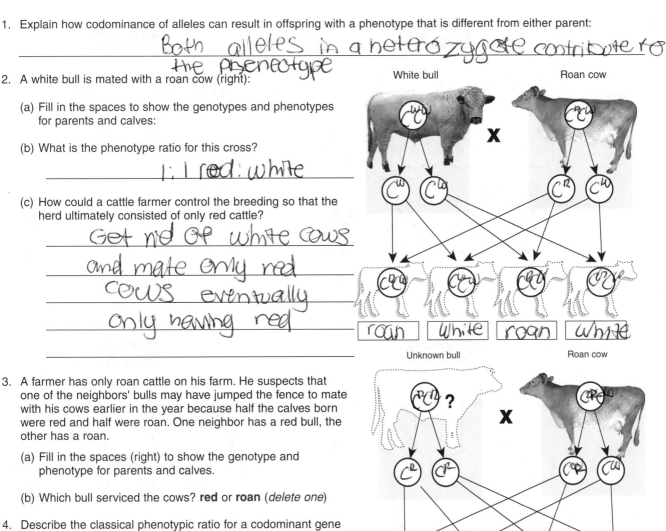

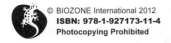

Codominance in Multiple Allele Systems

The four common blood groups of the human 'ABO blood group system' are determined by three alleles: *A*, *B*, and *O* (also represented in some textbooks as: I^A, I^B, and i^O or just i). This is an example of a **multiple allele** system for a gene. The ABO antigens consist of sugars attached to the surface of red blood cells. The alleles code for enzymes (proteins) that join these sugars together. The allele *O* produces a non-functioning enzyme that is unable to make any changes to the basic antigen (sugar) molecule. The other two alleles *(A, B)* are **codominant** and are expressed equally. They each produce a different functional enzyme that adds a different, specific sugar to the basic sugar molecule. The blood group A and B antigens are able to react with antibodies present in the blood from other people and must be matched for transfusion.

Recessive allele:	*O*	produces a non-functioning protein
Dominant allele:	*A*	produces an enzyme which forms **A antigen**
Dominant allele:	*B*	produces an enzyme which forms **B antigen**

Blood group (phenotype)	Possible genotypes	Frequency*		
		White	Black	Native American
O	*OO*	45%	49%	79%
A	*AA AO*	40%	27%	16%
B		11%	20%	4%
AB		4%	4%	1%

* Frequency is based on North American population
Source: www.kcom.edu/faculty/chamberlain/Website/MSTUART/Lect13.htm

If a person has the *AO* allele combination then their blood group will be group **A**. The presence of the recessive allele has no effect on the blood group in the presence of a dominant allele. Another possible allele combination that can create the same blood group is *AA*.

1. Use the information above to complete the table for the possible genotypes for blood group B and group AB.

2. Below are four crosses possible between couples of various blood group types. The first example has been completed for you. Complete the genotype and phenotype for the other five crosses below:

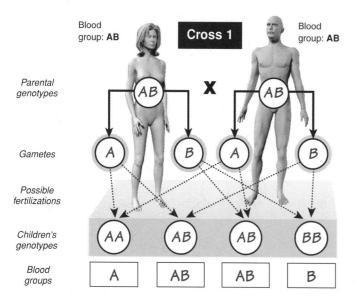

Cross 1 — Blood group: **AB** × Blood group: **AB**

Parental genotypes: AB, AB
Gametes: A, B, A, B
Children's genotypes: AA, AB, AB, BB
Blood groups: A, AB, AB, B

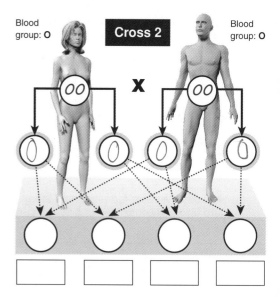

Cross 2 — Blood group: **O** × Blood group: **O**

Parental genotypes: OO, OO
Gametes: O, O, O, O

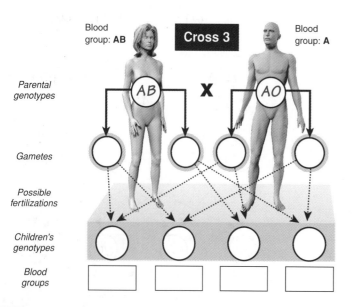

Cross 3 — Blood group: **AB** × Blood group: **A**

Parental genotypes: AB, AO

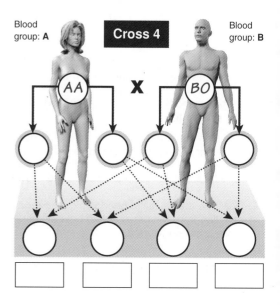

Cross 4 — Blood group: **A** × Blood group: **B**

Parental genotypes: AA, BO

Related activities: Codominance

A 1

Incomplete Dominance

Incomplete dominance refers to the situation where the action of one allele does not completely mask the action of the other and neither allele has dominant control over the trait. The heterozygous offspring are intermediate in phenotype between the contrasting homozygous parental phenotypes. In crosses involving incomplete dominance the phenotype and genotype ratios are identical. Examples of incomplete dominance includes flower color in snapdragons (*Antirrhinum*) and four o'clocks (*Mirabilis*) (below). In this type of inheritance the phenotype of the offspring results from the partial influence of both alleles.

Pure breeding snapdragons produce red or white flowers (left). When red and white-flowered parent plants are crossed a pink-flowered offspring is produced. If the offspring (F_1 generation) are then crossed together all three phenotypes (red, pink, and white) are produced in the F_2 generation.

Four o'clocks (above) are also known to have flower colors controlled by incompletely dominate alleles. Pure breeding four o'clocks produce crimson, yellow or white flowers. Crimson flowers crossed with yellow flowers produced reddish-orange flowers, while crimson flowers crossed with white flowers produce magenta (reddish-pink) flowers.

Red flower $C^R C^R$ **X** White flower $C^W C^W$
Parents

Gametes: C^R C^R C^W C^W

Offspring: $C^R C^W$ $C^R C^W$ $C^R C^W$ $C^R C^W$

Pink   Pink   Pink   Pink

1. Explain how incomplete dominance of alleles differs from complete dominance: *In incomplete dominance alleles are equally influenced and they blend*

2. A plant breeder wanted to produce snapdragons for sale that were only pink or white (i.e. no red). Determine the phenotypes of the two parents necessary to produce these desired offspring. Use the Punnett square (right) to help you:

   1 : 2
   red roun : white
   $C^R C^W \times C^W C^W$

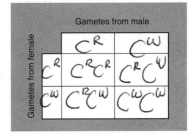

	Gametes from male	
	C^R	C^W
C^R	$C^R C^R$	$C^R C^W$
C^W	$C^R C^W$	$C^W C^W$

Gametes from female

3. Another plant breeder crossed two four o'clocks, known to have its flower color controlled by a gene which possesses incomplete dominant alleles. Pollen from a magenta flowered plant was placed on the stigma of a crimson flowered plant.

   (a) Fill in the spaces on the diagram on the right to show the genotype and phenotype for parents and offspring.

   (b) State the phenotype ratio:

   1 : 1
   red : pink

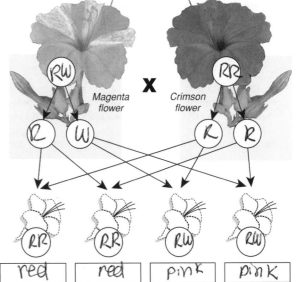

Magenta flower RW   **X** Crimson flower RR

Gametes: R W   R R

Possible fertilizations

Offspring: RR   RR   RW   RW

Phenotypes: red   red   pink   pink

Lethal Alleles

Lethal alleles are mutations of a gene that produce a gene product, which is not only nonfunctional, but may affect the organism's survival. Some lethal alleles are fully dominant and kill in one dose in the heterozygote. Others, such as in the **Manx** cat and yellow mice (below), produce viable offspring with a recognizable phenotype in the heterozygote. In some lethal alleles, the lethality is fully recessive and the alleles produce no detectable effect in the heterozygote at all. Furthermore, lethal alleles may take effect at different stages in development (e.g. in juveniles or, as in **Huntington's disease**, in adults).

When **Lucien Cuenot** investigated inheritance of coat color in yellow mice in 1905, he reported a peculiar pattern. When he mated two yellow mice, about 2/3 of their offspring were yellow, and 1/3 were non-yellow. This was a departure from the expected Mendelian ratio of 3:1. A test cross of the yellow offspring showed that they were all heterozygous.

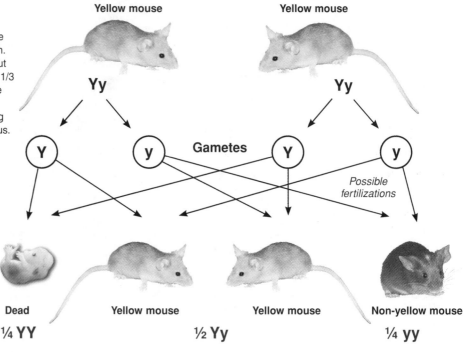

Yellow mouse · Yellow mouse

Yy · Yy

Gametes

Y · y · Y · y

Possible fertilizations

Dead
¼ **YY**

Yellow mouse · Yellow mouse
½ **Yy**

Non-yellow mouse
¼ **yy**

Cats possess a gene for producing a tail. The tailless **Manx** phenotype in cats is produced by the presence of an allele (**M^L**) that is lethal in the homozygous state (**M^L M^L**). The Manx allele (M^L) severely interferes with normal spinal development. In heterozygotes (**M^L M**), this results in the absence of a tail. In M^L M^L homozygotes, the double dose of the gene produces an extremely abnormal embryo, which does not survive.

Manx cats are born without a tail

Normal cats are born with a tail

Normal cat · **Manx cat**

1. Distinguish between recessive lethal alleles and dominant lethal alleles: _____

_____

_____

2. In Manx cats, the allele for taillessness (**M^L**) is incompletely dominant over the recessive allele for normal tail (**M**). Tailless Manx cats are heterozygous (**M^L M**) and carry a recessive allele for normal tail. Normal tailed cats are **MM**. A cross between two Manx (tailless) cats, produces two Manx to every one normal tailed cat (not a regular 3 to 1 Mendelian ratio).

   (a) State the genotypes arising from this type of cross: _____

   (b) State the phenotype ratio of Manx to normal cats and explain why it is not the expected 3:1 ratio: _____

   _____

3. Explain why Huntington disease persists in the human population when it is caused by a lethal, dominant allele:

   _____

   _____

Problems Involving Monohybrid Inheritance

The following problems involve Mendelian crosses. The alleles involved are associated with various phenotypic traits controlled by a single gene. The problems are to give you practice in problem solving using Mendelian genetics.

1. A dominant gene (**W**) produces wire-haired texture in dogs; its recessive allele (**w**) produces smooth hair. A group of heterozygous wire-haired individuals are crossed and their F_1 progeny are then test-crossed. Determine the expected genotypic and phenotypic ratios among the **test cross** progeny:

_____

_____

_____

_____

_____

2. In sheep, black wool is due to a recessive allele (**b**) and white wool to its dominant allele (**B**). A white ram is crossed to a white ewe. Both animals carry the black allele (b). They produce a white ram lamb, which is then back crossed to the female parent. Determine the probability of the **back cross** offspring being black:

_____

_____

_____

3. A recessive allele, a, is responsible for albinism, an inability to produce or deposit melanin in tissues. Humans and a variety of other animals can exhibit this phenotype. In each of the following cases, determine the possible genotypes of the mother and father, and of their children:

(a) Both parents have normal phenotypes; some of their children are albino and others are unaffected: _____

_____

(b) Both parents are albino and have only albino children: _____

(c) The woman is unaffected, the man is albino, and they have one albino child and three unaffected children:

_____

_____

4. Chickens with shortened wings and legs are called creepers. When creepers are mated to normal birds, they produce creepers and normals with equal frequency. When creepers are mated to creepers they produce two creepers to one normal. Crosses between normal birds produce only normal progeny. Explain these results:

_____

_____

5. In a dispute over parentage, the mother of a child with blood group O identifies a male with blood group A as the father. The mother is blood group B. Draw Punnett squares to show possible genotype/phenotype outcomes to determine if the male is the father and the reasons (if any) for further dispute:

_____

_____

_____

_____

_____

_____

_____

Related activities: Monhybrid Cross, Lethal Alleles, Codominance in Multiple Allele Systems

The Chromosomal Basis of Inheritance

A 2

Dihybrid Cross

A cross (or mating) between two organisms where the inheritance patterns of **two genes** are studied is called a **dihybrid cross** (compared with the study of one gene in a monohybrid cross). There are a greater number of gamete types produced when two genes are considered (four types). Remember that the genes described are being carried by separate chromosomes and are sorted independently of each other during meiosis (that is why you get four kinds of gamete). The two genes below control two unrelated characteristics, **hair color** and **coat length**. Black and short are dominant.

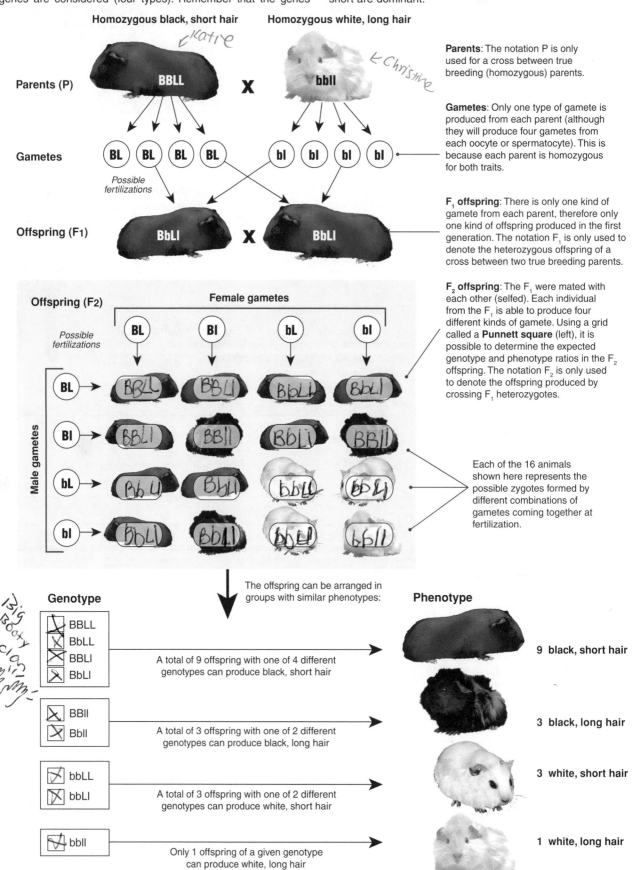

Parents: The notation P is only used for a cross between true breeding (homozygous) parents.

Gametes: Only one type of gamete is produced from each parent (although they will produce four gametes from each oocyte or spermatocyte). This is because each parent is homozygous for both traits.

F$_1$ offspring: There is only one kind of gamete from each parent, therefore only one kind of offspring produced in the first generation. The notation F$_1$ is only used to denote the heterozygous offspring of a cross between two true breeding parents.

F$_2$ offspring: The F$_1$ were mated with each other (selfed). Each individual from the F$_1$ is able to produce four different kinds of gamete. Using a grid called a **Punnett square** (left), it is possible to determine the expected genotype and phenotype ratios in the F$_2$ offspring. The notation F$_2$ is only used to denote the offspring produced by crossing F$_1$ heterozygotes.

Each of the 16 animals shown here represents the possible zygotes formed by different combinations of gametes coming together at fertilization.

The offspring can be arranged in groups with similar phenotypes:

Genotype

BBLL
BbLL
BBLl
BbLl

A total of 9 offspring with one of 4 different genotypes can produce black, short hair

BBll
Bbll

A total of 3 offspring with one of 2 different genotypes can produce black, long hair

bbLL
bbLl

A total of 3 offspring with one of 2 different genotypes can produce white, short hair

bbll

Only 1 offspring of a given genotype can produce white, long hair

Phenotype

9 black, short hair

3 black, long hair

3 white, short hair

1 white, long hair

1. Complete the Punnett square above and use it to fill in the number of each genotype in the boxes (above left).

Related activities: Basic Genetic Crosses
Weblinks: Drag and Drop Genetics

© BIOZONE International 2012
ISBN: 978-1-927173-11-4
Photocopying Prohibited

Inheritance of Linked Genes

Linkage refers to genes that are located on the same chromosome. Linked genes tend to be inherited together, and the extent of crossing over depends on how close together they are on the chromosome. Linkage generally reduces the variety of offspring that can be produced. In genetic crosses, linkage is indicated when a greater proportion of the offspring resulting from a cross are of the parental type (than would be expected if the alleles were assorting independently). If the genes in question had been on separate chromosomes, there would have been more genetic variation in the gametes and therefore in the offspring. Note that in the hypothetical example below there are only two possible genotype outcomes, both the same as the parent type. If the alleles were assorting independently (on different chromosomes) there would be four outcomes.

Overview of Linkage

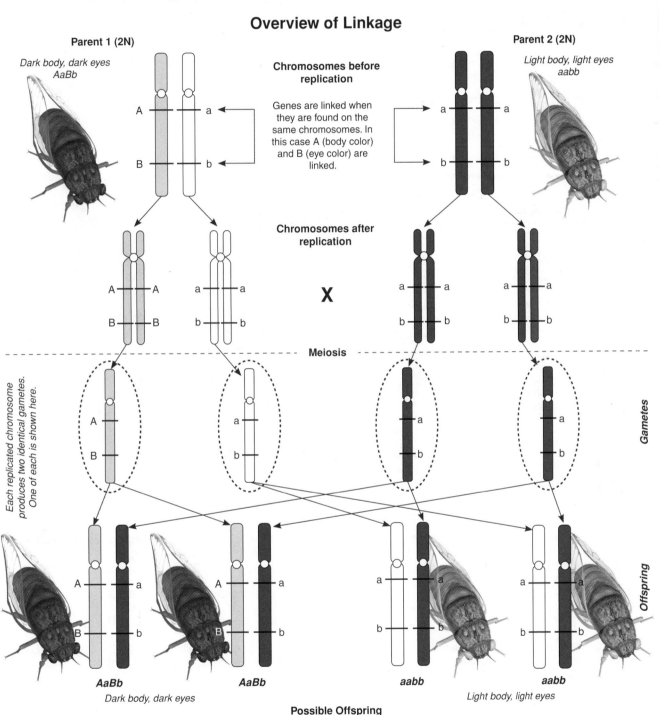

Possible Offspring
Only two kinds of genotype combinations are possible. They are the same as the parent genotype.

1. What is the effect of **linkage** on the inheritance of genes? _Linkage reduces the variety of offspring that can be produced_

2. Explain how linkage decreases the amount of genetic variation in the offspring: _tend to be inherited together, and the extent of crossing over depends on how close together they are in the chromosome._

Related activities: Recombination and Dihybrid Inheritance

A 2

An Example of Linked Genes in *Drosophila*

The genes for wing shape and body color are linked (they are on the same chromosome).

Parent	**Wild type female**	**Mutant male**
Phenotype	Straight wing Gray body	Curled wing Ebony body
Genotype	Cucu Ebeb	cucu ebeb
Linkage	*Cu* *Eb* / *cu* *eb*	*cu* *eb* / *cu* *eb*

-------- **Meiosis** --------

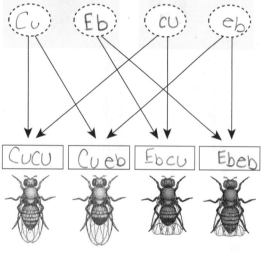

Gametes from female fly (N)   *Gametes from male fly (N)*

(Cu)   (Eb)   (cu)   (eb)

CuCu	Cu eb	Ebcu	Ebeb

Sex of offspring is irrelevant in this case

Contact **Newbyte Educational Software** for details of their superb *Drosophila Genetics* software package which includes coverage of linkage and recombination. *Drosophila* images © Newbyte Educational Software.

Drosophila and linked genes

In the example shown left, wild type alleles are dominant and are given an upper case symbol of the mutant phenotype (Cu or Eb). This symbology used for *Drosophila* departs from the convention of using the dominant gene to provide the symbol. This is necessary because there are many mutant alternative phenotypes to the wild type (e.g. curled and vestigial wings). A lower case symbol of the wild type (e.g. ss for straight wing) would not indicate the mutant phenotype involved.

Drosophila melanogaster is known as a model organism. Model organisms are used to study particular biological phenomena, such as mutation. *Drosophila melanogaster* is particularly useful because it produces such a wide range of heritable mutations. Its short reproduction cycle, high offspring production, and low maintenance make it ideal for studying in the lab.

Drosophila melanogaster examples showing variations in eye and body color. The wild type is marked with a w in the photo above.

3. Complete the linkage diagram above by adding the gametes in the ovals and offspring genotypes in the rectangles.

4. (a) List the possible genotypes in the offspring (above, left) if genes Cu and Eb had been on **separate chromosomes**:

_____

   (b) If the female *Drosophila* had been homozygous for the dominant wild type alleles (CuCu EbEb), state:

   The genotype(s) of the F_1: ___CuCu EbEb___   The phenotype(s) of the F_1: ___Female___

5. A second pair of *Drosophila* are mated. The female genotype is Vgvg EbEb (straight wings, gray body), while the male genotype is vgvg ebeb (straight wings, ebony body). Assuming the genes are linked, carry out the cross and list the genotypes and phenotypes of the offspring. Note vg = vestigial (no) wings:

   _____ Vg vg EbEb _____ Vgvg Ebeb _____

   _____ Vg vg EbEb _____ Vg vg Eb Eb _____

   _____ VgVg ebeb _____ Vg vg ebeb _____

   The genotype(s) of the F_1: ___Vgvg Ebeb___   The phenotype(s) of the F_1: ___straight wings, grey body___

6. Explain why *Drosophila* are often used as model organisms in the study of genetics: _____

   ___Produce a wide range of heritable mutations, high offspring production, short reproductive cycle___

Recombination and Dihybrid Inheritance

Genetic recombination refers to the exchange of alleles between homologous chromosomes as a result of **crossing over**. The alleles of parental linkage groups separate and new associations of alleles are formed in the gametes. Offspring formed from these gametes show new combinations of characteristics and are known as **recombinants** (offspring with genotypes unlike either parent). The proportion of recombinants in the offspring can be used to calculate the frequency of recombination (crossover value).

These values are fairly constant for any given pair of alleles and can be used to produce gene maps indicating the relative positions of genes on a chromosome. In contrast to linkage, recombination increases genetic variation. Recombination between the alleles of parental linkage groups is indicated by the appearance of non-parental types in the offspring, although not in the numbers that would be expected had the alleles been on separate chromosomes (independent assortment).

Overview of Recombination

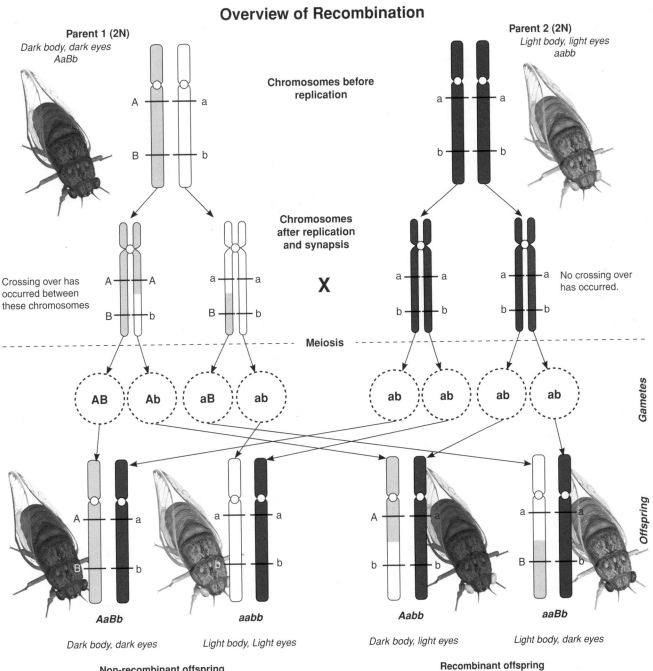

Parent 1 (2N)
Dark body, dark eyes
AaBb

Chromosomes before replication

Parent 2 (2N)
Light body, light eyes
aabb

Chromosomes after replication and synapsis

Crossing over has occurred between these chromosomes

No crossing over has occurred.

X

Meiosis

Gametes: **AB** **Ab** **aB** **ab**   **ab** **ab** **ab** **ab**

Gametes

Offspring

AaBb — *Dark body, dark eyes*
aabb — *Light body, Light eyes*
Aabb — *Dark body, light eyes*
aaBb — *Light body, dark eyes*

Non-recombinant offspring
These two offspring exhibit allele combinations that are expected as a result of independent assortment during meiosis. Also called parental types.

Recombinant offspring
These two offspring exhibit unexpected allele combinations. They can only arise if one of the parent's chromosomes has undergone crossing over.

Possible Offspring
Offspring with four kinds of genotype combinations are produced instead of the two kinds expected (AaBb and aabb) if no crossing over had occurred.

1. Describe the effect of **recombination** on the inheritance of genes: _Offsprings formed from these gametes show new cobinations of characteristics. Alleles of parental linkage groups seperate and new associations of alleles are formed in the gametes._

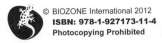

Related activities: Inheritance of Linked Genes

A 3

An Example of Recombination

In the female parent, crossing over occurs between the linked genes for wing shape and body color

	Wild type female	Mutant male
Parent		
Phenotype	Straight wing Gray body	Curled wing Ebony body
Genotype	Cucu Ebeb	cucu ebeb
Linkage		

Cu   *Eb*   *cu*   *eb*

cu   *eb*   *cu*   *eb*

---- *Meiosis* ----

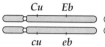

Gametes from female fly (N)   **Gametes from male fly (N)**

Crossing over has occurred, giving four types of gametes

Only one type of gamete is produced in this case

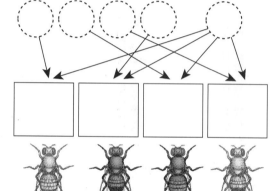

Imani

Non-recombinant offspring   **Recombinant offspring**

The sex of the offspring is irrelevant in this case

Contact **Newbyte Educational Software** for details of their superb *Drosophila Genetics* software package which includes coverage of linkage and recombination. *Drosophila* images © Newbyte Educational Software.

The cross (left) uses the same genotypes as the previous activity but, in this case, crossing over occurs between the alleles in a linkage group in one parent. The symbology used is the same.

Recombination Produces Variation

If crossing over does not occur, the possible combinations in the gametes remains limited. Crossing over and recombination increase the variation in the offspring. In humans, even without crossing over, there are approximately $(2^{23})^2$ or 70 trillion genetically different zygotes that could form for every couple. Taking crossing over and recombination into account produces $(4^{23})^2$ or 5000 trillion trillion genetically different zygotes for every couple.

Family members may resemble each other, but they'll never be identical (except for identical twins).

Using Recombination

Analysing recombination gave geneticists a way to map the genes on a chromosome. Crossing over is less likely to occur in genes that are close together on a chromosome than in genes that are far apart. By counting the number of offspring of each phenotype, you can calculate the **frequency of recombination**. The higher the frequency of recombination between two genes, the further apart they must be on the chromosome.

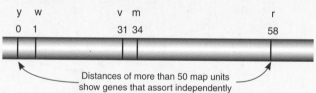

y   w   v   m   r

0   1   31   34   58

Distances of more than 50 map units show genes that assort independently

Map of the X chromosome of *Drosophila*, showing the relative distances between five different genes (in map units).

---

2. Complete the recombination diagram above, adding the gametes in the ovals and offspring genotypes and phenotypes in the rectangles:

3. Explain how recombination increases the amount of genetic variation in offspring: <u>One offspring would recieve a possible recombinant chromosome from each parent, while another offspring would recieve a different one.</u>

4. Explain why it is not possible to have a recombination frequency of greater than 50% (half recombinant progeny): <u>The higher the frequency of recombination between two genes, the further apart they must be on the chromosome</u>

5. A second pair of *Drosophila* are mated. The female is Cucu YY (straight wing, gray body), while the male is Cucu yy (straight wing, yellow body). Assuming recombination, perform the cross and list the offspring genotypes and phenotypes:

<u>Cucu YY   Cucu Yy</u>
<u>CuCu yy   Cucu YY</u>
<u>Cucu Yy   cucu yy</u>

Problems Involving Dihybrid Inheritance

Dihybrid inheritance can involve genes in which there is no interaction between them (such as genes for the wrinkliness and color of pea seeds). Other dihybrid crosses can involve genes that do interact with each other and the combination of dominant and recessive alleles can have an outcome on a single phenotype.

1. In cats, the following alleles are present for coat characteristics: black (B), brown (b), short (L), long (l), tabby (T), blotched tabby (tb). Use the information to complete the dihybrid crosses below:

Katie

(a) A black short haired (BBLl) male is crossed with a black long haired (Bbll) female. Determine the genotypic and phenotypic ratios of the offspring:

Genotype ratio: _____

_____

Phenotype ratio: _____

_____

_____

Christine

(b) A tabby, short haired male (TtbLl) is crossed with a blotched tabby, short haired (tbtbLl) female. Determine ratios of the offspring:

Genotype ratio: _____

_____

Phenotype ratio: _____

_____

_____

2. A plant with orange-striped flowers was cultivated from seeds. The plant was self-pollinated and the F_1 progeny appeared in the following ratios: 89 orange with stripes, 29 yellow with stripes, 32 orange without stripes, 9 yellow without stripes.

(a) Describe the dominance relationships of the alleles responsible for the phenotypes observed: _____

_____

_____

(b) Determine the genotype of the original plant with orange striped flowers: _____

_____

_____

3. In rabbits, spotted coat **S** is dominant to solid color **s**, while for coat color, black **B** is dominant to brown **b**. A brown spotted rabbit is mated with a solid black one and all the offspring are black spotted (the genes are not linked).

(a) State the genotypes:

Parent 1: _____

Parent 2: _____

Offspring: _____

(b) Use the Punnett square to show the outcome of a cross between the F_1 (the F_2):

(c) Using ratios, state the phenotypes of the F_2 generation: _____

_____

_____

_____

The Chromosomal Basis of Inheritance

© BIOZONE International 2012
ISBN: 978-1-927173-11-4
Photocopying Prohibited

Related activities: Dihybrid Cross, Recombination and Dihybrid Inheritance

A 2

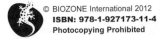

Sex Determination

The determination of the sex of an organism is controlled in most cases by the sex chromosomes provided by each parent. These have evolved to regulate the ratios of males and females produced and preserve the genetic differences between the sexes. In humans, males are the **heterogametic sex** because each somatic cell has one X and one Y chromosome. The determination of sex is based on the presence or absence of the Y chromosome; without it, an individual will develop into a female. In mammals, the male is always the heterogametic sex, but this is not necessarily the case in other taxa. In birds and butterflies, the female is the heterogametic sex, and in some insects the male is simply X whereas the female is XX.

Sex Determination in Humans

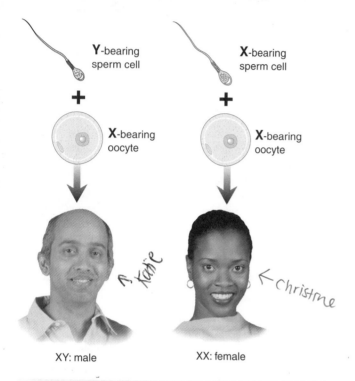

Y-bearing sperm cell

X-bearing sperm cell

+

+

X-bearing oocyte

X-bearing oocyte

↑ Kane

← Christine

XY: male

XX: female

XY Sex Determination

Female: **XX**   Male: **XY**

Examples: Humans (and all mammals), fruit flies (*Drosophila*), some dioecious plants (having separate male and female plants) such as kiwifruit.

In humans the female is the **homogametic sex** and has two similar sex chromosomes (XX), whereas the male is the heterogametic sex with two unlike chromosomes (XY). The primary sexual characteristics are initiated by special genes on the X chromosomes. Females must have a double dose (2X chromosomes). Maleness is determined by the presence of the Y chromosome.

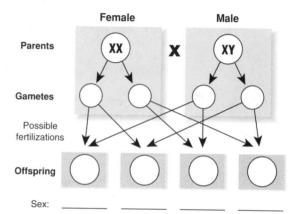

Female   Male

Parents   XX   X   XY

Gametes

Possible fertilizations

Offspring

Sex: _____ _____ _____ _____

Sex Determination in *Drosophila*

In *Drosophila*, maleness is determined by the ratio of X chromosomes to autosomes, not the presence of the Y chromosome. Two X chromosomes in a diploid cell produce a female fly, but one X chromosome in a diploid cell produces a male fly. The Y chromosome is not involved in determining sex, but does contains genes involved in the production of sperm in adult males. Thus XO in *Drosophila* is a sterile male, while in mammals it is a sterile female.

Temperature Dependent Sex Determination

In some vertebrate species, mostly reptiles, sex is determined by the temperature at which the eggs are incubated. In turtles, males are produced at lower incubation temperatures than females (22°C-27°C as opposed to 30°C). The hormone testosterone may be converted to estradiol at higher temperatures to produce females, but how temperature triggers gene expression and the pathway for genetic sex determination is poorly understood.

1. (a) Complete the diagram above, to show the resulting gametes, genotype and sex of the offspring:

   (b) Determine the probability of a conception producing a male child: _____

   (c) Determine the probability of a second conception producing a female child: _____

2. Explain what determines the sex of the offspring at the moment of conception in humans: _____
   _____

3. Explain why many genes on the X chromosome in males will be expressed regardless of their dominance status:
   _____
   _____

4. Estimate the ratio of males to females produced when a clutch of turtle eggs is incubated at 28.5°C: _____

Periodicals:
The Y chromosome:
it's a man thing

© BIOZONE International 2012
ISBN: 978-1-927173-11-4
Photocopying Prohibited

Sex Linkage

Sex linkage occurs when a gene is located on a sex chromosome (usually the X). The result of this is that the character encoded by the gene is usually seen only in one sex (the heterogametic sex) and occurs rarely in the homogametic sex. In humans, recessive sex linked genes are responsible for a number of heritable disorders in males, e.g. hemophilia. Women who have the recessive alleles on their chromosomes are said to be **carriers**. One of the gene loci controlling coat color in cats is sex-linked. The two alleles, red and non-red (or black), are found only on the X-chromosome.

Allele types

X_o = Non-red (=black)
X_O = Red

Genotypes

X_oX_o, X_oY = Black coated female, male
X_OX_O, X_OY = Orange coated female, male
X_OX_o = Tortoiseshell (intermingled black and orange in fur) in female cats only

Phenotypes

1. An owner of a cat is thinking of mating her black female cat with an orange male cat. Before she does this, she would like to know what possible coat colors could result from such a cross. Use the symbols above to fill in the diagram on the right. Summarize the possible genotypes and phenotypes of the kittens in the tables below.

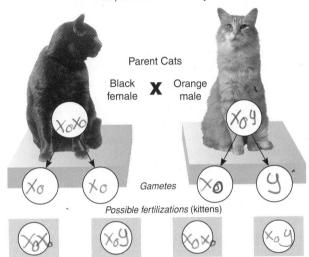

Parent Cats

Black female **X** Orange male

X_oX_o X_oY

Gametes X_o X_o X_O y

Possible fertilizations (kittens)

X_oX_o X_oy X_oX_o X_oy

	Genotypes	Phenotypes
Male kittens	X_oY	black coated male
	X_OY	orange coated male

Female kittens	X_oX_o	black coated female
	X_OX_O	orange coated female

2. A female tortoiseshell cat mated with an unknown male cat in the neighborhood and has given birth to a litter of six kittens. The owner of this female cat wants to know what the appearance and the genotype of the father was of these kittens. Use the symbols above to fill in the diagram on the right. Also show the possible fertilizations by placing appropriate arrows.

Describe the father cat's:

(a) Genotype: _____ $X_O y$ _____

(b) Phenotype: _____ Orange male _____

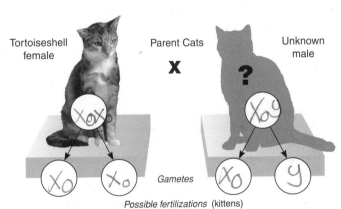

Tortoiseshell female **X** Parent Cats Unknown male

X_OX_o X_Oy

X_O X_o Gametes X_O y

Possible fertilizations (kittens)

X_OX_o X_OX_o X_oy X_Oy

2 orange females   1 tortoiseshell female   1 black male   2 orange males

3. The owner of another cat, a black female, also wants to know which cat fathered her two tortoiseshell female and two black male kittens. Use the symbols above to fill in the diagram on the right. Show the possible fertilizations by placing appropriate arrows.

Describe the father cat's:

(a) Genotype: _____ $X_O y$ _____

(b) Phenotype: _____ Orange male _____

(c) Was it the same male cat that fathered both this litter and the one above?

YES / NO   (*delete one*)

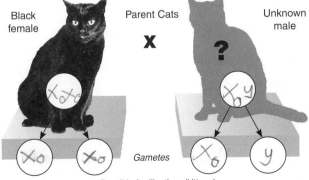

Black female **X** Parent Cats Unknown male

X_oX_o X_Oy

X_o X_o Gametes X_O y

Possible fertilizations (kittens)

X_oX_o X_oX_o X_oy X_oy

1 tortoiseshell female   1 tortoiseshell female   1 black male   1 black male

© BIOZONE International 2012
ISBN: 978-1-927173-11-4
Photocopying Prohibited

Related activities: *Inheritance Patterns, Pedigree Analysis*
Weblinks: *X Linked Inheritance*

Dominant allele in humans

A rare form of rickets in humans is determined by a **dominant** allele of a gene on the **X chromosome** (it is not found on the Y chromosome). This condition is not successfully treated with vitamin D therapy. The allele types, genotypes, and phenotypes are as follows:

Allele types		Genotypes	Phenotypes
X_R	= affected by rickets	X_RX_R, X_RX =	Affected female
X	= normal	X_RY =	Affected male
		XX, XY =	Normal female, male

As a genetic counselor you are presented with a married couple where one of them has a family history of this disease. The husband is affected by this disease and the wife is normal. The couple, who are thinking of starting a family, would like to know what their chances are of having a child born with this condition. They would also like to know what the probabilities are of having an affected boy or affected girl. Use the symbols above to complete the diagram right and determine the probabilities stated below (expressed as a proportion or percentage).

4. Determine the probability of having:

(a) Affected children: ___$\frac{1}{2}$___

(b) An affected girl: ___$\frac{1}{2}$___

(c) An affected boy: ___0___

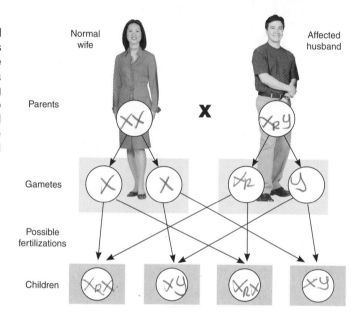

Another couple with a family history of the same disease also come in to see you to obtain genetic counseling. In this case the husband is normal and the wife is affected. The wife's father was not affected by this disease. Determine what their chances are of having a child born with this condition. They would also like to know what the probabilities are of having an affected boy or affected girl. Use the symbols above to complete the diagram right and determine the probabilities stated below (expressed as a proportion or percentage).

5. Determine the probability of having:

(a) Affected children: ___1___

(b) An affected girl: ___$\frac{1}{2}$___

(c) An affected boy: ___$\frac{1}{2}$___

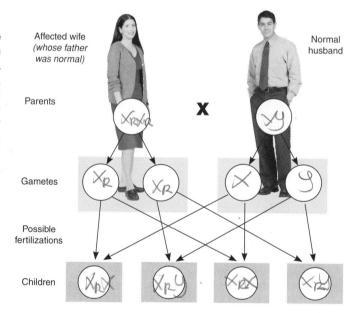

6. Describing examples other than those above, discuss the role of **sex linkage** in the inheritance of genetic disorders:

Some disorders are inherited other can just occur on their own.

Sex Limited Characteristics

The majority of our phenotypic traits are influenced by genes on **autosomes**. Some of these genes, however, are only expressed in one gender and are therefore **sex limited**. This can be due to responses to the different hormone levels in males and females. Sex influenced traits are traits that may be displayed in both genders but are more developed or pronounced in one.

Sex Limited Traits

Sex limited traits are commonly associated with secondary sexual characteristics. Lactation in mammals is an example. The secondary development of the breast tissue in females during puberty and preparation for milk production during pregnancy are both the result of the hormones estrogen and progesterone, produced by the ovaries.

Traits such as this become important in the breeding of dairy herds. Bulls may carry genes for high milk production and quality and crossing them with cows that carry similar genes is important in increasing milk production and quality in the offspring. DNA tests are able to identify these genes at birth, allowing farmers to save on the huge cost of raising bulls to breeding age to see if they sire good milk-producing offspring.

1. Describe the difference between a sex limited trait and a sex influenced trait:

_____

_____

_____

_____

_____

_____

_____

2. Why is the bull's genotype important when breeding high quality milk producing cows?

_____

_____

_____

_____

_____

3. (a) On the first Punnett square above right, use a blue pen to circle the genotype of the F_1 generation that would produce horned males and red pen to circle the genotype of horned females:

   (b) Repeat this for the F_2 generation on the second Punnett square:

Sex Influenced Traits

Two well known examples of sex influenced traits are male pattern baldness in humans and the appearance of horns in Dorset sheep.

Horned Sheep

The appearance of horns on Dorset sheep was first studied by Arkell and Davenport in 1912. Both sexes in Dorset sheep are horned. In the Suffolk breed, both sexes are hornless. In a cross of these two breeds, F_1 males have horns and F_1 females are hornless. If F_1 individuals are crossed, F_2 male offspring occur in a 3:1 ratio of horned to hornless rams, while the female ratio is the opposite.

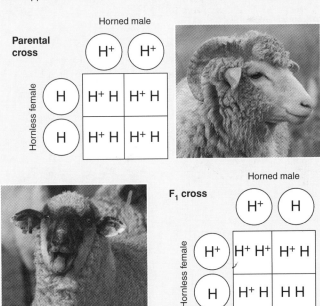

Parental cross

Horned male

	H^+	H^+
Hornless female H	H^+H	H^+H
H	H^+H	H^+H

F_1 cross

Horned male

	H^+	H
Hornless female H^+	H^+H^+	H^+H
H	H^+H	HH

Male Pattern Baldness

Male pattern baldness (**androgenic alopecia**) is often cited as an example of a sex influenced trait which acts as a dominant trait in males and as a recessive trait in females. Certainly, it is more pronounced in males, being less severe and occurring later in life in females. However, it appears to be controlled by many genes, rather than one, and it is still not clear how these genes interact.

Research in 2005 confirmed that the recessive allele of the androgen receptor gene (found on the X chromosome) played a role in pattern baldness. In this case, males require just one recessive allele while females require two. However, many other genes and their products are involved. Levels of the enzyme 5-alpha reductase are higher in males with pattern baldness. Drugs that interfere with this enzyme greatly reduce the progress of pattern baldness. Other genes on the Y chromosome and on chromosome 3 also appear to play a role. Sons whose fathers have hair loss are 2.5 times more likely to experience it than sons with fathers with no hair loss.

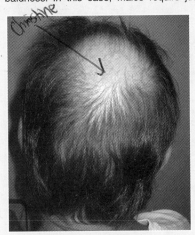

Non-nuclear Inheritance

Not all genes are encoded within nuclear DNA. Both mitochondria and chloroplasts contain their own plasmid-like copies of DNA, (denoted **mtDNA** and **cpDNA** respectively). Mitochondrial and chloroplast DNA replicates independently of the main genome and contains genes essential for that organelle's function, e.g. mitochondrial DNA contains genes for producing some of the proteins associated with the electron transport chain. Because mitochondria and chloroplasts are distributed randomly in gametes, traits determined by their genes do not follow simple Mendelian rules.

Mitochondrial and Chloroplast DNA

In most multicellular organisms, the mtDNA and cpDNA are inherited from the maternal line. Mitochondria in the cytoplasm are randomly divided between dividing primary oocytes so that the final egg cell has a random selection of mitochondria from the original cell. Mitochondria are numerous in sperm cells, providing the energy to power the tail. However, they do not enter the egg cell, or are destroyed by enzymes within it.

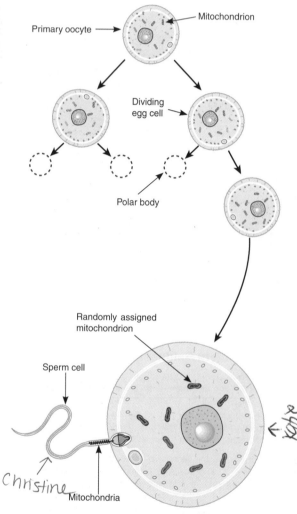

Using mtDNA and cpDNA

Because mitochondria and chloroplasts are passed through the maternal line they can be used to trace maternal lineage. Mutations occur in mtDNA at a rate of fewer than one per 100 people. This means that the mtDNA of any one person is probably the same as their direct maternal ancestor back many generations. The same applies for plant cpDNA and mtDNA. In humans, this concept has been used to trace our mitochondrial ancestor, a single female from Africa. This does not mean this person is the ancestor of all humans, rather that no other lineages produced daughters through which the mtDNA was passed to our generation.

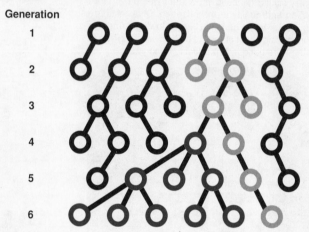

Experimental Evidence for Maternal Inheritance of cpDNA

In 1909, Carl Correns, a German geneticist, noticed that the branches of the four o'clock (*Mirabilis jalapa*) developed leaves that could be green, white, or variegated (a combination of green and white). After carrying out a series of crosses, Correns noticed that seeds from branches with solid green leaves grew into plants with solid green leaves, even if the pollen was from a branch with white or variegated leaves. Seeds from branches with variegated leaves produced plants with all three leaf types. Again, where the pollen came from did not matter. In addition, the appearance of these three types was never in any predictable Mendelian ratio. Correns concluded that the trait was being passed via the maternal line. We now know a defect in the cpDNA of the four o'clock causes the white color and these, like mitochondria, are passed randomly to the offspring.

1. (a) Why are mtDNA and cpDNA only passed along the maternal line? _____

_____

_____

(b) Why do the inheritance patterns of mtDNA and cpDNA not follow predictable Mendelian rules? _____

_____

2. (a) In which generation of the mtDNA diagram (blue box above) did a mutation in the mtDNA occur? _____

(b) How can this be used to help trace the lineage? _____

_____

_____

Genomic Imprinting

The phenotypic effects of some mammalian genes depend on whether they were inherited from the mother or the father. This phenomenon, called **genomic imprinting** (or parental imprinting), is part of **epigenetics**, the study of the heritable changes in gene function that occur without involving changes in the DNA sequence. Just as cells inherit genes, they also inherit the instructions that communicate to the genes when to become active, in which tissue, and to what extent. Epigenetic phenomena are important because they regulate when and at what level genes are expressed.

Genomic Imprinting

Genomic imprinting describes how a small subset of the genes in the genome are expressed according to their parent of origin. 'Imprints' can act as silencers or activators for imprinted genes. A mammal inherits two sets of chromosomes, one from the mother and one from the father. In this way the imprinted gene expression is balanced, a prerequisite for a viable offspring in mammals.

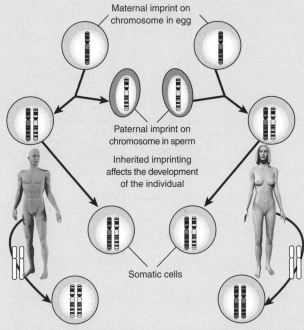

Maternal imprint on chromosome in egg

Paternal imprint on chromosome in sperm

Inherited imprinting affects the development of the individual

Somatic cells

Maternal and paternal chromosomes are differently imprinted. Chromosomes are newly imprinted (reprogrammed) each generation.

Imprinted Genes Are Different

Some imprinted genes are expressed from a maternally inherited chromosome and silenced on the paternal chromosome, while other imprinted genes show the opposite expression pattern and are only expressed from a paternally inherited chromosome. Evidence of this is seen in two human genetic disorders. Both are caused by the same mutation: a specific deletion on chromosome 15. The disorder expressed depends on whether the mutation is inherited from the father or the mother.

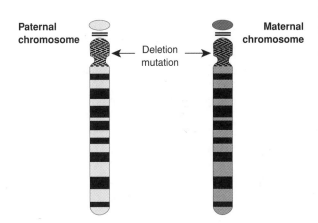

Paternal chromosome — Deletion mutation → Maternal chromosome

Inherited from the father: **Prader-Willi syndrome**

Phenotype: Mental retardation, obesity, short stature, unusually small hands and feet

Inherited from the mother **Angelman syndrome**

Phenotype Uncontrollable laughter, jerky movements, motor and mental abnormalities

How Are Genes Silenced?

■ In many instances, **gene silencing** is achieved through **methylation** of the DNA of genes or regulatory sequences, which results in the gene not being expressed.

■ **Methylation** turns off gene expression by adding a methyl group to cytosines in the DNA. This changes the state of the chromatin so that the expression of any genes in the methylated region is inhibited. Methylation is also important in X-inactivation.

■ In other instances, phosphorylation or other chemical modification of histone proteins appears to lead to silencing.

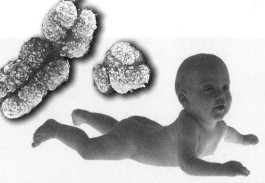

Which genes did you inherit from your mother and which from your father? For some genes, imprinting will affect phenotypic expression.

1. (a) What is meant by genomic imprinting? _____

(b) Describe one of the mechanisms by which imprinting is achieved: _____

2. Explain the significance of imprinting to the inheritance of genes: _____

Inheritance Patterns

Complete the following monohybrid crosses for different types of inheritance patterns in humans: autosomal recessive, autosomal dominant, sex linked recessive, and sex linked dominant inheritance.

1. Inheritance of autosomal recessive traits

Example: *Albinism*

Albinism (lack of pigment in hair, eyes and skin) is inherited as an autosomal recessive allele (not sex-linked).

Using the codes: **PP** (normal)
Pp (carrier)
pp (albino)

(a) Enter the parent phenotypes and complete the Punnett square for a cross between two carrier genotypes.

(b) Give the ratios for the phenotypes from this cross.

Phenotype ratios: $\frac{1}{4} : \frac{1}{2} : \frac{1}{4}$

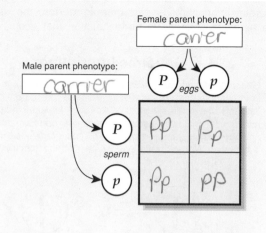

Female parent phenotype: carrier

Male parent phenotype: carrier

P eggs p

P | PP | Pp
p | Pp | pp

sperm

2. Inheritance of autosomal dominant traits

Example: *Woolly hair*

Woolly hair is inherited as an autosomal dominant allele. Each affected individual will have at least one affected parent.

Using the codes: **WW** (woolly hair)
Ww (woolly hair, heterozygous)
ww (normal hair)

(a) Enter the parent phenotypes and complete the Punnett square for a cross between two heterozygous individuals.

(b) Give the ratios for the phenotypes from this cross.

Phenotype ratios: $\frac{1}{4} : \frac{1}{2} : \frac{1}{4}$

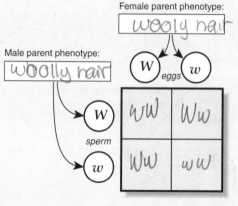

Female parent phenotype: wooly hair

Male parent phenotype: woolly hair

W eggs w

W | WW | Ww
w | Ww | ww

sperm

3. Inheritance of sex linked recessive traits

Example: *Hemophilia*

Inheritance of hemophilia is sex linked. Males with the recessive (hemophilia) allele, are affected. Females can be carriers.

Using the codes: **XX** (normal female)
XX$_h$ (carrier female)
X$_h$X$_h$ (hemophiliac female)
XY (normal male)
X$_h$Y (hemophiliac male)

(a) Enter the parent phenotypes and complete the Punnett square for a cross between a normal male and a carrier female.

(b) Give the ratios for the phenotypes from this cross.

Phenotype ratios: $\frac{1}{4} : \frac{1}{4} : \frac{1}{4} : \frac{1}{4}$

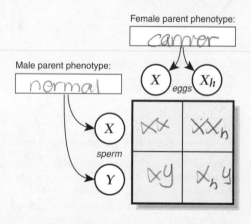

Female parent phenotype: carrier

Male parent phenotype: normal

X eggs X$_h$

X | XX | XX$_h$
Y | XY | X$_h$Y

sperm

4. Inheritance of sex linked dominant traits

Example: *Sex linked form of rickets*

A rare form of rickets is inherited on the X chromosome.

Using the codes: **XX** (normal female); **XY** (normal male)
X$_R$X (affected heterozygote female)
X$_R$X$_R$ (affected female)
X$_R$Y (affected male)

(a) Enter the parent phenotypes and complete the Punnett square for a cross between an affected male and heterozygous female.

(b) Give the ratios for the phenotypes from this cross.

Phenotype ratios: $\frac{1}{4} : \frac{1}{4} : \frac{1}{4} : \frac{1}{4}$

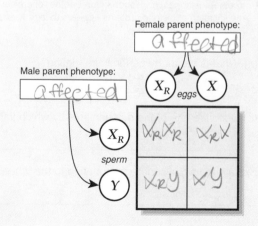

Female parent phenotype: affected

Male parent phenotype: affected

X$_R$ eggs X

X$_R$ | X$_R$X$_R$ | X$_R$X
Y | X$_R$Y | XY

sperm

Pedigree Analysis

Sample Pedigree Chart

Pedigree charts are a way of graphically illustrating inheritance patterns over a number of generations. They are used to study the inheritance of genetic disorders. The key (below the chart) should be consulted to make sense of the various symbols. Particular individuals are identified by their generation number and their order number in that generation. For example, **II-6** is the sixth person in the second row. The arrow indicates the **propositus**; the person through whom the pedigree was discovered (i.e. who reported the condition).

If the chart on the right were illustrating a human family tree, it would represent three generations: grandparents (I-1 and I-2) with three sons and one daughter. Two of the sons (II-3 and II-4) are identical twins, but did not marry or have any children. The other son (II-1) married and had a daughter and another child (sex unknown). The daughter (II-5) married and had two sons and two daughters (plus a child that died in infancy).

For the particular trait being studied, the grandfather was expressing the phenotype (showing the trait) and the grandmother was a carrier. One of their sons and one of their daughters also show the trait, together with one of their granddaughters.

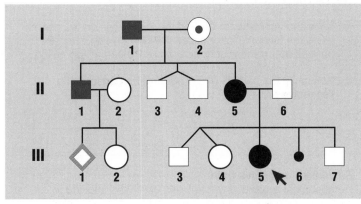

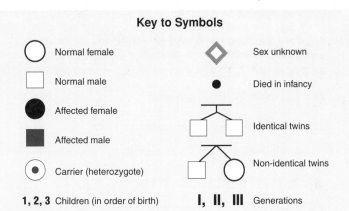

Key to Symbols

○	Normal female	◇	Sex unknown
□	Normal male	●	Died in infancy
●	Affected female		
■	Affected male	⊓	Identical twins
⊙	Carrier (heterozygote)	⋀	Non-identical twins
1, 2, 3 Children (in order of birth)		**I, II, III** Generations	

1. **Pedigree chart of your family**
   Using the symbols in the key above and the example illustrated as a guide, construct a pedigree chart of your own family (or one that you know of) starting with the parents of your mother and/or father on the first line. Your parents will appear on the second line (II) and you will appear on the third line (III). There may be a fourth generation line (IV) if one of your brothers or sisters has had a child. Use a ruler to draw up the chart carefully.

2. **Autosomal recessive traits**
   Albinos lack pigment in the hair, skin and eyes. This trait is inherited as an autosomal recessive allele (i.e. it is not carried on the sex chromosome).

   (a) Write the genotype for each of the individuals on the chart using the following letter codes: **PP** normal skin color; **P-** normal, but unknown if homozygous; **Pp** carrier; **pp** albino.

   (b) Why must the parents (II-3) and (II-4) be **carriers** of a **recessive** allele:

   _____

   _____

Albinism in humans

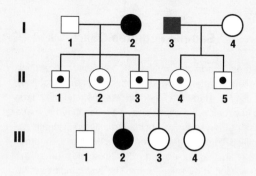

3. **Sex linked recessive traits**
   Hemophilia is a disease where blood clotting is affected. A person can die from a simple bruise (which is internal bleeding). The clotting factor gene is carried on the X chromosome.

   (a) Write the genotype for each of the individuals on the chart using the codes: **XY** normal male; X_hY affected male; **XX** normal female; X_hX female carrier; X_hX_h affected female:

   (b) Why can males never be carriers?

   _____

   _____

   _____

   _____

Hemophilia in humans

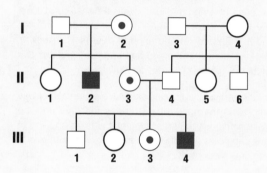

4. **Autosomal dominant traits**
   An unusual trait found in some humans is woolly hair (not to be confused with curly hair). Each affected individual will have at least one affected parent.

   (a) Write the genotype for each of the individuals on the chart using the following letter codes:
   WW woolly hair; **Ww** woolly hair (heterozygous); **W-** woolly hair, but unknown if homozygous; **ww** normal hair

   (b) Describe a feature of this inheritance pattern that suggests the trait is the result of a **dominant** allele:

   _____

   _____

   _____

Woolly hair in humans

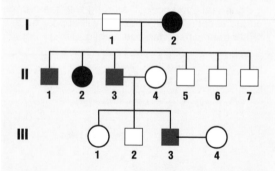

5. **Sex linked dominant traits**
   A rare form of rickets is inherited on the X chromosome. All daughters of affected males will be affected. More females than males will show the trait.

   (a) Write the genotype for each of the individuals on the chart using the following letter codes:
   XY normal male; X_RY affected male; **XX** normal female; X_{R-} female (unknown if homozygous); X_RX_R affected female.

   (b) Why will more females than males be affected?

   _____

   _____

   _____

   _____

A rare form of rickets in humans

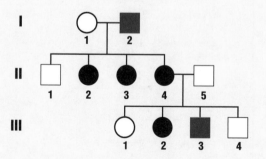

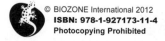

6. The pedigree chart below illustrates the inheritance of a trait (darker symbols) in two families joined in marriage.

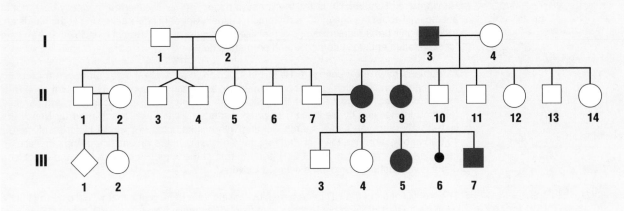

(a) State whether the trait is **dominant** or **recessive**, and explain your reasoning: _____

_____

_____

_____

(b) State whether the trait is **sex linked** or not, and explain your reasoning: _____

_____

_____

_____

7. The recessive sex linked gene (h) prolongs the blood-clotting time, resulting in the genetically inherited disease called hemophilia. From the information in the pedigree chart (right), answer the following questions:

Hemophilia in humans

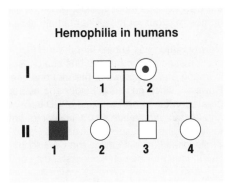

(a) If **II2** marries a normal man, determine the probability of her first child being a hemophiliac:

_____

(b) Suppose her first child is actually a hemophiliac. Determine the chance that her second child will be a hemophiliac boy:

_____

(c) If **II4** has children with a hemophilic man, determine the probability of her first child being phenotypically normal:

_____

(d) If the mother of **I2** was phenotypically normal, state the phenotype of her father: _____

8. The phenotypic expression of a dominant gene in Ayrshire cattle is a notch in the tips of the ears. In the pedigree chart on the right, notched animals are represented by the solid symbols.

Ear notches in Ayrshire cattle

Determine the probability of notched offspring being produced from the following matings:

(a) III1 x III3 _____

(b) III3 x III2 _____

(c) III3 x III4 _____

(d) III1 x III5 _____

(e) III2 x III5 _____

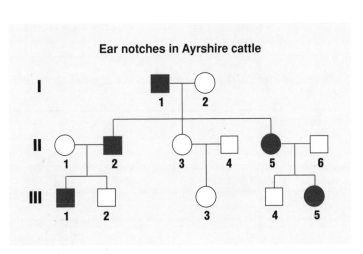

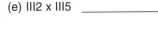

A Gene That Can Tell Your Future

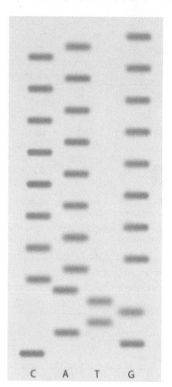

Huntington's disease (HD) is a genetic neuro-degenerative disease that normally does not affect people until about the age of 40. Its symptoms usually appear first as a shaking of the hands and an awkward gait. Later manifestations of the disease include serious loss of muscle control and mental function, often ending in dementia and premature death.

All humans have the huntingtin (**HTT**) gene, which in its normal state produces a protein with roles in gene transcription, synaptic transmission, and brain cell survival. The mutant gene (**mHTT**) causes changes to and death of the cells of the cerebrum, the hippocampus, and cerebellum, resulting in the atrophy (reduction) of brain matter. The gene was discovered by Nancy Wexler in 1983 after ten years of research working with cell samples and family histories of more than 10,000 people from the town of San Luis in Venezuela, where around 1% of the population have the disease (compared to about 0.01% in the rest of the world). Ten years later the exact location of the gene on the chromosome 4 was discovered.

The identification of the HD gene began by looking for a gene probe that would bind to the DNA of people who had HD, and not to those who didn't. Eventually a marker for HD, called **G8**, was found. The next step was to find which chromosome carried the marker and where on the chromosome it was. The researchers hybridized human cells with those of mice so that each cell contained only one human chromosome, a different chromosome in each cell. The hybrid cell with chromosome 4 was the one with the G8 marker. They then found a marker that overlapped G8 and then another marker that overlapped that marker. By repeating this many times, they produced a map of the genes on chromosome 4. The researchers then sequenced the genes and found people who had HD had one gene that was considerably longer than people who did not have HD. Moreover the increase in length was caused by the repetition of the base sequence CAG.

The HD mutation (mHTT) is called a trinucleotide repeat expansion. In the case of mHTT, the base sequence CAG is repeated multiple times on the short arm of chromosome 4. The normal number of CAG repeats is between 6 and 30. The mHTT gene causes the repeat number to be 35 or more and the size of the repeat often increases from generation to generation, with the severity of the disease increasing with the number of repeats. Individuals who have 27 to 35 CAG repeats in the HTT gene do not develop Huntington disease, but they are at risk of having children who will develop the disorder. The mutant allele, mHTT, is also dominant, so those who are homozygous or heterozygous for the allele are both at risk of developing HD.

Christine

New research has shown that the mHTT gene activates an enzyme called JNK3, which is expressed only in the neurons and causes a drop in nerve cell activity. While a person is young and still growing, the neurons can compensate for the accumulation of JNK3. However, when people get older and neuron growth stops, the effects of JNK3 become greater and the physical signs of HD become apparent. Because of mHTT's dominance, an affected person has a 50% chance of having offspring who are also affected. Genetic testing for the disease is relatively easy now that the genetic cause of the disease is known. While locating and counting the CAG repeats does not give a date for the occurrence of HD, it does provide some understanding of the chances of passing on the disease.

1. Describe the physical effects of Huntington's disease: _____

_____

2. Describe how the mHTT gene was discovered: _____

_____

_____

3. Discuss the cause of Huntington's disease and its pattern of increasing severity with each generation: _____

_____

_____

_____

_____

_____

_____

_____

Related activities: Lethal Alleles

Weblinks: Huntington's Disease

© BIOZONE International 2012
ISBN: 978-1-927173-11-4
Photocopying Prohibited

Genetic Counseling

Genetic counseling provides an analysis of the risk of producing offspring with known gene defects within a family. Counselors identify families at risk, investigate the problem present in the family, interpret information about the disorder, analyze inheritance patterns and risks of recurrence, and review available options with the family. Increasingly, there are DNA tests for the identification of specific defective genes. People usually consider genetic counseling if they have a family history of a genetic disorder, or if a routine prenatal screening test yields an unexpected result. While screening for many genetic disorders is now recommended, the use of presymptomatic tests for adult-onset disorders, such as Alzheimer's, is still controversial.

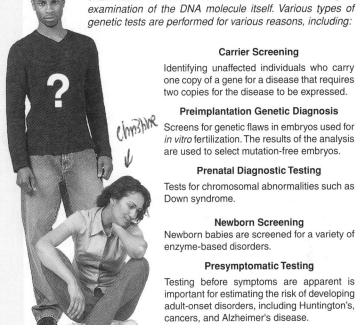

Autosomal Recessive Conditions

Common inherited disorders caused by recessive alleles on autosomes. Recessive conditions are evident only in homozygous recessive genotypes.

Cystic fibrosis: Malfunction of the pancreas and other glands; thick mucus leads to pneumonia and emphysema. Death usually occurs in childhood. CF is the most frequent lethal genetic disorder in childhood (about 1 case in 3700 live births).

Maple syrup urine disease: Mental and physical retardation produced by a block in amino acid metabolism. Isoleucine in the urine produces the characteristic odor.

Tay-Sachs disease: A lipid storage disease which causes progressive developmental paralysis, mental deterioration, and blindness. Death usually occurs by three years of age.

Autosomal Dominant Conditions

Inherited disorders caused by dominant alleles on autosomes. Dominant conditions are evident both in heterozygotes and in homozygous dominant individuals.

Huntington's disease: Involuntary movements of the face and limbs with later general mental deterioration. The beginning of symptoms is highly variable, but occurs usually between 30 to 40 years of age.

Genetic testing may involve biochemical tests for gene products such as enzymes and other proteins, microscopic examination of stained or fluorescent chromosomes, or examination of the DNA molecule itself. Various types of genetic tests are performed for various reasons, including:

Carrier Screening
Identifying unaffected individuals who carry one copy of a gene for a disease that requires two copies for the disease to be expressed.

Preimplantation Genetic Diagnosis
Screens for genetic flaws in embryos used for *in vitro* fertilization. The results of the analysis are used to select mutation-free embryos.

Prenatal Diagnostic Testing
Tests for chromosomal abnormalities such as Down syndrome.

Newborn Screening
Newborn babies are screened for a variety of enzyme-based disorders.

Presymptomatic Testing
Testing before symptoms are apparent is important for estimating the risk of developing adult-onset disorders, including Huntington's, cancers, and Alzheimer's disease.

Auditory test

About half of the cases of childhood deafness are the result of an autosomal recessive disorder. Early identification of the problem prepares families and allows early appropriate treatment.

Genetic counseling provides information to families who have members with birth defects or genetic disorders, and to families who may be at risk for a variety of inherited conditions.

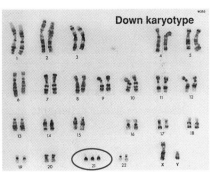

Down karyotype

Most pregnant women in developed countries will have a prenatal test to detect chromosomal abnormalities such as trisomy 21 (Down syndrome).

Cytogenetics Dept., Waikato Hospital

1. Outline the benefits of **carrier screening** to a couple with a family history of a genetic disorder:

_____

_____

2. (a) Suggest why Huntington's disease persists in the human population when it is caused by a lethal, dominant allele:

_____

_____

(b) How could presymptomatic genetic testing change this? _____

_____

_____

Related activities: Inherited Metabolic Disorders, Lethal Alleles, Aneuploidy in Humans, Trisomy in Human Autosomes **Weblinks**: *Genetic Disorders Library*

EA 2

Understanding how genes influence one's health and possible medical future raises many beneficial options and ethical issues. For example, a person diagnosed with a high risk of developing a treatable genetic disorder can plan for its occurrence or make lifestyle changes to delay its onset. A person diagnosed at high risk for an untreatable disorder may face years of anxiety over something that may not even occur. Moreover, how will health insurers react? Will they deny insurance cover or charge exorbitantly for cover?

Benefits and ethical issues arising from genetic testing

Medical benefits

- Improved **diagnosis** of disease and predisposition to disease by genetic testing.

- Better identification of disease carriers, through genetic testing.

- Better **drugs** can be designed using knowledge of protein structure (from gene sequence information) rather than by trial and error.

- Greater possibility of successfully using **gene therapy** to correct genetic disorders.

Non-medical benefits

- Greater knowledge of **family relationships** through genetic testing, e.g. paternity testing in family courts.

- Advances **forensic science** through analysis of DNA at crime scenes.

- Improved knowledge of the evolutionary relationships between humans and other organisms, which will help to develop better, more accurate classification systems.

Possible ethical issues

- It is unclear whether third parties, e.g. health insurers, have rights to genetic test results.

- If treatment is unavailable for a disease, genetic knowledge about it may have no use.

- Genetic tests are costly, and there is no easy answer as to who should pay for them.

- Genetic information is hereditary so knowledge of an individual's own genome has implications for members of their family.

Couples can already have a limited range of genetic tests to determine the risk of having offspring with some disease-causing mutations.

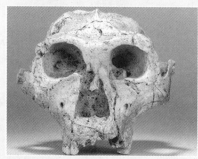

When DNA sequences are available for humans and their ancestors, comparative analysis may provide clues about human evolution.

Legislation is needed to ensure that there is no discrimination on the basis of genetic information, e.g. at work or for health insurance.

3. Huntington's disease can be tested for by looking for the number of CAG repeats on the HTT gene. Explain why someone with a family history of the disease may or may not want to be tested for their own disease risk:

_____

_____

_____

_____

4. Describe two possible **benefits** of genetic testing:

(a) Medical: _____

_____

_____

(b) Non-medical: _____

_____

_____

5. Suggest two possible points of view for one of the **ethical issues** described in the list above:

(a) _____

_____

_____

(b) _____

_____

_____

Polygenes

Some phenotypes (e.g. kernel color in maize and skin color in humans) are determined by more than one gene and show **continuous variation** in a population. The production of the skin pigment melanin in humans is controlled by at least three genes. The amount of melanin produced is directly proportional to the number of dominant alleles for either gene (from 0 to 6).

Very pale	Light	Medium light	Medium	Medium dark	Dark	Black
0	1	2	3	4	5	6

A light-skinned person   A dark-skinned person

There are seven shades skin color ranging from very dark to very pale, with most individuals being somewhat intermediate in skin color. No dominant allele results in a lack of dark pigment (aabbcc). Full pigmentation (black) requires six dominant alleles (AABBCC).

1. Complete the Punnett square for the F_2 generation (below) by entering the genotypes and the number of dark alleles resulting from a cross between two individuals of intermediate skin color. Color-code the offspring appropriately for easy reference.

   (a) How many of the 64 possible offspring of this cross will have darker skin than their parents:

   none

   (b) How many genotypes are possible for this type of gene interaction:

   $\frac{1}{2}$

2. Explain why we see many more than seven shades of skin color in reality:

   an affected gamete

Chnothe ↓   *Kahe* ↓

Parental generation

X

Black (AABBCC)   Pale (aabbcc)

cratie/Bufanda → Medium (AaBbCc)

F_2 generation (AaBbCc X AaBbCc)

GAMETES	ABC	ABc	AbC	Abc	aBC	aBc	abC	abc
ABC	AABBCC	AABBCc	AABbCC	AABbCc	AaBBCC	AaBBCc	AaBbCC	AaBbCc
ABc	AABBCc	AABBcc	AABbCc	AABbcc	AaBBCc	AaBBcc	AaBbCc	AaBbcc
AbC	AABbCC	AABbCc	AAbbCC	AAbbCc	AaBbCC	AaBbCc	AabbCC	AabbCc
Abc	AABbCc	AABbcc	AAbbCc	AAbbcc	AaBbCc	AaBbcc	AabbCc	Aabbcc
aBC	AaBBCC	AaBBCc	AaBbCC	AaBbCc	aaBBCC	aaBBCc	aaBbCC	aaBbCc
aBc	AaBBCc	AaBBcc	AaBbCc	AaBbcc	aaBBCc	aaBBcc	aaBbCc	aaBbcc
abC	AaBbCC	AaBbCc	AabbCC	AabbCc	aaBbCC	aaBbCc	aabbCC	aabbCc
abc	AaBbCc	AaBbcc	AabbCc	Aabbcc	aaBbCc	aaBbcc	aabbCc	aabbcc

Periodicals: The color code

Related activities: Selection for Skin Color in Humans
Weblinks: Polygenic Inheritance

3. Discuss the differences between **continuous** and **discontinuous** variation, giving examples to illustrate your answer:

_____

_____

_____

_____

_____

4 From a sample of no less than 30 adults, collect data (by request or measurement) for one continuous variable (e.g. height, weight, shoe size, or hand span). Record and tabulate your results in the space below, and then plot a frequency histogram of the data on the grid below:

Raw data

Tally Chart (frequency table)

Height	Total
0	1
1	6
2	15
3	20
4	15
5	6
6	1

Variable: _____

(Frequency histogram grid, x-axis labelled 0 1 2 3 4 5 6, y-axis Frequency 0 to 20)

(a) Calculate each of the following for your data. See *Descriptive Statistics* if you need help and attach your working:

Mean: _____ Mode: _____ Median: _____

Standard deviation: _____

(b) Describe the pattern of distribution shown by the graph, giving a reason for your answer: _____

_____

(c) What is the genetic basis of this distribution? _____

_____

(d) What is the importance of a large sample size when gathering data relating to a continuous variable?

_____

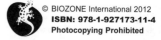

Gene-Environment Interactions

External environmental factors, such as diet, temperature, altitude or latitude, can affect the phenotype encoded by genes. **Phenotypic plasticity** describes the capacity of a single genotype to produce variable phenotypes in different environments. **Plasticity** is an adaptive response and is a function of flexibility in expression of the genotype. It is particularly common in insects, fish, and plants and can be induced by virtually any biotic or abiotic environmental factor.

Sources of Variation in Organisms

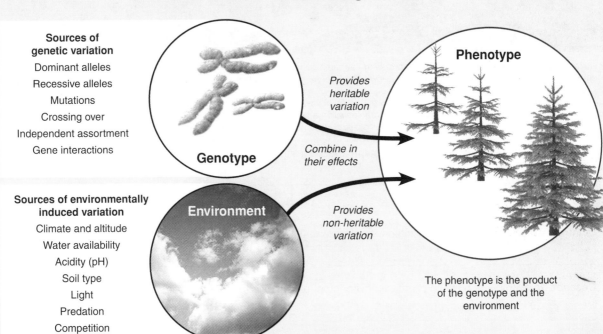

Sources of genetic variation
Dominant alleles
Recessive alleles
Mutations
Crossing over
Independent assortment
Gene interactions

Genotype

Provides heritable variation

Combine in their effects

Phenotype

Sources of environmentally induced variation
Climate and altitude
Water availability
Acidity (pH)
Soil type
Light
Predation
Competition

Environment

Provides non-heritable variation

The phenotype is the product of the genotype and the environment

The Effect of Temperature

The sex of some animals is determined by the incubation temperature during their embryonic development. Examples include turtles, crocodiles, and the American alligator. In some species, high incubation temperatures produce males and low temperatures produce females. In other species, the opposite is true. Temperature regulated sex determination may be advantageous by preventing inbreeding (since all siblings will tend to be of the same sex).

Color-pointing in breeds of cats and rabbits (e.g. Siamese, Himalayan) is a result of a temperature sensitive mutation in one of the enzymes in the metabolic pathway from tyrosine to melanin. The dark pigment is only produced in the cooler areas of the body (face, ears, feet, and tail), while the rest of the body is a paler version of the same color, or white.

The Effect of Other Organisms

Female

Male

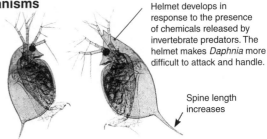

Helmet develops in response to the presence of chemicals released by invertebrate predators. The helmet makes *Daphnia* more difficult to attack and handle.

Spine length increases

Non-helmeted form   **Helmeted form with long tail spine**

The presence of other individuals of the same species may control sex determination for some animals. Some fish species, including some in the wrasse family (e.g. *Coris sandageri*, above), show this phenomenon. The fish live in groups consisting of a single male with attendant females and juveniles. In the presence of a male, all juvenile fish of this species grow into females. When the male dies, the dominant female will undergo physiological changes to become a male. The male has distinctive bands, whereas the female is pale in color and has very faint markings.

Some organisms respond to the presence of other, potentially harmful, organisms by changing their morphology or body shape. Invertebrates such as *Daphnia* will grow a large helmet when a predatory midge larva is present. Such responses are usually mediated through chemicals produced by the predator (or competitor), and are common in plants as well as animals.

Periodicals:
What is variation?

Related activities: Sex Determination

RA 2

The Effect of Altitude

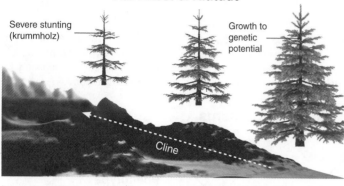

Severe stunting
(krummholz)

Growth to
genetic
potential

Cline

The Effect of Chemical Environment

Increasing altitude can stunt the phenotype of plants with the same genotype. In some conifers, e.g. Engelmann spruce, plants at low altitude grow to their full genetic potential, but become progressively more stunted as elevation increases, forming krummholz (gnarled bushy growth forms) at the highest, most severe sites. A continuous gradation in a phenotypic character within a species, associated with a change in an environmental variable, is called a **cline**.

The chemical environment can influence the expressed phenotype in both plants and animals. In hydrangeas, flower color varies according to soil pH. Flowers are blue in more acid soils (pH 5.0-5.5), but pink in more alkaline soils (pH 6.0-6.5). The blue color is due to the presence of aluminium compounds in the flowers and aluminium is more readily available when the soil pH is low.

1. Describe an example to illustrate how genotype and environment contribute to phenotype: _____

_____

_____

2. What are the physical factors associated with altitude that could affect plant phenotype? _____

_____

3. Describe an example of how the chemical environment of a plant can influence phenotype: _____

_____

_____

_____

4. Explain why the darker patches of fur in color-pointed cats and rabbits are found only on the face, paws and tail:

_____

_____

5. There has been much amusement over the size of record-breaking vegetables, such as enormous pumpkins, produced for competitions. Explain how you could improve the chance that a vegetable would reach its maximum genetic potential:

_____

_____

6. (a) What is a **cline**? _____

_____

(b) On a windswept portion of a coast, two different species of plant (species A and species B) were found growing together. Both had a low growing (prostrate) phenotype. One of each plant type was transferred to a greenhouse where "ideal" conditions were provided to allow maximum growth. In this controlled environment, species B continued to grow in its original prostrate form, but species A changed its growing pattern and became erect in form. Identify the **cause** of the prostrate phenotype in each of the coastal grown plant species and explain your answer:

Plant species A: _____

_____

_____

Plant species B: _____

_____

_____

(c) Which of these species (A or B) would be most likely to exhibit clinal variation? _____

KEY TERMS: Mix and Match

INSTRUCTIONS: Test your vocabulary by matching each term to its definition, as identified by its preceding letter code.

allele ....

autosome ....

codominance ....

continuous variation ......

cross ....

dihybrid cross ....

discontinuous variation ...

dominant ....

genetic counseling ....

genomic imprinting ....

genotype ....

heterozygous ....

homozygous ....

incomplete dominance .....

lethal allele ....

linkage ....

locus ....

monohybrid cross ....

multiple alleles ....

non-disjunction ....

non-nuclear inheritance ...

pedigree analysis ....

phenotype ....

phenotypic plasticity .....

polygenes (=multiple genes) ...

Punnett square ....

recessive ....

segregation ....

sex chromosome ....

sex-limited traits ....

sex linked gene ....

test cross ....

trait ....

A One of the two chromosomes that jointly determine the sex of an organism.

B Allele that expresses its characteristics only when in the homozygous condition.

C Non-allelic genes that together influence a phenotypic trait.

D A particular phenotypic character. Refers to the physical appearance as opposed to the mode of appearance e.g. blue eyes rather than eye color.

E The inheritance of genetic information from sources other than the nuclear chromosomes.

F Variation across a complete range of measurements from one extreme to the other, often as the result of polygenic inheritance.

G The specific allele combination of an organism.

H One of two or more forms of a particular gene.

I A chromosome that is not a sex chromosome.

J The physical appearance of the genotype.

K Possessing identical alleles for a particular gene.

L A diagnostic cross involving breeding with a recessive with known genotype.

M A deliberate breeding of two different individuals that results in offspring that carry part of the genetic material of each parent.

N Possessing two different forms of a particular gene.

O Dominance in which the action of one allele does not completely mask the action of the other is called this.

P Alleles that are fatal only when present in homozygous condition.

Q The position of a gene on a chromosome.

R The situation in which genes are located on the same chromosome.

S The failure of chromosome pairs to separate properly during meiosis I or II.

T A diagram used to predict an outcome of a particular cross or breeding experiment.

U Genetic cross between two individuals that differ in two traits of particular interest.

V A genetic phenomenon in which the expression of a gene depends on parent-of-origin.

W Allele that expresses its characteristics when in both the homozygous and the heterozygous condition is called this.

X Inheritance pattern in which both alleles in the heterozygote contribute to the phenotype.

Y Genetic cross between two individuals that differ in one trait of particular interest.

Z The capacity of a single genotype to exhibit variable phenotypes in different environments.

AA Genes located on one of the sex chromosomes (X or Y) but not the other.

BB The analysis of the family history or inheritance of an organism or trait.

CC The process by which individuals or relatives, at risk of an inherited disorder, are advised about the relevant aspects of their disorder and its inheritance.

DD Alleles for a gene where there are more than two alleles in the population.

EE Traits that are visible only within one sex.

FF Variation in which individuals fall into a number of distinct classes or categories.

GG Separation of alleles during cell division so that each gamete receives only one copy (allele) for each trait.

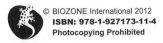

Regulation of Gene Expression

Christine
Katie
Alice Wiegand

Key concepts

▶ Changes in gene expression result in cell differentiation.

▶ Transcription factors regulate gene expression.

▶ Programmed cell death is important in development.

▶ Gene induction and gene repression are important in bacteria and viruses, but some genes are constitutive.

▶ Gene expression in eukaryotes involves regulatory sequences and transcription factors.

▶ Signal transmission mediates cell function.

Key terms

activator

apoptosis (=programmed cell death)

cancer

cell differentiation

constitutive gene

cytokine

embryonic induction

gene expression

gene induction

gene repression

genetic transplantation

gibberellin

homeotic genes

Hox genes

inducible genes

lac operon

microRNA (miRNA)

morphogenesis

morphogens

mutation

necrosis

p53

regulatory sequence

repressor

signal transmission

small interfering RNA (siRNA)

small regulatory RNA

transcription factors

Essential Knowledge

☐ 1. Use the **KEY TERMS** to compile a glossary for this topic.

Timing and Coordination of Development *(2.E.1)*   pages 100, 193-201

☐ 2. Explain how changes in **gene expression** result in **cell differentiation**.

☐ 3. Understand that the induction of **transcription factors** during development results in the sequential expression of genes.

☐ 4. Explain the role of **homeotic genes**, including the *Hox* genes, in developmental patterning in organisms. Suggest why these gene sequences are highly conserved across animal phyla.

☐ 5. Explain how **embryonic induction** during development results in developmental events occurring at the correct time in the correct sequence.

☐ 6. Explain the role of temperature and water availability in seed germination in most plants. Include reference to the role of **gibberellin** (gibberellic acid) as an intracellular signaling molecule during seed germination (*cross ref. 3.B.2*).

☐ 7. Describe an example of a **mutation** that results in abnormal development and explain the mechanism by which it operates.

☐ 8. Explain how **genetic transplantation** experiments provide evidence for the link between gene expression and normal development.

☐ 9. Explain the role of **microRNAs** (miRNAs) and **small interfering RNAs** (siRNAs) in regulating cellular activity.

☐ 10. Define the term **apoptosis** and distinguish it from **necrosis**. Explain the role of apoptosis in normal cell differentiation and **morphogenesis**. Describe the consequences of excessive or insufficient apoptosis during development.

Gene Regulation *(3.B.1-3.B.2)*   pages 74, 202-207

☐ 11. Distinguish between **regulatory sequences**, **regulatory genes**, and **small regulatory RNA**s. Outline the role of these in controlling gene expression.

☐ 12. Recognize both positive and negative control mechanisms are involved in the regulation of gene expression in bacteria and viruses.

☐ 13. Describe the structure and function of a prokaryote **operon**. Explain **gene induction** in prokaryotes, as illustrated by the *lac* operon in *E. coli*.

☐ 14. Explain **gene repression** in a prokaryote. Know that some genes (e.g. ribosomal genes) are continuously expressed (they are constitutive) and suggest why.

☐ 15. Explain the regulation of gene expression in eukaryotes, including the role of **transcription factors**. Recall that differences in gene regulation account for some of the phenotypic differences between genomically similar organisms.

☐ 16. Using an example, explain how **signal transmission** within and between cells mediates gene expression and cell function.

Periodicals:
Listings for this chapter are on page 383

Weblinks:
www.thebiozone.com/
weblink/AP1-3114.html

BIOZONE APP:
Student Review Series
The Nature of Genes

Cellular Differentiation

A zygote commences development by dividing into a small ball of a few dozen identical cells called **embryonic stem cells**. These cells start to take different developmental paths to become specialized cells such as nerve cells, which means they can no longer produce any other type of cell. **Differentiation** is cell specialization that occurs at the end of a developmental pathway.

Regulation of Gene Expression

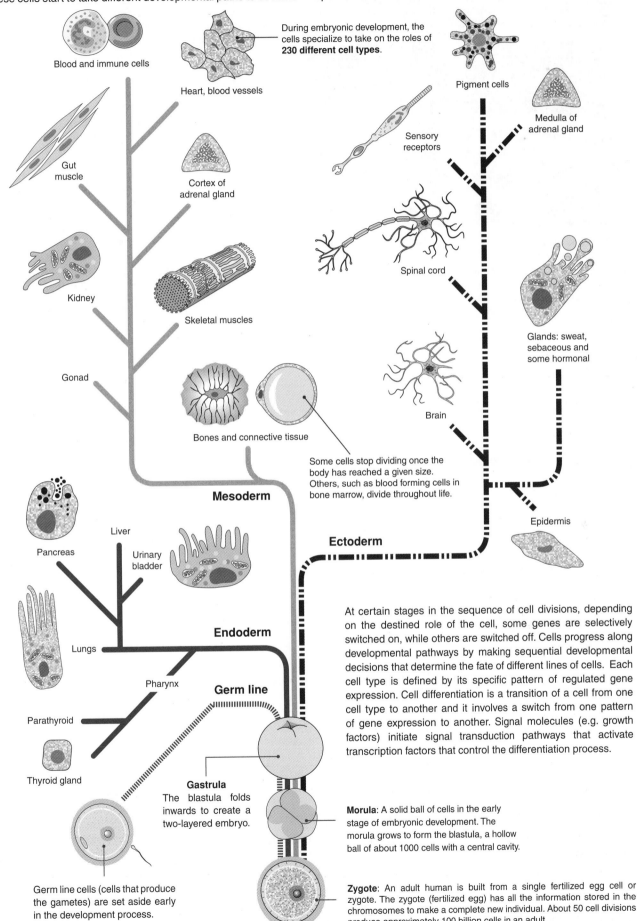

Blood and immune cells

During embryonic development, the cells specialize to take on the roles of **230 different cell types**.

Heart, blood vessels

Pigment cells

Medulla of adrenal gland

Gut muscle

Cortex of adrenal gland

Sensory receptors

Kidney

Skeletal muscles

Spinal cord

Glands: sweat, sebaceous and some hormonal

Gonad

Bones and connective tissue

Brain

Some cells stop dividing once the body has reached a given size. Others, such as blood forming cells in bone marrow, divide throughout life.

Epidermis

Mesoderm

Liver

Pancreas

Urinary bladder

Ectoderm

Lungs

Endoderm

Pharynx

Germ line

Parathyroid

Thyroid gland

At certain stages in the sequence of cell divisions, depending on the destined role of the cell, some genes are selectively switched on, while others are switched off. Cells progress along developmental pathways by making sequential developmental decisions that determine the fate of different lines of cells. Each cell type is defined by its specific pattern of regulated gene expression. Cell differentiation is a transition of a cell from one cell type to another and it involves a switch from one pattern of gene expression to another. Signal molecules (e.g. growth factors) initiate signal transduction pathways that activate transcription factors that control the differentiation process.

Gastrula
The blastula folds inwards to create a two-layered embryo.

Morula: A solid ball of cells in the early stage of embryonic development. The morula grows to form the blastula, a hollow ball of about 1000 cells with a central cavity.

Germ line cells (cells that produce the gametes) are set aside early in the development process.

Zygote: An adult human is built from a single fertilized egg cell or zygote. The zygote (fertilized egg) has all the information stored in the chromosomes to make a complete new individual. About 50 cell divisions produce approximately 100 billion cells in an adult.

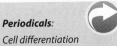

Development

Development is the process of progressive change through the lifetime of an organism. It involves growth (increase in size) and cell division (to generate the multicellular body). **Cellular differentiation** (the generation of specialized cells from more generalized ones) and **morphogenesis** (the creation of the shape and form of the body) are also part of development.

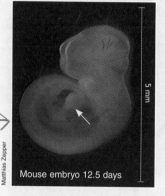 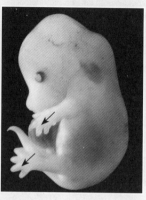

5 mm

Matthias Zepper

Mouse embryo 12.5 days

Katie →

Selective cell proliferation combined with selective apoptosis sculpts the tissues and structures in all vertebrates. In the mouse embryo, mesoderm forms between the toes giving the appearance of a webbed, paddle like structure (above left). As the embryo develops, this superfluous webbing is selectively destroyed by apoptosis so that each of the individual digits are visible (above right, arrowed).

Control Over Genes

Gene expression is the process by which a cell regulates the production of gene products (DNA or RNA) to meet the cell's requirements. Regulation is achieved by switching the transcription of specific genes on and off, starting or stopping the production of the gene product so that it is produced only when it is needed. Gene activity is regulated by cell type, cell function, chemical signals, and signals from the environment.

Many insect larvae have large **polytene chromosomes** (chromosomes which have undergone repeated rounds of DNA replication without cell division so that many sister chromatids remain synapsed together). They contain many copies of the same gene, so have a high level of gene expression. For example, *Chironomus* larvae produce large amounts of saliva to aid digestion of detritus. As they molt and grow, salivary gland development must keep pace with their increased food requirements. The molting hormone, **ecdysone**, acts as the **signal** to turn on the gene regulating the development of the salivary glands. The regions of the chromosome with this gene puff out (left) when the genes are being transcribed.

Chromosome puff

Ell

1. Germ line cells diverge (become isolated) from other cells at a very early stage in embryonic development:

   (a) Explain what the **germ line** is: _a cell lineage which consists of cells which are used to pass down genetic inheritance to the next generation_

   (b) Explain why it is necessary for the germ line to become separated at such an early stage of development: _make gametocytes which is used to make eggs and sperm_

2. Explain the genetic events that enable so many different cell types to arise from one unspecialized cell (the zygote): _Can produce different cell types provokes a particular set of protiens_

3. Using an example, describe the role of signal molecules in activating gene expression at the appropriate time:
   - An immediate damage of metabolism the cell
   - charge in electrical charge across plasma membrane
   - change in gene expression

4. Explain briefly the importance of gene regulation in development: _Gene regulation is a process in which a cell determines which genes it will express when._

Homeotic Genes and Development

Homeotic genes are genes that control the pattern of body formation in embryonic development. Homeotic gene sequences contain an evolutionarily highly conserved region, called the **homeobox**, which directs development in organisms as diverse as fungi, plants, and animals. A homeobox is about 180 base pairs long and encodes transcription factors responsible for switching on gene expression. The **Hox** genes are a particular group of homeobox genes in animals. They are found in clusters and determine where particular body segments and limbs grow. Very disparate organisms share this same tool kit of genes, but regulate them differently. This means that large changes in morphology or function are attributable to changes in gene regulation, rather than the evolution of new genes, and natural selection associated with gene switches plays a major role in evolution.

The Role of *Hox* Genes

Hox genes control the development of back and front parts of the body. The same genes (or homologous ones) are present in essentially all animals, including humans.

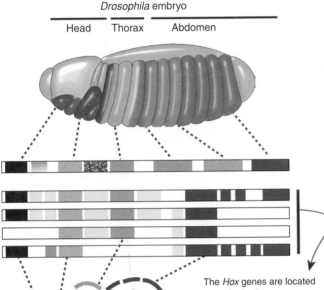

Drosophila embryo

Head | Thorax | Abdomen

The *Hox* genes are located on a single chromosome in *Drosophila*, and on four separate chromosomes in mice. The different shading indicates where in the body the genes are expressed.

The Evolution of Novel Forms

Even very small changes (mutations) in the *Hox* genes can have a profound effect on morphology. Such changes to the genes controlling development have almost certainly been important in the evolution of novel structures and body plans. Four principles underlie the role of developmental genes in the evolution of novel forms:

- **Evolution works with what is already present**: New structures are modifications of pre-existing structures.

- **Multifunctionality** and **redundancy**: Functional redundancy in any part of a multifunctional structure allows for specialization and division of labor through the development of two separate structures.

  Example: the diversity of appendages (including mouthparts) in arthropods.

- **Modularity**: Modular architecture in animals (arthropods, vertebrates) allows for the modification and specialization of individual body parts. Genetic switches allow changes in one part of a structure, independent of other parts.

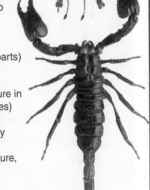

Regulation of Gene Expression

Shifting *Hox* Expression

Huge diversity in morphology in organisms within and across phyla could have arisen through small changes in the genes controlling development.

Differences in neck length in vertebrates provide a good example of how changes in gene expression can bring about changes in morphology. Different vertebrate species have different numbers of neck vertebrae (denoted by the black ovals on the diagram). The boundary between neck and trunk vertebrae is marked by expression of the **Hox c6** gene (c6 denotes the sixth cervical or neck vertebra) in all cases, but the position varies in each animal relative to the overall body. The forelimb (arrow) arises at this boundary in all four-legged vertebrates. In snakes, the boundary is shifted forward to the base of the skull and no limbs develop. As a result of these differences in expression, mice have a short neck, geese a long neck, and snakes, no neck at all.

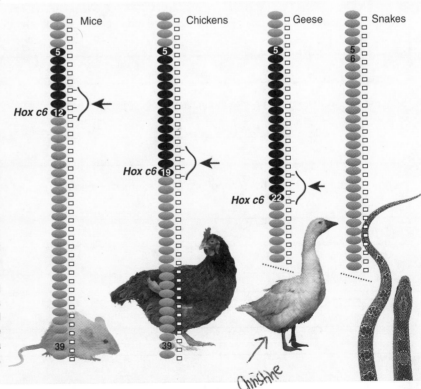

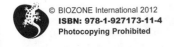

Periodicals:
Regulating evolution

Related activities: The Rate of Evolutionary Change
Weblinks: Genetic Tool Kit

EA 3

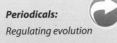

Genetic Switches in Evolution

The *Hox* genes are just part of the collection of genes that make up the genetic tool kit for animal development. The genes in the tool kit act as switches, shaping development by affecting how other genes are turned on or off. The distribution of genes in the tool kit indicates that it is ancient and was in place before the evolution of most types of animals. Differences in form arise through changes in genetic switches. One example is the evolution of eyespots on the wings of butterflies:

■ The **Distal-less** gene is one of the important **master body-building genes** in the genetic tool kit. Switches in the *Distal-less* gene control expression in the embryo (E), larval legs (L), and wing (W) in flies and butterflies, but butterflies have also evolved an extra switch (S) to control eyespot development.

■ Once spots evolved, changes in *Distal-less* expression (through changes in the switch) produced more or fewer spots.

Three switches in a fly

E // L // W → gene

A new switch, S, evolved in butterflies

E // L // W S → gene

Changes in *Distal-less* regulation were probably achieved by changing specific sequences of the *Distal-less* gene eyespot switch. The result? Changes in eyespot size and number.

Same Gene, New Tricks

Stichophthalma camadeva   *Junonia coenia (buckeye)*   *Taenaris macrops*

■ The action of a tool kit protein depends on context: where particular cells are located at the time when the gene is switched on.

■ Changes in the DNA sequence of a genetic switch can change the zone of gene expression without disrupting the function of the tool kit protein itself.

■ The spectacular **eyespots** on butterfly wings (arrowed above) represent different degrees of a basic pattern, from virtually all eyespot elements expressed (*Stichophthalma*) to very few (*Taenaris*).

1. Briefly describe the role of homeobox genes in development: _Control the expression_ _of dozens or even hundreds of other_ _genes that influence animal morphology_

2. (a) What does it mean when the homeobox genes are said to be highly conserved? _the gene develops_ _differently depending on the species_

(b) What does this tell you about the evolution and the importance of the homeobox sequences? _Needs to be different for each_ _organism_

3. Suggest why the *Hox* genes are found in clusters (i.e. grouped tightly together): _enhancing sharing between adjacent genes_

4. Using an example, discuss how changes in gene expression can bring about changes in morphology:

_____

_____

_____

_____

_____

_____

_____

_____

© BIOZONE International 2012
ISBN: 978-1-927173-11-4
Photocopying Prohibited

The Timing of Development

Timing is extremely important in embryonic development if the body is to develop normally. Cells must differentiate into the correct cell at the correct time and in the correct place. This is achieved by cell signaling, either by producing a gradient of molecules through a section of the embryo or by signaling a specific cell that then signals other cells in sequence. Cell signals regulate specific transcription factors that in turn control the differential gene expression that shapes embryos. The time taken from the start of this **embryonic induction** to the response is between hours and days, while the time a cell can respond to an inducing signal is strictly limited. The state (readiness) of the responding tissue is also important in determining the timing of the response.

Development in *Drosophila*

Morphogen concentration and cell fate

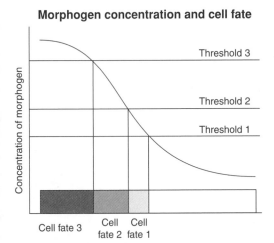

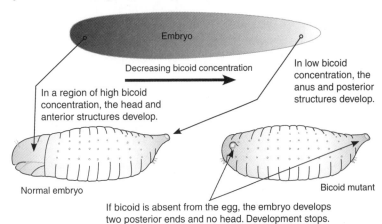

Embryo

Decreasing bicoid concentration

In a region of high bicoid concentration, the head and anterior structures develop.

In low bicoid concentration, the anus and posterior structures develop.

Normal embryo

Bicoid mutant

If bicoid is absent from the egg, the embryo develops two posterior ends and no head. Development stops.

Morphogens are signal molecules that govern the pattern of tissue development. Morphogens induce cells in different positions to adopt different fates by diffusing from an area of production through a field of cells.

The **bicoid** gene plays an important part in the development of *Drosophila*. After fertilization, bicoid mRNA from the mother is passed to the egg where it is translated into bicoid protein. Bicoid protein forms a gradient in the developing embryo, its concentration in certain regions determine where the anterior and posterior develop.

Regulation of Gene Expression

Development in *Caenorhabditis elegans*

A second way of inducing change in embryonic cells is through **sequential induction**. A signal molecule reaches cell A, which responds by developing and producing signal B and so on.

The fate of all of *C. elegans*' 1090 somatic cells are known. 131 somatic cells undergo apoptosis (programmed cell death). The genes that control the death of these cells are *ced-3* and *ced-4*. These genes are themselves regulated by the gene *ced-9*. There are three waves of apoptosis during development of *C. elegans*. The first removes 113 of 628 cells, the second removes another 18 cells, and the third removes half of the developing oocytes.

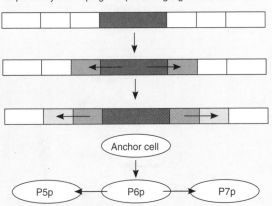

Anchor cell

P5p — P6p — P7p

The nematode worm, *Caenorhabditis elegans*, is a **model organism** that is often used in developmental studies. The cells that form the vulva (ventral opening for copulation and egg laying) display sequential induction. Signals from the anchor cell induce a change in cell P6p. Cell P6p then signals cells P5p and P7p to develop.

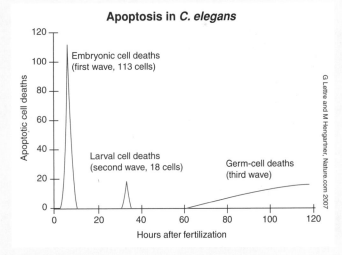

Apoptosis in *C. elegans*

Embryonic cell deaths (first wave, 113 cells)

Larval cell deaths (second wave, 18 cells)

Germ-cell deaths (third wave)

Apoptotic cell deaths

Hours after fertilization

G Lettre and M Hengartner, Nature.com 2007

1. (a) Describe the role of **morphogens**: _signal molecules that govern the pattern of tissue development_

(b) Explain the purpose of the **bicoid protein** in *Drosophila*: _forms a gradient in the developing embryo_

2. Describe **sequential induction** in *C. elegans*: _____

Related activities: Cellular Differentiation
Weblinks: Master Genes Control Basic Body Plans

RA 2

Factors Regulating Seed Germination

Seed germination refers to the beginning of seed growth. It involves rehydration of the seed, where a mature seed begins to take up water through the micropyle and testa, and reactivation of normal metabolism, in which the food stored in the seed is hydrolyzed to produce substrates for respiration (e.g. glucose), often after a period of low temperature. Uptake of water causes the seed to swell and the testa to split. Germinating seeds have a high oxygen requirement and respire rapidly. Water is essential

to the germination process. It enables expansion of the growing cells and activates the germination enzymes. The hormone, **gibberellic acid** is activated during germination and initiates production of the enzyme α-amylase. α-amylase breaks down the starch stored in the endosperm into simpler sugars by hydrolysis. Water is also required for the hydrolysis of stored starch and for the translocation of the mobilized food from the endosperm to the sites of growth.

Germination and Enzyme Activity

Summary of early germination events in a monocot seed

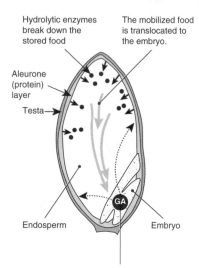

Hydrolytic enzymes break down the stored food

The mobilized food is translocated to the embryo.

Aleurone (protein) layer

Testa

GA

Endosperm

Embryo

The hormone gibberellic acid (GA) is produced in the embryo. It diffuses to the aleurone layer and stimulates the production of α-amylase.

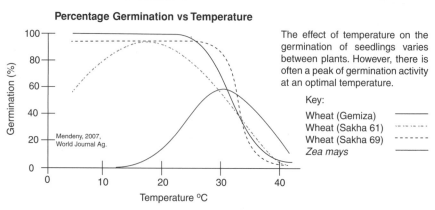

α-amylase Activity vs Temperature

Ching, 1975, Plant Physiology

The synthesis and activity of α-amylase is temperature dependent, peaking between 20°C and 30°C (there is some variation between plants).

α-amylase

Percentage Germination vs Temperature

Mendeny, 2007, World Journal Ag.

The effect of temperature on the germination of seedlings varies between plants. However, there is often a peak of germination activity at an optimal temperature.

Key:

Wheat (Gemiza)	————
Wheat (Sakha 61)	—·—·—
Wheat (Sakha 69)	– – – –
Zea mays	————

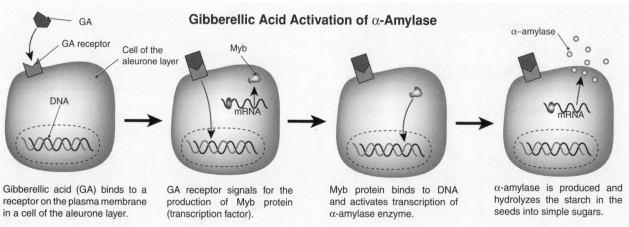

Gibberellic Acid Activation of α-Amylase

GA

GA receptor

Cell of the aleurone layer

Myb

DNA

mRNA

α–amylase

mRNA

Gibberellic acid (GA) binds to a receptor on the plasma membrane in a cell of the aleurone layer.

GA receptor signals for the production of Myb protein (transcription factor).

Myb protein binds to DNA and activates transcription of α-amylase enzyme.

α-amylase is produced and hydrolyzes the starch in the seeds into simple sugars.

1. Describe how temperature affects α-amylase activity: _____

_____

2. Describe how temperature affects germination in plants: _____

_____

3. How do temperature and gibberellic acid affect the breakdown of the endosperm and germination in a seedling?

_____

_____

_____

Mutations and Development

Mutation of the DNA molecule can lead to abnormal development in the individual. Every time a DNA molecule is copied (DNA replication), there is a chance that a base or series of bases will be copied incorrectly. Some changes in development can be mild and have little effect, others can be of greater significance. DNA replication has a low **error rate**, with only one mistake for every billion base pairs copied. Errors that have no effect on the organism or its offspring are called **neutral mutations**. Other errors may create new **alleles**, some of which may be beneficial, although most will be detrimental to development. One example is the most common form of genetic hearing loss (called NSRD), which accounts for up to 50% of childhood deafness.

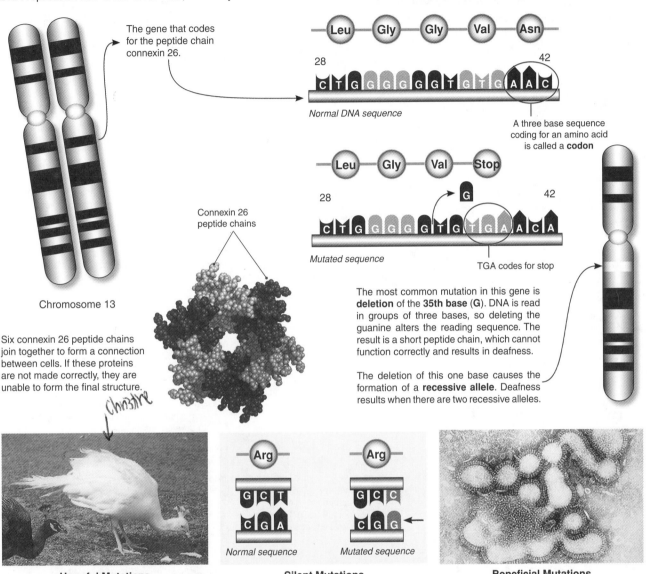

The gene that codes for the peptide chain connexin 26.

Normal DNA sequence

A three base sequence coding for an amino acid is called a **codon**

Mutated sequence

TGA codes for stop

Chromosome 13

Connexin 26 peptide chains

Six connexin 26 peptide chains join together to form a connection between cells. If these proteins are not made correctly, they are unable to form the final structure.

Christie

The most common mutation in this gene is **deletion** of the **35th base** (**G**). DNA is read in groups of three bases, so deleting the guanine alters the reading sequence. The result is a short peptide chain, which cannot function correctly and results in deafness.

The deletion of this one base causes the formation of a **recessive allele**. Deafness results when there are two recessive alleles.

Regulation of Gene Expression

Harmful Mutations
Most mutations cause harmful effects, usually because they stop or alter the production of a protein (often an enzyme). Albinism (above) is one of the more common mutations in nature, and leaves an animal with no pigmentation.

Silent Mutations
Silent mutations do not change the amino acid sequence nor the final protein. In the genetic code, several codons may code for the same amino acid. Silent mutations are also **neutral** if they do not alter the fitness of the organism.

Normal sequence

Mutated sequence

Beneficial Mutations
Sometimes mutations help the survival of an organism. In viruses (such as the *Influenzavirus* above) genes coding for the glycoprotein coat are constantly mutating, producing new strains that avoid detection by the host's immune system.

1. How can changes in a DNA sequence occur? _____

_____

2. How can a mutation in a single base be as damaging as a mutation in a sequence of bases? _____

_____

_____

_____

3. Explain how mutation can be harmful or beneficial: _____

_____

_____

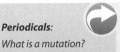

Periodicals: What is a mutation?

Related activities: The Genetic Code, Mutagens
Weblinks: Damage to DNA

Gene Transplantation Experiments

Developmental biology studies the way organisms grow and develop. Part of this is understanding how genes regulate development. We can study some types of gene regulation by investigating the effect of particular gene mutations on the development of an individual. We can also analyze the effect of **gene transplantation** (i.e. inserting DNA from one cell type into another). Experiments on myoblasts and cavefish have shown clear links between gene expression and cell development.

Genes and the Fate of Cells

As an embryo develops, populations of undifferentiated cells pass a point where they become committed to becoming a certain type of cell. An understanding of what causes this was developed after gene experiments by Harold Weintraub were carried out (below).

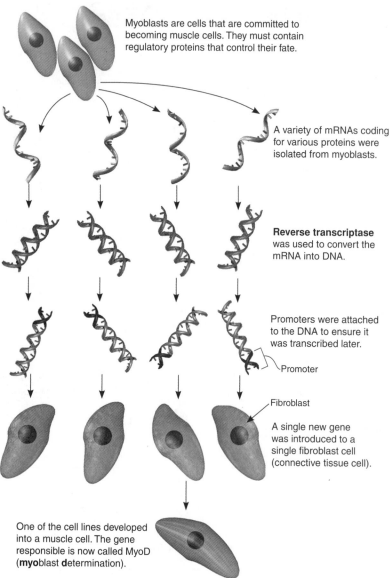

Myoblasts are cells that are committed to becoming muscle cells. They must contain regulatory proteins that control their fate.

A variety of mRNAs coding for various proteins were isolated from myoblasts.

Reverse transcriptase was used to convert the mRNA into DNA.

Promoters were attached to the DNA to ensure it was transcribed later.

Promoter

Fibroblast

A single new gene was introduced to a single fibroblast cell (connective tissue cell).

One of the cell lines developed into a muscle cell. The gene responsible is now called MyoD (**myo**blast **d**etermination).

Genes and Cavefish

The Mexican tetra or blind cavefish (*Astyanax mexicanus*) is a freshwater fish that develops distinct forms, an eyed form living in lighted areas and various blind forms geographically isolated by their cave environments. The cave living form develops no eyes or pigmentation while the fish living in lit areas does. Despite their differences, all forms can interbreed.

The eyes of both forms begin development in the early embryo stage, but development stops soon after in the blind form. Researchers reasoned that a gene in the eye cells might be regulating the eye's development.

The lens from an eyed *Astyanax* embryo was transplanted into a blind embryo. The eye in the blind embryo then developed normally, showing the cells in the lens were producing the substances required for eye development. The transplant also showed that the cells in the eye of blind cave fish embryos have the capacity to develop into correctly functioning eyes if they are given the correct signals at the correct time.

Breeding experiments using two different and geographically isolated populations of cavefish found that when they were bred together the hybrid offspring developed functioning eyes. This was explained by the fact that eye development is controlled by numerous genes. Each isolated population of cavefish developed blindness through mutations to different genes. The hybrid offspring gained a working version of each gene, one from each parent, enabling their eyes to develop normally.

Ltshears

1. (a) Give a brief description of how myoblasts form muscle cells: _____

_____

(b) Explain why transferring the MyoD gene into non-muscle cells caused them to develop into muscle cells:

_____

_____

2. Explain what the results of the lens transplantation experiment in cavefish tell us about genes and development:

_____

_____

_____

Related activities: The Timing of the Development

© BIOZONE International 2012
ISBN: 978-1-927173-11-4
Photocopying Prohibited

Apoptosis: Programmed Cell Death

Apoptosis or programmed cell death (PCD) is a normal and necessary mechanism in multicellular organisms to trigger the death of a cell. Apoptosis has a number of crucial roles in the body, including the maintenance of adult cell numbers, and defense against damaged or dangerous cells, such as virus-infected cells and cells with DNA damage. Apoptosis also has a role in "sculpting" embryonic tissue during its development, e.g. in the formation of fingers and toes in a developing human embryo. Programmed cell death involves an orderly series of biochemical events that result in set changes in cell morphology and end in cell death. The process is carried out in such a way as to safely dispose of cell remains and fragments. This is in contrast to another type of cell death, called **necrosis**, in which traumatic damage to the cell results in premature cell death and spillage of the cell contents. Apoptosis is tightly regulated by a balance between the factors that promote cell survival and those that trigger cell death. An imbalance between these regulating factors leads to defective apoptotic processes and is implicated in an extensive variety of diseases. For example, low rates of apoptosis result in uncontrolled proliferation of cells and cancers.

Stages in Apoptosis

Apoptosis is a normal cell suicide process in response to particular cell signals. It characterized by an overall compaction (shrinking) of the cell and its nucleus, and the orderly dissection of chromatin by endonucleases. Death is finalized by a rapid engulfment of the dying cell by phagocytosis. The cell contents remain membrane-bound and there is no inflammation.

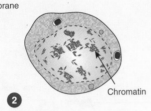

1 The cell shrinks and loses contact with neighboring cells. The chromatin condenses and begins to degrade.

2 The nuclear membrane degrades. The cell loses volume. The chromatin clumps into **chromatin bodies**.

3 **Zeiosis**: The plasma membrane forms bubble like **blebs** on its surface.

4 The nucleus collapses, but many membrane-bound organelles are not affected.

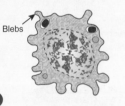
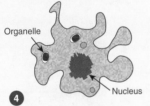

5 The nucleus breaks up into spheres and the DNA breaks up into small fragments.

6 The cell breaks into numerous **apoptotic bodies**, which are quickly resorbed by phagocytosis.

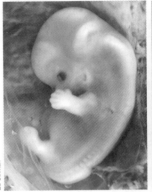
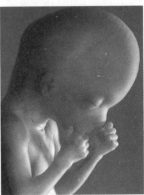

Ed Uthman

In humans, the mesoderm initially formed between the fingers and toes is removed by apoptosis. Forty one days after fertilization (top left), the digits of the hands and feet are webbed, making them look like small paddles. Apoptosis selectively destroys this superfluous webbing, sculpting them into digits when can be seen later in development (top right).

Regulating Apoptosis

Apoptosis is a complicated and tightly controlled process, distinct from cell necrosis (uncontrolled cell death), when the cell contents are spilled. Apoptosis is regulated through both:

Positive signals, which prevent apoptosis and allow a cell to function normally. They include:
▶ interleukin-2
▶ bcl-2 protein and growth factors

Interleukin-2 is a positive signal for cell survival. Like other signaling molecules, it binds to cell surface receptors to regulate metabolism.

Negative signals (death activators), which trigger the changes leading to cell death. They include:
▶ inducer signals generated from within the cell itself in response to stress, e.g. DNA damage or cell starvation.
▶ signalling proteins and peptides such as lymphotoxin.

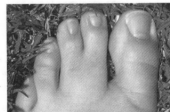

1. The photograph (right) shows a condition called syndactyly. Explain what might have happened during development to result in this condition:

_____

_____

2. Describe one difference between apoptosis and necrosis: _____

_____

3. Describe two situations, other than digit formation in development, in which apoptosis plays a crucial role:

(a) _____

_____

(b) _____

_____

Periodicals: What is cell suicide?

Related activities: Defective Gene Regulation
Weblinks: Apoptosis: Dance of Death

A 2

Regulation of Gene Expression

Gene Induction in Prokaryotes

Jacob and Monod proposed the operon mechanism to explain how prokaryotes regulate gene activity to meet their metabolic requirements. Their work was carried out with the bacterium *Escherichia coli,* and the model can only be applied to prokaryotes because genes in eukaryotic cells are not found as operons. An **operon** consists of a group of closely linked genes that act together and code for the enzymes that control a particular metabolic pathway. Structural genes, a promoter, and operator sites make up an operon. Structural genes encode enzymes used in the metabolic pathway. They are transcribed as a single **transcription unit** (a DNA sequence that constitutes a gene). Structural genes are controlled by a **promoter**, which initiates the formation of the mRNA, and a region of the DNA in front of the structural genes called the **operator**. A gene

outside the operon, called the **regulator gene**, produces a **repressor** molecule that can bind to the operator, and block the transcription of the structural genes. It is the repressor that switches the structural genes on or off and controls the metabolic pathway. Two mechanisms operate in the operon model: gene induction and gene repression. **Gene induction** occurs when genes are switched on by an inducer binding to the repressor molecule and deactivating it. In the *lac* **operon model**, lactose acts as the **inducer**, binding to the repressor and permitting transcription of the structural genes for the utilization of lactose (an infrequently encountered substrate). **Gene repression** occurs when genes that are normally switched on (e.g. genes for synthesis of an amino acid) are switched off by activation of the repressor.

Control of Gene Expression Through Induction: the *lac* Operon

Structure of the operon

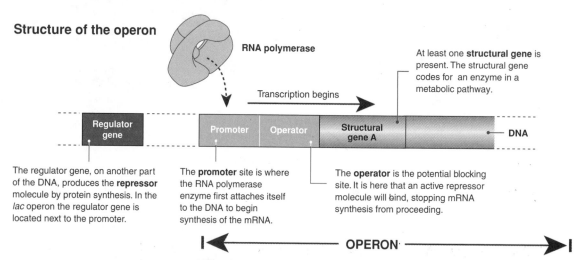

RNA polymerase

Transcription begins

At least one **structural gene** is present. The structural gene codes for an enzyme in a metabolic pathway.

Regulator gene — Promoter — Operator — Structural gene A — DNA

The regulator gene, on another part of the DNA, produces the **repressor** molecule by protein synthesis. In the *lac* operon the regulator gene is located next to the promoter.

The **promoter** site is where the RNA polymerase enzyme first attaches itself to the DNA to begin synthesis of the mRNA.

The **operator** is the potential blocking site. It is here that an active repressor molecule will bind, stopping mRNA synthesis from proceeding.

OPERON

The operon consists of the structural genes and the promoter and operator sites

Structural genes switched off

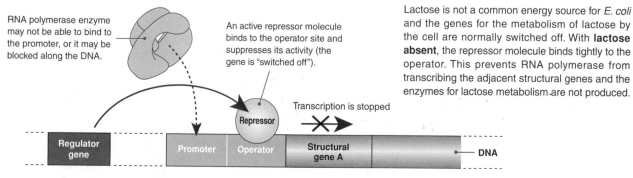

RNA polymerase enzyme may not be able to bind to the promoter, or it may be blocked along the DNA.

An active repressor molecule binds to the operator site and suppresses its activity (the gene is "switched off").

Transcription is stopped

Repressor

Regulator gene — Promoter — Operator — Structural gene A — DNA

Lactose is not a common energy source for *E. coli* and the genes for the metabolism of lactose by the cell are normally switched off. With **lactose absent**, the repressor molecule binds tightly to the operator. This prevents RNA polymerase from transcribing the adjacent structural genes and the enzymes for lactose metabolism are not produced.

Gene induction

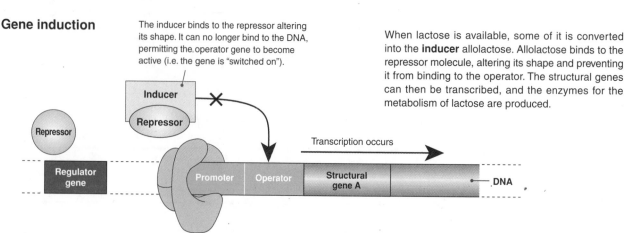

The inducer binds to the repressor altering its shape. It can no longer bind to the DNA, permitting the operator gene to become active (i.e. the gene is "switched on").

Inducer
Repressor
Repressor

Transcription occurs

Regulator gene — Promoter — Operator — Structural gene A — DNA

When lactose is available, some of it is converted into the **inducer** allolactose. Allolactose binds to the repressor molecule, altering its shape and preventing it from binding to the operator. The structural genes can then be transcribed, and the enzymes for the metabolism of lactose are produced.

Related activities: Gene Control in Eukaryotes
Weblinks: Induction of the lac Operon

Periodicals:
Gene structure and
expression

© BIOZONE International 2012
ISBN: 978-1-927173-11-4
Photocopying Prohibited

Diauxie in *E.coli*

Diauxie means two growth phases. **Diauxic growth** describes how microbes grown on a mixed sugar source in batch culture will preferentially metabolize one sugar source before moving on to the second. This sequential metabolism results in two distinct growth phases (right).

In the example (right) *E.coli* is grown on a mixed substrate of glucose and lactose. Diauxie occurs because the presence of glucose in excess suppresses the *lac* operon so that only the enzymes required for glucose metabolism are produced. As the glucose supply diminishes, the *lac* operon becomes activated, and *E.coli* begins to metabolize lactose.

The lag period represents the time taken for the *lac* operon to become active and the synthesis of the enzymes required for lactose metabolism to begin. This mechanism allows *E.coli* to preferentially metabolize the substrate it can grow fastest on, before moving to the second substrate.

The diauxic growth curve of *E.coli* when grown on glucose and lactose

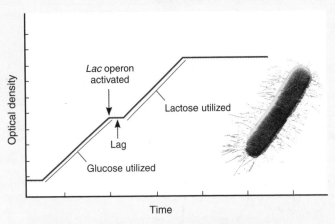

Jacques Monod discovered diauxic growth in 1941 prior to the discovery of the lac operon, which explained the lag phenomenon at the genetic level.

1. Explain the functional role of each of the following in relation to gene regulation in a prokaryote, e.g. *E. coli*:

   (a) Operon: _a group of closely linked genes that act together and code for enzymes that control metabolic pathways_

   (b) Regulator gene: _produces a repressor molecule that can block the transcription of the structural gene_

   (c) Operator: _initiates the formation of the mRNA, and a region of DNA in front of the structural genes_

   (d) Promoter: _controls structural genes_

   (e) Structural genes: _encode enzymes used in the metabolic pathway_

2. (a) Explain the advantage in having an inducible enzyme system that is regulated by the presence of a substrate: _Alter the shape of the repressor preventing it from binding to the Operater_

   (b) Suggest when it would not be adaptive to have an inducible system for metabolism of a substrate: _When lactose is absent_

   (c) Suggest how gene control in a non-inducible system might be achieved:

3. Explain how the operon model explains the diauxic growth of bacteria on two sugar substrates: _it allows glucose and lactose metabolism_

Regulation of Gene Expression

Gene Repression in Prokaryotes

In *E. coli*, the enzyme tryptophan synthetase synthesizes the amino acid tryptophan. The gene for producing this enzyme is normally switched on. When tryptophan is present in excess, some of it acts as an effector (also called a co-repressor). The effector activates the repressor, and they bind to the operator gene, preventing any further transcription of the structural gene. Once transcription stops, the enzyme tryptophan synthetase is no longer produced. This is an example of end-product inhibition (feedback inhibition).

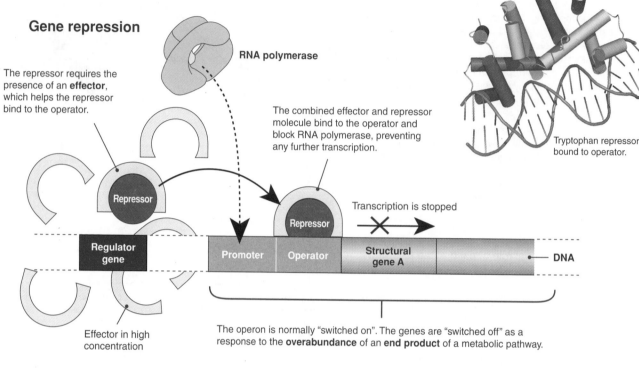

Gene repression

RNA polymerase

The repressor requires the presence of an **effector**, which helps the repressor bind to the operator.

The combined effector and repressor molecule bind to the operator and block RNA polymerase, preventing any further transcription.

Tryptophan repressor bound to operator.

Repressor

Repressor

Transcription is stopped

Regulator gene

Promoter   Operator   Structural gene A   DNA

Effector in high concentration

The operon is normally "switched on". The genes are "switched off" as a response to the **overabundance** of an **end product** of a metabolic pathway.

1. (a) Giving an example, outline how gene control is achieved through **gene repression**:

   Tryptophan activates the repressor and binds to the operater gene, preventing any transcription of the structural gene

   (b) Explain how the tryptophan operon forms a negative feedback loop: _Prevents the further transcription of the structural gene._

2. Explain why the operator is always downstream of the promoter: _____

3. Explain how gene repression is different from gene induction: _____

   Gene repression represses the transcription of transcription and gene induction encourages it.

Related activities: Gene Induction and the lac Operon

Gene Control in Eukaryotes

All the cells in your body contain identical copies of your genetic instructions. Yet these cells appear very different (e.g. muscle, nerve, and epithelial cells have little in common). These morphological differences reflect profound differences in the expression of genes during the cell's development. For example, muscle cells express the genes for the proteins that make up the contractile elements of the muscle fiber. This diversity of cell structure and function reflects precise control over the time, location, and extent of expression of a huge variety of genes. The physical state of the DNA in or near a gene is important in helping to control whether the gene is even available for transcription. When the **heterochromatin** is condensed, the transcription proteins cannot reach the DNA and the gene is not expressed. To be transcribed, a gene must first be unpacked from its condensed state. Once unpacked, control of gene expression involves the interaction of **transcription factors** with DNA sequences that control the specific gene. Initiation of transcription is the most important and universally used control point in gene expression. A simplified summary of this process is outlined below. Note the differences between this model and the operon model, described earlier, which is not applicable to eukaryotes because eukaryotic genes are not found as operons.

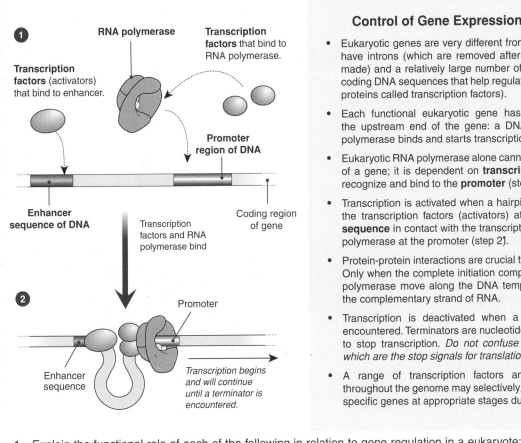

1 Transcription factors (activators) that bind to enhancer. RNA polymerase. Transcription factors that bind to RNA polymerase. Promoter region of DNA. Enhancer sequence of DNA. Transcription factors and RNA polymerase bind. Coding region of gene.

2 Enhancer sequence. Promoter. Transcription begins and will continue until a terminator is encountered.

Control of Gene Expression in Eukaryotes

- Eukaryotic genes are very different from prokaryotic genes: they have introns (which are removed after the primary transcript is made) and a relatively large number of **control elements** (non-coding DNA sequences that help regulate transcription by binding proteins called transcription factors).

- Each functional eukaryotic gene has a **promoter region** at the upstream end of the gene: a DNA sequence where RNA polymerase binds and starts transcription.

- Eukaryotic RNA polymerase alone cannot initiate the transcription of a gene; it is dependent on **transcription factors** in order to recognize and bind to the **promoter** (step 1).

- Transcription is activated when a hairpin loop in the DNA brings the transcription factors (activators) attached to the **enhancer sequence** in contact with the transcription factors bound to RNA polymerase at the promoter (step 2).

- Protein-protein interactions are crucial to eukaryotic transcription. Only when the complete initiation complex is assembled can the polymerase move along the DNA template strand and produce the complementary strand of RNA.

- Transcription is deactivated when a terminator sequence is encountered. Terminators are nucleotide sequences that function to stop transcription. *Do not confuse these with stop codons, which are the stop signals for translation.*

- A range of transcription factors and enhancer sequences throughout the genome may selectively activate the expression of specific genes at appropriate stages during cell development.

Regulation of Gene Expression

1. Explain the functional role of each of the following in relation to gene regulation in a eukaryote:

(a) Promoter: _a DNA sequence where RNA polymerase binds and starts transcription_

(b) Transcription factors: _bind to RNA polymerase to initiate transcription of a gene_

(c) Enhancer sequence: _contacts transcription factors bound to RNA polymerase at the promoter_

(d) RNA polymerase: _transcribes DNA sequences_

(e) Terminator sequence: _deactivates transcription_

2. Identify one difference between the mechanisms of gene control in eukaryotes and prokaryotes:
   Operons are used in Prokaryotes

Related activities: Gene Induction and the lac Operon   Weblinks: Control of Gene Expression in Eukaryotes, Regulation of Eukaryotic DNA Transcription

RA 3

Defective Gene Regulation in Cancer

Normal cells do not live forever; they are programmed to die under certain circumstances, particularly during development. Cells that become damaged beyond repair will normally undergo this programmed cell death (called **apoptosis** or cell suicide). Cancer cells evade this control and become immortal, continuing to divide regardless of any damage incurred. **Carcinogens** are agents capable of causing cancer. Roughly 90% of carcinogens are also mutagens, i.e. they damage DNA. Chronic exposure to carcinogens accelerates the rate at which dividing cells make errors. Susceptibility to cancer is also influenced by genetic make-up. Any one or a number of cancer-causing factors (including defective genes) may interact to induce cancer.

Cancer: Cells out of Control

Cancerous transformation results from changes in the genes controlling normal cell growth and division. The resulting cells become immortal and no longer carry out their functional role. Two types of gene are normally involved in controlling the cell cycle: **proto-oncogenes**, which start the cell division process and are essential for normal cell development, and **tumor-suppressor genes**, which switch off cell division. In their normal form, both kinds of genes work as a team, enabling the body to perform vital tasks such as repairing defective cells and replacing dead ones. But mutations in these genes can disrupt these finely tuned checks and balances. Proto-oncogenes, through mutation, can give rise to **oncogenes**; genes that lead to uncontrollable cell division. Mutations to tumor-suppressor genes initiate most human cancers. The best studied tumor-suppressor gene is **p53**, which encodes a protein that halts the cell cycle so that DNA can be repaired before division.

The panel, right, shows the mutagenic action of some selected carcinogens on four of five codons of the **p53 gene**.

Features of Cancer Cells

The diagram right shows a single **lung cell** that has become cancerous. It no longer carries out the role of a lung cell, and instead takes on a parasitic lifestyle, taking from the body what it needs in the way of nutrients and contributing nothing in return. The rate of cell division is greater than in normal cells in the same tissue because there is no *resting phase* between divisions.

A mutation in one or two of the controlling genes causes a **benign** (nonmalignant) **tumor.** As the number of controlling genes with mutations increases, so too does the loss of control until the cell becomes cancerous.

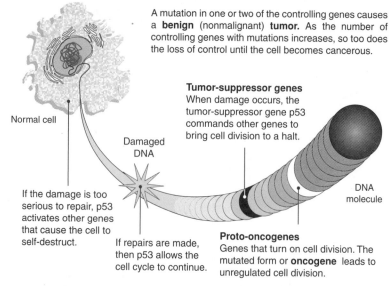

Normal cell

Damaged DNA

Tumor-suppressor genes
When damage occurs, the tumor-suppressor gene p53 commands other genes to bring cell division to a halt.

DNA molecule

If the damage is too serious to repair, p53 activates other genes that cause the cell to self-destruct.

If repairs are made, then p53 allows the cell cycle to continue.

Proto-oncogenes
Genes that turn on cell division. The mutated form or **oncogene** leads to unregulated cell division.

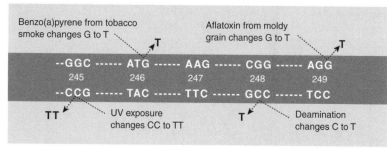

Benzo(a)pyrene from tobacco smoke changes G to T

Aflatoxin from moldy grain changes G to T

```
--GGC ------ ATG ------ AAG ------ CGG ------ AGG
   245        246        247        248        249
--CCG ------ TAC ------ TTC ------ GCC ------ TCC
```

UV exposure changes CC to TT

Deamination changes C to T

Given a continual supply of nutrients, cancer cells can go on dividing indefinitely and are said to be immortal.

Cancer cells may have unusual numbers of chromosomes.

The bloated, lumpy shape is readily distinguishable from a healthy cell, which has a flat, scaly appearance.

Metabolism is disrupted and the cell ceases to function constructively.

Cancerous cells lose their attachments to neighboring cells.

1. How do cancerous cells differ from normal cells? _Continues to go through mitosis_

2. Explain how the cell cycle is normally controlled, including reference to the role of **tumor-suppressor genes**:
   The tumor-suppresor genes = switch off cell division.

3. With reference to the role of **oncogenes**, explain how the normal controls over the cell cycle can be lost:
   Oncogenes lead to uncontrolable cell division

Related activities: Apoptosis: Programmed Cell Death
Web links: p53

Periodicals:
Living with the enemy

© BIOZONE International 2012
ISBN: 978-1-927173-11-4
Photocopying Prohibited

Estrogen, Transcription, and Cancer

Estrogen, a hormone found in high levels in premenopausal women, has long been implicated in immune system function. **Autoimmune diseases**, in which the body attacks its own tissues, tend to be more common in women than in men and normal immune responses to infection are slightly faster in women than in men. Both of these responses have been linked to levels of estrogen in the blood. We now know that estrogen plays an important role in the immune system by acting as a switch to turn on the gene involved in antibody production. As it happens, activation of that gene is also linked to immune system cancer.

1 Estrogen binds to **estrogen receptors** and acts as a **transcription factor** which binds to the **A**ctivation **I**nduced **D**eaminase (AID) gene.

2 The estrogen ER transcription factor activates the AID gene and results in the production of the AID protein.

3 AID's normal role is to cause somatic hypermutation in the DNA of the **B cells** of the immune system. This allows them to produce hundreds of novel antibodies in readiness for unknown antigens.

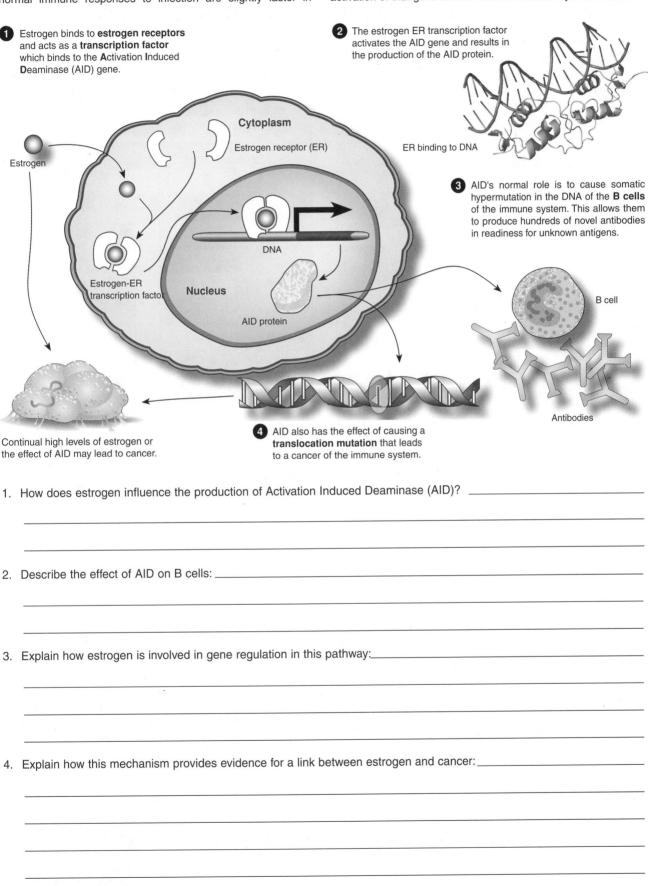

Cytoplasm
Estrogen receptor (ER)
Estrogen
ER binding to DNA
Estrogen-ER transcription factor
Nucleus
DNA
AID protein
B cell
Antibodies

4 AID also has the effect of causing a **translocation mutation** that leads to a cancer of the immune system.

Continual high levels of estrogen or the effect of AID may lead to cancer.

Regulation of Gene Expression

1. How does estrogen influence the production of Activation Induced Deaminase (AID)? _____

2. Describe the effect of AID on B cells: _____

3. Explain how estrogen is involved in gene regulation in this pathway:_____

4. Explain how this mechanism provides evidence for a link between estrogen and cancer:_____

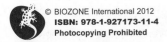
© BIOZONE International 2012
ISBN: 978-1-927173-11-4
Photocopying Prohibited

Periodicals:
Living with the enemy

Related activities: Defective Gene Regulation

A 3

KEY TERMS: Crossword

Complete the crossword below, which will test your understanding of key terms in this chapter and their meanings.

CLUES ACROSS

1. Activation of an inactive gene to carry out transcription (2 words: 4, 9).
3. A signal molecule that controls and directs the development of cells and tissues.
7. This occurs when genes that are normally switched on are switched off by the activation of a repressor (2 words: 4, 10).
9. In negative control in a prokaryote operon, this molecule binds to the operator to prevent transcription.
10. A change to the DNA sequence of an organism.
11. An important group of highly conserved developmental genes (2 words: 3, 5).
12. Small cell signaling proteins used in intercellular communication.
14. The process by which a less specialized cell becomes a more specialized cell type (2 words: 4, 15).
16. A condition where mutations in genes regulating cell growth and division allow cells to grow and divide in an uncontrolled fashion resulting in a tumor.
17. A cluster of genes in *E. coli*, under the control of a single regulatory promoter, in which lactose acts as the inducer (2 words: 3, 6).
18. The creation of the shape and form of the body.
19. Proteins that bind to specific DNA sequences to regulate transcription. (2 words: 13, 7).

CLUES DOWN

2. The process by which the presence of one tissue influences the development of others (2 words: 9, 9).
4. A class of phytohormones involved in breaking seed dormancy through activation of α-amylase.
5. The process of transferring the information encoded in a gene to its functional gene product (2 words: 4, 10).
6. A short RNA molecule derived from exogenous genetic material that has a role in gene regulation.
8. A short RNA molecule made from endogenous DNA that has a role in gene regulation.
13. Programmed cell death is also known as this.
15. Premature death of cells or tissues through injury or disease.

Sources of **Variation**

Key concepts

▶ Biological systems have several mechanisms to increase genetic variation.

▶ Mutations are changes to the DNA sequence. The may be positive, negative, or neutral in their effect.

▶ Some mutations result in phenotypic changes that have a selective advantage in some environments.

▶ Specific features of viral replication increase variation.

▶ Gene duplication may enable the evolution of new traits.

Key terms

aneuploidy
background
    mutation rate
bacteriophage
Barr body
beneficial (positive) mutation
fertilization
gametic mutation
harmful mutation
heterozygote advantage
induced mutation
lysogenic phase
lytic phase
maternal age effect
meiosis
mutagen
mutation
neutral mutation
non-disjunction
pathogenicity
polyploidy
sexual reproduction
silent (=synonymous) mutation
somatic mutation
spontaneous mutation
syndrome
transposition
trisomy
X-inactivation

Essential Knowledge

☐ 1. Use the **KEY TERMS** to compile a glossary for this topic.

Changes in Genotype *(3.C.1, 4.C.1: b)*   pages 212-226, 231, 255, 265

☐ 2. Describe, in a general way, how **mutation** can lead to changes in phenotype.

☐ 3. Distinguish between **beneficial (positive)**, **harmful (negative)**, and **silent** or synonymous **mutations** and describe examples. Note that recent research has shown that silent mutations are not necessarily neutral as was once thought. Comment on the evolutionary potential of **neutral mutations**.

☐ 4. Describe and explain the effects of **mutagens** on DNA. Distinguish between **spontaneous** and **induced mutations**. Understand that there is a species-specific naturally occurring **background mutation rate** and explain how it arises.

☐ 5. Using examples, explain how errors during cell division (e.g. non-disjunction) can result in changes in phenotype. Examples could include **polyploidy** in crop plants, and human aneuploidies, such as trisomy 21 (see #6 below).

☐ 6. Explain **aneuploidy** arising as a result of non-disjunction during meiosis. Explain the **maternal age effect**. Explain the significance of **Barr bodies** to the expression of human aneuploidies. Describe and explain examples of aneuploidy in human sex chromosomes (e.g. Turner syndrome) or autosomes (e.g **trisomy** 21).

☐ 7. Use examples to explain how genetic changes that result in new phenotypes may enhance survival and reproduction in particular environments. Examples include:
    (a) Antibiotic resistance in bacteria and pesticide resistance in insects.
    (b) Antigenic variability (antigenic drifts and antigenic shifts) in *Influenzavirus*.
    (c) Sickle cell mutation and **heterozygote advantage** (also see #8).

☐ 8. Explain how multiple copies of alleles or genes (**gene duplication**) may provide phenotypes that have a selective advantage in certain environments.

Genetic Variation *(3.C.2)*   pages 149-150, 210-211, 227,231, 257, 325

☐ 9. Describe the mechanisms by which biological system increase genetic variation:
    (a) Errors in DNA replication and repair.
    (b) Horizontal gene transfer by **transformation**, **transduction**, **conjugation**, and **transposition** (transfer of transposable elements).
    (c) **Sexual reproduction** (**meiosis** and **fertilization**).

Viral Replication *(3.C.3)*   pages 228-231, 326

☐ 10. Describe the nature of replication in viruses. Explain how the features of viral replication enable rapid evolution and acquisition of new phenotypes.

☐ 11. Describe examples of replication in viruses. Examples include:
    (a) Bacteriophage with **lytic** and **lysogenic phases**. Explain how latent viral genome can contribute to increase **pathogenicity** in the host bacterial cell.
    (b) Replication in RNA viruses (e.g. HIV). Explain how rapid evolution of the virus and the property of latency contribute to pathogenicity of the viral infection.

Periodicals:
Listings for this
chapter are on page 383

Weblinks:
www.thebiozone.com/
weblink/AP1-3114.html

BIOZONE APP:
Student Review Series
Mutations

Sources of Genetic Variation

Variation refers to the diversity within and between species. The genetic variation in species is largely due to meiosis and sexual reproduction, which shuffles existing genetic material into new combinations as it is passed from generation to generation. **Mutation** is also a source of variation as it may create new alleles. Variation gives species more opportunity to adapt to a changing environment because, at any one time, some individuals will have higher fitness (leave more offspring)

than others. Variation in a population can be continuous or discontinuous. Traits determined by a single gene (e.g. ABO blood groups) show **discontinuous variation**, with a very limited number of variants present in the population. In contrast, traits determined by a large number of genes (e.g. skin color) show **continuous variation**, and the number of phenotypes is very large. Environmental influences (differences in diet for example) also contribute to the observable variation in a population.

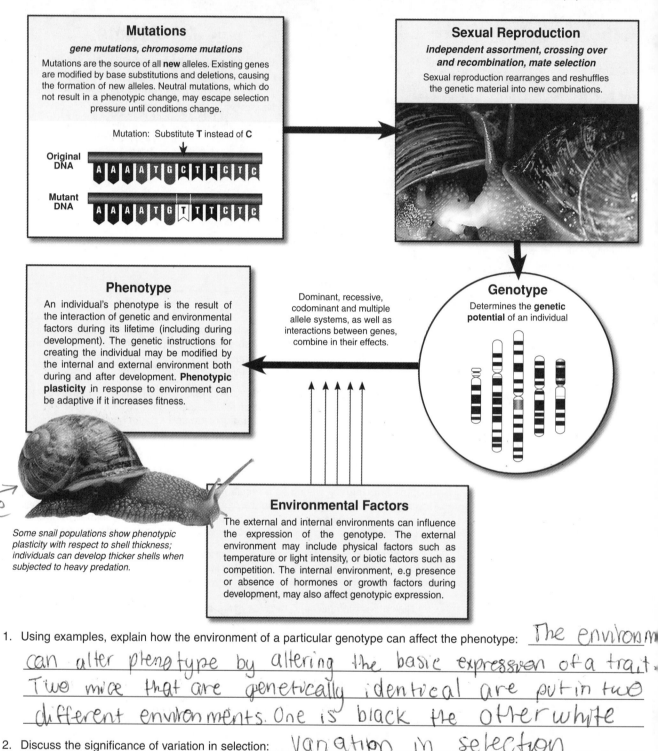

Mutations
gene mutations, chromosome mutations

Mutations are the source of all **new** alleles. Existing genes are modified by base substitutions and deletions, causing the formation of new alleles. Neutral mutations, which do not result in a phenotypic change, may escape selection pressure until conditions change.

Mutation: Substitute **T** instead of **C**

Original DNA: A A A A T G C T T T C T C

Mutant DNA: A A A A T G T T T T C T C

Sexual Reproduction
independent assortment, crossing over and recombination, mate selection

Sexual reproduction rearranges and reshuffles the genetic material into new combinations.

Phenotype

An individual's phenotype is the result of the interaction of genetic and environmental factors during its lifetime (including during development). The genetic instructions for creating the individual may be modified by the internal and external environment both during and after development. **Phenotypic plasticity** in response to environment can be adaptive if it increases fitness.

Dominant, recessive, codominant and multiple allele systems, as well as interactions between genes, combine in their effects.

Genotype

Determines the **genetic potential** of an individual

Some snail populations show phenotypic plasticity with respect to shell thickness; individuals can develop thicker shells when subjected to heavy predation.

Katie

Environmental Factors

The external and internal environments can influence the expression of the genotype. The external environment may include physical factors such as temperature or light intensity, or biotic factors such as competition. The internal environment, e.g presence or absence of hormones or growth factors during development, may also affect genotypic expression.

1. Using examples, explain how the environment of a particular genotype can affect the phenotype: *The environm can alter phenotype by altering the basic expression of a trait. Two mice that are genetically identical are put in two different environments. One is black the other white*

2. Discuss the significance of variation in selection: *Variation in selection can create new phenotype and new genotype in an organism causing more variation*

Related activities: Meiosis, Gene-Environment Interactions, The Nature of Mutation

Periodicals: What is variation?

© BIOZONE International 2012
ISBN: 978-1-927173-11-4
Photocopying Prohibited

Albinism (above) is the result of the inheritance of recessive alleles for melanin production. Those with the albino phenotype lack melanin pigment in the eyes, skin, and hair.

Comb shape in poultry is a **qualitative trait** and birds have one of four phenotypes depending on which combination of four alleles they inherit. The dash (missing allele) indicates that the allele may be recessive or dominant.

Quantitative traits are characterized by **continuous variation**, with individuals falling somewhere on a normal distribution curve of the phenotypic range. Typical examples include skin color and height in humans (left), grain yield in corn (above), growth in pigs (above, left), and milk production in cattle (far left). Quantitative traits are determined by genes at many loci (polygenic) but most are also influenced by environmental factors.

Single comb	Walnut comb	Pea comb	Rose comb
rrpp	**R_P_**	**rrP_**	**R_pp**

Flower color in snapdragons (right) is also a **qualitative trait** determined by two alleles. (red and white) The alleles show incomplete dominance and the heterozygote (C^RC^W) exhibits an intermediate phenotype between the two homozygotes.

C^RC^R

C^WC^W

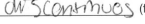
Christine

3. Describe three ways in which sexual reproduction can provide genetic variation in individuals:

(a) By providing an assertment of different gametes as a result of independent assertment

(b) Crossing over in chromosomes

(c) Mate selection will bring together the genes of two different individuals

4. (a) What is a **neutral mutation**? not harmful or beneficial

(b) What is the significance of neutral mutations? It benefit or disadventage on individual in the future

5. Describe the differences between **continuous** and **discontinuous** variation, giving examples to illustrate your answer:

Discontinuous variation is limited in the number of variants while in continuos it isnt

6. Identify each of the following phenotypic traits as continuous (quantitative) or discontinuous (qualitative):

(a) Wool production in sheep: Continuous

(b) Hand span in humans: Continuous

(c) Blood groups in humans: discontinuos

(d) Albinism in mammals: Continuous

(e) Body weight in mice: continuous

(f) Flower color in snapdragons: discontinuos

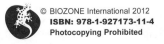

Mutagens

Mutations occur in all organisms spontaneously. The natural rate at which a gene will undergo a change is normally very low, but this rate can be increased by environmental factors such as ionizing radiation and mutagenic chemicals (e.g. benzene). Only mutations taking place in cells producing gametes (**gametic** **mutations**) will be inherited. If they occur in a body cell after the organism has begun to develop beyond the zygote (fertilized egg cell) stage, they are called **somatic mutations**. In some cases a mutation may trigger the onset of **cancer**, if the normal controls over gene regulation and expression are disrupted.

1. Describe examples of environmental factors that induce mutations under the following headings:

   (a) Radiation: _Nuclear radiation_
   exposure to
   radioisotopes,
   UV radiation, Xrays,
   Gamma rays

   (b) Chemical agents: _Poisons_
   and Irritants. Diets
   Alcohol and
   Tobacco smoke

2. Explain how **mutagens** cause mutations:

   They increase
   the chance of
   a mutation
   to occur.

3. Distinguish between **gametic** and **somatic** **mutations** and comment on the significance of the difference:

   Somatic is in the
   body and can't be
   passed down. Gametic
   can be passed down
   to future generations

Mutagen and Effect

Ionizing Radiation

Nuclear radiation from nuclear fallout or exposure to radioisotopes. Ultraviolet radiation from the sun and tanning lamps. X-rays and gamma rays from medical diagnosis and treatment. Ionizing radiation is associated with the development of cancers, e.g. thyroid cancers, and skin cancer (from high exposure to ultraviolet). Fair skinned people at low latitudes are at risk from ultraviolet radiation. Safer equipment has considerably reduced the risks to those working with ionizing radiation (e.g. radiographers).

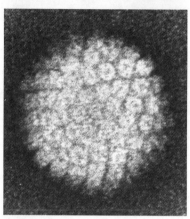

Viruses and Microorganisms

Some viruses integrate into the human chromosome, upsetting genes and triggering cancers. Examples include hepatitis B virus (liver cancer), HIV (Kaposi's sarcoma), and Epstein-Barr virus (Burkitt's lymphoma, Hodgkin's disease), and HPV (left) which is implicated in cervical cancer. Aflatoxins produced by the fungus *Aspergillus flavus* are potent inducers of liver cancer.

Those at higher risk of viral infections include intravenous drug users and those with unsafe sex practices

Poisons and Irritants

Many chemicals are mutagenic. Synthetic and natural examples include organic solvents such as benzene, asbestos, formaldehyde, tobacco tar, vinyl chlorides, coal tars, some dyes, and nitrites. Those most at risk include workers in the chemicals industries, including the glue, paint, rubber, resin, and leather industries, petrol pump attendants, and those in the coal and other mining industries.

Photo left: Firefighters and those involved in environmental clean-up of toxic spills are at high risk of exposure to mutagens.

Diet, Alcohol and Tobacco Smoke

Diets high in fat, especially those containing burned or fatty, highly preserved meat, slow the passage of food through the gut giving time for mutagenic irritants to form in the lower bowel.

High alcohol intake increases the risk of some cancers and increases susceptibility to tobacco-smoking related cancers.

Tobacco tar is one of the most damaging constituents of tobacco smoke. Tobacco tars contain at least 17 known carcinogens (cancer inducing mutagens) that cause chronic irritation of the gas exchange system and cause cancer in smokers.

Related activities: Changes to the DNA Sequence
Weblinks: Damage to DNA

Periodicals:
What is a mutation?

The Nature of Mutation

A **mutation** is a change in the genetic sequence of a genome. Mutations may occur spontaneously, as a result of errors occurring during meiosis or DNA replication, or they may be induced. The rate of spontaneous or natural mutation forms the **background mutation rate**, and it provides a means by which scientists can determine the divergence of taxa during evolution. Induced mutations are the result of agents called **mutagens**, which increase the natural mutation rate. These agents include ionizing radiation, some viruses, and chemicals such as formaldehyde, asbestos, coal tar, and tobacco smoke. While changes to DNA are likely to be harmful, there are many documented cases of mutations conferring a survival advantage. These **beneficial mutations** occur most frequently in organisms with short generation times, such as viruses, bacteria, and insects. Mutations that cause no change in the amino acid sequence are called **silent**. Until recently, it was supposed that these could be carried without effect until subjected to selection pressure at a later time. However, recent research indicates that even these silent mutations can influence function in unpredictable ways by altering mRNA stability and the accuracy of protein synthesis.

Harmful Mutations

There are many well-documented examples of mutations that cause harmful effects. Examples are the mutations giving rise to cystic fibrosis (CF) and sickle cell disease. The sickle cell mutation involves a change to only one base in the DNA sequence, whereas the CF mutation involves the loss of a single triplet (three nucleotides). The malformed proteins that result from these mutations cannot carry out their normal biological functions (although interestingly, the CF mutation, which disrupts chloride transport, makes carriers less susceptible to cholera). Albinism is caused by a mutation in the gene that produces an enzyme in the metabolic pathway to produce melanin. It is common in most vertebrate taxa. Albinos are more visible to predators, and are also more susceptible to the damaging effects of ultraviolet radiation because they lack protective pigmentation.

Beneficial Mutations

Salmonella

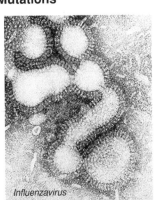
Influenzavirus

Bacteria reproduce asexually by binary fission. They are susceptible to antibiotics (substances that harm them or inhibit their growth) but are well-known for acquiring **antibiotic resistance** through mutation. The genes for bacterial resistance can be transferred within or even between bacterial species. New, multi-resistant bacterial superbugs have arisen in this way.

Viruses, including HIV and *Influenzavirus*, have membrane envelopes coated with glycoproteins. These are used by the host to identify the virus so that it can be destroyed. The genes coding for these glycoproteins on the virus are constantly mutating. The result is that each new viral 'strain' goes undetected by the immune system until well after the infection is established.

Are Silent Mutations Really Silent?

So-called **silent** or **synonymous mutations** are those that result in no change in the sequence of amino acids making up a protein. The redundancy of the genetic code provides a buffer against the effect of DNA changes that affect the third base. Such mutations have routinely been assumed to be neutral, meaning that, at that time, they have no effect on the phenotype or fitness of the individual carrying the mutation.

Right: A change to the third base of a codon may not change the amino acid encoded, but it does change the exonic sequence.

However, so-called silent changes still affect transcription, splicing, and mRNA stability, even though they do not change the codon information. Disruptions to RNA splicing sequences can cause exon skipping and lead to RNA not being processed properly. Synonymous variations have been associated with a number of diseases including cystic fibrosis and mental disorders. For example, one silent mutation causes the dopamine receptor D2 gene to be less stable and degrade faster, under-expressing the gene. Experimental evidence with the CFTR gene also shows that synonymous mutations cause exon skipping, yielding a short CFTR protein.

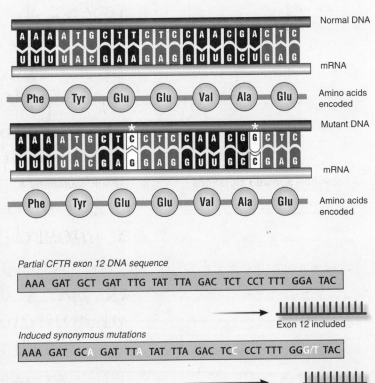

© BIOZONE International 2012
ISBN: 978-1-927173-11-4
Photocopying Prohibited

Periodicals:
What is a mutation?
The price of silent mutations

Related activities: Gene Mutations, Evolution of Drug Resistance
Web links: Evolution in E.coli

Sources of Variation

Only mutations taking place in the cells that produce gametes are inherited. If they occur in a body cell after the organism has begun to develop beyond the zygote stage, then they may give rise to a **chimera** (an organism with a mix of genetically different cells). In some cases, mutations trigger the onset of **cancer** through the disruption of the normal controls regulating cell division.

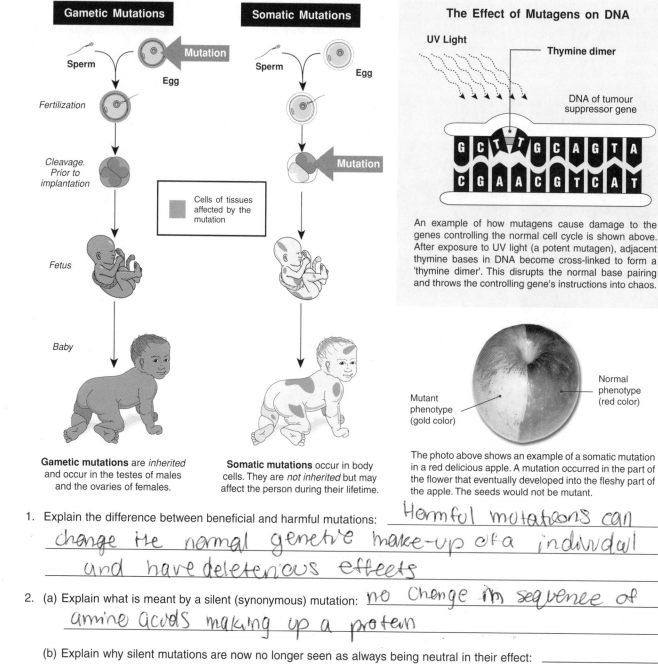

Gametic Mutations

Sperm

Egg

Mutation

Fertilization

Cleavage. Prior to implantation

Cells of tissues affected by the mutation

Fetus

Baby

Gametic mutations are *inherited* and occur in the testes of males and the ovaries of females.

Somatic Mutations

Sperm

Egg

Mutation

Somatic mutations occur in body cells. They are *not inherited* but may affect the person during their lifetime.

The Effect of Mutagens on DNA

UV Light

Thymine dimer

DNA of tumour suppressor gene

G C T T G C A G T A
C G A A C G T C A T

An example of how mutagens cause damage to the genes controlling the normal cell cycle is shown above. After exposure to UV light (a potent mutagen), adjacent thymine bases in DNA become cross-linked to form a 'thymine dimer'. This disrupts the normal base pairing and throws the controlling gene's instructions into chaos.

Mutant phenotype (gold color)

Normal phenotype (red color)

The photo above shows an example of a somatic mutation in a red delicious apple. A mutation occurred in the part of the flower that eventually developed into the fleshy part of the apple. The seeds would not be mutant.

1. Explain the difference between beneficial and harmful mutations: *Harmful mutations can change the normal genetic make-up of a individual and have deleterious effects*

2. (a) Explain what is meant by a silent (synonymous) mutation: *no change in sequence of amine acids making up a protein*

   (b) Explain why silent mutations are now no longer seen as always being neutral in their effect: *Still effect transcription, splicing mRNA stability*

3. Explain how **somatic mutations** differ from **gametic mutations** and comment on the significance of the difference: *Somatic mutation occur in the body and gametic mutation can be inherited*

4. Explain why organisms such as *Drosophila* and bacteria are frequently used in the study of mutations: *They have short life spans and cumulative effects of mutations over several generation can be feasibly studied*

5. Explain why the mutation seen in the red delicious apple (above right) will not be inherited: *It only happens in the body*

A Beneficial Mutation

High blood cholesterol in humans can be fatal. It is associated with various cardiovascular diseases such as atherosclerosis, which can result in a myocardial infarction (heart attack). In 1979, a railway employee born in the town of Limone sul Garda in Italy, but living in Milan, had a routine medical checkup. A blood test showed levels of cholesterol and triglycerides that amazed his doctors and sparked investigations into one of the more famous, **beneficial mutations** in humans ever studied.

The blood tests from the man from Limone showed he had very high levels of blood cholesterol, but he had no signs of damage to his heart or arteries. Further tests found his blood contained an anomalous protein similar to **Apolipoprotein A-1** (ApoA1), the protein that aids clearance of cholesterol from the blood. The protein, now called **Apolipoprotein A-1 Milan** (ApoA1 Milan), was able to remove cholesterol from the blood many times more efficiently than ApoA1.

Steep mountains behind the village

Lake Garda

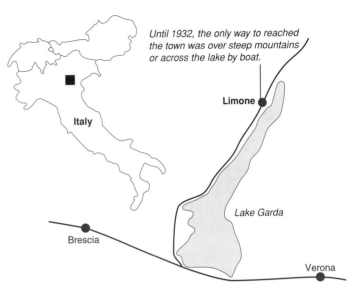

Until 1932, the only way to reached the town was over steep mountains or across the lake by boat.

Limone

Italy

Lake Garda

Brescia

Verona

The protein ApoA1 is a major component of high density lipoprotein (HDL) in blood plasma. HDL transports cholesterol to the liver where it is excreted into the bile (and then into the small intestine). Structurally it is composed of eight α-helices that form a twisted loop (below). Studies have shown ApoA1 Milan is characterized by a change in a single amino acid, from arginine at position 173 to cysteine. This small change makes it ten times more efficient at removing excess cholesterol to the liver.

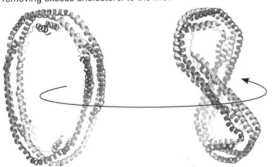

Researchers tested the blood of the man's family members and found the blood of his father and daughter both contained ApoA1 Milan. Blood tests were then conducted on all Limone's residents (the population at the time being around 1000), and found that many also had the mutated protein in their blood. The genealogy of the town was constructed using town and church records and found all the people who carried the ApoA1 Milan gene were related to Cristoforo Pomaroli and Rosa Giovanelli, who were married in 1644. Because the town was relatively isolated from the outside world until 1932, the gene spread through the population with little dilution from outside genes.

The bacterium *E. coli* was engineered to produce the ApoA1 Milan protein in order to try to produce a useful cholesterol reducing drug. In 2003 researchers in the U.S. produced an experimental drug for human testing. It was administered to 47 patients with heart disease. After a six week period there was an average 4.2% reduction in plaque along the arteries. Shortly after, Pfizer paid US$1.3 billion for the rights to the drug's production. However, due to technical difficulties little progress was made and Pfizer sold the rights for US$160 million in 2008.

E. coli

Sources of Variation

1. (a) What is the health effect of high cholesterol levels in the blood? ___can cause cardiovascular diseases___

   (b) Describe the function of ApoA1: ___protein that aids clearance of cholesterol from the blood___

   (c) How is the structure of ApoA1 Milan different to ApoA1? ___ApoA1 has a single change in amino acid___

2. Explain why people with the ApoA1 Milan gene show no sign of heart disease, even though they may have high levels of blood cholesterol:
   ___It is ten times more efficient at removing excess cholesterol to the liver___

3. Explain why Limone sul Garda's isolation may have help the spread of the gene in the population: _____

_____

_____

Related activities: Nature of Mutation

A 2

Gene Mutations

Gene mutations are small, localized changes in the structure of a DNA strand. These changes may be induced by a **mutagen** or arise spontaneously as a result of errors during DNA replication. The changes may involve a single nucleotide (often called **point mutations**), or changes to a triplet (e.g. triplet deletion or triplet repeat). The diagrams below show how point mutations can occur by substitution, insertion, or deletion. These alterations in the DNA are at the **nucleotide** level where individual **codons** are affected. Alteration of the precise nucleotide sequence of a coded gene in turn alters the mRNA transcribed from the mutated DNA and may affect the polypeptide chain that it creates. Note that mutations do not always result in altered proteins, because more than one codon may code for the same amino acid. Most of this **degeneracy** in the code occurs at the third base of a codon.

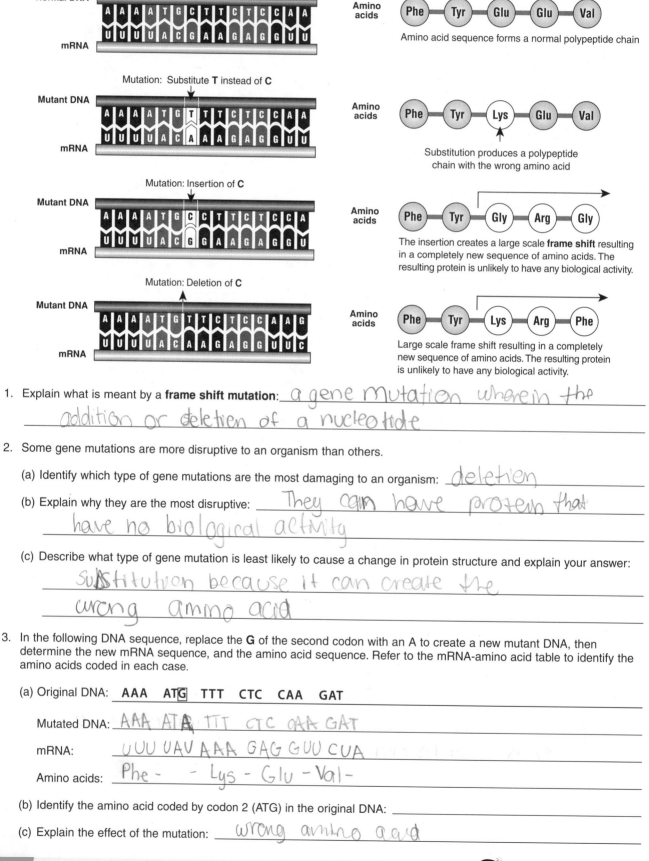

1. Explain what is meant by a **frame shift mutation**: a gene mutation wherein the addition or deletion of a nucleotide

2. Some gene mutations are more disruptive to an organism than others.

   (a) Identify which type of gene mutations are the most damaging to an organism: deletion

   (b) Explain why they are the most disruptive: They can have protein that have no biological activity

   (c) Describe what type of gene mutation is least likely to cause a change in protein structure and explain your answer: Substitution because it can create the wrong amino acid

3. In the following DNA sequence, replace the **G** of the second codon with an A to create a new mutant DNA, then determine the new mRNA sequence, and the amino acid sequence. Refer to the mRNA-amino acid table to identify the amino acids coded in each case.

   (a) Original DNA: **AAA   ATG   TTT   CTC   CAA   GAT**

   Mutated DNA: AAA ATA TTT CTC CAA GAT

   mRNA: UUU UAU AAA GAG GUU CUA

   Amino acids: Phe -   - Lys - Glu - Val -

   (b) Identify the amino acid coded by codon 2 (ATG) in the original DNA: _____

   (c) Explain the effect of the mutation: wrong amino acid

Related activities: The Genetic Code, Cystic Fibrosis, Mutation, Sickle Cell Mutation
Web links: Mutation by Base Substitution, Addition and Deletion Mutations

Inherited Metabolic Disorders

Humans have more than 6000 physiological diseases attributed to mutations in single genes and over one hundred syndromes known to be caused by chromosomal abnormality. The number of genetic disorders identified increases every year. Rapid progress of the Human Genome Project is enabling the identification of the genetic basis of these disorders. This will facilitate the development of new drug therapies and gene therapies. Four genetic disorders are summarized below.

Sickle Cell Disease	β-Thalassemia	Cystic Fibrosis	Huntington Disease
Synonym: Sickle cell anemia	**Synonyms**: Cooley anemia, Mediterranean anemia	**Synonyms**: Mucoviscidosis, CF	**Synonyms**: Huntington's chorea, HD (abbreviated)
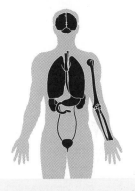	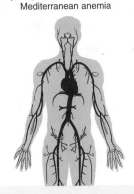	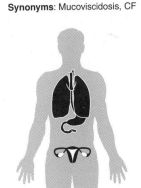	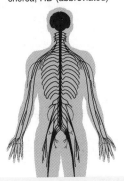
Incidence: Occurs most commonly in people of African ancestry. West Africans: 1% (10-45% carriers) West Indians: 0.5%	**Incidence**: Most common type of thalassemia affecting 1% of some populations. More common in Asia, Middle East and Mediterranean.	**Incidence**: Varies with populations: United States: 1 in 1000 (0.1%) Asians in England: 1 in 10,000 Caucasians: 1 in 20-28 are carriers.	**Incidence**: An uncommon disease present in one in 20,000.
Gene type: Autosomal mutation which results in the substitution of a single nucleotide in the HBB gene that codes for the beta hemoglobin chain. The allele is codominant.	**Gene type**: Autosomal recessive mutation of the HBB gene coding for the hemoglobin beta chain. It may arise through a gene deletion or a nucleotide deletion or insertion.	**Gene type**: Autosomal recessive. Over 500 different recessive mutations (deletions, missense, nonsense, terminator codon) of the CFTR gene have been identified.	**Gene type**: An autosomal dominant mutation of the HD gene (IT15) caused by an increase in the length (36-125) of a CAG repeat region (normal range is 11-30 repeats).
Gene location: Chromosome 11	**Gene location**: Chromosome 11	**Gene location**: Chromosome 7	**Gene location**: Chromosome 4
HBB	HBB	CFTR	IT15
Symptoms: Include pain, ranging from mild to severe, in the chest, joints, back, or abdomen; swollen hands and feet; jaundice; repeated infections, particularly pneumonia or meningitis; kidney failure; gallstones (at an early age); strokes (at an early age), anemia.	**Symptoms**: The result of hemoglobin with few or no beta chains, causes a severe anemia during the first few years of life. People with this condition are tired and pale because not enough oxygen reaches the cells.	**Symptoms**: Disruption of glands: the pancreas, intestinal glands, biliary tree (biliary cirrhosis), bronchial glands (chronic lung infections), and sweat glands (high salt content of which becomes depleted in a hot environment). Infertility occurs in males/females.	**Symptoms**: Mutant gene forms defective protein: **huntingtin**. Progressive, selective nerve cell death associated with chorea (jerky, involuntary movements), psychiatric disorders, and dementia (memory loss, disorientation, impaired ability to reason, and personality changes).
Treatment and outlook: Patients are given folic acid. Acute episodes may require oxygen therapy, intravenous infusions of fluid, and antibiotic drugs. Experimental therapies include bone marrow transplants and gene therapy.	**Treatment and outlook**: Patients require frequent blood transfusions. This causes iron build-up in the organs, which is treated with drugs. Bone marrow transplants and gene therapy hold promise and are probable future treatments.	**Treatment and outlook**: Conventional: chest physiotherapy, a modified diet, and the use of TOBI antibiotic to control lung infections. Outlook: Gene therapy inserting normal CFTR gene using adenoviral vectors and liposomes.	**Treatment and outlook**: Surgical treatment may be possible. Research is underway to discover drugs that interfere with huntingtin protein. Genetic counselling coupled with genetic screening of embryos may be developed in the future.

Sources of Variation

1. For each of the genetic disorder below, indicate the following:

    (a) Sickle cell disease: Gene name: **HBB** Chromosome: **11** Mutation type: **Substitution**

    (b) β-thalassemia: Gene name: *HBB* Chromosome: *11* Mutation type: *deletion*

    (c) Cystic fibrosis: Gene name: *CFTR* Chromosome: *7* Mutation type: *recessive*

    (d) Huntington disease: Gene name: *ITIS* Chromosome: *4* Mutation type: *autosomal*

2. Explain the cause of the symptoms for people suffering from β-thalassemia: *Hemoglobin with few or no beta chains*

3. Suggest a reason for the differences in the country-specific incidence rates for some genetic disorders:

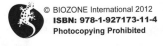

Related activities: Sickle Cell Mutation, Cystic Fibrosis Mutation

A 2

Sickle Cell Mutation

Sickle cell disease (formerly called sickle cell anemia) is an inherited disorder caused by a gene mutation which codes for a faulty beta (β) chain hemoglobin (Hb) protein. This in turn causes the red blood cells to deform causing a whole range of medical problems. The DNA sequence below is the beginning of the transcribing sequence for the **normal** β-chain Hb molecule.

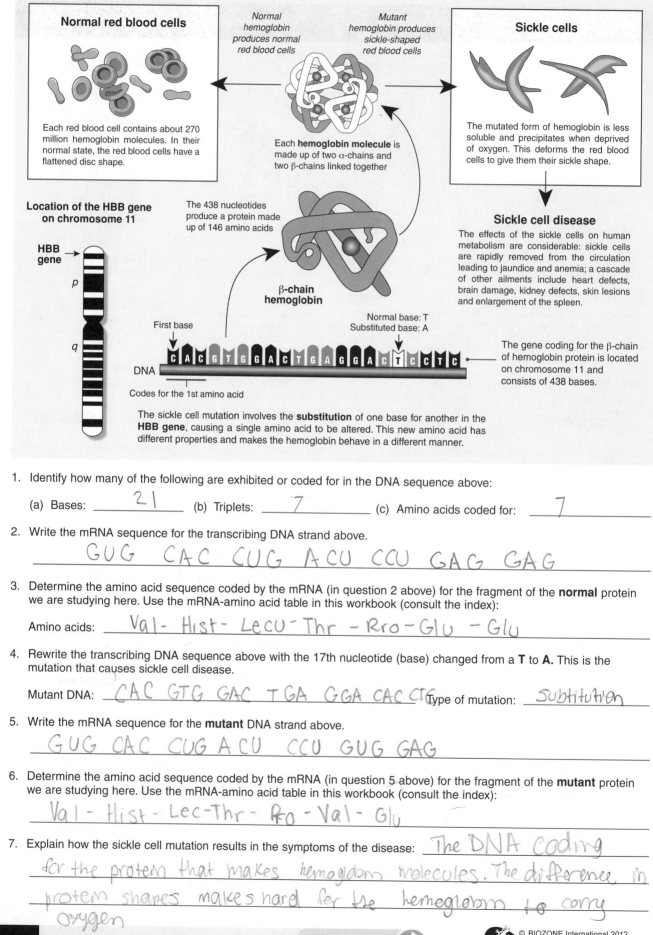

Normal red blood cells

Each red blood cell contains about 270 million hemoglobin molecules. In their normal state, the red blood cells have a flattened disc shape.

Normal hemoglobin produces normal red blood cells

Mutant hemoglobin produces sickle-shaped red blood cells

Each **hemoglobin molecule** is made up of two α-chains and two β-chains linked together

Sickle cells

The mutated form of hemoglobin is less soluble and precipitates when deprived of oxygen. This deforms the red blood cells to give them their sickle shape.

Location of the HBB gene on chromosome 11

The 438 nucleotides produce a protein made up of 146 amino acids

HBB gene

p

β-chain hemoglobin

Sickle cell disease

The effects of the sickle cells on human metabolism are considerable: sickle cells are rapidly removed from the circulation leading to jaundice and anemia; a cascade of other ailments include heart defects, brain damage, kidney defects, skin lesions and enlargement of the spleen.

q

First base

Normal base: T
Substituted base: A

DNA

C A C G T G G A C T G A G G A C T C C T C

Codes for the 1st amino acid

The gene coding for the β-chain of hemoglobin protein is located on chromosome 11 and consists of 438 bases.

The sickle cell mutation involves the **substitution** of one base for another in the **HBB gene**, causing a single amino acid to be altered. This new amino acid has different properties and makes the hemoglobin behave in a different manner.

1. Identify how many of the following are exhibited or coded for in the DNA sequence above:

   (a) Bases: _____21_____   (b) Triplets: _____7_____   (c) Amino acids coded for: _____7_____

2. Write the mRNA sequence for the transcribing DNA strand above.

   GUG   CAC   CUG   ACU   CCU   GAG   GAG

3. Determine the amino acid sequence coded by the mRNA (in question 2 above) for the fragment of the **normal** protein we are studying here. Use the mRNA-amino acid table in this workbook (consult the index):

   Amino acids: Val- Hist- Lecu-Thr -Rro-Glu -Glu

4. Rewrite the transcribing DNA sequence above with the 17th nucleotide (base) changed from a **T** to **A**. This is the mutation that causes sickle cell disease.

   Mutant DNA: CAC GTG GAC TGA GGA CAC CTG   Type of mutation: Subtitution

5. Write the mRNA sequence for the **mutant** DNA strand above.

   GUG CAC CUG ACU CCU GUG GAG

6. Determine the amino acid sequence coded by the mRNA (in question 5 above) for the fragment of the **mutant** protein we are studying here. Use the mRNA-amino acid table in this workbook (consult the index):

   Val - Hist- Lec-Thr- Rro -Val- Glu

7. Explain how the sickle cell mutation results in the symptoms of the disease: The DNA coding for the protein that makes hemoglobin molecules. The difference in protein shapes makes hard for the hemoglobin to carry oxygen

Related activities: The Genetic Code
Weblinks: Sickle Cell Disease

Periodicals: Genetics of sickle cell anemia

© BIOZONE International 2012
ISBN: 978-1-927173-11-4
Photocopying Prohibited

Cystic Fibrosis Mutation

Cystic fibrosis an inherited disorder caused by a mutation of the **CF gene**. It is one of the most common lethal autosomal recessive conditions affecting caucasians, with an incidence of 1 in 2500 live births and a **carrier frequency** of 4%. It is uncommon in Asians and Africans. The CF gene's protein product, **CFTR**, is a membrane-based protein with a function in regulating the transport of chloride across the membrane. A faulty gene in turn codes for a faulty CFTR. More than 500 mutations of the CF gene

have been described, giving rise to disease symptoms of varying severity. One mutation is particularly common and accounts for more than 70% of all defective CF genes. This mutation, called δ(delta)F508, leads to the absence of CFTR from its correct position in the membrane (below). Another CF mutation, R117H, which is also relatively common, produces a partially functional CFTR protein. The DNA sequence below is part of the transcribing sequence for the **normal** CF gene.

Normal CFTR *(1480 amino acids)*
Correctly controls chloride ion balance in the cell

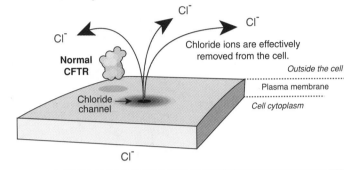

Abnormal CFTR *(1479 amino acids)*
Unable to control chloride ion balance in the cell

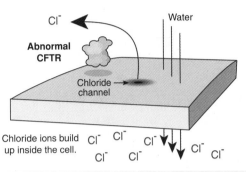

The CF gene on chromosome 7

The CF gene is located on chromosome 7. The δF508 mutation of the CF gene describes a deletion of the 508th triplet, which in turn causes the loss of a single **amino acid** from the gene's protein product, the cystic fibrosis transmembrane conductance regulator (CFTR). This protein normally regulates the chloride channels in cell membranes, but the mutant form fails to achieve this. The portion of the DNA containing the mutation site is shown below:

The CFTR protein consists of 1480 amino acids

CFTR protein

The δF508 mutant form of CFTR fails to take up its position in the membrane. Its absence results in defective chloride transport and leads to a net increase in water absorption by the cell. This accounts for the symptoms of cystic fibrosis, where mucus-secreting glands, particularly in the lungs and pancreas, become fibrous and produce abnormally thick mucus. The widespread presence of CFTR throughout the body also explains why CF is a multisystem condition affecting many organs.

Base 1630

← CFTR gene

DNA: C C G T G G T A A T T T C T T T T A T A G T A G A A A C C A C C A

This triplet codes for the 500th amino acid

The 508th triplet is absent in the form with the δF508 mutation

1. (a) Write the mRNA sequence for the transcribing DNA strand above:

   GGC ACC AUU AAA GAA AAU AUC AUC UUU GGU GGU

   (b) Use the mRNA-amino acid table earlier in this workbook to determine the amino acid sequence coded by the mRNA for the fragment of the normal protein we are studying here:

   Gly - Thr - Iso - Lys - Glu - Asp - Iso - Iso - Gly · Gly

2. (a) Rewrite the mRNA sequence for the mutant DNA strand:

   CCG - TGG - TAA - TTT - CTT - TTA - TAG - TAG - CCA - CCA

   (b) State what kind of mutation δF508 is: deletion mutation

   (c) Determine the amino acid sequence coded by the mRNA for the fragment of the δF508 mutant protein:

   GGC - ACC - AUU - AA - GAA - AAU - AUC - AUC - GGU - GGU

   (d) Identify the amino acid that has been removed from the protein by this mutation: Phenylalanine

3. Suggest why cystic fibrosis is a disease with varying degrees of severity:

© BIOZONE International 2012
ISBN: 978-1-927173-11-4
Photocopying Prohibited

Related activities: The Genetic Code, Inherited Metabolic Disorders
Weblinks: Cystic Fibrosis

RA 3

Sources of Variation

Chromosome Mutations

The diagrams below show the different types of chromosome mutation that can occur only during **meiosis**. These mutations (sometimes also called block mutations) involve the rearrangement of blocks of genes, rather than individual bases within a gene. Each type of mutation results in an alteration in the number and/or sequence of whole sets of genes (represented by letters) on the chromosome. In humans, **translocations** occur with varying frequency (several rare types of Down syndrome occur in this way). Individuals with a **balanced translocation** have the correct amount of genetic material and appear phenotypically normal but have an increased chance

of producing faulty gametes. Translocation may sometimes involve the fusion of whole chromosomes, thereby reducing the chromosome number of an organism. This is thought to be an important mechanism by which **instant speciation** can occur. There is good evidence that a translocation has happened in human evolution. Humans have 23 pairs of chromosomes, while the rest of the apes have 24. Evidence based on centromeres and telomeres in chromosome 2 show that this was once two separate chromosomes. A similar situation is found between Przewalski's wild horse, which has 33 pairs of chromosomes, and domestic horses, which have 32.

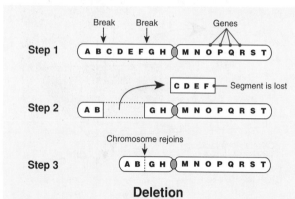

Deletion

A break may occur at two points on the chromosome and the middle piece of the chromosome falls out. The two ends then rejoin to form a chromosome deficient in some genes. Alternatively, the end of a chromosome may break off and is lost.

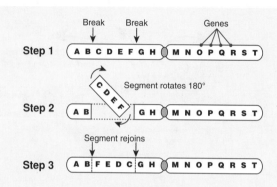

Inversion

The middle piece of the chromosome falls out and rotates through 180° and then rejoins. There is no loss of genetic material. The genes will be in a reverse order for this segment of the chromosome.

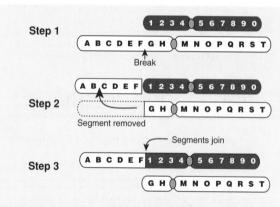

Translocation

Translocation involves a group of genes moving between chromosomes. The large chromosome (white) and the small chromosome (blue) are not homologous. A piece of one chromosome breaks off and joins to another. This will cause major problems when the chromosomes are passed to the gametes. Some will receive extra genes, while some will be deficient.

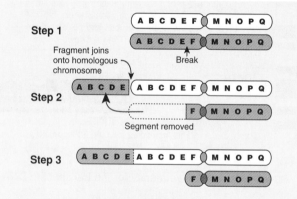

Duplication

A segment is lost from one chromosome and is added to its homologue. In this diagram, the darker chromosome on the bottom is the 'donor' of the duplicated piece of chromosome. The chromosome with the segment removed is deficient in genes. Some gametes will receive double the genes while others will have no genes for the affected segment.

1. For each of the chromosome (block) mutations illustrated above, write the original gene sequence and the new gene sequence after the mutation has occurred (the first one has been done for you):

	Original sequence(s)	Mutated sequence(s)
(a) Deletion:	A B C D E F G H M N O P Q R S T	A B G H M N O P Q R S T
(b) Inversion:	ABCDEF GH MNOPQRST	AB FEDC GH MNOPQRST
(c) Translocation:	ABCDEF GHMNO PQRST	ABCDEF 1234567890
	1234567890	GH MNOPQRST
(d) Duplication:	ABCDEF MNOPQST	ABCDE ABCDEF MNOPQ
	AB CDEF MNOPQST	FMNOPQ

Related activities: Trisomy in Human Autosomes
Weblinks: Cri-du-chat Syndrome

Periodicals:
What is a chromosome mutation?

© BIOZONE International 2012
ISBN: 978-1-927173-11-4
Photocopying Prohibited

Gene Duplication and Evolution

Over the last four decades, gene duplication as an evolutionary process has been hypothesized. Duplication creates more genetic information, which can then be influenced by selection pressures. Studies have shown that gene duplication has occurred in virtually every species and often more than once. It is well known that many important crop plants (e.g. wheat) have duplicated not just one gene, but their entire genome. The duplication of a gene frees one of those genes to develop a new function, while the original continues as it was. In the case of genes that produce proteins with a tendency or ability to perform two functions, there may be **adaptive conflict**, in which its ability to perform one function compromises its ability to perform another. Gene duplication solves this problem by allowing natural selection to act on the genes so that they follow different evolutionary paths.

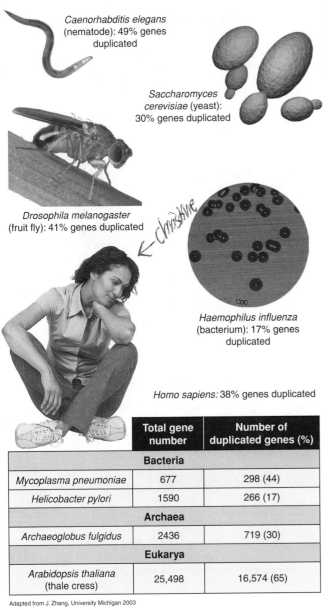

Caenorhabditis elegans (nematode): 49% genes duplicated

Saccharomyces cerevisiae (yeast): 30% genes duplicated

Drosophila melanogaster (fruit fly): 41% genes duplicated

← christine

Haemophilus influenza (bacterium): 17% genes duplicated

CDC

Homo sapiens: 38% genes duplicated

ART G CC 2.0

Gene duplication in colobine monkeys has enabled the production of enzymes that optimally perform similar functions in different body environments. The primary food source of colobines, unlike most other primates, is leaves. The leaves are fermented in the gut by bacteria, which are then digested with the assistance of an enzyme produced by RNase genes.

In colobines there are two forms of the RNase genes, **RNase1** and **RNase1B**, while in other primates there is only RNase1.

The optimal pH for the enzyme RNase1 is 7.4. For RNase1B, the optimal pH is 6.3. In colobine monkeys, the pH of the digestive system is 6-7, but in other primates it is 7.4-8. RNase1B is six times more efficient at degrading RNA in the gut of colobines than RNase1. RNase1 is also expressed in cells outside the digestive system where it degrades double stranded RNA and may assist in defense against viral infection. RNaseB1 is 300 times less efficient at this function.

Fish living in the near freezing waters of the Antarctic must have a way of ensuring their blood remains ice free. In many species, this is done by employing proteins with antifreeze properties. There are four major antifreeze proteins used by fish (labeled AFP types I - IV). The gene for the protein AFP III, found in Antarctic eelpout, is very similar to the gene that

AFP III

produces sialic acid synthase (SAS) (also found in humans). Molecular studies have found that a slight modification in the SAS gene causes the production and secretion of AFP III. More importantly, the SAS gene also shows ice binding capabilities. It appears that a duplication of the SAS gene produced a new gene that was selected for its ice binding capabilities and thus diverged to become the AFP III gene in Antarctic eelpout.

Total gene number	Number of duplicated genes (%)
Bacteria	
Mycoplasma pneumoniae 677	298 (44)
Helicobacter pylori 1590	266 (17)
Archaea	
Archaeoglobus fulgidus 2436	719 (30)
Eukarya	
Arabidopsis thaliana (thale cress) 25,498	16,574 (65)

Adapted from J. Zhang, University Michigan 2003

2. Which of the types of block mutation is likely to be the least damaging to an organism? Explain your answer:

_____

_____

3. (a) Use an example to explain how gene duplication allows for the evolution of new gene function:

_____

_____

_____

(b) Explain how duplication can resolve adaptive conflict: _____

_____

_____

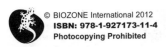
Sources of Variation

Aneuploidy in Humans

Euploidy is the condition of having an exact multiple of the haploid number of chromosomes. Normal euploid humans have 46 chromosomes (2N). **Aneuploidy** is the condition where the chromosome number is not an exact multiple of the normal haploid set for the species (the number may be more, e.g. 2N+2, or less, e.g. 2N–1). **Polysomy** is aneuploidy involving reduplication of some of the chromosomes beyond the normal diploid number (e.g. 2N+1). Aneuploidy usually results from the **non-disjunction** (failure to separate) of homologous chromosomes during meiosis. The two most common forms are monosomy (e.g. Turner syndrome) and trisomy (e.g. Down and Klinefelter syndrome) as outlined on the next few pages.

Faulty Egg Production	Faulty Sperm Production
The male has produced normal gametes, but the female has not. The two X-sex chromosomes failed to separate during the first division of meiosis.	The female has produced normal gametes, while the male has had an error during the first division of meiosis. The two sex chromosomes (**X** and **Y**) failed to separate during the first division.

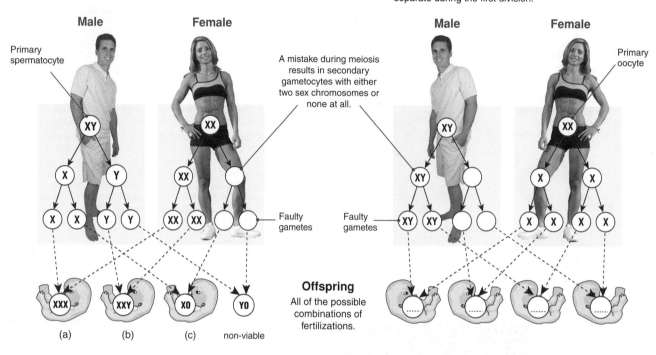

1. Identify the sex chromosomes in each of the unlabeled embryos (above, right):

2. Using the next activity, identify the syndrome for each of the offspring labeled (a) to (g), above:

   (a) _____   (e) _____

   (b) _____   (f) _____

   (c) _____   (g) _____

   (d) _____

3. Explain why the YO configuration (above) is non-viable (i.e. there is no embryonic development): _____

   _____

4. Discuss the nature and causes of aneuploidy, describing its causative role in one named **syndrome**:

   _____

   _____

   _____

   _____

   _____

   _____

   _____

   _____

Related activities: Meiosis, Aneuploidy in Sex Chromosomes

Aneuploidy in Sex Chromosomes

Turner Syndrome (XO)

Turner syndrome results from the non-disjunction of the sex chromosomes during meiosis (see the activity "Aneuploidy in Humans" for an explanation of the process). The individual has only **one sex chromosome (X)**, totally lacking either another **X** or a **Y**. The karyotype (pictured here) has a total of 45 chromosomes (one less than the normal 46). The incidence rate is 1 in 5000 live female births.

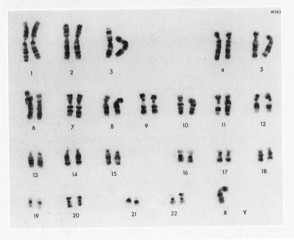

Klinefelter Syndrome (XXY)

Klinefelter syndrome also results from the non-disjunction of the sex chromosomes during meiosis. The individual has an extra sex chromosome (**X**), to produce a total complement of **XXY**. The karyotype (pictured here) of a Klinefelter syndrome patient shows a total complement of **47** including **XXY** sex chromosomes. The incidence rate is an average of 1 in 1000 live male births, with a maternal age effect.

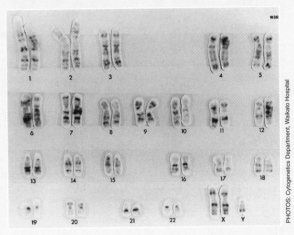

PHOTOS: Cytogenetics Department, Waikato Hospital

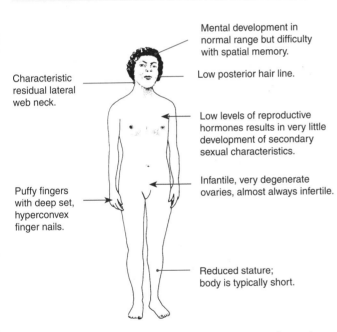

Mental development in normal range but difficulty with spatial memory.

Characteristic residual lateral web neck.

Low posterior hair line.

Low levels of reproductive hormones results in very little development of secondary sexual characteristics.

Infantile, very degenerate ovaries, almost always infertile.

Puffy fingers with deep set, hyperconvex finger nails.

Reduced stature; body is typically short.

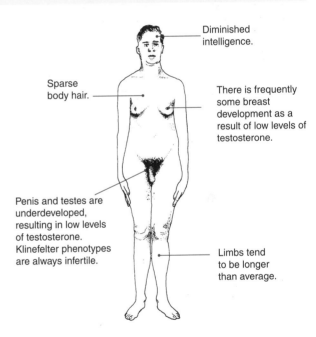

Diminished intelligence.

Sparse body hair.

There is frequently some breast development as a result of low levels of testosterone.

Penis and testes are underdeveloped, resulting in low levels of testosterone. Klinefelter phenotypes are always infertile.

Limbs tend to be longer than average.

1. Study the Klinefelter syndrome and Turner syndrome karyotypes above.

   (a) Place a circle around the sex chromosomes of each.

   (b) For Turner syndrome: State the chromosome configuration: _____ Sex: _____

   (c) For Klinefelter syndrome: State the chromosome configuration: _____ Sex: _____

2. Describe the features of Turner syndrome, identifying the traits associated with low levels of female hormones:

   _____

   _____

3. Describe the features of Klinefelter syndrome, identifying the traits associated with low levels of testosterone:

   _____

   _____

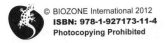
Related activities: Meiosis, Aneuploidy in Sex Chromosomes
Weblinks: Klinefelter Syndrome, Turner Syndrome

A 2

Sources of variation

Examples of Aneuploidy in Human Sex Chromosomes

Sex chromosomes and chromosome condition	Apparent sex	Phenotype
XO, monosomic	Female	Turner syndrome
XX, disomic	Female	**Normal female**
XXX, trisomic	Female	Metafemale. Most appear normal; they have a greater tendency to criminality
XXXX, trisomic	Female	Rather like Down syndrome, low fertility and intelligence
XY, disomic	Male	**Normal male**
XYY, trisomic	Male	Jacob syndrome, apparently normal male, tall, aggressive
XXY, trisomic	Male	Klinefelter syndrome (infertile). Incidence rate 1 in 1000 live male births, with a maternal age effect.
XXXY, tetrasomic	Male	Extreme Klinefelter, mentally retarded

ABOVE: Features of selected aneuploidies in humans. Note that this list represents only a small sample of the possible sex chromosome aneuploidies in humans.

RIGHT: Symbolic representation of Barr body occurrence in various human karyotypes. The chromosome number is given first, and the inactive X chromosomes (Xi) are framed by a black box. Note that in aneuploid syndromes, such as those described here, all but one of the X chromosomes are inactivated, regardless of the number present.

If extra copies of X are inactivated, why do extra copies still produce the aneuploidy syndromes? This is because some of the genes on the Xi escape inactivation so the dosage of these non-silenced genes will differ as they escape inactivation.

Barr Bodies

In the nucleus of any non-dividing somatic cell, one of the X chromosomes condenses to form a visible piece of chromatin, called a **Barr body**. This chromosome is inactivated (Xi), so that only one X chromosome in a cell ever has its genes expressed. The inactivation is random, so Xi may be either the maternal homologue (from the mother) or the paternal homologue (from the father).

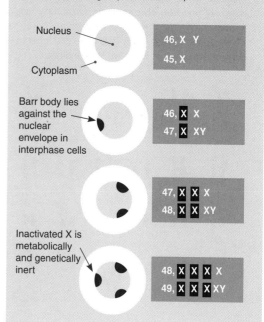

4. State how many Barr bodies are present in each somatic cell for each of the following syndromes:

   (a) Jacob syndrome: _____ (b) Klinefelter syndrome: _____ (c) Turner syndrome: _____

5. Explain the consequence of X-chromosome inactivation in terms of the proteins encoded by the X chromosome genes:

   _____

   _____

6. State how many chromosomes for each set of homologues are present for the following forms of aneuploidy:

   (a) Nullisomy: _____ (c) Trisomy: _____

   (b) Monosomy: _____ (d) Polysomy: _____

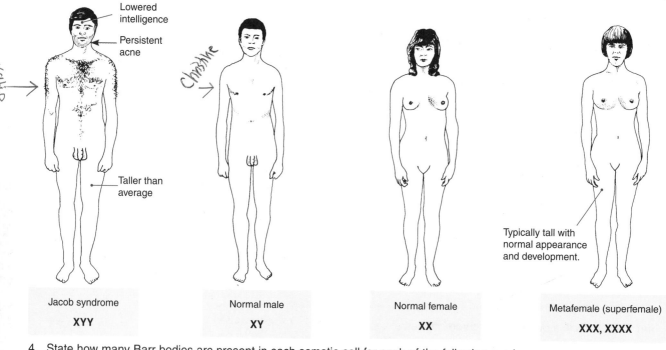

Trisomy in Humans Autosomes

Trisomy is a form of **polysomy** where the nucleus of the cells have one chromosome pair represented by three chromosomes (2N+1). The extra chromosome disturbs the overall chromosomal balance causing abnormalities or death. In humans, about 50% of all spontaneous abortions result from chromosomal abnormalities, and trisomies are responsible for about half of these (25% of all spontaneous abortions). About 6% of live births involve children with chromosomal abnormalities. Autosomal trisomies make up only 0.1% of all pregnancies. Of the three autosomal trisomies surviving to birth, trisomy 21 (**Down** syndrome) is the most common. The other two, **Edward** and **Patau**, show severe physical and mental abnormalities. Trisomies in other autosomes are rare.

Down Syndrome (Trisomy 21)

Down syndrome is the most common of the human aneuploidies. The incidence rate in humans is about 1 in 800 births for women aged 30 to 31 years, with a maternal age effect (the rate increases rapidly with maternal age). The most common form of this condition arises when meiosis fails to separate the pair of chromosome number 21s in the eggs that are forming in the woman's ovaries (it is apparently rare for males to be the cause of this condition). In addition to growth failure and mental retardation, there are a number of well known phenotypic traits (see diagram right).

Down syndrome may arise from several causes:

Non-disjunction: Nearly all cases (approximately 95%) result from **non-disjunction** of chromosome 21 during **meiosis**. When this happens, a gamete (usually the oocyte) ends up with 24 rather than 23 chromosomes, and fertilization produces a trisomic offspring (see the karyotype photo, above right).

Translocation: One in twenty cases of Down syndrome (fewer than 3-4%) arise from a **translocation** mutation where one parent is a translocation carrier (chromosome 21 is fused to another chromosome, usually number 14).

Mitotic errors: A very small proportion of cases (fewer than 3%) arise from the failure of the pair of chromosomes 21 to separate during **mitosis** at an early embryonic stage. The resulting individual is a **mosaic** in which two cell lines exist, one of which is trisomic. If the mitotic abnormality occurs very early in development, a large number of cells are affected and the full Down syndrome is expressed. If only a few cells are affected, there are only mild expressions of the syndrome.

A child showing features of the Down syndrome phenotype.

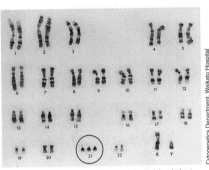
The karyotype of a trisomic 21 individual that produces the phenotype known as Down syndrome.

Cytogenetics Department, Waikato Hospital

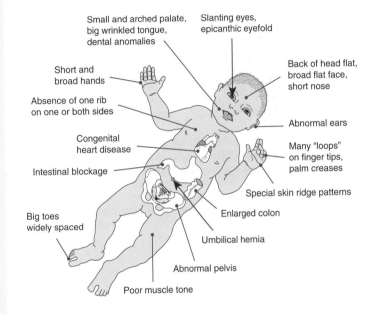

Small and arched palate, big wrinkled tongue, dental anomalies

Slanting eyes, epicanthic eyefold

Short and broad hands

Absence of one rib on one or both sides

Congenital heart disease

Intestinal blockage

Big toes widely spaced

Back of head flat, broad flat face, short nose

Abnormal ears

Many "loops" on finger tips, palm creases

Special skin ridge patterns

Enlarged colon

Umbilical hernia

Abnormal pelvis

Poor muscle tone

1. Distinguish between an autosomal aneuploidy and one involving the sex chromosomes: _____

_____

2. (a) Suggest a possible reason why the presence of an extra chromosome causes such a profound effect on the development of a person's phenotype:

_____

(b) With reference to Down syndrome, explain what you understand by the term **syndrome**: _____

_____

3. (a) Describe the main cause of Down syndrome: _____

_____

(b) Describe the features of the Down phenotype: _____

_____

4. Describe one other cause of Down syndrome: _____

_____

5. State how many chromosomes would be present in the somatic cells of an individual with Down syndrome: _____

Related activities: Aneuploidy in Humans
Web links: Down Syndrome

A 2

Sources of Variation

Polyploidy as a Source of Variation

Polyploidy is a condition in which a cell or organism contains three or more times the haploid number of chromosomes (**3N** or more). Polyploidy is rare in animals, but more common in plants. Natural polyploid species occur in animals particularly among hermaphrodites (those having both male and female sex organs) such as flatworms and earthworms. Animals with **parthenogenetic** females (which produce viable offspring without fertilization) can be polyploid (e.g. some beetles, moths, shrimp, goldfish, and salamanders). Polyploid species occur in all major groups of plants. They are common in ferns and occur in about 47% of all flowering plants. There are two types of polyploidy recognized: **allopolyploidy** and **autopolyploidy**.

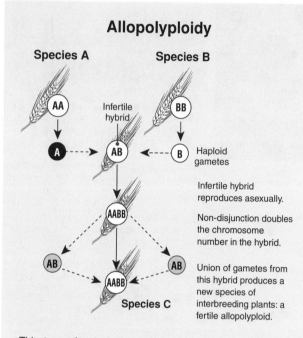

Allopolyploidy

Species A   Species B

AA   Infertile hybrid   BB

A - - -> AB <- - - B   Haploid gametes

Infertile hybrid reproduces asexually.

AABB

Non-disjunction doubles the chromosome number in the hybrid.

AB   AB

AABB

Species C

Union of gametes from this hybrid produces a new species of interbreeding plants: a fertile allopolyploid.

This type of polyploidy results from mating between **two species**. The resulting **hybrid**, with chromosomes from each of the parent species, may be sterile. Non-disjunction in the sterile hybrid can result in all chromosomes having a homologue with which to pair during meiosis. Self-fertilization may then produce a viable, fertile hybrid. Allopolyploids show greater heterozygosity and hybrid vigor than autopolyploids.

Autopolyploidy

Same species   Same species

AA   AA   AA   AA

A   AA   AA <- Diploid gametes - AA

Normal haploid gamete   Diploid gamete

AAA   Sterile hybrid   AAAA   Fertile hybrid

Autopolyploidy refers to the multiplication of one basic set of chromosomes. It occurs when chromosomes fail to separate during meiosis or from the failure of the cell to divide after the chromatids have separated. The karyotype possesses chromosomes only from **one species**. Two forms are shown above, one sterile, the other fertile. The total chromosome complement is represented by a multiple of identical sets.

Many commercial plant varieties are allopolyploids. Their increased hybrid vigor relative to the parental genotypes is the result of many factors, including epigenetic influences and masking of deleterious alleles. Polyploidy also results in gene redundancy and so provides the ability to diversify gene function so that extra copies of genes might end up being used in entirely different ways.

Common wheat
6N = 42

Tobacco
4N = 48

Banana
3N = 27

Boysenberry
7N = 49

Strawberry
8N = 56

1. Distinguish between **polyploidy** and **polysomy** (a type of aneuploidy): _____
   Organisms contain 3 or mor times the haploid # of chromosomes

2. Explain why allopolyploids show greater heterozygosity than autopolyploids: _____
   _____

3. Explain why polyploids might have a selective advantage over the parental genotypes: _____
   Polyploidy allow for more protein combinations as a result, polyploid plants may have the ability to survive in a wider variety of environment

Related activities: Breeding Modern Wheat

The Genetic Basis of Resistance in Bacteria

Bacteria are well known for their ability to become resistant to antibiotics. This can be done by a random mutation in a gene that gives the bacteria resistance. This can then be passed on to the next generation during binary fusion. However, even though bacteria have a short generation time and can produce enormous numbers in a very short time, acquiring resistance in this way is inefficient and does not allow the resistance to spread through the population outside the original strain. A more important way for acquiring resistance in bacteria is by horizontal gene transfer (HGT). This is the transfer of genes between individuals which are not related (i.e. parents and offspring). In this way bacteria can acquire a resistant gene from another bacteria or from the environment, allowing the gene to more quickly spread through many strains of bacteria.

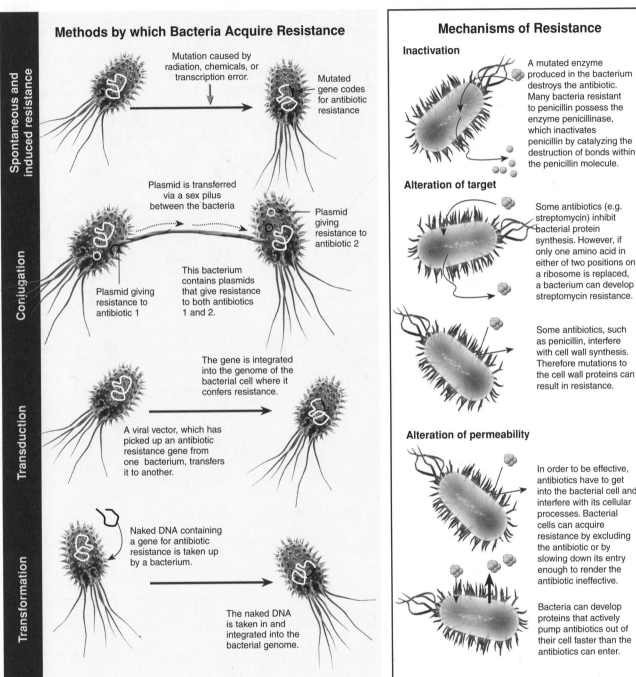

Methods by which Bacteria Acquire Resistance

Spontaneous and induced resistance

Mutation caused by radiation, chemicals, or transcription error.

Mutated gene codes for antibiotic resistance

Conjugation

Plasmid is transferred via a sex pilus between the bacteria

Plasmid giving resistance to antibiotic 2

Plasmid giving resistance to antibiotic 1

This bacterium contains plasmids that give resistance to both antibiotics 1 and 2.

Transduction

The gene is integrated into the genome of the bacterial cell where it confers resistance.

A viral vector, which has picked up an antibiotic resistance gene from one bacterium, transfers it to another.

Transformation

Naked DNA containing a gene for antibiotic resistance is taken up by a bacterium.

The naked DNA is taken in and integrated into the bacterial genome.

Mechanisms of Resistance

Inactivation

A mutated enzyme produced in the bacterium destroys the antibiotic. Many bacteria resistant to penicillin possess the enzyme penicillinase, which inactivates penicillin by catalyzing the destruction of bonds within the penicillin molecule.

Alteration of target

Some antibiotics (e.g. streptomycin) inhibit bacterial protein synthesis. However, if only one amino acid in either of two positions on a ribosome is replaced, a bacterium can develop streptomycin resistance.

Some antibiotics, such as penicillin, interfere with cell wall synthesis. Therefore mutations to the cell wall proteins can result in resistance.

Alteration of permeability

In order to be effective, antibiotics have to get into the bacterial cell and interfere with its cellular processes. Bacterial cells can acquire resistance by excluding the antibiotic or by slowing down its entry enough to render the antibiotic ineffective.

Bacteria can develop proteins that actively pump antibiotics out of their cell faster than the antibiotics can enter.

Sources of Variation

1. Which of the methods in the blue panel above show horizontal gene transfer? _____

_____

2. Discuss how HGT contributes to the rapid spread of drug resistance in bacteria: _____

_____

_____

_____

_____

© BIOZONE International 2012
ISBN: 978-1-927173-11-4
Photocopying Prohibited

Related activities: *The Evolution of Antibiotic Resistance*
Weblinks: *Transduction, Bacterial Conjugation, Transposons*

A 2

Replication in Bacteriophages

Viruses infect living cells, commanding the metabolism of the host cell and producing new viral particles. In viruses that use bacterial cells as a host (**bacteriophages**), this process may not immediately follow infection. Instead, the virus may integrate its nucleic acid into the host cell's DNA, forming a provirus or **prophage**. This type of cycle, called **lysogenic** or **temperate**, does not kill the host cell outright. Instead, the host cell is occupied by the virus and used to replicate the viral genes. During this time, the viral infection is said to be **latent**. The virus may be **transduced** into becoming active again, entering the **lytic cycle** and utilizing the host's cellular mechanisms to produce new virions. The lytic cycle results in death of the host cell through **cell lysis**. Although the multiplication of animal viruses follows a similar pattern to that of bacteriophage multiplication there are notable differences. Animal viruses have different mechanisms by which they enter host cells and, once inside the cell, the production of new virions is different. This is partly because of differences in host cell structure and metabolism and partly because the structure of animal viruses themselves is very variable.

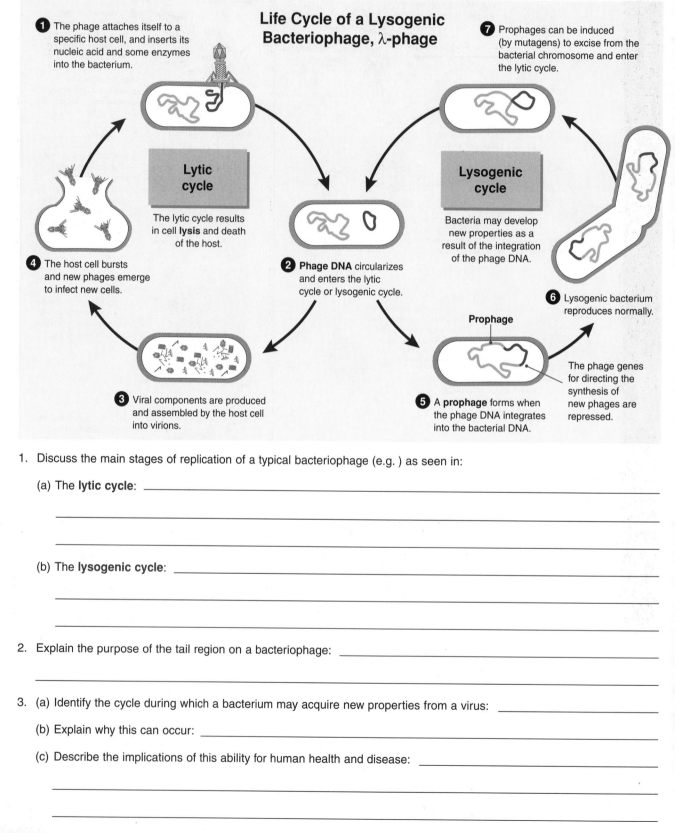

Life Cycle of a Lysogenic Bacteriophage, λ-phage

1 The phage attaches itself to a specific host cell, and inserts its nucleic acid and some enzymes into the bacterium.

7 Prophages can be induced (by mutagens) to excise from the bacterial chromosome and enter the lytic cycle.

Lytic cycle

The lytic cycle results in cell **lysis** and death of the host.

Lysogenic cycle

Bacteria may develop new properties as a result of the integration of the phage DNA.

4 The host cell bursts and new phages emerge to infect new cells.

2 Phage DNA circularizes and enters the lytic cycle or lysogenic cycle.

6 Lysogenic bacterium reproduces normally.

Prophage

The phage genes for directing the synthesis of new phages are repressed.

3 Viral components are produced and assembled by the host cell into virions.

5 A **prophage** forms when the phage DNA integrates into the bacterial DNA.

1. Discuss the main stages of replication of a typical bacteriophage (e.g.) as seen in:

   (a) The **lytic cycle**: _____

   _____

   _____

   (b) The **lysogenic cycle**: _____

   _____

   _____

2. Explain the purpose of the tail region on a bacteriophage: _____

   _____

3. (a) Identify the cycle during which a bacterium may acquire new properties from a virus: _____

   (b) Explain why this can occur: _____

   (c) Describe the implications of this ability for human health and disease: _____

   _____

   _____

Web links: Lytic Cycle, Lysogeny

Replication in Animal Viruses

Animal viruses are more complex and varied in structure than the viruses that infect bacteria. Likewise, animal host cells are more diverse in structure and metabolism than bacterial cells. Consequently, animal viruses exhibit a number of different mechanisms for **replicating**, i.e. entering a host cell and producing and releasing new virions. Enveloped viruses bud out from the host cell, whereas those without an envelope are released by rupture of the cell membrane. Three processes (attachment, penetration, and uncoating) are shared by both DNA- and RNA-containing animal viruses but the methods of

biosynthesis vary between these two major groups. Generally, **DNA viruses** replicate their DNA in the nucleus of the host cell using viral enzymes, and synthesize their capsid and other proteins in the cytoplasm using the host cell's enzymes. This is outlined below for a typical enveloped DNA virus. **RNA viruses** are more variable in their methods of biosynthesis. The example on the next page describes replication in the retrovirus HIV, where the virus uses its own reverse transcriptase to synthesize viral DNA and produce either **latent proviruses** or active, mature retroviruses.

Entry of an Enveloped Virus into a Cell

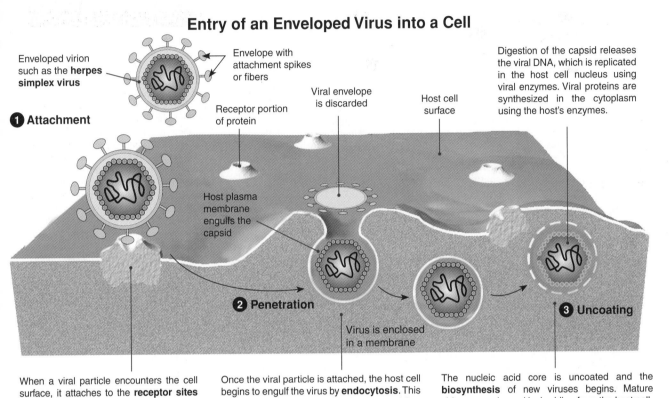

Enveloped virion such as the **herpes simplex virus**

Envelope with attachment spikes or fibers

Viral envelope is discarded

Host cell surface

Digestion of the capsid releases the viral DNA, which is replicated in the host cell nucleus using viral enzymes. Viral proteins are synthesized in the cytoplasm using the host's enzymes.

1 Attachment

Receptor portion of protein

Host plasma membrane engulfs the capsid

2 Penetration

Virus is enclosed in a membrane

3 Uncoating

When a viral particle encounters the cell surface, it attaches to the **receptor sites** of proteins on the cell's plasma membrane.

Once the viral particle is attached, the host cell begins to engulf the virus by **endocytosis**. This is the cell's usual response to foreign particles.

The nucleic acid core is uncoated and the **biosynthesis** of new viruses begins. Mature virions are released by budding from the host cell.

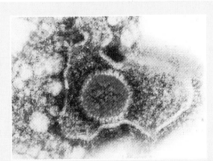

Coronaviruses are irregularly shaped viruses associated with upper respiratory infections and SARS. The envelope bears distinctive projections.

Herpesviruses are medium-sized enveloped viruses that cause various diseases including fever blisters, chickenpox, shingles, and herpes.

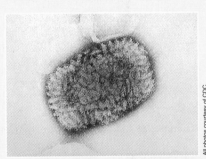

This **Vaccinia** virus belongs to the family of pox viruses; large (200-350 nm), enveloped DNA viruses that cause diseases such as smallpox.

All photos courtesy of CDC

Sources of Variation

1. Describe the purpose of the glycoprotein spikes found on some enveloped viruses: _____

_____

2. (a) Explain the significance of endocytosis to the entry of an enveloped virus into an animal cell: _____

_____

_____

(b) State where an enveloped virus replicates its viral DNA: _____

(c) State where an enveloped virus synthesizes its proteins: _____

Related activities: The Structure of Viruses, Replication in Bacteriophages
Web links: HSV Infection and Replication, HIV Life Cycle, Retrovirus Life Cycle

RA 2

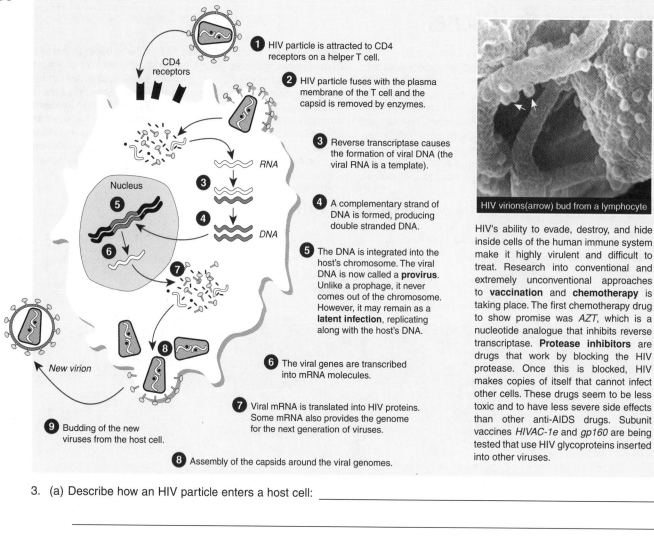

1 HIV particle is attracted to CD4 receptors on a helper T cell.

2 HIV particle fuses with the plasma membrane of the T cell and the capsid is removed by enzymes.

3 Reverse transcriptase causes the formation of viral DNA (the viral RNA is a template).

4 A complementary strand of DNA is formed, producing double stranded DNA.

5 The DNA is integrated into the host's chromosome. The viral DNA is now called a **provirus**. Unlike a prophage, it never comes out of the chromosome. However, it may remain as a **latent infection**, replicating along with the host's DNA.

6 The viral genes are transcribed into mRNA molecules.

7 Viral mRNA is translated into HIV proteins. Some mRNA also provides the genome for the next generation of viruses.

8 Assembly of the capsids around the viral genomes.

9 Budding of the new viruses from the host cell.

CD4 receptors

Nucleus

RNA

DNA

New virion

HIV virions(arrow) bud from a lymphocyte

CDC

HIV's ability to evade, destroy, and hide inside cells of the human immune system make it highly virulent and difficult to treat. Research into conventional and extremely unconventional approaches to **vaccination** and **chemotherapy** is taking place. The first chemotherapy drug to show promise was *AZT*, which is a nucleotide analogue that inhibits reverse transcriptase. **Protease inhibitors** are drugs that work by blocking the HIV protease. Once this is blocked, HIV makes copies of itself that cannot infect other cells. These drugs seem to be less toxic and to have less severe side effects than other anti-AIDS drugs. Subunit vaccines *HIVAC-1e* and *gp160* are being tested that use HIV glycoproteins inserted into other viruses.

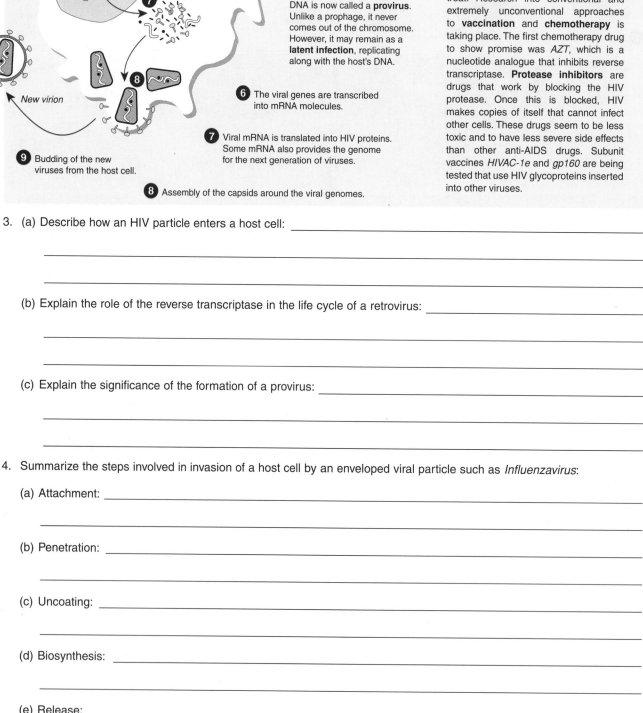

3. (a) Describe how an HIV particle enters a host cell: _____

_____

_____

(b) Explain the role of the reverse transcriptase in the life cycle of a retrovirus: _____

_____

_____

(c) Explain the significance of the formation of a provirus: _____

_____

_____

4. Summarize the steps involved in invasion of a host cell by an enveloped viral particle such as *Influenzavirus*:

(a) Attachment: _____

_____

(b) Penetration: _____

_____

(c) Uncoating: _____

_____

(d) Biosynthesis: _____

_____

(e) Release: _____

_____

Antigenic Variability in Pathogens

Influenza (flu) is a disease of the upper respiratory tract caused by the viral genus *Influenzavirus*. Globally, up to 500,000 people die from influenza every year. Three types of *Influenzavirus* affect humans. They are simply named *Influenzavirus* A, B, and C. The most common and most virulent of these strains is *Influenzavirus* A, which is discussed in more detail below. Influenza viruses are constantly undergoing genetic changes. **Antigenic drifts** are small changes in the virus which happen continually over time. Such changes mean that the influenza vaccine must be adjusted each year to include the most recently circulating influenza viruses. **Antigenic shift** occurs when two or more different viral strains (or different viruses) combine to form a new subtype. The changes are large and sudden and most people lack immunity to the new subtype. New influenza viruses arising from antigenic shift have caused influenza pandemics that have killed millions people over the last century. *Influenzavirus* A is considered the most dangerous to human health because it is capable of antigenic shift.

Structure of *Influenzavirus*

Viral strains are identified by the variation in their H and N surface antigens. Viruses are able to combine and readily rearrange their RNA segments, which alters the protein composition of their H and N glycoprotein spikes.

The *influenzavirus* is surrounded by an **envelope** containing protein and lipids.

The genetic material is actually closely surrounded by protein capsomeres (these have been omitted here and below right in order to illustrate the changes in the RNA more clearly).

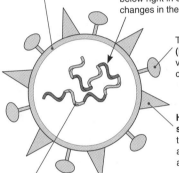

The **neuraminidase (N) spikes** help the virus to detach from the cell after infection.

Hemagglutinin (H) spikes allow the virus to recognize and attach to cells before attacking them.

The viral genome is contained on **eight RNA segments**, which enables the exchange of genes between different viral strains.

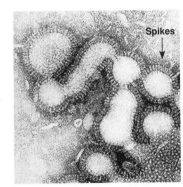

Spikes

Photo right: *Electron micrograph of Influenzavirus showing the glycoprotein spikes projecting from the viral envelope*

Antigenic Shift in *Influenzavirus*

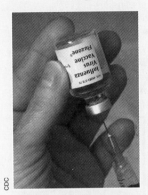

CDC

Influenza vaccination is the primary method for preventing influenza and is 75% effective. The ability of the virus to recombine its RNA enables it to change each year, so that different strains dominate in any one season. The 'flu' vaccination is updated annually to incorporate the antigenic properties of currently circulating strains. Three strains are chosen for each year's vaccination. Selection is based on estimates of which strains will be predominant in the following year.

CDC

H1N1, H1N2, and H3N2 (below) are the known *Influenza A* viral subtypes currently circulating among humans. Although the body will have acquired antibodies from previous flu strains, the new combination of N and H spikes is sufficiently different to enable new viral strains to avoid detection by the immune system. The World Health Organization coordinates strain selection for each year's influenza vaccine.

| H1N1 | H1N2 | H3N2 |

1. The *Influenzavirus* is able to mutate readily and alter the composition of H and N spikes on its surface.

    (a) Why can the virus mutate so readily? _____

    _____

    _____

    (b) How does this affect the ability of the immune system to recognize and respond to the virus? _____

    _____

    _____

    _____

2. Discuss why a virus capable of antigenic shift is more dangerous to humans than a virus undergoing antigenic drift:

    _____

    _____

    _____

Sources of Variation

The Origin of Living Systems

Prebiotic Earth	• Conditions on prebiotic Earth • RNA and the first cells
The first cells	• The role of prokaryotic life

Evidence for Biological Evolution

Fossil and molecular evidence	• Fossils and stratigraphy • Comparative molecular analyses • Comparative anatomy
Biogeography	• Biogeography and continental drift

Connect 1.A.4 with 1.D.2 — *Scientific evidence from many disciplines supports models for the origin of life.*

Genetic Change in Populations

The modern synthesis	• The Darwin-Wallace theory • Neo-Darwinism: updating the picture • Adaptation and fitness
Mechanisms of evolution	• Mutation and gene flow • Natural selection and genetic drift • Changing selection pressures
Evolution in real time	• Drug and insecticide resistance • Skin color in humans - what is race? • Melanism, birth weight, and sickle cells
Artificial selection	• Selective breeding in animals • Selective breeding in crop plants

Connect 1.A.2 with 3.C — *Evolutionary processes work with natural variability, arising through mutation and sexual reproduction.*

Living systems evolved by natural processes. The evidence for evolution comes from many fields science.

Natural selection operates in the current environment to establish adaptive phenotypes. Random processes influence the rate of evolution.

Andrew Dunn

Evolution

Important in this section...

- *Recognize the wealth of evidence for evolution*
- *Understand evolutionary fitness*
- *Explain changes in allele frequencies in gene pools*
- *Explain the role of evolution in producing diversity*

Structural and functional evidence shows that organisms are linked by lines of descent from common ancestors.

Speciation and extinction have occurred throughout Earth's history and continue today.

Carolina parakeet, recently extinct

Fritz Geller-Grimm cc2.5

Descent and common ancestry	• The importance of synapomorphies • Drawing and interpreting cladograms
Recognizing diversity	• The features of taxonomic groups • Classification system • Different classification schemes

Connect 1.B.1 with 3.D.1 — *Common ancestry of organisms is evident in conserved molecules and processes.*

Connect 1.B.1 with 2.D.2 — *Homeostatic mechanisms reflect common ancestry and adaptation to environment.*

The Relatedness of Organisms

The species concept	• The biological species concept • Ring species, and cryptic species
Speciation	• Isolation and adaptation • Allopatric and sympatric speciation
Rates of evolution	• Gradualism (phyletic gradualism) • Punctuated equilibria
The evolution of diversity	• Convergent evolution • Coevolution • Adaptive radiation

Connect 1.C.1 with 4.C.3 — *Species extinction rates increase during times of ecological stress.*

Speciation and Extinction

Genetic Change
in Populations

Key concepts

▶ Populations are rarely in genetic equilibrium, therefore their allele frequencies change over time.

▶ The Hardy-Weinberg equation can be used to calculate and predict the changes in the allele frequencies of populations.

▶ Changes in the allele frequencies in populations are determined by many processes, including genetic drift, mutation, migration, and natural selection.

▶ These processes are relatively more important in small populations and in a changing environment.

Key terms

adaptation
allele frequency
artificial selection
Darwin
differential survival
directional selection
disruptive selection
evolution
evolutionary fitness
fixation (of alleles)
founder effect
gene flow
gene pool
genetic bottleneck
genetic drift
genetic equilibrium
genetic variation
genotype
Hardy-Weinberg equation
hybrid
migration
modern synthesis
mutation
natural selection
phenotype
polymorphism
population
random mating

Essential Knowledge

☐ 1. Use the **KEY TERMS** to compile a glossary for this topic.

Natural Selection and Evolution *(1.A.1)*   pages 234-248

☐ 2. Briefly describe Darwin's theory of **natural selection** and its significance. Understand the terms **adaptation** and **evolutionary fitness**.

☐ 3. Understand the concept of the **gene pool** and state the principle of **genetic equilibrium**. Understand that the condition for genetic equilibrium are seldom, if ever, met and explain the consequences of this.

☐ 4. Evaluate evidence to qualitatively and quantitatively investigate the role of natural selection in **evolution**.

☐ 5. Explain the basis of the **Hardy-Weinberg equation**. Use the Hardy-Weinberg equation to calculate change in **allele frequencies** for a population over time.

☐ 6. Describe the role of chance and random events in evolutionary processes. Comment on why these are relatively more important in small populations.

The Role of Variation *(1.A.2)*   pages 249-257, 261-266

☐ 7. Describe examples to show how the selective environment can change. Examples could include soot pollution in the Industrial Revolution and climate change on a global and/or local scale.

☐ 8. Describe the role of **genetic variation** in providing the raw material on which natural selection can act. Use specific examples, e.g. sickle cell disease, to relate phenotypic variation to fitness.

☐ 9. Describe and evaluate examples of genetic change in real populations over time. Suitable examples include peppered moths in the UK or USA, sickle cell disease, and/or DDT resistance in insects.

☐ 10. Describe the impact of human activity on variation in other species. Examples could include **artificial selection** in crops or livestock, or antibiotic misuse creating a selective environment for resistance.

The Role of Random Processes *(1.A.3)*   pages 258-260, 265-266

☐ 11. Use the Hardy-Weinberg equation to predict future changes in the **allele frequencies** for a population given certain events, e.g. **founder effect**, **genetic bottleneck**, and **migration**.

☐ 12. Analyze **genetic drift** and the effects of selection in the evolution of specific populations. Recognize the importance of genetic drift in small populations. Justify the use of mathematical models to make these analyses.

☐ 13. Make predictions about the effects of genetic drift, migration, and **artificial selection** on the genetic makeup of a population. Explain your rationale.

Periodicals:
Listings for this chapter are on page 383

Weblinks:
www.thebiozone.com/
weblink/AP1-3114.html

BIOZONE APP:
Student Review Series
Evolution

A Pictorial History of Evolutionary Thought

Although **Charles Darwin** is largely credited with the development of the theory of evolution by natural selection, his ideas did not develop in isolation, but within the context of the work of others before him. The **modern synthesis of evolution** (below) has a long history with contributors from all fields of science. The diagram below summarizes just some of the important players in the story of evolutionary biology. This is not to say they were collaborators or always agreed. However, the work of many has contributed to a deeper understanding of evolutionary processes. This understanding continues to develop in the light of increasingly sophisticated molecular techniques and collaborative work between scientists across many disciplines.

Find out more!
This timeline has been adapted from the University of California, Berkeley's excellent *Evolution 101* website. Go to the Weblink indicated at the bottom of the page to find out more about the events and the people described.

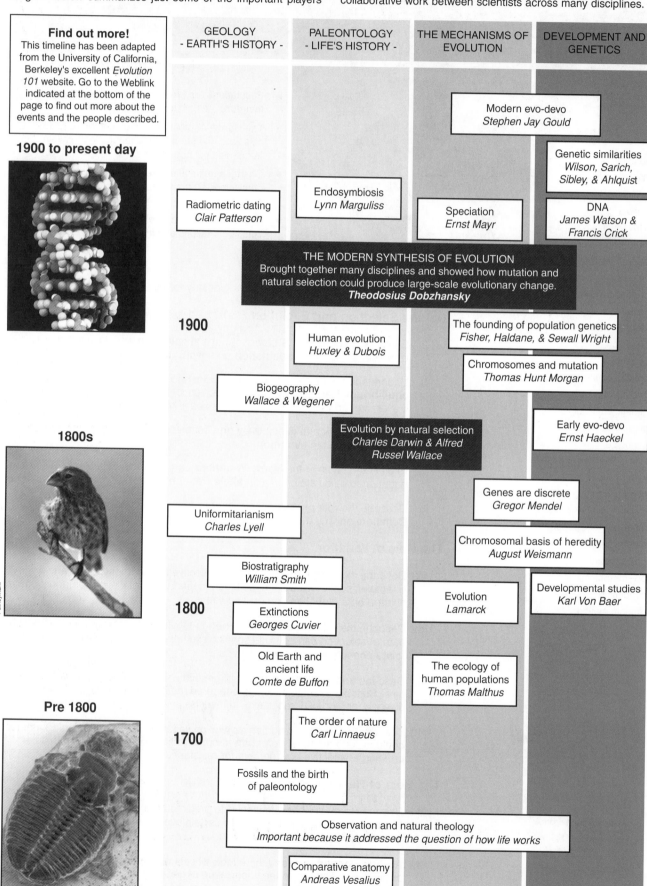

1900 to present day

1800s

Putneymark

Pre 1800

GEOLOGY
- EARTH'S HISTORY -

PALEONTOLOGY
- LIFE'S HISTORY -

THE MECHANISMS OF EVOLUTION

DEVELOPMENT AND GENETICS

Modern evo-devo
Stephen Jay Gould

Genetic similarities
Wilson, Sarich, Sibley, & Ahlquist

Radiometric dating
Clair Patterson

Endosymbiosis
Lynn Marguliss

Speciation
Ernst Mayr

DNA
James Watson & Francis Crick

THE MODERN SYNTHESIS OF EVOLUTION
Brought together many disciplines and showed how mutation and natural selection could produce large-scale evolutionary change.
Theodosius Dobzhansky

1900

Human evolution
Huxley & Dubois

The founding of population genetics
Fisher, Haldane, & Sewall Wright

Chromosomes and mutation
Thomas Hunt Morgan

Biogeography
Wallace & Wegener

Evolution by natural selection
Charles Darwin & Alfred Russel Wallace

Early evo-devo
Ernst Haeckel

Genes are discrete
Gregor Mendel

Uniformitarianism
Charles Lyell

Chromosomal basis of heredity
August Weismann

Biostratigraphy
William Smith

1800

Extinctions
Georges Cuvier

Evolution
Lamarck

Developmental studies
Karl Von Baer

Old Earth and ancient life
Comte de Buffon

The ecology of human populations
Thomas Malthus

The order of nature
Carl Linnaeus

1700

Fossils and the birth of paleontology

Observation and natural theology
Important because it addressed the question of how life works

Comparative anatomy
Andreas Vesalius

Related activities: Darwin's Theory
Weblinks: The History of Evolutionary Thought

The Development of the Modern Synthesis

Christine ↓

Darwin

Wallace

Charles Darwin (1809-1882) and **Alfred Russel Wallace** (1823-1913) jointly and independently proposed the theory of evolution by natural selection. Both amassed large amounts of supporting evidence, Darwin from his voyages aboard the Beagle and in the Galápagos Islands and Wallace from his studies in the Amazon and the Malay archipelago. Wallace wrote to Darwin of his ideas on evolution by natural selection, spurring Darwin to publish *The Origin of Species*.

Gregor Mendel (1822-1884) developed ideas of the genetic basis of inheritance. Mendel's *particulate model of inheritance* was recognized decades later as providing the means by which natural selection could occur.

Katie ↓

"An inordinate fondness for beetles"

Display: Oxford University Museum of Natural History

Theodosius Dobzhansky (1900-1975) was a Ukrainian who synthesized the ideas of genetics and evolutionary biology and defined evolution as "a change in the frequency of an allele within a gene pool". Dobzhansky worked on the genetics of wild *Drosophila* species and was famously quoted as saying "Nothing in biology makes sense except in the light of evolution".

Ernst Mayr (1904-2005) was a German evolutionary biologist who collaborated with Dobzhansky to formulate the modern evolutionary synthesis. He worked on and defined various mechanisms of speciation and proposed the existence of rapid speciation events, which became important for later ideas about punctuated equilibrium.

Ronald Fisher, **JBS Haldane**, and **Sewall Wright** founded population genetics, building sophisticated mathematical models of genetic change in populations. Their models, together with the work of others like Mayr and Dobzhansky contributed to a refinement and development of Darwin's theory into the **modern synthesis**. Haldane was quoted as saying that the Creator must have "an inordinate fondness for beetles".

The Modern Synthesis Today

James Watson and **Francis Crick**'s discovery of DNA's structure in 1953 revolutionized evolutionary biology. The genetic code could be understood and deciphered, and the role of mutation as the source of new alleles was realized.

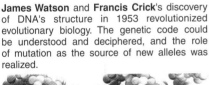

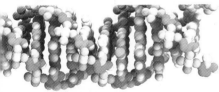

Stephen Jay Gould (1941-2002)

After Haeckel's flawed work on embryology fell out of favor, the evolutionary study of embryos was largely abandoned for decades. However, in the 1970s, **Stephen Jay Gould**'s work on the genetic triggers for developmental change brought studies of embryological development back into the forefront. Today evo-devo is providing some of the strongest evidence for how novel forms can rapidly arise.

In recent decades, DNA hybridization studies, DNA sequencing and protein analyses have revolutionized our understanding of phylogeny. **Allan Wilson** was one of a small group of pioneers in this field, using molecular approaches to understand evolutionary change and reconstruct phylogenies, including those of human ancestors.

1. Using a separate sheet, research and then write a 150 word account of the development of evolutionary thought and the importance of contributors from many scientific disciplines in shaping what became the modern synthesis. You should choose specific examples to illustrate your points of discussion.

Darwin's Theory

In 1859, Darwin and Wallace jointly proposed that new species could develop by a process of natural selection. Natural selection is the term given to the mechanism by which better adapted organisms survive to produce a greater number of viable offspring. This has the effect of increasing their proportion in the population so that they become more common. It is Darwin who is best remembered for the theory of evolution by natural selection through his famous book: **'On the origin of species by means of natural selection'**, written 23 years after returning from his voyage on the Beagle, from which much of the evidence for his theory was accumulated. Although Darwin could not explain the origin of variation nor the mechanism of its transmission (this was provided later by Mendel's work), his basic theory of evolution by natural selection (outlined below) is widely accepted today. The study of population genetics has greatly improved our understanding of evolutionary processes, which are now seen largely as a (frequently gradual) change in allele frequencies within a population. Be aware that scientific debate on the subject of evolution centers around the relative merits of various alternative hypotheses about the nature of evolutionary processes. The debate is not about the existence of the phenomenon of evolution itself.

Darwin's Theory of Evolution by Natural Selection

Overproduction
Populations produce too many young: many must die

Populations tend to produce more offspring than are needed to replace the parents. Natural populations normally maintain constant numbers. There must therefore be a certain number that die without producing offspring.

Variation
Individuals show variation: some are more favorable than others

Individuals in a population vary in their phenotype and therefore, their genotype. Some variants are better suited to the prevailing environment and have greater survival and reproductive success.

Natural Selection
Natural selection favors the best suited at the time

The struggle for survival amongst individuals competing for limited resources will favor those with the most favorable variations. Relatively more of those without favorable variations will die.

Inherited
Variations are inherited: the best suited variants leave more offspring

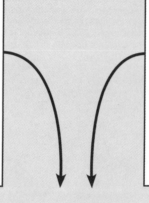

The variations (both favorable and unfavorable) are passed on to offspring. Each new generation will contain proportionally more descendants of individuals with favorable characters.

The banded or grove snail, *Cepaea nemoralis*, is famous for the highly variable colors and banding patterns of its shell. These **polymorphisms** are thought to have a role in differential survival in different regions, associated with both the risk of predation and maintenance of body temperature. Dark brown grove snails are more abundant in dark woodlands, whilst snails with light yellow shells and thin banding are more commonly found in grasslands.

Andrew Dunn www.andrewdunnphoto.com

1. In your own words, describe how Darwin's theory of evolution by natural selection provides an explanation for the change in the appearance of a species over time:

Survival of the fittest; if a certain genotype has led to survival that genotype is more likely to mate and leave offsprings with faverable traits

Related activities: A Pictorial History of Evolutionary Thought
Weblinks: Variation: Snails

Periodicals:
Was Darwin wrong?

© BIOZONE International 2012
ISBN: 978-1-927173-11-4
Photocopying Prohibited

Adaptation and Fitness

An **adaptation**, is any heritable trait that equips an organism to its functional position in the environment (its niche). These traits may be structural, physiological, or behavioral and reflect ancestry as well as adaptation. Adaptation is important in an evolutionary sense because adaptive features promote fitness. **Fitness** is a measure of an organism's ability to maximize the numbers of offspring surviving to reproductive age. Genetic adaptation must not be confused with **physiological adjustment** (acclimatization), which refers to an organism's ability to adapt during its lifetime to changing environmental conditions (e.g. a person's acclimatization to altitude). Examples of adaptive features arising through evolution are illustrated below.

Ear Length in Rabbits and Hares

The external ears of many mammals are used as important organs to assist in thermoregulation (controlling loss and gain of body heat). The ears of rabbits and hares native to hot, dry climates, such as the jack rabbit of south-western USA and northern Mexico, are relatively very large. The Arctic hare lives in the tundra zone of Alaska, northern Canada and Greenland, and has ears that are relatively short. This reduction in the size of the extremities (ears, limbs, and noses) is typical of cold adapted species.

Arctic hare: *Lepus arcticus*   Black-tail jackrabbit: *Lepus californicus*

Body Size in Relation to Climate

Regulation of body temperature requires a large amount of energy and mammals exhibit a variety of structural and physiological adaptations to increase the effectiveness of this process. Heat production in any endotherm depends on body volume (heat generating metabolism), whereas the rate of heat loss depends on surface area. Increasing body size minimizes heat loss to the environment by reducing the surface area to volume ratio. Animals in colder regions therefore tend to be larger overall than those living in hot climates. This relationship is know as **Bergman's rule** and it is well documented in many mammalian species. Cold adapted species also tend to have more compact bodies and shorter extremities than related species in hot climates.

The **fennec fox** of the Sahara illustrates the adaptations typical of mammals living in hot climates: a small body size and lightweight fur, and long ears, legs, and nose. These features facilitate heat dissipation and reduce heat gain.

The **Arctic fox** shows the physical characteristics typical of cold adapted mammals: a stocky, compact body shape with small ears, short legs and nose, and dense fur. These features reduce heat loss to the environment.

Number of Horns in Rhinoceroses

Not all differences between species can be convincingly interpreted as adaptations to particular environments. Rhinoceroses charge rival males and predators, and the horn(s), when combined with the head-down posture, add effectiveness to this behavior. Horns are obviously adaptive, but it is not clear if having one (Indian rhino) or two (black rhino) horns is related to the functionality in the environment or a reflection of evolution from a small hornless ancestor.

1. Distinguish between adaptive features (genetic) and acclimatization: _Adaptive features are any heritable traits that equips an organism to its functional position in the environment_

2. Explain the nature of the relationship between the length of extremities (such as limbs and ears) and climate: _It seems that hotter temperature cause limbs and ears to be longer_

3. Explain the adaptive value of a larger body size in a colder climate: _To keep themselves warm in the cold climate._

Related activities: Genes, Inheritance, and Evolution
Weblinks: Natural Selection and Adaptation

A 2

Natural Selection

Natural selection operates on the phenotypes of individuals, produced by their particular combinations of alleles (genotypes). The differential survival of some genotypes over others is called **natural selection** and, as a result of it, organisms with phenotypes most suited to the prevailing environment will be relatively more successful and so leave relatively more offspring. Favorable phenotypes become more numerous while unfavorable phenotypes become less common or disappear. Natural selection is not a static phenomenon; it is always linked to phenotypic suitability in the prevailing environment. It may favor existing phenotypes or shift the phenotypic median one way or another, as is shown below. The top row of diagrams represents the population phenotypic spread before selection, and the bottom row the spread afterwards. Note that balancing selection is similar to disruptive selection, but the balanced polymorphism that results is not associated with phenotypic extremes. Balanced polymorphism can occur as a result of heterozygous advantage or frequency dependent predation.

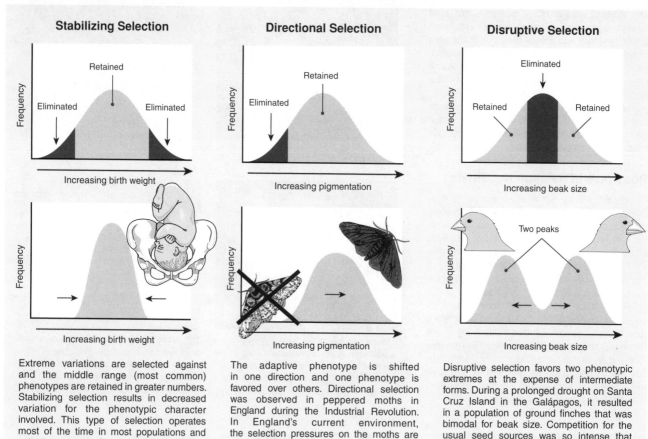

Stabilizing Selection

Extreme variations are selected against and the middle range (most common) phenotypes are retained in greater numbers. Stabilizing selection results in decreased variation for the phenotypic character involved. This type of selection operates most of the time in most populations and acts to prevent divergence of form and function, e.g. birth weight of human infants.

Directional Selection

The adaptive phenotype is shifted in one direction and one phenotype is favored over others. Directional selection was observed in peppered moths in England during the Industrial Revolution. In England's current environment, the selection pressures on the moths are more typically balanced, and the proportions of each form vary regionally.

Disruptive Selection

Disruptive selection favors two phenotypic extremes at the expense of intermediate forms. During a prolonged drought on Santa Cruz Island in the Galápagos, it resulted in a population of ground finches that was bimodal for beak size. Competition for the usual seed sources was so intense that birds able to exploit either small or large seeds were favored, although intermediate phenotypes remained in low numbers.

1. Explain why fluctuating (as opposed to stable) environments favor disruptive selection:

_____

_____

_____

2. Disruptive selection can be important in the formation of new species:

(a) Describe the evidence from the ground finches on Santa Cruz Island that provides support for this statement:

Competition for ~~the~~ usual seed sources was so intense that
birds able to exploit small or large seeds were favored

(b) The ground finches on Santa Cruz Island are one interbreeding population with a strongly bimodal distribution for the character of beak size. Suggest what conditions could lead to the two phenotypic extremes diverging further:

_____

_____

(c) Predict the consequences of the end of the drought and an increased abundance of medium size seeds as food:

_____

A 3

Related activities: Selection for Human Birth Weight, Industrial Melanism,
Disruptive Selection in Darwin's Finches **Weblinks:** *Variation: Snails*

Gene Pool Exercise

Cut out each of the beetles on this page and use them to reenact different events within a gene pool as described in this topic

(pages: *Gene Pools and Evolution, Changes in a Gene Pool, The Founder Effect, Population Bottlenecks, Genetic Drift*).

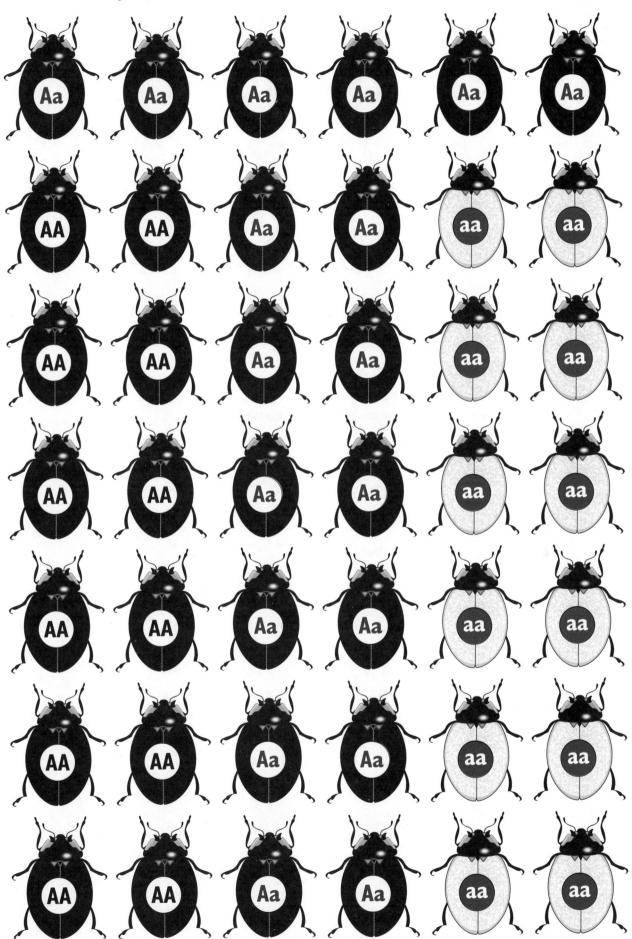

Related activities: *Gene Pools and Evolution, Changes in a Gene Pool, The Founder Effect, Population Bottlenecks, Genetic Drift*

This page has deliberately been left blank

Gene Pools and Evolution

This activity illustrates the dynamic nature of **gene pools**. It portrays two populations of one hypothetical beetle species. Each beetle is a 'carrier' of genetic information, represented here by the alleles (A and a) for a single gene that controls color and has a dominant/recessive expression pattern. There are normally two phenotypes: black and pale. Mutations may create other versions of the phenotype. Some of the microevolutionary processes that can affect the genetic composition (**allele frequencies**) of the gene pool are illustrated.

Mutations: Spontaneous mutations can develop that alter the allele frequencies of the gene pool, and even create new alleles. Mutation is very important to evolution, because it is the original source of genetic variation that provides new material for natural selection.

Immigration: Populations can gain alleles when they are introduced from other gene pools. Immigration is one aspect of gene flow.

Emigration: Genes may be lost to other gene pools. Emigration is an aspect of gene flow.

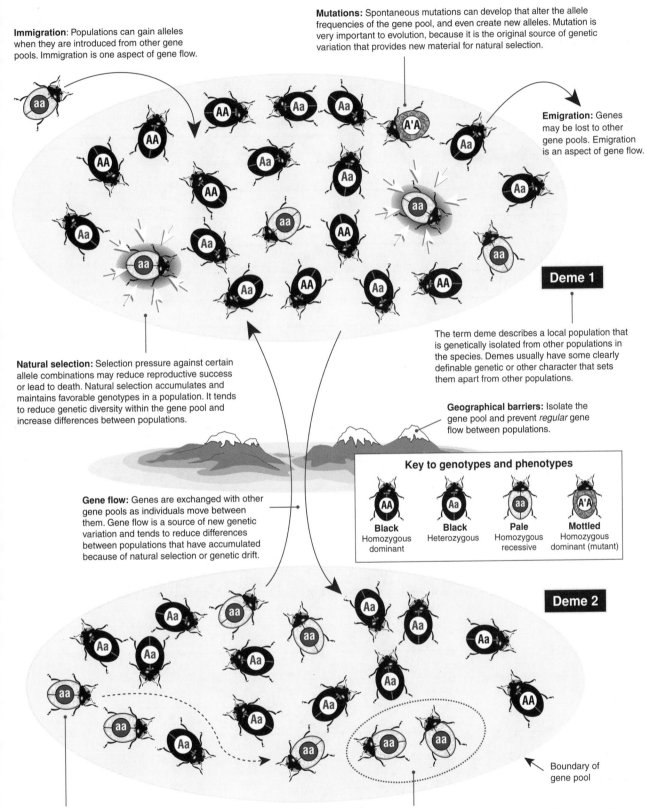

Natural selection: Selection pressure against certain allele combinations may reduce reproductive success or lead to death. Natural selection accumulates and maintains favorable genotypes in a population. It tends to reduce genetic diversity within the gene pool and increase differences between populations.

The term deme describes a local population that is genetically isolated from other populations in the species. Demes usually have some clearly definable genetic or other character that sets them apart from other populations.

Geographical barriers: Isolate the gene pool and prevent *regular* gene flow between populations.

Gene flow: Genes are exchanged with other gene pools as individuals move between them. Gene flow is a source of new genetic variation and tends to reduce differences between populations that have accumulated because of natural selection or genetic drift.

Key to genotypes and phenotypes

Black Homozygous dominant | **Black** Heterozygous | **Pale** Homozygous recessive | **Mottled** Homozygous dominant (mutant)

Boundary of gene pool

Mate choice (non-random mating): Individuals may not select their mate randomly and may seek out particular phenotypes, increasing the frequency of these "favored" alleles in the population.

Genetic drift: Chance events can cause the allele frequencies of small populations to "drift" (change) randomly from generation to generation. Genetic drift can play a significant role in the microevolution of very small populations. The two situations most often leading to populations small enough for genetic drift to be significant are the **bottleneck effect** (where the population size is dramatically reduced by a catastrophic event) and the **founder effect** (where a small number of individuals colonize a new area).

Related activities: The Founder Effect, Population Bottlenecks
Weblinks: Mechanisms of Evolution, Processes in Gene Pools, Speciation

A 2

1. Study the diagram on the previous page and review the factors affecting gene pools. For each of the factors listed below, describe how and why the allele frequency in the gene pool is affected:

    (a) Population size (large population vs small population): _____

    _____

    _____

    _____

    (b) Mate choice (random vs non-random mating): _____

    _____

    _____

    _____

    (c) Gene flow between populations: _____

    _____

    _____

    _____

    (d) Mutations: _____

    _____

    _____

    _____

    (e) Natural selection: _____

    _____

    _____

    _____

    (f) Genetic drift: _____

    _____

    _____

    _____

2. Using the information in your answers above, discuss the role of microevolutionary processes in evolution:

    _____

    _____

    _____

    _____

    _____

    _____

    _____

    _____

3. Identify the factors that tend to:

    (a) Increase genetic variation in gene pools: _____

    (b) Decrease genetic variation in gene pools: _____

Sexual Selection

The success of an individual is measured not only by the number of offspring it leaves, but also by the quality or likely reproductive success of those offspring. This means that it becomes important who its mate will be. It was Darwin (1871) who first introduced the concept of sexual selection, a special type of natural selection that produces anatomical and behavioral traits that affect an individual's ability to acquire mates. Biologists today recognize two types: **intrasexual selection** (usually male-male competition) and **intersexual selection** or mate choice. One result of either type is the evolution of **sexual dimorphism**.

Intrasexual selection

Intrasexual selection involves competition within one sex (usually males) with the winner gaining access to the opposite sex. Competition often takes place before mating, and males compete to establish dominance or secure a territory for breeding or mating. This occurs in many species of ungulates (**deer, antelope, cattle**) and in many birds. In deer and other ungulates, the males typically engage in highly ritualized battles with horns or antlers. The winners of these battles gain dominance over rival males and do most of the mating.

Katie

In other species, males compete for territories. These may contain resources or they may consist of an isolated area within a special arena used for communal courtship display (a **lek**). In lek species, males with the best territories on a lek (the dominant males) have more chances to mate with females. In some species of grouse (right), this form of sexual selection can be difficult to distinguish from intersexual selection, because once males establish their positions on the lek the females then choose among them. In species where access to females is limited and females are promiscuous, **sperm competition** may also be a feature of male-male competition.

Christine

Intersexual selection

In intersexual selection (or **mate choice**), individuals of one sex (usually the males) advertise themselves as potential mates and members of the other sex (usually the females) choose among them. Intersexual selection results in development of exaggerated ornamentation, such as elaborate plumages. Female preference for elaborate male ornaments is well supported by both anecdotal and experimental evidence. For example, in the **long-tailed widow bird** (*Euplectes progne*), females prefer males with long tails. When tails are artificially shortened or lengthened, females still prefer males with the longest tails; they therefore select for long tails, not another trait correlated with long tails.

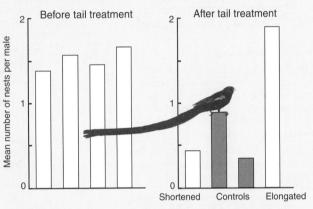

As shown above, there was no significant difference in breeding success between the groups before the tails were altered. When the tails were cut or lengthened, breeding success went down and up respectively relative to the unaltered controls.

In male-male competition for mates, ornamentation is used primarily to advertise superiority to rival males, and not to mortally wound opponents. However, injuries do occur, most often between closely matched rivals, where dominance must be tested and established through the aggressive use of their weaponry rather than mere ritual duels.

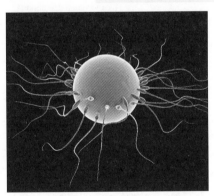

Sperm competition occurs when females remate within a relatively short space of time. The outcome of sperm competition may be determined by mating order. In some species, including those that guard their mates, the first male has the advantage, but in many the advantage accrues to the sperm of the second or subsequent males.

How do male features, such as the extravagant plumage of the peacock, persist when increasingly elaborate plumage must become detrimental to survival at some point? At first, preference for such traits must confer a survival advantage. Male adornment and female preference then advance together until a stable strategy is achieved.

1. Explain the difference between **intrasexual selection** and **mate choice**, identifying the features associated with each:

In intrasexual selection the male compete against each other

2. Suggest how sexual selection results in marked **sexual dimorphism**: _____

Periodicals:
Animal attraction

Related activities: Gene Pools and Evolution
Weblinks: Pheasant Sexual Selection

in a Gene Pool

low shows an hypothetical population of beetles
ges as it is subjected to two 'events'. The three
t a progression in time (i.e. the same gene pool,
nge). The beetles have two phenotypes (black and pale) determined by the amount of pigment deposited in the cuticle. The gene controlling this character is represented by two alleles **A** and **a**. Your task is to analyze the gene pool as it undergoes changes.

Phase 1: Initial gene pool

Calculate the frequencies of the allele types and allele combinations by counting the actual numbers, then working them out as percentages.

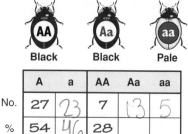

AA Black   **Aa** Black   **aa** Pale

	A	a	AA	Aa	aa
No.	27	23	7	13	5
%	54	46	28		

Allele types   Allele combinations

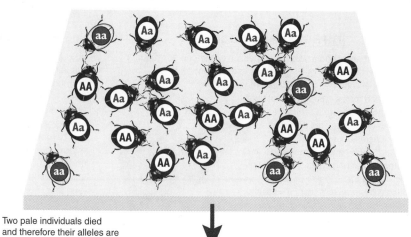

Phase 2: Natural selection

In the same gene pool at a later time there was a change in the allele frequencies. This was the result of the loss of certain allele combinations due to natural selection. Some of those with a genotype of aa were eliminated (poor fitness).

Calculate as for above. Do not include the individuals surrounded by small white arrows in your calculations: they are dead!

Two pale individuals died and therefore their alleles are removed from the gene pool.

	A	a	AA	Aa	aa	
No.	27	19	7	13	3	6
%	59	41	30	56	13	

(above table: handwritten "14 13" above A/a columns)

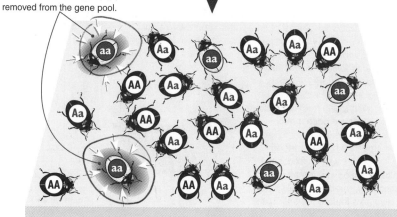

Phase 3: Immigration and emigration

This particular kind of beetle exhibits wandering behavior. The allele frequencies change again due to the introduction and departure of individual beetles, each carrying certain allele combinations.

Calculate as above. In your calculations, include the individual coming into the gene pool (AA), but remove the one leaving (aa).

This individual is entering the population and will add its alleles to the gene pool.

This individual is leaving the population, removing its alleles from the gene pool.

	A	a	AA	Aa	aa
No.	29	17	8	13	2
%	63	37	35	57	8

(above table: handwritten "16 13" above A/a columns)

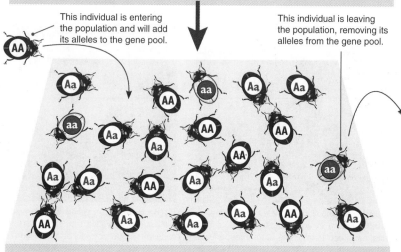

1. Explain how the number of dominant alleles (A) in the genotype of a beetle affects its phenotype:

_____

_____

2. For each phase in the gene pool above (place your answers in the tables provided; some have been done for you):

   (a) Determine the relative frequencies of the two alleles: A and a. Simply total the **A** alleles and **a** alleles separately.
   (b) Determine the frequency of how the alleles come together as allele pair combinations in the gene pool (AA, Aa and aa). Count the number of each type of combination.
   (c) For each of the above, work out the frequencies as percentages:

$$Allele\ frequency\ =\ No.\ counted\ alleles\ \div\ Total\ no.\ of\ alleles\ \times\ 100$$

Related activities: Gene Pool Exercise, Natural Selection
Weblinks: Natural Selection in Populations

Calculating Allele Frequencies in Populations

The **Hardy-Weinberg equation** provides a simple mathematical model of genetic equilibrium in a gene pool, but its main application in population genetics is in calculating allele and genotype frequencies in populations, particularly a [...] of studying changes and measuring their rate. The u[...] Hardy-Weinberg equation is described below.

Punnett square

A   a

Frequency of allele combination **AA** in the population is represented as p^2

Frequency of allele combination **aa** in the population is represented as q^2

Frequency of allele combination **Aa** in the population (add these together to get **2pq**)

$$(p + q)^2 = p^2 + 2pq + q^2 = 1$$

Frequency of allele types

p = Frequency of allele A
q = Frequency of allele a

Frequency of allele combinations

p^2 = Frequency of AA (homozygous dominant)
2pq = Frequency of Aa (heterozygous)
q^2 = Frequency of aa (homozygous recessive)

The Hardy-Weinberg equation is applied to populations with a simple genetic situation: dominant and recessive alleles controlling a single trait. The frequency of all of the dominant (A) and recessive alleles (a) equals the total genetic complement, and adds up to 1 or 100% of the alleles present.

How To Solve Hardy-Weinberg Problems

In most populations, the frequency of two alleles of interest is calculated from the proportion of homozygous recessives (q^2), as this is the only genotype identifiable directly from its phenotype. If only the dominant phenotype is known, q^2 may be calculated (1 – the frequency of the dominant phenotype). The following steps outline the procedure for solving a Hardy-Weinberg problem:

Remember that all calculations must be carried out using proportions, NOT PERCENTAGES!

1. Examine the question to determine what piece of information you have been given about the population. In most cases, this is the percentage or frequency of the homozygous recessive phenotype q^2, or the dominant phenotype $p^2 + 2pq$ (see note above).

2. The first objective is to find out the value of p or q, If this is achieved, then every other value in the equation can be determined by simple calculation.

3. Take the square root of q^2 to find q.

4. Determine p by subtracting q from 1 (i.e. p = 1 – q).

5. Determine p^2 by multiplying p by itself (i.e. $p^2 = p \times p$).

6. Determine 2pq by multiplying p times q times 2.

7. Check that your calculations are correct by adding up the values for $p^2 + q^2 + 2pq$ (the sum should equal 1 or 100%).

Worked example

Among Caucasians in the USA, approximately 70% of people can taste the chemical phenylthiocarbamide (PTC) (the dominant phenotype), while 30% are non-tasters (the recessive phenotype).

Determine the frequency of:	Answers
(a) Homozygous recessive phenotype(q^2).	30% - provided
(b) The dominant allele (**p**).	45.2%
(c) Homozygous tasters (**p^2**).	20.5%
(d) Heterozygous tasters (**2pq**).	49.5%

Data: The frequency of the dominant phenotype (70% tasters) and recessive phenotype (30% non-tasters) are provided.

Working:

Recessive phenotype: **q^2** = 30%
use 0.30 for calculation

therefore: **q** = 0.5477
square root of 0.30

therefore: **p** = 0.4523
1 – q = p
1 – 0.5477 = 0.4523

Use p and q in the equation (top) to solve any unknown:

Homozygous dominant   **p^2** = 0.2046
(p x p = 0.4523 x 0.4523)

Heterozygous: **2pq** = 0.4953

---

1. A population of hamsters has a gene consisting of 90% M alleles (black) and 10% m alleles (gray). Mating is random.

   Data: Frequency of recessive allele (10% m) and dominant allele (90% M).

   Determine the proportion of offspring that will be black and the proportion that will be gray (show your working):

   1% will be grey   _99%_

Recessive allele:	q =	.1
Dominant allele:	p =	.9
Recessive phenotype:	q^2 =	.01
Homozygous dominant:	p^2 =	.81
Heterozygous:	2pq =	.18

Periodicals:
The Hardy-Weinberg principle

Related activities: Analysis of a Squirrel Gene Pool

Genetic Cha... 245
as a means
se of the

...ea plants and found 36 plants out of 400 were dwarf.
...essive phenotype (36 out of 400 = 9%)

...ncy of the tall gene: _____ 49%

...er of heterozygous pea plants:

...8 heterozygous pea plant

Recessive allele:	q =	.3
Dominant allele:	p =	.7
Recessive phenotype:	q^2 =	.09
Homozygous dominant:	p^2 =	.49
Heterozygous:	2pq =	.42

3. In humans, the ability to taste the chemical phenylthiocarbamide (PTC) is inherited as a simple dominant characteristic. Suppose you found out that 360 out of 1000 college students could not taste the chemical.
   Data: Frequency of recessive phenotype (360 out of 1000).

   (a) State the frequency of the gene for tasting PTC:

   40%

   (b) Determine the number of heterozygous students in this population:

   480

Recessive allele:	q =	.6
Dominant allele:	p =	.4
Recessive phenotype:	q^2 =	.36
Homozygous dominant:	p^2 =	.16
Heterozygous:	2pq =	.48

4. A type of deformity appears in 4% of a large herd of cattle. Assume the deformity was caused by a recessive gene.
   Data: Frequency of recessive phenotype (4% deformity).

   (a) Calculate the percentage of the herd that are carriers of the gene:

   32 %

   (b) Determine the frequency of the dominant gene in this case:

   20%

Recessive allele:	q =	.2
Dominant allele:	p =	.8
Recessive phenotype:	q^2 =	.04
Homozygous dominant:	p^2 =	.64
Heterozygous:	2pq =	.32

5. Assume you placed 50 pure bred black guinea pigs (dominant allele) with 50 albino guinea pigs (recessive allele) and allowed the population to attain genetic equilibrium (several generations have passed).
   Data: Frequency of recessive allele (50%) and dominant allele (50%).

   Determine the proportion (%) of the population that becomes white:

Recessive allele:	q =	.5
Dominant allele:	p =	.5
Recessive phenotype:	q^2 =	.25
Homozygous dominant:	p^2 =	.25
Heterozygous:	2pq =	.50

6. It is known that 64% of a large population exhibit the recessive trait of a characteristic controlled by two alleles (one is dominant over the other).
   Data: Frequency of recessive phenotype (64%). Determine the following:

   q = .8
   p = .2
   q^2 = .64
   p^2 = .04
   2pq = .32

   (a) The frequency of the recessive allele: _____ 80%

   (b) The percentage that are heterozygous for this trait: _____ 32%

   (c) The percentage that exhibit the dominant trait: _____ 36%

   (d) The percentage that are homozygous for the dominant trait: _____ 4%

   (e) The percentage that has one or more recessive alleles: _____

7. Albinism is recessive to normal pigmentation in humans. The frequency of the albino allele was 10% in a population.
   Data: Frequency of recessive allele (10% albino allele).

   Determine the proportion of people that you would expect to be albino:

Recessive allele:	q =	
Dominant allele:	p =	
Recessive phenotype:	q^2 =	
Homozygous dominant:	p^2 =	
Heterozygous:	2pq =	

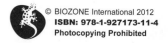

© BIOZONE International 2012
ISBN: 978-1-927173-11-4
Photocopying Prohibited

Analysis of a Squirrel Gene Pool

In Olney, Illinois, there is a unique population of albino (white) and gray squirrels. Between 1977 and 1990, students at Olney Central College carried out a study of this population. They recorded the frequency of gray and albino squirrels. The albinos displayed a mutant allele expressed as an albino phenotype only in the homozygous recessive condition. The data they collected are provided in the table below. Using the **Hardy-Weinberg equation** for calculating genotype frequencies, it was possible to estimate the frequency of the normal 'wild' allele (G) providing gray fur coloring, and the frequency of the mutant albino allele (g) producing white squirrels when homozygous. This study provided real, first hand, data that students could use to see how genotype frequencies can change in a real population.

Thanks to **Dr. John Stencel**, Olney Central College, Olney, Illinois, US, for providing the data for this exercise.

Gray squirrel, usual color form

Albino form of gray squirrel

Population of gray and white squirrels in Olney, Illinois (1977-1990)

Year	Gray	White	Total	GG	Gg	gg	Freq. of g	Freq. of G
1977	602	182	784	26.85	49.93	23.21	48.18	51.82
1978	511	172	683	24.82	50.00	25.18	50.18	49.82
1979	482	134	616	28.47	49.77	21.75	46.64	53.36
1980	489	133	622	28.90	49.72	21.38	46.24	53.76
1981	536	163	699	26.74	49.94	23.32	48.29	51.71
1982	618	151	769	31.01	49.35	19.64	44.31	55.69
1983	419	141	560	24.82	50.00	25.18	50.18	49.82
1984	378	106	484	28.30	49.79	21.90	46.80	53.20
1985	448	125	573	28.40	49.78	21.82	46.71	53.29
1986	536	155	691	27.71	49.86	22.43	47.36	52.64
1987	No data collected this year							
1988	652	122	774	36.36	47.88	15.76	39.70	60.30
1989	552	146	698	29.45	49.64	20.92	45.74	54.26
1990	603	111	714	36.69	47.76	15.55	39.43	60.57

1. **Graph population changes**: Use the data in the first 3 columns of the table above to plot a line graph. This will show changes in the phenotypes: numbers of gray and white (albino) squirrels, as well as changes in the total population. Plot: **gray**, **white**, and **total** for each year:

(a) Determine by how much (as a %) total population numbers have fluctuated over the sampling period:

_____

(b) Describe the overall trend in total population numbers and any pattern that may exist:

_____

_____

_____

_____

_____

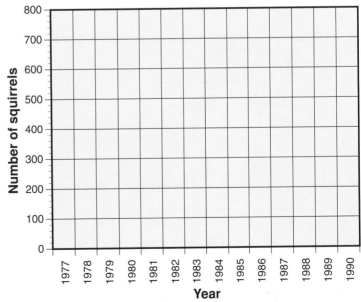

Periodicals: Evolution at a snail's pace

Related activities: *Calculating Allele Frequencies in Populations*

2. **Graph genotype changes**: Use the data in the genotype columns of the table on the opposite page to plot a line graph. This will show changes in the allele combinations (**GG**, **Gg**, **gg**). Plot: **GG**, **Gg**, and **gg** for each year:

Describe the overall trend in the frequency of:

(a) Homozygous dominant (**GG**) genotype:

_____

_____

(b) Heterozygous (**Gg**) genotype:

_____

_____

(c) Homozygous recessive (**gg**) genotype:

_____

_____

_____

Graph: Percentage frequency of genotype (y-axis, 0 to 60) vs Year (x-axis, 1977 to 1990). Empty grid.

3. **Graph allele changes**: Use the data in the last two columns of the table on the previous page to plot a line graph. This will show changes in the *allele frequencies* for each of the dominant (**G**) and recessive (**g**) alleles.
Plot: the frequency of **G** and the frequency of **g**:

(a) Describe the overall trend in the frequency of the dominant allele (**G**):

_____

_____

_____

(b) Describe the overall trend in the frequency of the recessive allele (**g**):

_____

_____

_____

_____

_____

_____

Graph: Percentage frequency of allele (y-axis, 0 to 70) vs Year (x-axis, 1977 to 1990). Empty grid.

4. (a) State which of the three graphs best indicates that a significant change may be taking place in the gene pool of this population of squirrels:

_____

(b) Give a reason for your answer: _____

_____

_____

5. Describe a possible cause of the changes in allele frequencies over the sampling period: _____

_____

_____

_____

_____

Industrial Melanism

Natural selection may act on the frequencies of phenotypes (and hence genotypes) in populations in one of three different ways (through stabilizing, directional, or disruptive selection). Over time, natural selection may lead to a permanent change in the genetic makeup of a population. The increased prevalence of melanic (dark) forms of the peppered moth, *Biston betularia*, during the Industrial Revolution in the UK is one of the best known examples of directional selection following a change in environmental conditions. Although the protocols used in the central experiments on *Biston*, and the conclusions drawn from these experiments, have been called into question, the collections of moths do provide documented evidence of phenotypic change.

Industrial melanism in peppered moths, *Biston betularia*

The **peppered moth**, *Biston betularia*, occurs in two forms (morphs): the gray mottled form, and a dark melanic form. Changes in the relative abundance of these two forms was hypothesized to be the result of selective predation by birds, with pale forms suffering higher mortality in industrial areas because they are more visible. The results of experiments by H.D. Kettlewell supported this hypothesis but did not confirm it, since selective predation by birds was observed but not quantified. Other research indicates that predation by birds is not the only factor determining the relative abundance of the different color morphs.

Gray or mottled morph:
vulnerable to predation in industrial areas where the trees are dark.

Melanic or carbonaria morph:
dark color makes it less vulnerable to predation in industrial areas.

> The gene controlling color in the peppered moth, is located on a single locus. The allele for the melanic (dark) form (**M**) is dominant over the allele for the gray (light) form (**m**).

1850   1900

Museum collections of the peppered moth made over the last 150 years show a marked change in the frequency of the melanic form. Moths collected in 1850 (above left), prior to the major onset of the Industrial Revolution in England, were predominantly pale. Fifty years later (above right) the frequency of the darker melanic forms had greatly increased. Even as late as the mid 20th century, coal-based industries predominated in some centers, and the melanic form was relatively more common in these areas (see map, right).

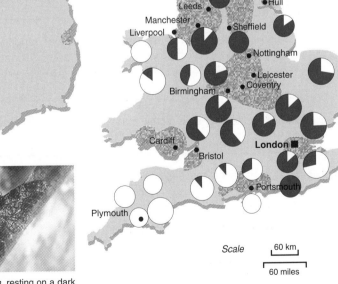

Glasgow
Belfast
Newcastle
Middlesbrough
Leeds
Hull
Manchester
Liverpool
Sheffield
Nottingham
Leicester
Coventry
Birmingham
Cardiff
Bristol
London
Portsmouth
Plymouth

Map: Relative frequencies of the two forms of peppered moth in the UK in 1950, a time when coal-based industries still predominated in some major centers.

Scale
60 km
60 miles

Key to frequency graphs

Gray or speckled form

Melanic or carbonaria form

Industrial areas

Non-industrial areas

A gray (mottled) form of *Biston*, camouflaged against a lichen covered bark surface. In the absence of soot pollution, mottled forms appear to have the selective advantage.

A melanic form of *Biston*, resting on a dark branch, so that it appears as part of the branch. Note that the background has been faded out so that the moth can be seen.

Periodicals:
The moths of war

Related activities: Adaptations and Fitness
Weblinks: Butterfly Shows Evolution at Work

RDA 2

Changes in Frequency of Melanic Peppered Moths

Frequency of melanic peppered moth related to reduced air pollution

In the 1940s and 1950s, coal burning was still at intense levels around the industrial centers of Manchester and Liverpool. During this time, the melanic form of the moth was still very dominant. In the rural areas further south and west of these industrial centers, the occurrence of the gray form increased dramatically. Subsequently (1960-1980), air quality improved with the decline of coal burning factories and the introduction of the Clean Air Act in cities. Sulfur dioxide and smoke levels dropped to a fraction of their previous levels. This coincided with a decline in the relative numbers of melanic moths (right).

1. The populations of peppered moth in England have undergone changes in the frequency of an obvious phenotypic character over the last 150 years. Describe the phenotypic character that changed in its frequency:

   _____

2. (a) Identify the (proposed) selective agent for phenotypic change in *Biston*: _____

   (b) How has the selection pressure on the light colored morph changed with changing environmental conditions over the last 150 years?

   _____

   _____

3. The industrial centers for England in 1950 were located around London, Birmingham, Liverpool, Manchester, and Leeds. Glasgow in Scotland also had a large industrial base. Comment on how the relative frequencies of the two forms of peppered moth were affected by the geographic location of industrial regions:

   _____

   _____

   _____

   _____

4. The level of pollution dropped around Manchester and Liverpool between 1960 and 1985.

   (a) State how much the pollution dropped by: _____

   (b) Describe how the frequency of the darker melanic form responded to this reduced pollution: _____

   _____

   _____

5. In the example of the peppered moths, state whether the selection pressure is disruptive, stabilizing, or directional:

   _____

6. Outline the key difference between natural and artificial selection: _____

   _____

7. Discuss the statement "the environment directs natural selection": _____

   _____

   _____

   _____

© BIOZONE International 2012
ISBN: 978-1-927173-11-4
Photocopying Prohibited

Selection for Skin Color in Humans

Pigmented skin of varying tones is a feature of humans that evolved after early hominins lost the majority of their body hair. However, the distribution of skin color globally is not random; people native to equatorial regions have darker skin tones than people from higher latitudes. For many years, biologists postulated that this was because darker skins had evolved to protect against skin cancer. The problem with this explanation was that skin cancer is not tied to evolutionary fitness because it affects post-reproductive individuals and cannot therefore provide a mechanism for selection. More complex analyses of the physiological and epidemiological evidence has shown a more complex picture in which selection pressures on skin color are finely balanced to produce a skin tone that regulates the effects of the sun's ultraviolet radiation on the nutrients vitamin D and folate, both of which are crucial to successful human reproduction, and therefore evolutionary fitness.

Skin Color in Humans: A Product of Natural Selection

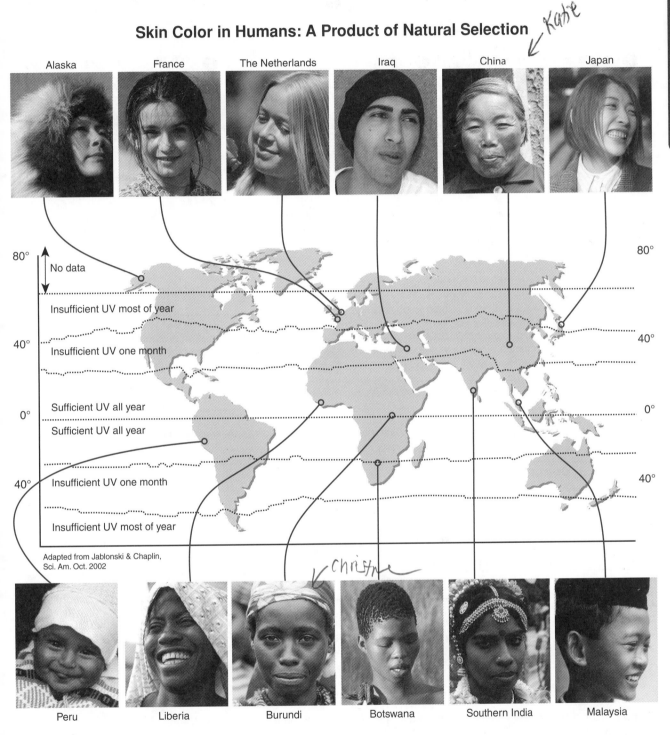

Alaska   France   The Netherlands   Iraq   China   Japan

80°   80°

No data

Insufficient UV most of year

40°   40°

Insufficient UV one month

Sufficient UV all year

0°   0°

Sufficient UV all year

Insufficient UV one month

40°   40°

Insufficient UV most of year

Adapted from Jablonski & Chaplin, Sci. Am. Oct. 2002

Peru   Liberia   Burundi   Botswana   Southern India   Malaysia

Human skin color is the result of two opposing selection pressures. Skin pigmentation has evolved to protect against destruction of folate by ultraviolet light, but the skin must also be light enough to receive the light required to synthesise vitamin D. Vitamin D synthesis is a process that begins in the skin and is inhibited by dark pigment. Folate is needed for healthy neural development in humans and a deficiency is associated with fatal neural tube defects. Vitamin D is required for the absorption of calcium from the diet and therefore normal skeletal development.

Women also have a high requirement for calcium during pregnancy and lactation. Populations that live in the tropics receive enough ultraviolet (UV) radiation to synthesise vitamin D all year long. Those that live in northern or southern latitudes do not. In temperate zones, people lack sufficient UV light to make vitamin D for one month of the year. Those nearer the poles lack enough UV light for vitamin D synthesis most of the year (above). Their lighter skins reflect their need to maximize UV absorption (the photos show skin color in people from different latitudes).

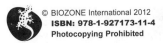

Periodicals:
Skin deep, Fair enough

Related activities: Adaptations and Fitness
Weblinks: Nina Jablonski Breaks the Illusion of Skin Color

252

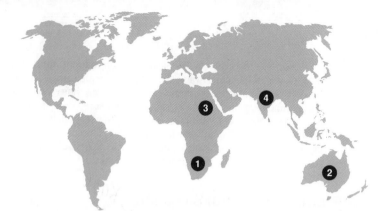

Long-term resident   Recent immigrant

1 Southern Africa: ~ 20-30°S

Khoisan-Namibia   *Zulu: 1000 years ago*

2 Australia: ~ 10-35°S

Aborigine   *European: 300 years ago*

3 Banks of the Red Sea: ~ 15-30°N

Nuba-Sudan   *Arab: 2000 years ago*

4 India: ~ 10-30°S

West Bengal   *Tamil: ~100 years ago*

The skin of people who have inhabited particular regions for millennia has adapted to allow sufficient vitamin D production while still protecting folate stores. In the photos above, some of these original inhabitants are illustrated to the left of each pair and compared with the skin tones of more recent immigrants (to the right of each pair, with the number of years since immigration). The numbered locations are on the map.

1. (a) Describe the role of folate in human physiology: _____

   (b) Describe the role of vitamin D in human physiology: _____

2. (a) Early hypotheses to explain skin color linked pigmentation level only to the degree of protection it gave from UV-induced skin cancer. Explain why this hypothesis was inadequate in accounting for how skin color evolved:

   _____

   _____

   (b) Explain how the new hypothesis for the evolution of skin color overcomes these deficiencies: _____

   _____

   _____

   _____

3. Explain why, in any given geographical region, women tend to have lighter skins (by 3-4% on average) than men:

   _____

   _____

4. The Inuit people of Alaska and northern Canada have a diet rich in vitamin D and their skin color is darker than predicted on the basis of UV intensity at their latitude. Explain this observation:

   _____

   _____

5. (a) What health problems might be expected for people of African origin now living in Canada?

   _____

   _____

   (b) How could these people avoid these problems in their new higher latitude environment? _____

   _____

   _____

Selection for Human Birth Weight

Selection pressures operate on populations in such a way as to reduce mortality. For humans, giving birth is a special, but often traumatic, event. In a study of human birth weights it is possible to observe the effect of selection pressures operating to constrain human birth weight within certain limits. This is a good example of **stabilizing selection**. This activity explores the selection pressures acting on the birth weight of human babies. Carry out the steps below:

Step 1: Collect the birth weights from 100 birth notices from your local newspaper (or 50 if you are having difficulty getting enough; this should involve looking back through the last 2-3 weeks of birth notices). If you cannot obtain birth weights in your local newspaper, a set of 100 sample birth weights is provided in the Model Answers booklet.

Step 2: Group the weights into each of the 12 weight classes (of 0.5 kg increments). Determine what percentage (of the total sample) fall into each weight class (e.g. 17 babies weigh 2.5-3.0 kg out of the 100 sampled = 17%)

Step 3: Graph these in the form of a histogram for the 12 weight classes (use the graphing grid provided right). Be sure to use the scale provided on the left vertical (y) axis.

Step 4: Create a second graph by plotting percentage mortality of newborn babies in relation to their birth weight. Use the scale on the right y axis and data provided (below).

Step 5: Draw a line of 'best fit' through these points.

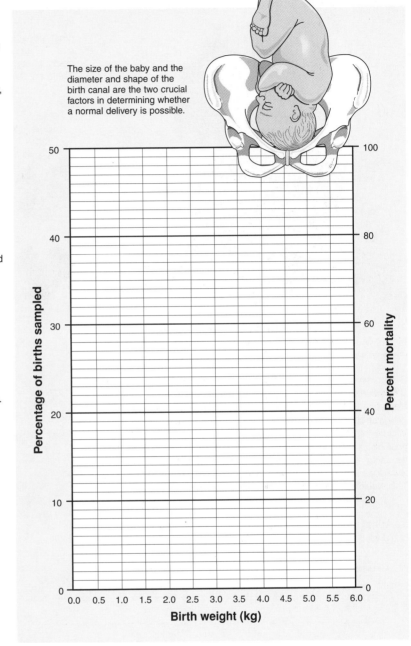

The size of the baby and the diameter and shape of the birth canal are the two crucial factors in determining whether a normal delivery is possible.

Mortality of newborn babies related to birth weight

Weight (kg)	Mortality (%)
1.0	80
1.5	30
2.0	12
2.5	4
3.0	3
3.5	2
4.0	3
4.5	7
5.0	15

Source: Biology: The Unity & Diversity
of Life (4th ed), by Starr and Taggart

1. Describe the shape of the histogram for birth weights: _____

2. What is the optimum birth weight in terms of the lowest newborn mortality? _____

3. Describe the relationship between newborn mortality and birth weight: _____

_____

4. Describe the selection pressures that are operating to control the range of birth weight: _____

_____

_____

5. How might have modern medical intervention during pregnancy and childbirth altered these selection pressures?

_____

_____

Related activities: Natural Selection   **PDA 2**

Black is the New Gray

The North American gray squirrel (*Sciurus carolinensis*) was introduced into the UK in the 19th century. By 1958 it had spread to most of England and Wales, displacing the smaller native red squirrel (*Sciurus vulgaris*). Now, a black variation of the gray squirrel is beginning to displace the original gray variety, and is common in four areas of Britain (below right). Observations suggest the black form is more aggressive than the gray. This behavior may give them a competitive advantage and may provide a characteristic for selection. This would help explain why the numbers of black squirrels are increasing locally.

Most gene mutations are recessive, but the mutation for black coat color shows incomplete dominance. This allows the black variant to spread more rapidly through the population. There are three genotypes for the gray squirrel:

M^GM^G: Gray coat
Two normal gray alleles

M^GM^B: Brown-black coat
One normal gray allele, one mutant black allele

M^BM^B: Jet black coat
Two mutant black alleles

Photo: D. Gordon E. Robertson

The black color (above) results from a major sequence mutation in the MC1R gene which regulates hair and skin color in mammals. The gene controls the amount of the black eumelanin pigment produced.

Black squirrel numbers have rapidly increased in Hertfordshire, Bedfordshire, Huntingdonshire, and Cambridgeshire. It is proposed that black coat color affords the black squirrels a competitive advantage over gray variants. Although none have yet been proven, likely advantages include higher levels of aggression than the gray varieties, and the possibility that the females find the black coat more attractive when selecting mates. A Canadian study has shown that black squirrels have a lower basal metabolic rate and lose heat more slowly than gray squirrels. This would benefit the black variant in cold conditions.

1. What features of the black variant of the gray squirrel appear to give it a competitive advantage?

_____

_____

_____

2. (a) Complete the Punnet square (left) and use it to help explain why the frequency of the black squirrel is increasing relative to the gray variety:

_____

_____

_____

_____

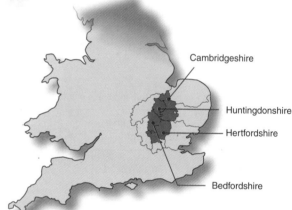

(b) Explain what would happen to the rate of change in black squirrel frequency if the mutation for black was recessive:

_____

_____

_____

Periodicals:
Black squirrels

Heterozygous Advantage

Natural selection operates on phenotypes (and therefore their genotypes) in the prevailing environment. For some phenotypic conditions controlled by a single gene with two alleles, a heterozygote may have a higher fitness than either of the homozygous conditions. This situation is called **heterozygous advantage**. In the case of the sickle cell allele outlined below, susceptibility to malaria is high in the homozygous dominant condition, but relatively low in the heterozygous condition.

Consequently, the heterozygote has a higher fitness in malaria-prone regions. Heterozygous advantage can result in the stable coexistence of different phenotypes in a population (a state called **balanced polymorphism**) and can account for the persistence of detrimental alleles. The maintenance of the sickle cell mutation in malaria-prone regions is one of the few well documented examples in which the evidence for heterozygous advantage is conclusive.

The Sickle Cell Allele (Hb^S)

Sickle cell disease is caused by a mutation in a gene that directs the production of the human blood protein called hemoglobin. The mutant allele is known as **Hb^S** and produces a form of hemoglobin that differs from the normal form by just one amino acid in the β-chain. This small change however causes a cascade of physiological problems in people with the allele. The red blood cells containing mutated hemoglobin alter their shape to become irregular and spiky: the so-called **sickle cells**.

Sickle cells have a tendency to clump together and work less efficiently. In people with just one sickle cell allele plus a normal allele (the heterozygote condition **Hb^{S}Hb**), there is a mixture of both red blood cell types and they are said to have the sickle cell trait. They are generally unaffected by the disease except in low oxygen environments (e.g. climbing at altitude). People with two Hb^S genes (**Hb^{S}Hb^S**) suffer severe illness and even premature death. HbS is therefore considered a **lethal gene**.

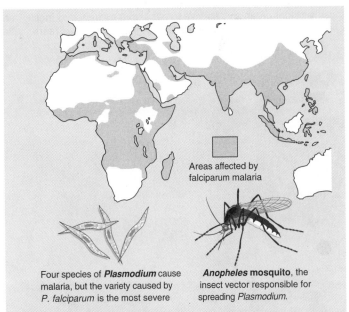

Four species of **Plasmodium** cause malaria, but the variety caused by *P. falciparum* is the most severe

Anopheles mosquito, the insect vector responsible for spreading *Plasmodium*.

Areas affected by falciparum malaria

Fig. 1: Incidence of falciparum malaria

Heterozygous Advantage in Malarial Regions

Falciparum malaria is widely distributed throughout central Africa, the Mediterranean, Middle East, and tropical and semi-tropical Asia (Fig. 1). It is transmitted by the *Anopheles* mosquito, which spreads the protozoan *Plasmodium falciparum* from person to person as it feeds on blood.

SYMPTOMS: These appear 1-2 weeks after being bitten, and include headache, shaking, chills, and fever. Falciparum malaria is more severe than other forms of malaria, with high fever, convulsions, and coma. It can be fatal within days of the first symptoms appearing.

THE PARADOX: The Hb^S allele offers considerable protection against malaria. Sickle cells have low potassium levels, which causes *Plasmodium* parasites inside these cells to die. Those with a normal phenotype are very susceptible to malaria, but heterozygotes (Hb^{S}Hb) are much less so. This situation, called **heterozygous advantage**, has resulted in the Hb^S allele being present in moderately high frequencies in parts of Africa and Asia despite its harmful effects (Fig. 2). This is a special case of balanced polymorphism, called a **balanced lethal system** because neither of the homozygotes produces a phenotype that survives, but the heterozygote is viable.

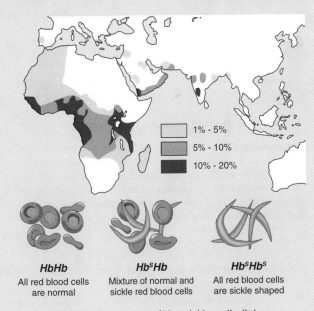

1% - 5%

5% - 10%

10% - 20%

HbHb
All red blood cells are normal

Hb^{s}Hb
Mixture of normal and sickle red blood cells

Hb^{s}Hb^s
All red blood cells are sickle shaped

Fig. 2: Frequency of the sickle cell allele

1. With respect to the sickle cell allele, explain how **heterozygous advantage** can lead to **balanced polymorphism**:

_____

_____

_____

_____

_____

Periodicals: Polymorphism, Genetics of sickle cell anemia

Related activities: Sickle Cell Mutation
Weblinks: A Mutation Story, Natural Selection in Humans

Insecticide Resistance

Insecticides are pesticides used to control insects considered harmful to humans, their livelihood, or environment. Insecticides have been used for hundreds of years, but their use has proliferated since the advent of synthetic insecticides (e.g. DDT) in the 1940s. When **insecticide resistance** develops the control agent will no longer control the target species. Insecticide resistance can arise through a combination of behavioral, anatomical, biochemical, and physiological mechanisms, but the underlying process is a form of **natural selection**, in which the most resistant organisms survive to pass on their genes to their offspring. To combat increasing resistance, higher doses of more potent pesticides are sometimes used. This drives the selection process, so that increasingly higher dose rates are required to combat rising resistance. This phenomenon is made worse by the development of multiple resistance in some pest species. High application rates may also kill non-target species, and persistent chemicals may remain in the environment and accumulate in food chains. These concerns have led to some insecticides being banned (DDT has been banned in most developed countries since the 1970s). Insecticides are used in medical, agricultural, and environmental applications, so the development of resistance has serious environmental and economic consequences.

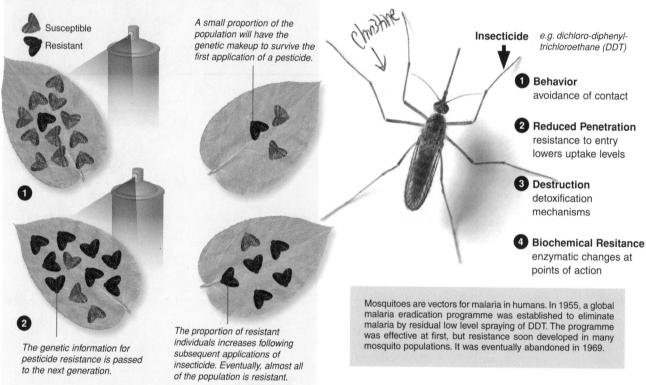

- Susceptible
- Resistant

A small proportion of the population will have the genetic makeup to survive the first application of a pesticide.

① The genetic information for pesticide resistance is passed to the next generation.

② The proportion of resistant individuals increases following subsequent applications of insecticide. Eventually, almost all of the population is resistant.

Insecticide e.g. dichloro-diphenyl-trichloroethane (DDT)

❶ Behavior avoidance of contact

❷ Reduced Penetration resistance to entry lowers uptake levels

❸ Destruction detoxification mechanisms

❹ Biochemical Resistance enzymatic changes at points of action

Mosquitoes are vectors for malaria in humans. In 1955, a global malaria eradication programme was established to eliminate malaria by residual low level spraying of DDT. The programme was effective at first, but resistance soon developed in many mosquito populations. It was eventually abandoned in 1969.

The Development of Resistance

The application of an insecticide can act as a potent selection pressure for resistance in pest insects. The insecticide acts as a selective agent, and only individuals with greater natural resistance survive the application to pass on their genes to the next generation. These genes (or combination of genes) may spread through all subsequent populations.

Mechanisms of Resistance in Insect Pests

Insecticide resistance in insects can arise through a combination of mechanisms. (1) Increased sensitivity to an insecticide will cause the pest to avoid a treated area. (2) Certain genes (e.g. the *PEN* gene) confer stronger physical barriers, decreasing the rate at which the chemical penetrates the cuticle. (3) Detoxification by enzymes within the insect's body can render the pesticide harmless, and (4) structural changes to the target enzymes make the pesticide ineffective. No single mechanism provides total immunity, but together they transform the effect from potentially lethal to insignificant.

1. Give two reasons why widespread insecticide resistance can develop very rapidly in insect populations:

   (a) _____

   (b) _____

2. Explain how repeated insecticide applications act as a selective agent for evolutionary change in insect populations:

   _____

   _____

3. With reference to synthetic insecticides, discuss the implications of insecticide resistance to human populations:

   _____

   _____

   _____

© BIOZONE International 2012
ISBN: 978-1-927173-11-4
Photocopying Prohibited

Related activities: The Evolution of Antibiotic Resistance, Chloroquine Resistance in Protozoa

The Evolution of Antibiotic Resistance

Antibiotic resistance arises when a genetic change allows bacteria to tolerate levels of antibiotic that would normally inhibit growth. This resistance may arise spontaneously, through mutation or copying error, or by transfer of genetic material between microbes. Genomic analyses from 30,000 year old permafrost sediments show that the genes for antibiotic resistance are not new. They have long been present in the bacterial genome, predating the modern selective pressure of antibiotic use. In the current selective environment, these genes have proliferated and antibiotic resistance has spread. For example, methicillin resistant strains of *Staphylococcus aureus* (MRSA) have acquired genes for resistance to all penicillins. Such strains are called superbugs.

The Evolution of Antibiotic Resistance in Bacteria

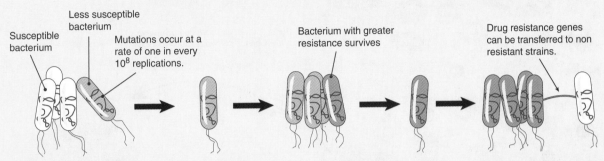

Susceptible bacterium — Less susceptible bacterium — Mutations occur at a rate of one in every 10^8 replications. — Bacterium with greater resistance survives — Drug resistance genes can be transferred to non resistant strains.

Any population, including bacterial populations, includes variants with unusual traits, in this case reduced sensitivity to an antibiotic. These variants arise as a result of mutations in the bacterial chromosome. Such mutations are well documented and some are ancient.

When a person takes an antibiotic, only the most susceptible bacteria will die. The more resistant cells remain and continue dividing. Note that the antibiotic does not create the resistance; it provides the environment in which selection for resistance can take place.

If the amount of antibiotic delivered is too low, or the course of antibiotics is not completed, a population of resistant bacteria develops. Within this population too, there will be variation in susceptibility. Some will survive higher antibiotic levels.

A highly resistant population has evolved. The resistant cells can exchange genetic material with other bacteria (via horizontal gene transmission), passing on the genes for resistance. The antibiotic initially used against this bacterial strain will now be ineffective.

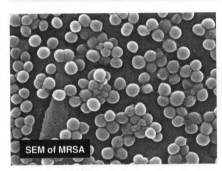

SEM of MRSA

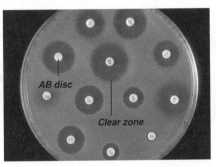

AB disc / Clear zone

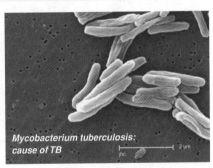

Mycobacterium tuberculosis: cause of TB   2 µm

Staphylococcus aureus is a common bacterium responsible for various minor skin infections in humans. MRSA is a variant strain that has evolved resistance to penicillin and related antibiotics. MRSA is troublesome in hospital-associated infections because patients with open wounds, invasive devices (e.g. catheters), or poor immunity are at greater risk for infection than the general public.

The photo above shows an antibiogram plate culture of *Enterobacter sakazakii*, a rare cause of invasive infections in infants. An antibiogram measures the biological resistance of disease-causing organisms to antibiotic agents. The bacterial lawn (growth) on the agar plate is treated with antibiotic discs, and the sensitivity to various antibiotics is measured by the extent of the clearance zone in the bacterial lawn.

TB is a disease that has experienced spectacular ups and downs. Drugs were developed to treat it, but then people became complacent when they thought the disease was beaten. TB has since resurged because patients stop their medication too soon and infect others. Today, one in seven new TB cases is resistant to the two drugs most commonly used as treatments, and 5% of these patients die.

1. Describe two ways in which antibiotic resistance can become widespread:

    (a) _____

    (b) _____

2. Genomic evidence indicates that the genes for antibiotic resistance are ancient:

    (a) How could these genes have arisen in the first place? _____

    (b) Why were they not lost from the bacterial genome? _____

    (c) Explain why these genes are proliferating now: _____

Periodicals: The enemy within

Related activities: Insecticide Resistance, Drug Resistance in HIV
Weblinks: Why Evolution Matters Now, The Rise in Antibiotic Resistance

A 2

The Founder Effect

Occasionally, a small number of individuals from a large population may migrate away from, or become isolated from, their original population. If this colonizing or 'founder' population is made up of only a few individuals, it will probably have a non-representative sample of alleles from the parent population's gene pool. As a consequence of this **founder effect**, the colonising population may evolve differently from that of the parent population, particularly since the environmental conditions for the isolated population may be different. In some cases, it may be possible for certain alleles to be missing altogether from the individuals in the isolated population. Future generations of this population will not have this allele.

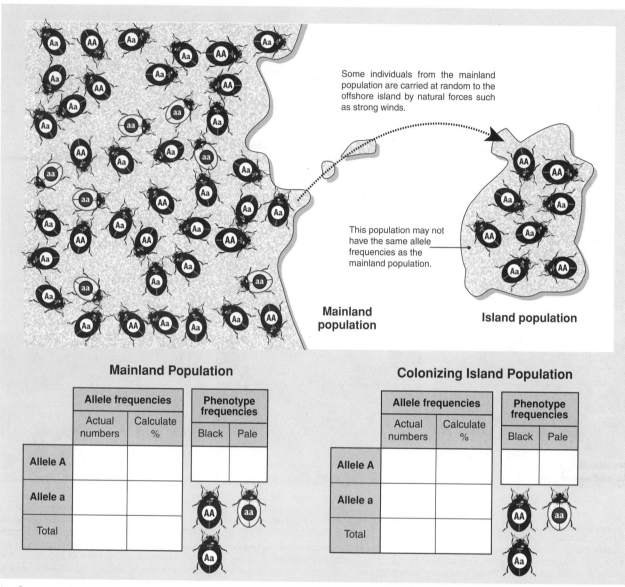

Some individuals from the mainland population are carried at random to the offshore island by natural forces such as strong winds.

This population may not have the same allele frequencies as the mainland population.

Mainland population

Island population

Mainland Population

	Allele frequencies		Phenotype frequencies	
	Actual numbers	Calculate %	Black	Pale
Allele A				
Allele a				
Total				

Colonizing Island Population

	Allele frequencies		Phenotype frequencies	
	Actual numbers	Calculate %	Black	Pale
Allele A				
Allele a				
Total				

1. Compare the mainland population to the population which ended up on the island (use the spaces in the tables above):
   (a) Count the **phenotype** numbers for the two populations (i.e. the number of black and pale beetles).
   (b) Count the **allele** numbers for the two populations: the number of dominant alleles (A) and recessive alleles (a). Calculate these as a percentage of the total number of alleles for each population.

2. How are the allele frequencies of the two populations different? _____

_____

3. Describe some possible ways in which various types of organism can be **carried** to an offshore island:

   (a) Plants: _____

   (b) Land animals: _____

   (c) Non-marine birds: _____

4. Founder populations are usually very small. What other process may act quite rapidly to further alter allele frequencies?

_____

_____

© BIOZONE International 2012
ISBN: 978-1-927173-11-4
Photocopying Prohibited

Related activities: Gene Pool Exercise, Genetic Drift, Oceanic Island Colonizers

Population Bottlenecks

Populations may sometimes be reduced to low numbers by predation, disease, or periods of climatic change. A population crash may not be 'selective': it may affect all phenotypes equally. Large scale catastrophic events, such as fire or volcanic eruption, are examples of such non-selective events. Humans may severely (and selectively) reduce the numbers of some species through hunting and/or habitat destruction. These populations may recover, having squeezed through a 'bottleneck' of low numbers.

The diagram below illustrates how population numbers may be reduced as a result of a catastrophic event. Following such an event, the small number of individuals contributing to the gene pool may not have a representative sample of the genes in the pre-catastrophe population, i.e. the allele frequencies in the remnant population may be severely altered. **Genetic drift** may cause further changes to allele frequencies. The small population may return to previous levels but with a reduced genetic diversity.

Population numbers

Low — High

Large population with plenty of genetic diversity.

Population crashes to a very low number and loses most of its genetic diversity.

Population grows to a large size again, but has lost much of its genetic diversity.

Time

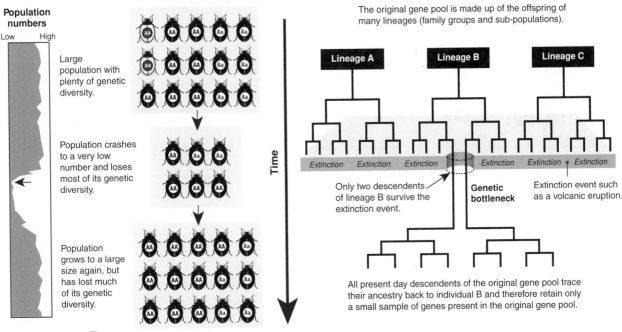

The original gene pool is made up of the offspring of many lineages (family groups and sub-populations).

| Lineage A | Lineage B | Lineage C |

Extinction   Extinction   Extinction   Extinction   Extinction   Extinction

Only two descendents of lineage B survive the extinction event.

Genetic bottleneck

Extinction event such as a volcanic eruption.

All present day descendents of the original gene pool trace their ancestry back to individual B and therefore retain only a small sample of genes present in the original gene pool.

Photo: Dept. of Natural Resources, Illinois

Modern Examples of Population Bottlenecks

Cheetahs: The world population of cheetahs currently stands at fewer than 20,000. Recent genetic analysis has found that the entire population exhibits very little genetic diversity. It appears that cheetahs may have narrowly escaped extinction at the end of the last ice age, about 10-20,000 years ago. If all modern cheetahs arose from a very limited genetic stock, this would explain their present lack of genetic diversity. The lack of genetic variation has resulted in a number of problems that threaten cheetah survival, including sperm abnormalities, decreased fecundity, high cub mortality, and sensitivity to disease.

Illinois prairie chicken: When Europeans first arrived in North America, there were millions of prairie chickens. As a result of hunting and habitat loss, the Illinois population of prairie chickens fell from about 100 million in 1900 to fewer than 50 in the 1990s. A comparison of the DNA from birds collected in the mid-twentieth century and DNA from the surviving population indicated that most of the genetic diversity has been lost.

1. Endangered species are often subjected to population bottlenecks. Explain how population bottlenecks affect the ability of a population of an endangered species to recover from its plight:

_____

_____

_____

2. Why has the lack of genetic diversity in cheetahs increased their sensitivity to disease? _____

_____

_____

3. Describe the effect of a population bottleneck on the potential of a species to adapt to changes (i.e. its ability to evolve):

_____

_____

Periodicals: The Cheetah: Losing the race?

Related activities: Gene Pool Exercise, Genetic Drift

Genetic Drift

Not all individuals, for various reasons, will be able to contribute their genes to the next generation. **Genetic drift** (also known as the Sewall-Wright Effect) refers to the *random changes in allele frequency* that occur in all populations, but are much more pronounced in small populations. In a small population, the effect of a few individuals not contributing their alleles to the next generation can have a great effect on allele frequencies. Alleles may even become **lost** from the gene pool altogether (frequency becomes 0%) or **fixed** as the only allele for the gene present (frequency becomes 100%).

The genetic makeup (allele frequencies) of the population changes randomly over a period of time

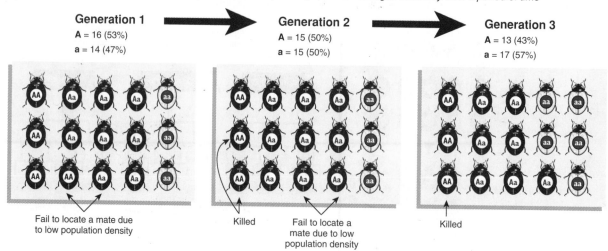

Generation 1
A = 16 (53%)
a = 14 (47%)

Generation 2
A = 15 (50%)
a = 15 (50%)

Generation 3
A = 13 (43%)
a = 17 (57%)

Fail to locate a mate due to low population density

Killed

Fail to locate a mate due to low population density

Killed

This diagram shows the gene pool of a hypothetical small population over three generations. For various reasons, not all individuals contribute alleles to the next generation. With the random loss of the alleles carried by these individuals, the allele frequency changes from one generation to the next. The change in frequency is directionless as there is no selecting force. The allele combinations for each successive generation are determined by how many alleles of each type are passed on from the preceding one.

Computer Simulation of Genetic Drift

Below are displayed the change in allele frequencies in a computer simulation showing random genetic drift. The breeding population progressively gets smaller from left to right. Each simulation was run for 140 generations.

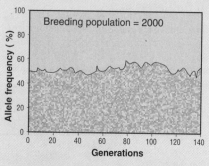

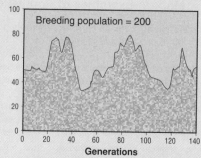

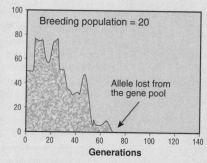

Allele lost from the gene pool

Large breeding population
Fluctuations are minimal in large breeding populations because the large numbers buffer the population against random loss of alleles. On average, losses for each allele type will be similar in frequency and little change occurs.

Small breeding population
Fluctuations are more severe in smaller breeding populations because random changes in a few alleles cause a greater percentage change in allele frequencies.

Very small breeding population
Fluctuations in very small breeding populations are so extreme that the allele can become fixed (frequency of 100%) or lost from the gene pool altogether (frequency of 0%).

1. What is **genetic drift**? _____

_____

2. What is the effect of genetic drift on the genetic variation present in very small populations? _____

_____

_____

_____

3. Name a small breeding population of animals or plants in your country in which genetic drift could be occurring:

_____

© BIOZONE International 2012
ISBN: 978-1-927173-11-4
Photocopying Prohibited

Related activities: Gene Pool Exercise, The Founder Effect
Weblinks: Genetic Drift Simulation

Artificial Selection in Animals

The domestication of livestock has a long history dating back at least 8000 years. Today's important stock breeds were all derived from wild ancestors that were domesticated by humans, who then used **artificial selection** (*aka* selective breeding) to produce livestock to meet specific requirements. Artificial selection of domesticated animals involves identifying desirable qualities (e.g. high wool production or meat yield), and breeding together individuals with those qualities so the trait is reliably passed on. Practices such as **inbreeding**, **line-breeding**, and **out-crossing** are used to select and 'fix' desirable traits in varieties. Today's reproductive technologies, such as artificial insemination, are

also widely used so that the desirable characteristics of one male can be passed on to many females. These new technologies refine the selection process and increase the rate at which stock improvements are made. Rates are predicted to accelerate further as new technologies, such as genomic selection, become more widely available and less costly. However, producing highly inbred lines of animals with specific traits can have disadvantages. **Homozygosity** for a number of desirable traits can cause physiological or physical problems to the animal itself. For example, animals bred specifically for rapid weight gain often grow so fast that they have skeletal and muscular difficulties.

The Origin of Domestic Animals

PIG
Wild ancestor: Boar
Origin: Anatolia, 9000 years BP
Now: More than 12 distinct modern breeds, including the Berkshire (meat) and Tamworth (hardiness).

DOMESTIC FOWL
Wild ancestor: Red jungle fowl
Origin: Indus Valley, 4000 BP
Now: More than 60 breeds including Rhode Island Red (meat) and Leghorn (egg production).

Each domesticated breed has been bred from the wild ancestor (pictured). The date indicates the earliest record of the domesticated form (years before present or BP). Different countries have different criteria for selection, based on their local environments and consumer preferences.

GOAT
Wild ancestor: Bezoar goat
Origin: Iraq, 10,000 years BP
Now: approx. 35 breeds including Spanish (meat), Angora (fibre) and Nubian (dairy).

SHEEP
Wild ancestor: Asiatic mouflon
Orign: Iran, Iraq, Levant, 10,000 years BP
Now: More than 200 breeds including Merino (wool), Suffolk (meat), Friesian (milk), and dual purpose (Romney).

CATTLE
Wild ancestor: Auroch (extinct)
Origin: SW Asia, 10,000 years BP
Now: 800 modern breeds including the Aberdeen Angus (meat), Friesian and Jersey (milk), and Zebu (draught).

1. Distinguish between inbreeding and out-crossing, explaining the significance of each technique in artificial selection:

Both are used to select and 'fix' desirable traits in varieties. They increase the rate of which stock improvements are made. Highly inbred populations can cause psycological and physical problems

2. How are new reproductive technologies contributing to rapid phenotypic change in populations?

Desireable traits of one male can be passed onto many female

Periodicals:
Taming the wild

Related activities: Artificial Selection in Crop Plants
Weblinks: Dogs and More Dogs

RA 2

Dogs provide a striking example of selective breeding, with more than 400 recognized breeds. Over centuries, humans have selected for desirable physical and behavioral traits. All breeds of dog are members of the same species, *Canis familiaris*. This species descended from a single wild species, the gray wolf *Canis lupus*, over 15,000 years ago. Five ancient dog breeds are recognized, from which all other breeds are thought to have descended by artificial selection.

Gray wolf *Canis lupus pallipes*

The gray wolf is distributed throughout Europe, North America, and Asia. Amongst members of this species, there is a lot of variation in coat coloration. This accounts for the large variation in coat colours of dogs today.

The Ancestor of Domestic Dogs

Until recently, it was unclear whether the ancestor to the modern domestic dogs was the desert wolf of the Middle East, the woolly wolf of central Asia, or the gray wolf of the Northern hemisphere. Recent genetic studies (mitochondrial DNA comparisons) now provide strong evidence that the ancestor of domestic dogs throughout the world is the gray wolf. It seems likely that this evolutionary change took place in a single region, probably China.

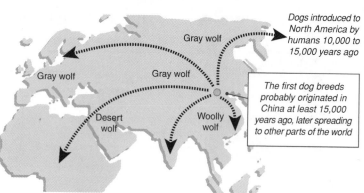

Dogs introduced to North America by humans 10,000 to 15,000 years ago

Gray wolf

Gray wolf

Gray wolf

Desert wolf

Woolly wolf

The first dog breeds probably originated in China at least 15,000 years ago, later spreading to other parts of the world

Mastiff-type
Canis familiaris inostranzevi

Originally from Tibet, the first records of this breed of dog go back to the Stone Age.

Grayhound
Canis familiaris leineri

Drawings of this breed on pottery dated from 8000 years ago in the Middle East make it one of the oldest.

Pointer-type
Canis familiaris intermedius

Probably derived from the grayhound breed for the purpose of hunting small game.

Sheepdog
Canis familiaris metris optimae

Originating in Europe, this breed has been used to guard flocks from predators for thousands of years.

Wolf-like
Canis familiaris palustris

Found in snow covered habitats in northern Europe, Asia (Siberia), and North America (Alaska).

3. How has selective breeding contributed to changes in the gene pool of domestic dogs? _____

Humans have selected dogs with the most desirable physical and behavioral traits.

4. Which behavioral characteristics of wolves predisposed them to successful domestication? _____

Evolutionary change took place in a single region

5. List the physical and behavioral traits that would be desirable (selected for) in the following uses of a dog:

(a) Hunting large game (e.g. boar and deer): _____

(b) Game fowl dog: _____

(c) Stock control (sheep/cattle dog): _____

(d) Family pet (house dog): _____

(e) Guard dog: _____

Artificial Selection in Crop Plants

For thousands of years, farmers have used the variation in wild and cultivated plants to develop crops. Genetic diversity gives species the ability to adapt to new environmental challenges, such as new pests, diseases, or growing conditions. The genetic diversity within different crop varieties provides options to develop, through selection, new and more productive crop plants. *Brassica* *oleracea* is a good example of the variety that can be produced by selectively growing plants with desirable traits. Not only are there six varieties of *Brassica oleracea*, but each of those has a number of sub varieties as well. Although brassicas have been cultivated for several thousand years, cauliflower, broccoli, and brussels sprouts appeared only in the last 500 years.

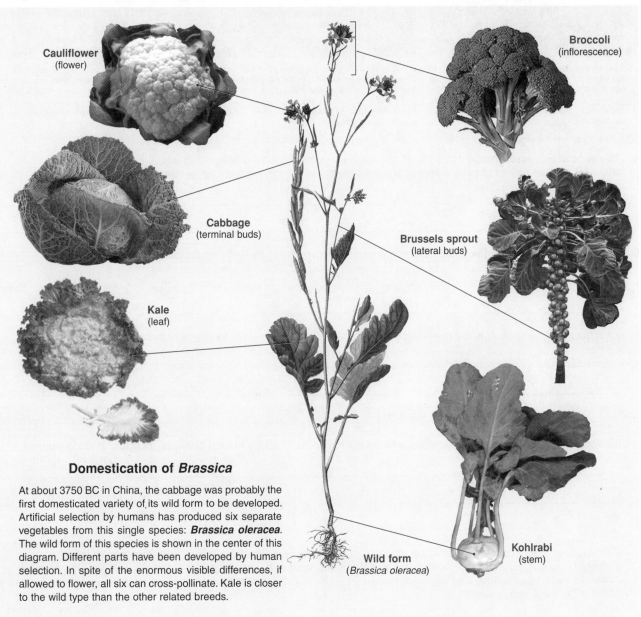

Cauliflower (flower)

Broccoli (inflorescence)

Cabbage (terminal buds)

Brussels sprout (lateral buds)

Kale (leaf)

Kohlrabi (stem)

Wild form (*Brassica oleracea*)

Domestication of *Brassica*

At about 3750 BC in China, the cabbage was probably the first domesticated variety of its wild form to be developed. Artificial selection by humans has produced six separate vegetables from this single species: ***Brassica oleracea***. The wild form of this species is shown in the center of this diagram. Different parts have been developed by human selection. In spite of the enormous visible differences, if allowed to flower, all six can cross-pollinate. Kale is closer to the wild type than the other related breeds.

1. Study the diagram above and identify which part of the plant has been selected for to produce each of the vegetables:

   (a) Cauliflower: _____flower_____
   (d) Brussels sprout: _____lateral buds_____

   (b) Kale: _____leaf_____
   (e) Cabbage: _____terminal buds_____

   (c) Broccoli: _____inflorenscence_____
   (f) Kohlrabi: _____stem_____

2. Describe the feature of these vegetables that suggests they are members of the same species: _____
   _____They are all green vegetables_____

3. Describe the method used to develop broccoli and the features one would look for when doing so: _____
   _____The method used to develop broccoli is artificial selection one may use color and size to get the desired broccoli_____

Related activities: Breeding Modern Wheat

A 3

The number of apple varieties is now a fraction of the many hundreds grown a century ago. Apples are native to Kazakhstan and breeders are now looking back to this center of diversity to develop apples resistant to the bacterial disease that causes fire blight.

In 18th-century Ireland, potatoes were the main source of food for about 30% of the population, and farmers relied almost entirely on one very fertile and productive variety. That variety proved susceptible to the potato blight fungus which resulted in a widespread famine.

Hybrid corn varieties have been bred to minimize harm inflicted by insect pests such as corn rootworm (above). Hybrids are important because they recombine the genetic characteristics of parental lines and show increased heterozygosity and hybrid vigor.

4. The genetic processes involved in artificial and natural selection are essentially no different. Explain how this has changed with the advent of genetic engineering technology and why it is particularly relevant to crop plants:

_____

_____

5. Describe a phenotypic characteristic that might be desirable in an apple tree and explain your choice:

_____

_____

6. (a) Explain why genetic diversity might decline during selective breeding for particular characteristics:

_____

_____

_____

(b) With reference to an example, discuss why retaining genetic diversity in crop plants is important for food security:

_____

_____

_____

7. Cultivated American cotton plants have a total of 52 chromosomes (2N = 52). In each cell there are 26 large chromosomes and 26 small chromosomes. Old World cotton plants have 26 chromosomes (2N = 26), all large. Wild American cotton plants have 26 chromosomes, all small. How might cultivated American cotton have originated from Old World cotton and wild American cotton?

_____

_____

8. The Cavendish is the variety of banana most commonly sold in world supermarkets. It is seedless, sterile, and under threat of extinction by Panama disease Race 4. Explain why Cavendish banana crops are so endangered by this fungus:

_____

_____

_____

9. Discuss the need to maintain the biodiversity of wild plants and ancient farm breeds: _____

_____

_____

_____

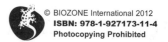

Breeding Modern Wheat

Wheat has been cultivated for more than 9000 years and has undergone many genetic changes during its domestication. The evolution of modern bread wheat from its wild ancestors (below) involved two natural **hybridization** events, accompanied by polyploidy. Once wheat became domesticated, artificial selection under cultivation emphasized characteristics such as high protein (gluten) content, high yield, and pest resistance to pests and disease. Hybrid vigor in wheat cultivars is produced by crossing inbred lines and selecting for desirable traits in the progeny, which can now be identified using genetic techniques such as marker assisted selection. This is an indirect selection process where a trait of interest is selected on the basis of a marker linked to it. Increasingly, research is focused on enhancing the genetic diversity of wheat to provide for future crop development. With this in mind, there is renewed interest in some of the lower yielding, ancient wheat varieties, such as wild emmer, Kamut®, and spelt, which pre-date common wheat. Not only do these older varieties offer a broader spectrum of nutrients than common and durum wheat, but they contain alleles no longer present in modern inbred varieties.

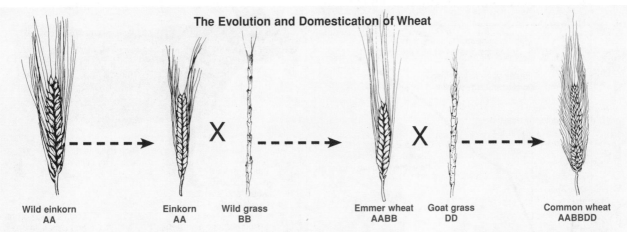

The Evolution and Domestication of Wheat

| Wild einkorn AA | Einkorn AA | Wild grass BB | Emmer wheat AABB | Goat grass DD | Common wheat AABBDD |

Wild einkorn becomes domesticated in the Middle East. There are slight changes to phenotype but not chromosome number.

A sterile hybrid between einkorn and wild grass undergoes a chromosome doubling to create fertile emmer wheat.

A sterile hybrid between emmer wheat and goat grass undergoes a chromosome doubling to create fertile common wheat.

Ancient cereal grasses had heads which shattered easily so that the seeds were widely scattered. In this more primitive morphology, the wheat ear breaks into spikelets when threshed, and milling or pounding is needed to remove the hulls and obtain the grain. Cultivation and repeated harvesting and sowing of the grains of wild grasses led to domestic strains with larger seeds and sturdier heads. Modern selection methods incorporate genetic techniques to identify and isolate beneficial genes, e.g. the RHt dwarfing gene, which gave rise to shorter stemmed modern wheat varieties.

Modern bread wheat has been selected for its non-shattering heads, high yield, and high gluten (protein) content. The grains are larger and the seeds (spikelets) remain attached to the ear by a toughened rachis during harvesting. On threshing, the chaff breaks up, releasing the grains. Selection for these traits by farmers might not necessarily have been deliberate, but occurred because these traits made it easier to gather the seeds. Such 'incidental' selection was an important part of crop domestication. **Hybrid vigor** in cultivars is generated by crossing inbred lines.

Durum wheat is a modern variety developed by artificial selection of the domesticated emmer wheat strains. Durum (also called hard wheat) has large firm kernels with a high protein content. These properties make it suitable for pasta production. As with all new wheat varieties, new cultivars are produced by crossing two lines using hand pollination, then selfing or **inbreeding** the progeny that combine the desirable traits of both parents. Progeny are evaluated for several years for the traits of interest, until they can be released as established varieties or cultivars.

1. Describe three phenotypic characteristics that would be desirable in a wheat plant:

   (a) _____ height

   (b) number of grains

   (c) grain size

2. Explain how both natural events and artificial selection have contributed to the high yielding modern wheat varieties:

   Natural events such as climate keep the best fit alive causing those to be breed more

Designer Herds

Most of the economically important traits in dairy cattle are quantitative traits, i.e. they are traits that are affected by many genes, as well as environment. The most important traits (below right) are expressed only in females, but the main opportunity for selection is in males. Intense selection of the best bulls, combined with their worldwide use through **artificial insemination** and frozen semen has seen a rapid genetic gain in dairy cattle since the 1970s. Bulls are assigned statistically-determined and unbiased **breeding values** based on the performance of their daughters and granddaughters. In this way, the bulls and cows with the best genetics can be selected to produce the next generation. More recent genetic techniques include **marker assisted selection**, in which a genetic marker is used to screen for particular alleles associated with traits of interest. Such techniques have enabled farmers to improve the accuracy of their herd records and the certainty with which they select their breeding stock.

Selective Breeding and Genetic Gain in Cattle

Improvements due to marker assisted selection and transgenics

Further gains using embryo transfer and sib-selection

Gradual replacement of cows with EMT heifers from selected lines

Gain due to more accurate assessment of genetic merit (breeding value)

Testing of the first clone lines

Steady progress using standard selection techniques and artificial insemination

Percentage genetic gain (y-axis: 0, 20, 40, 60, 80)

Years (x-axis: 0, 10, 20, 30, 40)

Sources: Breeds of Livestock, Oklahoma State University and Genetics Australia.

Cattle are selected on the basis of particular desirable traits (e.g. milk solids or muscle mass). Most of the genetic improvement in dairy cattle has relied on selection of high quality progeny from proven stock and extensive use of superior sires through artificial insemination (AI). Improved breeding techniques accelerate the **genetic gain**, i.e. the gain toward the desirable phenotype of a breed. The graph (above) shows the predicted gains based on artificial insemination and standard selection techniques (based on criteria such as production or temperament). These are compared with the predicted gains using breeding values (the value of the genes to the progeny) and reproductive technologies such as embryo multiplication and transfer (EMT) of standard and transgenic stock, marker assisted selection, and sib-selection (selecting bulls on the basis of their sisters' performance).

The Perfect Dairy Cow

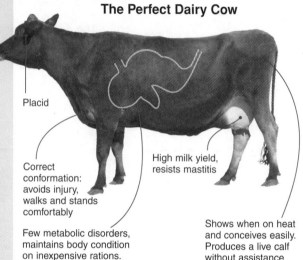

Placid

Correct conformation: avoids injury, walks and stands comfortably

High milk yield, resists mastitis

Few metabolic disorders, maintains body condition on inexpensive rations.

Shows when on heat and conceives easily. Produces a live calf without assistance

Breeding programs select not only for milk production, but also for fertility, udder characteristics, and good health. In addition, artificial selection can be based on milk composition, e.g. high butterfat content (a feature of the Jersey breed, above).

A2 milk, which contains the A2 form of the beta casein protein, has recently received worldwide attention for claims that its consumption lowers the risk of childhood diabetes and coronary heart disease. Selection for the A2 variant in **Holstein cattle** has increased the proportion of A2 milk produced in some regions. A2 milk commands a higher price than A1 milk, so there is a commercial incentive to farmers to produce it.

1. Explain why artificial selection can effect changes in phenotype much more rapidly than can natural selection:

   The most desirable looking organism will be preferred to produce offsprings in artificial selection

2. Suggest why selective breeding has proceeded particularly rapidly in dairy cattle:

   Based on milk composition and less expensive to take care off

3. Explain how molecular genetics has enhanced modern artificial selection techniques in the dairy cow:

   Enhance the gain toward the desirable phenotype of a bree.

© BIOZONE International 2012
ISBN: 978-1-927173-11-4
Photocopying Prohibited

Related activities: Artificial Selection in Animals

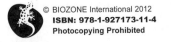

KEY TERMS: Word Find

Use the clues below to find the relevant key terms in the WORD FIND grid

```
T D I S R U P T I V E C X I W G E N E P O O L L O
I K P H E N O T Y P E B P X G G E N O T Y P E K V
B T M O D E R N S Y N T H E S I S S O E D B H P S
U Y X A R T I F I C I A L S E L E C T I O N J W L
P X F L I E U D F H Z I K D E V O L U T I O N T T
F O U N D E R E F F E C T L L U L J V T F M I H L
V X P Z F H Z J H I P O L Y M O R P H I S M N L B
F M F F W Z X L Y U M V M U Y U Y C Y J K V N J
V G H I R I X C Y Q R K G E N E F L O W I G X R V
G I G T Y G K I Q J I N F E F R E Q U E N C Y B L
E Y X N Y P F G E N E T I C D R I F T I D S M L S
Q D E E Q D I R E C T I O N A L F G D I Y P W P B
T E W S V V I Q P L L C P R O R R K W J I E D R L
A M E S I I S T A B I L I Z I N G O W J P C W D R
U E H Z R H Y B R I D Y V P P I O T D V X I C I F
G L G E N E T I C E Q U I L I B R I U M I E P I F
H A R D Y W E I N B E R G E Q U A T I O N S C V N
```

Selection in which the most common phenotype is favored and there is decreased variation in the phenotypic range for the character in question.

A group of organisms capable of breeding together to produce viable offspring.

The occurrence of several different forms within the same species usually caused by different allele combinations of the same gene.

The union of ideas from several areas of biology to provide today's widely accepted account of evolution. It arose as a consequence of population genetics and the realisation that Mendelian genetics was consistent with natural selection.

A population in which there are no destabilizing factors acting to change allele frequencies is said to be in this state (2 words).

An organism produced by the crossing of two unrelated species or strains.

The specific allele makeup of the individual.

The change in allele frequencies in a population simply as a result of random sampling. It is more pronounced in smaller populations
(2 words).

The collective group of genes in a population.

The movement of genes, as a result of mating or migration, within populations.

The number of times an allele appears in a population is termed the allele

_____.

A measure of the relative contribution of an individual's genes to the next generation.

Cumulative change in the characteristics and gene pool of a population over generations.

Selection favoring the differential survival and reproduction of two phenotypic extremes.

Selection in which the phenotypic norm is shifted in one direction.

A interbreeding part of a population that possesses clearly definable characteristics.

Allele frequencies in a gene pool can be calculated using this (3 words).

A process by which humans drive the direction of the phenotype in organisms by selecting and breeding from individuals with particular traits. Also called selective breeding
(2 words).

The loss of genetic variation that occurs when a new population is established by a very small number of individuals from a larger population
(2 words).

All the observable traits of an organism.

Evidence for Biological Evolution

Key concepts

▶ Overwhelming evidence for the fact of evolution comes from many fields of science.

▶ Molecular analyses have provided clarification, and sometimes prompted revision, of the phylogenies established using traditional methods.

▶ New methods in developmental biology provide compelling and current evidence for the mechanisms of evolution.

▶ Evolution is not a thing of the past; it continues and can be observed today.

Key terms

biogeography

chronometric dating
(*aka* absolute dating)

common ancestor

comparative anatomy

epoch

evo-devo

fossil

fossil record

geologic era

geologic period

geologic time scale

homologous structures

molecular clock

paleontology

phylogeny

radiometric dating

relative dating

rock strata

transitional fossil

vestigial organ

Essential Knowledge

☐ 1. Use the **KEY TERMS** to compile a glossary for this topic.

Molecules, Morphology, and Genes *(1.A.4)*   pages 195, 269-279

☐ 2. Use examples to show that biological evolution is supported by scientific evidence from many disciplines, including chemistry and mathematics.

☐ 3. Explain how **fossils** are formed. Recall how fossil-bearing rocks have provided the data for dividing the history of life on Earth into geologic periods.

☐ 4. Distinguish between **relative dating** and **chronometric dating** methods and explain why they are both important. Outline the variety of methods for dating rocks and fossils, including radiometric dating. Appreciate the degree of accuracy achieved by different dating methods.

☐ 5. Explain the significance of **transitional fossils**. Using examples, describe the trends that fossils indicate in the evolution of related taxonomic groups. Outline the **paleontological evidence** for evolution, e.g. in the evolution of horses.

☐ 6. Explain the biochemical evidence provided by the universality of DNA, amino acids, and protein structures (e.g. cytochrome *c*) for the common ancestry of living organisms. Describe how comparisons of specific molecules between species are used as an indication of relatedness or **phylogeny**.

☐ 7. Discuss how biochemical variations can be used as a **molecular clock** to determine probable dates of divergence from a **common ancestor**.

☐ 8. Explain how **comparative anatomy** and physiology have contributed to an understanding of evolutionary relationships.

☐ 9. Recall how evolutionary developmental biology (**evo-devo**) provides some of the strongest evidence for the mechanisms of evolution, particularly for the evolution of novel forms.

☐ 10. Describe the significance of **vestigial organs** and structures as indicators of evolutionary trends in some groups.

Biogeography and Extant Populations *(1.A.4)*   pages 280-286

☐ 11. Explain how the geographical distribution of living and extinct organisms provides evidence of dispersal from a point of origin.

☐ 12. Explain how mathematical models and simulations can be used to illustrate and support evolutionary concepts (e.g. modeling allopatric speciation events).

☐ 13. Appreciate that evidence for evolution is not confined to the past. There are many documented examples of evolution occurring in the present day. Modern examples include the evolution of drug resistance in pathogens, the evolution of pesticide resistance in insects, speciation in Hawaiian *Drosophila*, increase in black squirrels in England, and many others.

Periodicals:
Listings for this chapter are on page 384

Weblinks:
www.thebiozone.com/
weblink/AP1-3114.html

BIOZONE APP:
Student Review Series
Evolution

The History of Life on Earth

The history of life on Earth is one of diversification and extinction. Through fossil and biochemical evidence, it traces the processes by which organisms have evolved since life first originated, perhaps as early as 4000 mya. The biochemical similarities between all present day organisms indicate the presence of a universal common ancestor (an ancestral prokaryote) from which all known species have diverged. Autotrophic cyanobacteria were present as far back as 3500 mya and their photosynthetic activities and the buildup of free atmospheric oxygen were crucial to the later evolution of more complex life forms. Once multicellularity arose in the Precambrian, life diversified rapidly, with the Cambrian explosion being notable for the extraordinary number of new adaptive radiations from Precambrian forms. The rest of Earth's biological history is marked by major geologic and paleontological events, such as mass extinctions, which divide the record of biological diversity into geologic eras and periods.

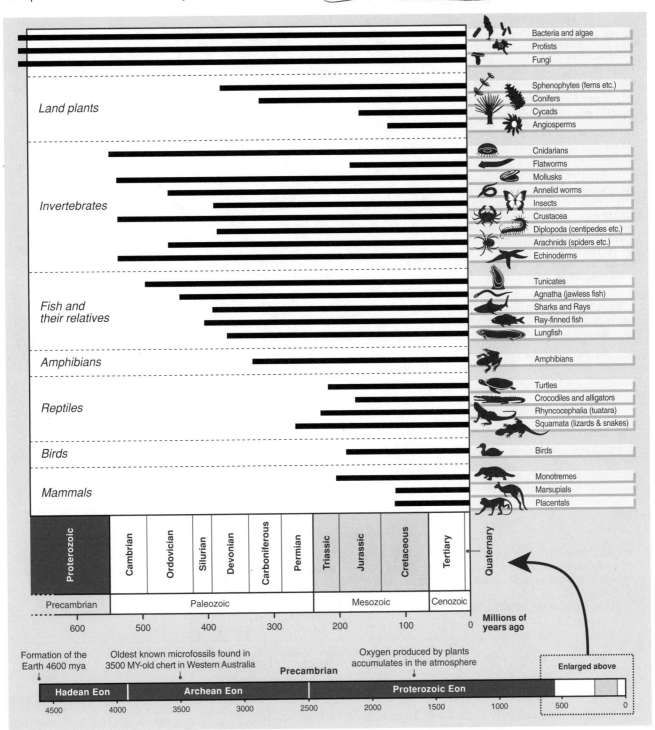

Evidence for Biological Evolution

1. What was the significance of the buildup of free oxygen in the atmosphere for the evolution of animal life?

   was more crucial for more complex life forms

2. Using the diagram above, determine how many millions of years ago the fossil record shows the first appearance of:

   (a) Invertebrates: *550*

   (b) Fish (ray-finned): *500*

   (c) Land plants: *400*

   (d) Reptiles: *300*

   (e) Birds: *200*

   (f) Mammals: *200*

Periodicals:
A cool early life

Related activities: Landmarks in Earth's History
Weblinks: Deep Time, A Brief History of Life

A 2

Cenozoic Era

1.65 mya: Modern humans evolve. Their activities, starting at the most recent ice age, are implicated in extinction of the megafauna.

3-5 mya: Early humans arise from ape-like ancestors.

65-1.65 mya: Major shifts in climate. Major adaptive radiations of angiosperms (flowering plants), insects, birds, and mammals.

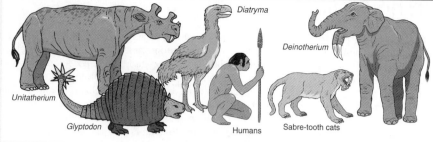
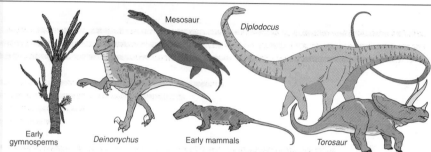

Unitatherium *Diatryma* *Deinotherium* *Glyptodon* *Humans* *Sabre-tooth cats*

Mesozoic Era

65 mya: Apparent asteroid impact implicated in the mass extinction of many marine species and all dinosaurs.

135-65 mya: Major radiations of dinosaurs, fishes, and insects. Origin of angiosperms.

181-135 mya: Major radiations of dinosaurs.

240-205 mya: Recoveries and adaptive radiation of marine invertebrates, dinosaurs, and fishes. Origin of mammals. Gymnosperms become dominant land plants.

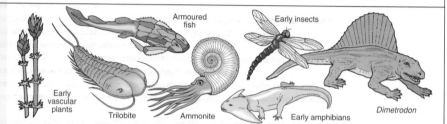

Mesosaur *Diplodocus* *Early gymnosperms* *Deinonychus* *Early mammals* *Torosaur*

Later Paleozoic Era

240 mya: Mass extinction of nearly all species on land and in the sea.

435-280 mya: Vast swamps with the first vascular plants. Origin and adaptive radiation of reptiles, insects, and spore bearing plants (including gymnosperms).

500-435 mya: Major adaptive radiations of marine invertebrates and early fishes.

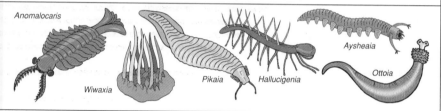

Armoured fish *Early insects* *Early vascular plants* *Trilobite* *Ammonite* *Early amphibians* *Dimetrodon*

Early Paleozoic Era (Cambrian)

550-500 mya: Origin of animals with hard parts (appear as fossils in rocks). Simple marine communities. A famous Canadian site with a rich collection of early Cambrian fossils is known as the Burgess Shale deposits; examples are shown on the right.

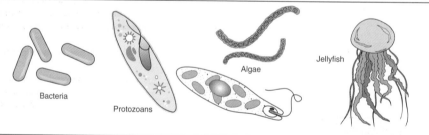

Anomalocaris *Aysheaia* *Ottoia* *Wiwaxia* *Pikaia* *Hallucigenia*

Precambrian

2500–570 mya: Origin of protists, fungi, algae, and animals.

3800–2500 mya: Origin of photosynthetic bacteria.

4600–3800 mya: Chemical and molecular evolution leading to origin of life; protocells to anaerobic bacteria.

4600 mya: Origin of Earth.

Bacteria *Protozoans* *Algae* *Jellyfish*

---

3. An important feature of the history of life is that it has not been a steady progression of change. There have been bursts of evolutionary change as newly evolved groups undergo **adaptive radiations** and greatly increase in biodiversity. Such events are often associated with the sudden mass extinction of other, unrelated groups.

(a) How were mass extinctions important in stimulating new biodiversity?

They evidence of mass extinction divide the record of biological diversity into geological eras and periods

(b) How has the biodiversity of the Earth changed since the origin of life?

The biodiversity of the Earth has changed since the origin of life with the complexity of organisms increasing.

Interpreting the Fossil Record

Relative dating establishes the sequential (relative) order of past events in a rock profile, but it can not provide an absolute date for an event. Each rock layer (stratum) is unique in terms of the type of rock (sedimentary or volcanic) and the type of fossils it contains. Rock layers (**strata**) are arranged in the order that they were deposited (unless they have been disturbed by geological events). The most recent layers are near the surface and the oldest are at the bottom. Strata from widespread locations can be correlated because a particular stratum at one location is the same age as the same stratum at a different location.

Profile with Sedimentary Rocks Containing Fossils

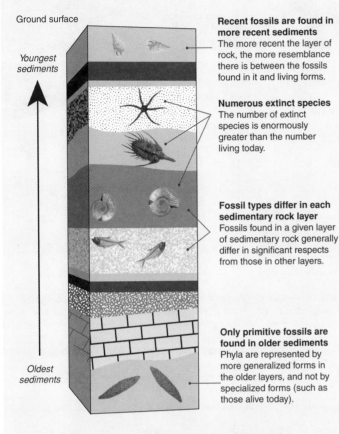

Ground surface

Youngest sediments

Recent fossils are found in more recent sediments
The more recent the layer of rock, the more resemblance there is between the fossils found in it and living forms.

Numerous extinct species
The number of extinct species is enormously greater than the number living today.

Fossil types differ in each sedimentary rock layer
Fossils found in a given layer of sedimentary rock generally differ in significant respects from those in other layers.

Oldest sediments

Only primitive fossils are found in older sediments
Phyla are represented by more generalized forms in the older layers, and not by specialized forms (such as those alive today).

New fossil types mark changes in environment
In the rocks marking the end of one geologic period, it is common to find many new fossils that become dominant in the next. Each geologic period had an environment very different from those before and after. Their boundaries coincided with drastic environmental changes and the appearance of new niches. These produced new selection pressures, resulting in new adaptive features in the surviving species as they responded to the changes.

The rate of evolution can vary

According to the fossil record, rates of evolutionary change seem to vary. There are bursts of species formation and long periods of relative stability within species (stasis). The occasional rapid evolution of new forms apparent in the fossil record, is probably a response to a changing environment. During periods of stable environmental conditions, evolutionary change may slow down.

The Fossil Record of Proboscidea

African and Indian elephants have descended from a diverse group of animals known as **proboscideans** (named for their long trunks). The first pig-sized, trunkless members of this group lived in Africa 40 million years ago. From Africa, their descendants invaded all continents except Antarctica and Australia. As the group evolved, they became larger, an effective evolutionary response to deter predators. Examples of extinct members of this group are illustrated below:

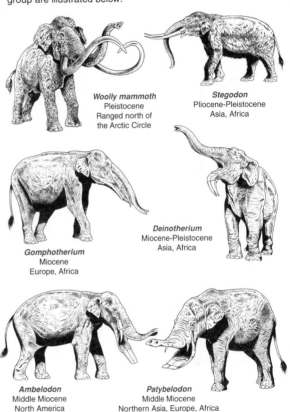

Woolly mammoth
Pleistocene
Ranged north of the Arctic Circle

Stegodon
Pliocene-Pleistocene
Asia, Africa

Gomphotherium
Miocene
Europe, Africa

Deinotherium
Miocene-Pleistocene
Asia, Africa

Ambelodon
Middle Miocene
North America

Patybelodon
Middle Miocene
Northern Asia, Europe, Africa

- **Modern day species can be traced:** The evolution of many present-day species can be very well reconstructed. For instance, the evolutionary history of the modern elephants is exceedingly well documented for the last 40 million years. The modern horse also has a well understood fossil record spanning the last 50 million years.

- **Fossil species are similar to but differ from today's species:** Most fossil animals and plants belong to the same major taxonomic groups as organisms living today. However, they do differ from the living species in many features.

---

1. Name an animal or plant taxon (e.g. family, genus, or species) that has:

   (a) A good fossil record of evolutionary development: _____ horse _____

   (b) Appeared to have changed very little over the last 100 million years or so: _____ dog _____

2. Discuss the importance of **fossils** as a record of evolutionary change over time: _____
   Show how a species physical phenotype changed over the course of evolution

---

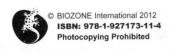

Periodicals:
How old is...
The quick and the dead

Related activities: Dating Fossils
Weblinks: Getting into the Fossil Record, 29+ Evidences for Macroevolution

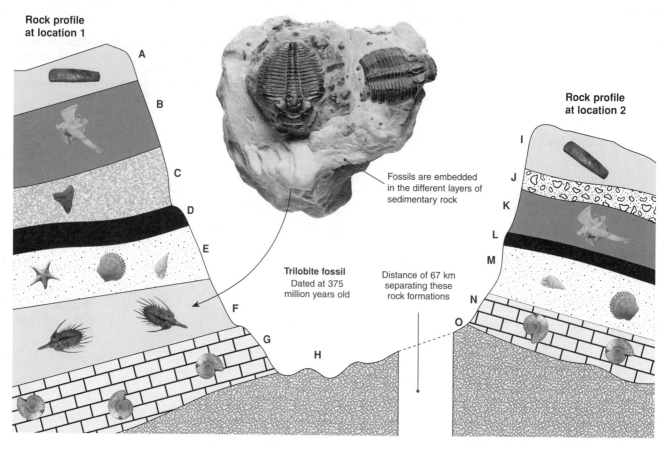

Rock profile at location 1

A
B
C
D
E
F
G
H

Fossils are embedded in the different layers of sedimentary rock

Trilobite fossil
Dated at 375 million years old

Distance of 67 km separating these rock formations

Rock profile at location 2

I
J
K
L
M
N
O

The questions below relate to the diagram above, showing a hypothetical rock profile from two locations separated by a distance of 67 km. There are some differences between the rock layers at the two locations. Apart from layers D and L which are volcanic ash deposits, all other layers comprise sedimentary rock.

3. Assuming there has been no geologic activity (e.g. tilting or folding), state in which rock layer (A-O) you would find:

   (a) The youngest rocks at location 1: _____ A _____   (c) The youngest rocks at location 2: _____ I _____

   (b) The oldest rocks at location 1: _____ H _____   (d) The oldest rocks at location 2: _____ O _____

4. (a) State which layer at location 1 is of the same age as layer M at location 2: _____ G + N _____

   (b) Explain the reason for your answer above: _____ Some types of fossils can be found _____

5. The rocks in layer H and O are sedimentary rocks. Explain why there are no visible fossils in these layers:

   _____ No living organisms were found then _____

6. (a) State which layers present at location 1 are missing at location 2: _____ C _____

   (b) State which layers present at location 2 are missing at location 1: _____ J _____

7. Describe three methods of dating rocks: _____ Examing fossils founds _____

8. Using radiometric dating, the trilobite fossil was determined to be approximately 375 million years old. The volcanic rock layer (D) was dated at 270 million years old, while rock layer B was dated at 80 million years old. Give the approximate **age range** (i.e. greater than, less than, or between given dates) of the rock layers listed below:

   $< x < L70$

   (a) Layer A: _____ less than _____   (d) Layer G: _____ greater than _____

   (b) Layer C: _____ between given dates _____   (e) Layer L: _____ between given dates _____

   (c) Layer E: _____ greater than _____   (f) Layer O: _____ greater than _____

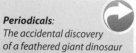

Periodicals:
The accidental discovery
of a feathered giant dinosaur

© BIOZONE International 2012
ISBN: 978-1-927173-11-4
Photocopying Prohibited

The Evolution of Horses

The evolution of the horse from the ancestral *Hyracotherium* to modern *Equus* is well documented in the fossil record. For this reason it is often used to illustrate the process of evolution. The rich fossil record, which includes numerous **transitional fossils**, has enabled scientists to develop a robust model of horse phylogeny. Although the evolution of the line was once considered to be a gradual straight line process, it has been radically revised to a complex tree-like lineage with many divergences (below). It showed no inherent direction, and a diverse array of species coexisted for some time over the 55 million year evolutionary period. The environmental transition from forest to grasslands drove many of the changes observed in the equid fossil record. These include reduction in toe number, increased size of cheek teeth, lengthening of the face, and increasing body size.

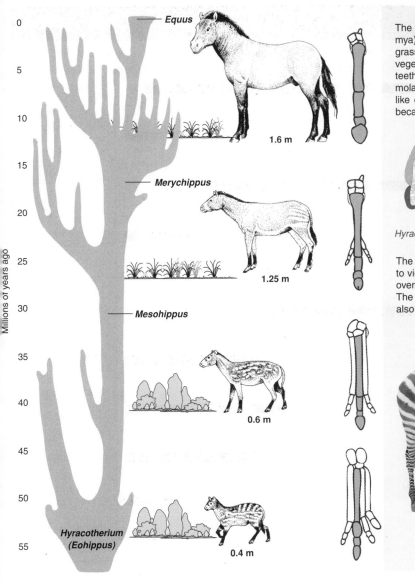

The cooler climates that prevailed in the Miocene (23 -5 mya) brought about a reduction in forested areas with grasslands becoming more abundant. The change in vegetation resulted in the equids developing more durable teeth to cope with the harsher diet. Over time the equid molar became longer and squarer with a hard cement-like covering to enable them to grind the grasses which became their primary diet.

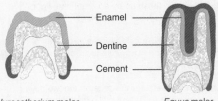

Hyracotherium molar   *Equus* molar

The equids also became taller and faster to enable them to view and escape their predators. This is evident in their overall increase in size and the elongation of their limbs. The reduction in the number of toes from four to one (left) also enabled them to run faster and more efficiently.

The majority of equid evolution took place in North America, although now extinct species did migrate to other areas of the globe at various times. During the late Pliocene (2.6 mya) *Equus* spread into the Old World and diversified into several species including the modern zebra of Africa and the true horse, *Equus caballus*. Ironically, the horse became extinct in the Americas about 11,000 years ago, and was reintroduced in the 16th century by Spanish explorers.

1. Explain how the environmental change from forest to grassland influenced the following aspects of equid evolution:

   (a) Change in tooth structure: _the tooth became flatter, change in diet_

   (b) Limb length: _limb length became longer, escape predators_

   (c) Reduction in number of toes: _to help escape predator faster_

2. In which way does the equid fossil record provide a good example of the evolutionary process? _shows how the environment of the equid changed the phenotype of the organism_

Related activities: History of Life on Earth, Fossil Formation
Weblinks: Horse Evolution, Fossil Horse Cybermuseum

Dating Fossils

Radiometric dating methods allow an **absolute date** to be assigned to fossils, usually by dating the rocks around the fossils. In the early days of developing these techniques, there were problems in producing dependable results, but the methods have been refined and often now provide dates with a high degree of certainty. Multiple dating methods for samples provides cross-referencing, which gives further confidence in a given date. Absolute, or **chronometric**, dating methods most often involve radiometric dating (e.g. **radiocarbon**, **potassium-argon**, **fission track**), which relies on the radioactive decay of an element. Non-radiometric methods (e.g. **tree-rings**, **paleomagnetism**) can be used in certain specific circumstances.

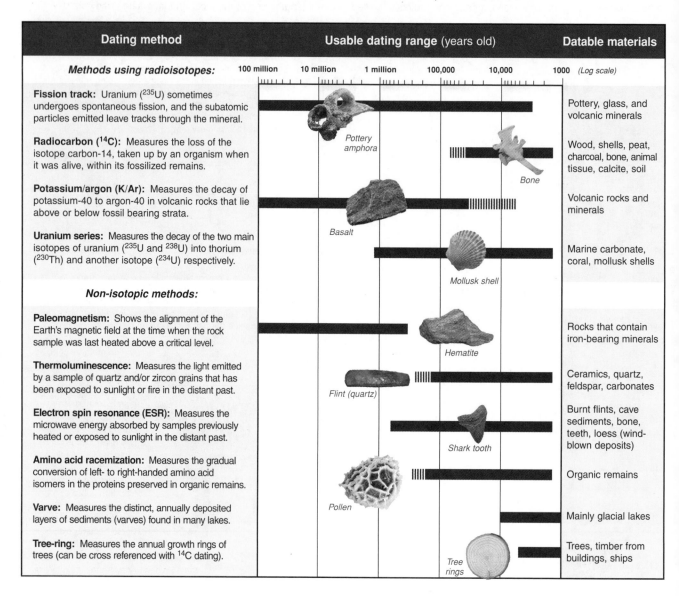

Dating method	Usable dating range (years old)	Datable materials

Methods using radioisotopes: (Log scale: 100 million, 10 million, 1 million, 100,000, 10,000, 1000)

Fission track: Uranium (^{235}U) sometimes undergoes spontaneous fission, and the subatomic particles emitted leave tracks through the mineral. — Pottery, glass, and volcanic minerals

Radiocarbon (^{14}C): Measures the loss of the isotope carbon-14, taken up by an organism when it was alive, within its fossilized remains. — Wood, shells, peat, charcoal, bone, animal tissue, calcite, soil

Potassium/argon (K/Ar): Measures the decay of potassium-40 to argon-40 in volcanic rocks that lie above or below fossil bearing strata. — Volcanic rocks and minerals

Uranium series: Measures the decay of the two main isotopes of uranium (^{235}U and ^{238}U) into thorium (^{230}Th) and another isotope (^{234}U) respectively. — Marine carbonate, coral, mollusk shells

Non-isotopic methods:

Paleomagnetism: Shows the alignment of the Earth's magnetic field at the time when the rock sample was last heated above a critical level. — Rocks that contain iron-bearing minerals

Thermoluminescence: Measures the light emitted by a sample of quartz and/or zircon grains that has been exposed to sunlight or fire in the distant past. — Ceramics, quartz, feldspar, carbonates

Electron spin resonance (ESR): Measures the microwave energy absorbed by samples previously heated or exposed to sunlight in the distant past. — Burnt flints, cave sediments, bone, teeth, loess (wind-blown deposits)

Amino acid racemization: Measures the gradual conversion of left- to right-handed amino acid isomers in the proteins preserved in organic remains. — Organic remains

Varve: Measures the distinct, annually deposited layers of sediments (varves) found in many lakes. — Mainly glacial lakes

Tree-ring: Measures the annual growth rings of trees (can be cross referenced with ^{14}C dating). — Trees, timber from buildings, ships

Image labels: Pottery amphora, Bone, Basalt, Mollusk shell, Hematite, Flint (quartz), Shark tooth, Pollen, Tree rings

1. Examine the diagram above and determine the approximate dating range (note the logarithmic time scale) and datable materials for each of the methods listed below:

	Dating Range	Datable Materials
(a) Potassium-argon method:	5,000 – 100 million	volcanic rocks
(b) Radiocarbon method:	2,000 – 50,000 yrs	woods, shells, charcoal, bone
(c) Tree-ring method:	6,000 – 1000 years	trees, timber, ships
(d) Thermoluminescence:	2,000 – 10,100 years	ceramics, quartz, feldspar

2. When the date of a sample has been determined, it is common practice to express it in the following manner: Example: **1.88 ± 0.02** million years old. Explain what the **± 0.02** means in this case:

error

3. Suggest a possible source of error that could account for an incorrect dating measurement using a radioisotope method:

human error

 © BIOZONE International 2012
ISBN: 978-1-927173-11-4
Photocopying Prohibited

Homologous Structures

The evolutionary relationship between groups of organisms is determined mainly by structural similarities called **homologous structures** (homologies), which suggest that they all descended from a common ancestor with that feature. The bones of the forelimb of air-breathing vertebrates are composed of similar bones arranged in a comparable pattern. This is indicative of common ancestry. The early land vertebrates were amphibians and possessed a limb structure called the **pentadactyl limb**: a limb with five fingers or toes (below left). All vertebrates that descended from these early amphibians, including reptiles, birds and mammals, have limbs that have evolved from this same basic pentadactyl pattern. They also illustrate the phenomenon known as **adaptive radiation**, since the basic limb plan has been adapted to meet the requirements of different niches.

Generalized Pentadactyl Limb

The forelimbs and hind limbs have the same arrangement of bones but they have different names. In many cases bones in different parts of the limb have been highly modified to give it a specialized locomotory function.

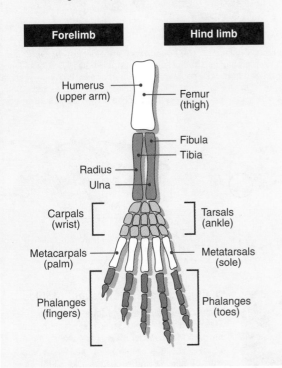

Forelimb	Hind limb

Humerus (upper arm) — Femur (thigh)
Fibula
Tibia
Radius
Ulna
Carpals (wrist) — Tarsals (ankle)
Metacarpals (palm) — Metatarsals (sole)
Phalanges (fingers) — Phalanges (toes)

Specializations of Pentadactyl Limbs

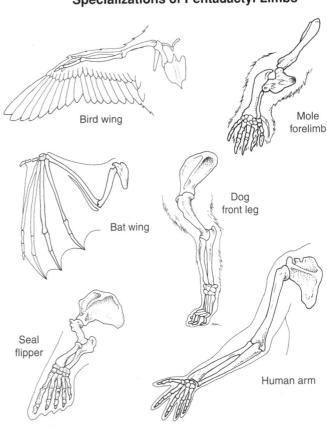

Bird wing

Mole forelimb

Bat wing

Dog front leg

Seal flipper

Human arm

1. Briefly describe the purpose of the major anatomical change that has taken place in each of the limb examples above:

   (a) Bird wing: _Highly modified for flight. Forelimb is shaped for aerodynamic lift and feather attachment._

   (b) Human arm: _lift things, hold things._

   (c) Seal flipper: _move around_

   (d) Dog front leg: _to walk, run, jump_

   (e) Mole forelimb: _to dig into the ground_

   (f) Bat wing: _to help fly_

2. Explain how homology in the pentadactyl limb is evidence for adaptive radiation: _Shew how limbs have evolved from the SAME pentadactyl basic limb pattern._

3. Homology in the behavior of animals (for example, sharing similar courtship or nesting rituals) is sometimes used to indicate the degree of relatedness between groups. How could behavior be used in this way:
   It shew how organism can share a common ancester / fetal development pattern

© BIOZONE International 2012
ISBN: 978-1-927173-11-4
Photocopying Prohibited

Periodicals:
A fin is a limb is a wing...

Weblinks: *All in the Family, Homologous Structures*

DA 2

Vestigial Characteristics

Some classes of characters are more valuable than others as reliable indicators of common ancestry. Often, the less any part of an animal is used for specialized purposes, the more important it becomes for classification. This is because common ancestry is easier to detect if a particular feature is unaffected by specific adaptations arising later during the evolution of the species. Vestigial organs are an example of this because, if they have no clear function and they are no longer subject to natural selection, they will remain unchanged through a lineage. It is sometimes argued that some vestigial organs are not truly vestigial, i.e. they may perform some small function. While this may be true in some cases, the features can still be considered vestigial if their new role is a minor one, unrelated to their original function.

Ancestors of Modern Whales

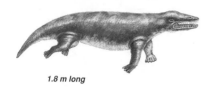

1.8 m long

2.5 m long

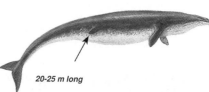

20-25 m long

Pakicetus (early Eocene). A carnivorous, four limbed, early Eocene whale ancestor, probably rather like a large otter. It was still partly terrestrial and not fully adapted for aquatic life.

Protocetus (mid Eocene). Much more whale-like than *Pakicetus*. The hind limbs were greatly reduced and although they still protruded from the body (arrowed), they were useless for swimming.

Basilosaurus (late Eocene). A very large ancestor of modern whales. The hind limbs contained all the leg bones, but were vestigial and located entirely within the main body, leaving a tissue flap on the surface (arrowed).

Vestigial organs are common in nature. The vestigial hind limbs of modern whales (right) provide anatomical evidence for their evolution from a carnivorous, four footed, terrestrial ancestor. The oldest known whale, *Pakicetus*, from the early Eocene (~54 mya) still had four limbs. By the late Eocene (~40 mya), whales were fully marine and had lost almost all traces of their former terrestrial life. For fossil evidence, see *Whale Origins* at: www.neoucom.edu/Depts/Anat/whaleorigins.htm

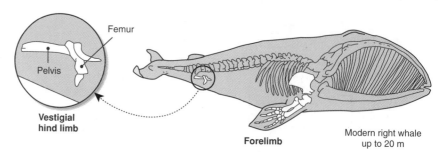

Femur

Pelvis

Vestigial hind limb

Forelimb

Modern right whale up to 20 m

RM-DoC

Vestigial organs in birds and reptiles

In all snakes (far left), one lobe of the lung is vestigial (there is not sufficient room in the narrow body cavity for it). In some snakes there are also vestiges of the pelvic girdle and hind limbs of their walking ancestors. Like all ratites, kiwis (left) are flightless. However, more than in other ratites, the wings of kiwis are reduced to tiny vestiges. Kiwis evolved in the absence of predators to a totally ground dwelling existence.

1. In terms of natural selection, explain how structures that were once useful to an organism could become vestigial:

   Over time as organizm evolved some organs were not of much use any more and become virtually useless

2. Suggest why a vestigial structure, once it has been reduced to a certain size, may not disappear altogether:

   It has no negative or positive effection on organism.

3. Whale evolution shows the presence of **transitional forms** (fossils that are intermediate between modern forms and very early ancestors). Suggest how vestigial structures indicate the common ancestry of these forms:

   The vestigial hind limbs of modern whales provide anatomical evidence for their evolution from a carnivorous, four rooted, terrestrial ancestor

Related activities: Homologous Structures
Weblinks: Vestigial Organs

Periodicals: A waste of space

© BIOZONE International 2012
ISBN: 978-1-927173-11-4
Photocopying Prohibited

Homologous Proteins

Traditionally, phylogenies were based largely on anatomical or behavioral traits and biologists attempted to determine the relationships between organisms based on overall degree of similarity or by tracing the appearance of key characteristics. With the advent of molecular techniques, homologies can now be studied at the molecular level as well and these can be compared to the phylogenies established using other methods. Protein sequencing provides an excellent tool for establishing **homologies** (similarities resulting from shared ancestry). Each

protein has a specific number of amino acids arranged in a specific order. Any differences in the sequence reflect changes in the DNA sequence. Commonly studied proteins include blood proteins, such as **hemoglobin** (below), and the respiratory protein **cytochrome c** (next page). Many of these proteins are **highly conserved**, meaning they change very little over time, presumably because mutations would be detrimental to basic function. Conservation of protein sequences is indicated by the identical amino acid residues at corresponding parts of proteins.

Amino Acid Differences in Hemoglobin

Human beta chain	0
Chimpanzee	0
Gorilla	1
Gibbon	2
Rhesus monkey	8
Squirrel monkey	9
Dog	15
Horse, cow	25
Mouse	27
Gray kangaroo	38
Chicken	45
Frog	67

When the sequence of the **beta hemoglobin chain** (right), which is 146 amino acids long, is compared between humans, five other primates, and six other vertebrates, the results support the phylogenies established using other methods. The numbers in the table (left) represent the number of amino acid differences between the beta chain of humans and those of other species. In general, the number of amino acid differences between the hemoglobins of different vertebrates is inversely proportional to genetic relatedness.

Shading indicates (from top) primates, non-primate placental mammals, marsupials, and non-mammals.

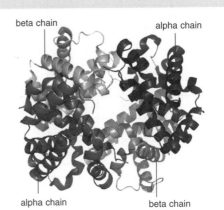

beta chain   alpha chain

alpha chain   beta chain

In most vertebrates, the oxygen-transporting blood protein hemoglobin is composed of four polypeptide chains, two alpha chains and two beta chains. Hemoglobin is derived from myoglobin, and ancestral species had just myoglobin for oxygen transport. When the amino acid sequences of myoglobin, the hemoglobin alpha chain, and the hemoglobin beta chain are compared, there are several amino acids that remain conserved between all three. These amino acid sequences must be essential for function because they have remained unchanged throughout evolution.

Using Immunology to Determine Phylogeny

The immune system of one species will recognise the blood proteins of another species as foreign and form antibodies against them. This property can be used to determine the extent of relatedness between species. Blood proteins, such as albumins, are used to prepare **antiserum** in rabbits. The antiserum contains antibodies against the test blood proteins (e.g. human) and will react to those proteins in any blood sample they are mixed with. The extent of the reaction indicates how similar the proteins are; the greater the reaction, the more similar the proteins. This principle is illustrated (right) for antiserum produced to human blood and its reaction with the blood of other primates and a rat.

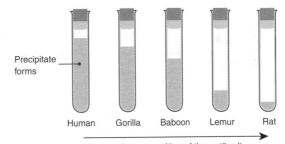

Precipitate forms

Human   Gorilla   Baboon   Lemur   Rat

Decreasing recognition of the antibodies against human blood proteins

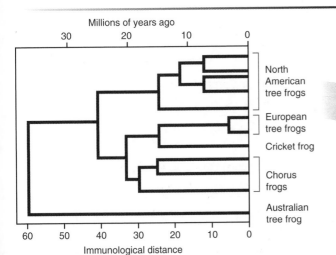

Millions of years ago

30   20   10   0

North American tree frogs

European tree frogs

Cricket frog

Chorus frogs

Australian tree frog

60   50   40   30   20   10   0

Immunological distance

The relationships among tree frogs have been established by immunological studies based on blood proteins such as immunoglobulins and albumins. The **immunological distance** is a measure of the number of amino acid substitutions between two groups. This, in turn, has been calibrated to provide a time scale showing when the various related groups diverged.

Related activities: DNA Homologies
Weblinks: Evidence for Evolution

DA 2

Cytochrome *c* and the Molecular Clock Theory

Evolutionary change at the molecular level occurs primarily through fixation of neutral mutations by genetic drift. The rate at which one neutral mutation replaces another depends on the mutation rate, which is fairly constant for any particular gene.

If the rate at which a protein evolves is roughly constant over time, the amount of molecular change that a protein shows can be used as a molecular clock to date evolutionary events, such as the divergence of species.

The molecular clock for each species, and each protein, may run at different rates, so scientists calibrate the molecular clock data with other evidence (morphological, molecular) to confirm phylogenetic relationships.

For example, 20 amino acid substitutions in a protein since two organisms diverged from a known common ancestor 400 mya indicates an average substitution rate of 5 substitutions per 100 my.

		1	2	3	4	5	6	7	8	9	10	11	12	13	14	15	16	17	18	19	20	21	22
Human		Gly	Asp	Val	Glu	Lys	Gly	Lys	Lys	Ile	Phe	Ile	Met	Lys	Cys	Ser	Gln	Cys	His	Thr	Val	Glu	Lys
Pig												Val	Gln			Ala							
Chicken				Ile						Val		Val	Gln			Ala							
Dogfish										Val		Val	Gln			Ala							Asn
Drosophila	<<									Leu		Val	Gln	Arg		Ala							Ala
Wheat	<<		Asn	Pro	Asp	Ala		Ala				Lys	Thr	Arg		Ala						Asp	Ala
Yeast	<<		Ser	Ala	Lys			Ala	Thr	Leu		Lys	Thr	Arg		Glu	Leu						

This table shows the N-terminal 22 amino acid residues of human cytochrome *c*, with corresponding sequences from other organisms aligned beneath. Sequences are aligned to give the most position matches. A shaded square indicates no change. In every case, the cytochrome's heme group is attached to the Cys-14 and Cys-17. In *Drosophila*, wheat, and yeast, arrows indicate that several amino acids precede the sequence shown.

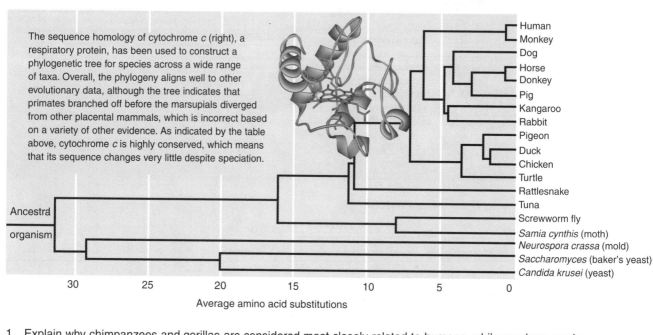

The sequence homology of cytochrome *c* (right), a respiratory protein, has been used to construct a phylogenetic tree for species across a wide range of taxa. Overall, the phylogeny aligns well to other evolutionary data, although the tree indicates that primates branched off before the marsupials diverged from other placental mammals, which is incorrect based on a variety of other evidence. As indicated by the table above, cytochrome *c* is highly conserved, which means that its sequence changes very little despite speciation.

Average amino acid substitutions

1. Explain why chimpanzees and gorillas are considered most closely related to humans, while monkeys are less so:

   They have more amino acid difference in hemoglobin

2. (a) Why would a respiratory protein like cytochrome *c* be highly conserved?

   Because mutations could cause respiratory problem

   (b) Why are highly conserved proteins good candidates for use in establishing protein homologies?

   They create very little change over time keeping the proteins the same

3. Discuss some of the limitations of using protein homology, specifically molecular clocks, to establish phylogeny:

   The molecular clocks are different within the organism's organs, the heart has different rhythms than brain

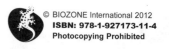

Homologous DNA Sequences

Establishing a phylogeny on the basis of homology in a protein such as cytochrome *c* is valuable, but it is also analogous to trying to see a complete picture through a small window. The technique of **DNA-DNA hybridization** provides a way to compare the total genomes of different species by measuring the degree of genetic similarity between pools of DNA sequences. It is usually used to determine the genetic distance between two species; the more closely two species are related, the fewer differences there will

be between their genomes. This is because there has been less time for the point mutations that will bring about these differences to occur. This technique gives a measure of 'relatedness', and can be calibrated as a **molecular clock** against known fossil dates. It has been applied to primate DNA samples to help determine the approximate date of human divergence from the apes, which has been estimated to be between 10 and 5 million years ago.

DNA Hybridization

1. DNA from the two species to be compared is extracted, purified and cut into short fragments (e.g. 600-800 base pairs).

2. The DNA of one species is mixed with the DNA of another.

3. The mixture is incubated to allow DNA strands to dissociate and reanneal, forming hybrid double-stranded DNA.

4. The hybridized sequences that are highly similar will bind more firmly. A measure of the heat energy required to separate the hybrid strands provides a measure of DNA relatedness.

DNA Homologies Today

DNA-DNA hybridization has been criticized because duplicated sequences within a single genome make it unreliable for comparisons between closely related species.

Today, DNA sequencing and computed comparisons are more widely used to compare genomes, although DNA-DNA hybridization is still used to help identify bacteria.

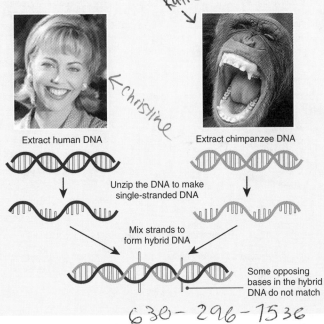

Katie

←Christine

Extract human DNA

Extract chimpanzee DNA

Unzip the DNA to make single-stranded DNA

Mix strands to form hybrid DNA

Some opposing bases in the hybrid DNA do not match

630-296-7536

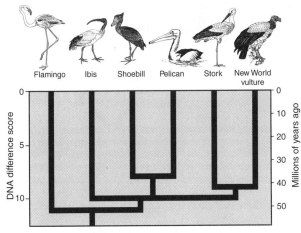

The relationships among the New World vultures and storks have been determined using DNA hybridization. It has been possible to estimate how long ago various members of the group shared a common ancestor.

Similarity of human DNA to that of other primates

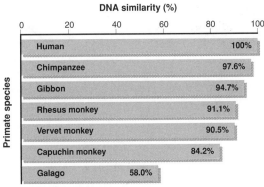

The genetic relationships among the primates has been investigated using DNA hybridization. Human DNA was compared with that of the other primates. It largely confirmed what was suspected from anatomical evidence.

Evidence for Biological Evolution

1. Explain how **DNA hybridization** can give a measure of genetic relatedness between species:

Provide a way to compare the total genomes of different species by measuring the degrees of genetic similarity between pools of DNA sequences

2. Study the graph showing the results of a DNA hybridization between human DNA and that of other primates.

(a) Which is the most closely related primate to humans? *chimpanzee*

(b) Which is the most distantly related primate to humans? *galago*

3. State the DNA difference score for: (a) Shoebills and pelicans: *7* (b) Storks and flamingos: *12*

4. On the basis of DNA hybridization, state how long ago the ibises and New World vultures shared a common ancestor:

for about a billion years

Periodicals: Uprooting the tree of life

Related activities: Protein Homologies

DA 2

Analyzing Data from Extant Populations

The Galápagos Islands are home to 13 species of finches descended from a common ancestor. They were much studied by Darwin and current studies of the island populations continue to provide evidence of evolutionary mechanisms in operation. A study during a long drought on Santa Cruz Island showed how **disruptive selection** can change the distribution of genotypes in a population. During the drought, large and small seed sizes were more abundant than the preferred intermediate seed sizes.

Measurements of the beak length, width, and depth were combined into one **single measure**.

CBU

Beak size vs fitness in *G. fortis*

Fitness is a measure of the reproductive success of each genotype.

Higher fitness

Higher fitness

Local fitness (2004-2006)

Beak size (single measure)

A.P. Hendry et. al 2009

*Fitness showed a **bimodal distribution** (arrowed) being highest for smaller and larger beak sizes.*

Beak sizes of the finch *Geospiza fortis* were measured over a three year period (2004-2006), at the start and end of each year. At the start of the year, individuals were captured, banded, and their beaks were measured.

The presence or absence of banded individuals was recorded at the end of the year when the birds were recaptured. Recaptured individuals had their beaks measured.

The proportion of banded individuals in the population at the end of the year gave a measure of fitness. Absent individuals were presumed dead (fitness = 0).

Fitness related to beak size showed a bimodal distribution (left) typical of disruptive selection.

Beak size and pairing in *G. fortis*

Pairing under extremely wet conditions

Male beak size (single measure)

Female beak size (single measure)

S.K. Huber *et al.* 2007

Pairing under moderately wet conditions

Male beak size (single measure)

Female beak size (single measure)

S.K. Huber *et al.* 2007

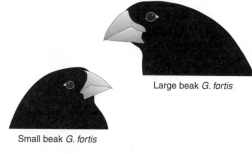

Large beak *G. fortis*

Small beak *G. fortis*

A 2007 study found that breeding pairs of birds had similar beak sizes. Male and females with small beaks tended to breed together, and males and females with large beaks tended to breed together. Mate selection maintained the biomodal distribution in the population during extremely wet conditions. If beak size wasn't a factor in mate selection, the beak size would even out.

1. (a) How did the drought affect seed size on Santa Cruz Island? _____

_____

(b) How did the change in seed size during the drought create a selection pressure for changes in beak size?

_____

_____

_____

2. How does beak size relate to fitness (differential reproductive success) in *G. fortis*? _____

_____

_____

_____

3. (a) Is mate selection in *G. fortis* random / non-random? (delete one)

(b) Give reasons for your answer: _____

_____

_____

Related activities: *Natural Selection*

Weblinks: *Darwin's Finches*

© BIOZONE International 2012
ISBN: 978-1-927173-11-4
Photocopying Prohibited

Oceanic Island Colonizers

The distribution of organisms around the world lends powerful support to the idea that modern forms evolved from ancestral populations. **Biogeography** is the study of the geographical distribution of species, both present-day and extinct. It stresses the role of dispersal of species from a point of origin across pre-existing barriers. Studies from island populations (below) indicate that flora and fauna of different islands are more closely related to adjacent continental species than to each other.

Galápagos and Cape Verde islands

Biologists did not fully appreciate the uniqueness and diversity of tropical island biota until explorers began to bring back samples of flora and fauna from their expeditions in the 19th century. The Galápagos Islands, the oldest of which arose 3-4 million years ago, had species similar to but distinct from those on the South American mainland. Similarly, in the Cape Verde Islands, species had close relatives on the West Africa mainland. This suggested to biologists that ancestral forms found their way from the mainland to the islands where they then underwent evolutionary changes.

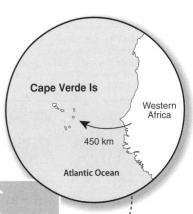

Evidence for Biological Evolution

Tristan da Cunha

The island of Tristan da Cunha in the South Atlantic Ocean is a great distance from any other land mass. Even though it is closer to Africa, there are more species closely related to South American species found there (see table on right). This is probably due to the predominant westerly trade winds from the direction of South America. The flowering plants of universal origin are found in both Africa and South America and could have been introduced from either land mass.

Tristan da Cunha species
South American origin
7 Flowering plants
5 Ferns
30 Liverworts

African origin
2 Flowering plants
2 Ferns
5 Liverworts

Universal origin
19 Flowering plants

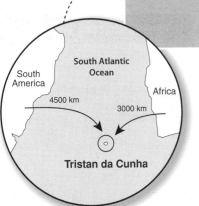

The flightless cormorant (above) is one of a number of bird species that lost the power of flight after becoming an island resident. Giant tortoises, such as the 11 subspecies remaining on the Galápagos today (center) were, until relatively recently, characteristic of many islands in the Indian Ocean including the Seychelles archipelago, Reunion,

Mauritius, Farquhar, and Diego Rodriguez. These were almost completely exterminated by early Western sailors, although a small population remains on the island of Aldabra. Another feature of oceanic islands is the adaptive radiation of colonizing species into different specialist forms. The three species of Galápagos iguana almost certainly arose,

through speciation, from a hardy traveller from the South American mainland. The marine iguana (above) feeds on shoreline seaweeds and is an adept swimmer. The two species of land iguana (not pictured) feed on cacti, which are numerous. One of these (the pink iguana) was identified as a separate species only in 2009.

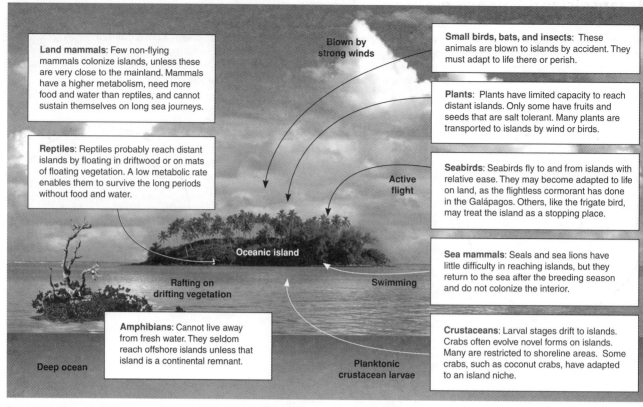

Land mammals: Few non-flying mammals colonize islands, unless these are very close to the mainland. Mammals have a higher metabolism, need more food and water than reptiles, and cannot sustain themselves on long sea journeys.

Reptiles: Reptiles probably reach distant islands by floating in driftwood or on mats of floating vegetation. A low metabolic rate enables them to survive the long periods without food and water.

Blown by strong winds

Small birds, bats, and insects: These animals are blown to islands by accident. They must adapt to life there or perish.

Plants: Plants have limited capacity to reach distant islands. Only some have fruits and seeds that are salt tolerant. Many plants are transported to islands by wind or birds.

Active flight

Seabirds: Seabirds fly to and from islands with relative ease. They may become adapted to life on land, as the flightless cormorant has done in the Galápagos. Others, like the frigate bird, may treat the island as a stopping place.

Oceanic island

Rafting on drifting vegetation

Swimming

Sea mammals: Seals and sea lions have little difficulty in reaching islands, but they return to the sea after the breeding season and do not colonize the interior.

Amphibians: Cannot live away from fresh water. They seldom reach offshore islands unless that island is a continental remnant.

Deep ocean

Planktonic crustacean larvae

Crustaceans: Larval stages drift to islands. Crabs often evolve novel forms on islands. Many are restricted to shoreline areas. Some crabs, such as coconut crabs, have adapted to an island niche.

The diversity and uniqueness of island biota is determined by migration to and from the island and extinctions and diversifications following colonization. These events are themselves affected by a number of other factors (see table, below right). The organisms that successfully colonize islands have to be marine in habit, or able to survive long periods at sea or in the air. The biota of the **Galápagos Islands** provide a good example of the results of such a colonization process. For example, all the subspecies of giant tortoise evolved in Galápagos from a common ancestor that arrived from the mainland, floating with the ocean currents.

1. The Galápagos and the Cape Verde Islands are both tropical islands close to the equator, yet their biotas are quite different. Explain why this is the case:

   Both have different migration to and from the island and extinctions and diversifications following colonization.

Factors affecting final biota
Degree of isolation
Length of time of isolation
Size of island
Climate (tropical/Arctic, arid/humid)
Location relative to ocean currents
Initial plant and animal composition
The species composition of earliest arrivals (if always isolated)
Serendipity (chance arrivals)

2. Explain why the majority of the plant species found on Tristan da Cunha originated from South America, despite its greater distance from the island:

   The plant seeds were carried by wind or birds.

3. The table (right) identifies some of the factors influencing the composition of island biota. Explain how each of the three following might affect the diversity and uniqueness of the biota found on an oceanic island:

   (a) Large island area: Can create large diversity

   (b) Long period of isolation from other land masses: Can create a change in gene due to change to adapt to environment

   (c) Relatively close to a continental land mass: Can travel back to continent

4. Describe one feature typical of an oceanic island colonizer and explain its significance:

   ridge / river

Continental Drift and Evolution

Continental drift is a measurable phenomenon; it has happened in the past and continues today. Movements of up to 2-11 cm a year have been recorded between continents using laser technology. The movements of the Earth's 12 major crustal plates are described by a geologic process known as **plate tectonics.** Some continents appear to be drifting apart while others are on a direct collision course. Various lines of evidence show that the modern continents were once joined together as 'supercontinents'. One supercontinent, called **Gondwana**, was made up of the southern continents some 200 million years ago. The diagram below shows some of the data that provide evidence of how the modern continents once fitted together.

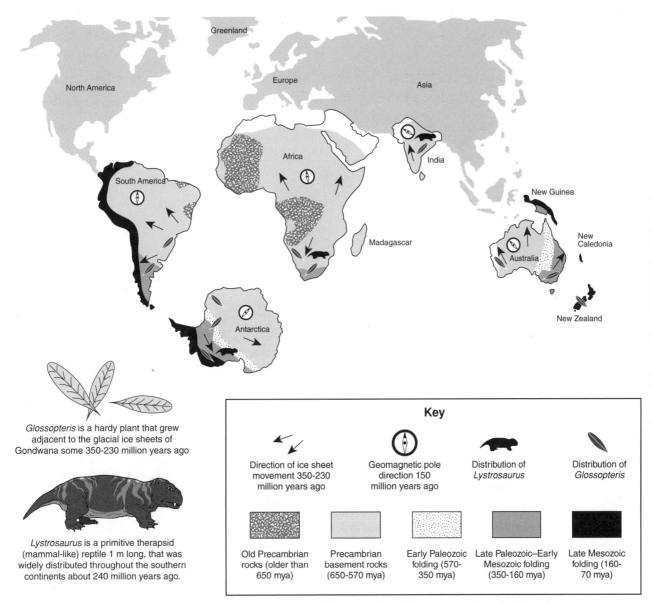

Glossopteris is a hardy plant that grew adjacent to the glacial ice sheets of Gondwana some 350-230 million years ago

Lystrosaurus is a primitive therapsid (mammal-like) reptile 1 m long, that was widely distributed throughout the southern continents about 240 million years ago.

Key

Direction of ice sheet movement 350-230 million years ago	Geomagnetic pole direction 150 million years ago	Distribution of *Lystrosaurus*	Distribution of *Glossopteris*	
Old Precambrian rocks (older than 650 mya)	Precambrian basement rocks (650-570 mya)	Early Paleozoic folding (570-350 mya)	Late Paleozoic–Early Mesozoic folding (350-160 mya)	Late Mesozoic folding (160-70 mya)

Evidence for Biological Evolution

1. Name the modern landmasses (continents and large islands) that made up the supercontinent of Gondwana:

_____

2. Cut out the southern continents on page 285 and arrange them to recreate the supercontinent of Gondwana. Take care to cut the shapes out close to the coastlines. When arranging them into the space showing the outline of Gondwana on the next page, take into account the following information:
   (a) The location of ancient rocks and periods of mountain folding during different geologic ages.
   (b) The direction of ancient ice sheet movements.
   (c) The geomagnetic orientation of old rocks (the way that magnetic crystals are lined up in ancient rock gives an indication of the direction the magnetic pole was at the time the rock was formed).
   (d) The distribution of fossils of ancient species such as *Lystrosaurus* and *Glossopteris*.

3. Once you have positioned the modern continents into the pattern of the supercontinent, mark on the diagram:
   (a) The likely position of the South Pole 350-230 million years ago (as indicated by the movement of the ice sheets).
   (b) The likely position of the geomagnetic South Pole 150 million years ago (as indicated by ancient geomagnetism).

4. State what general deduction you can make about the position of the polar regions with respect to land masses:

_____

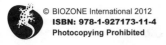
Related activities: Oceanic Island Colonizers, The Fossil Record
Weblinks: Evolution of the Continents

PDA 3

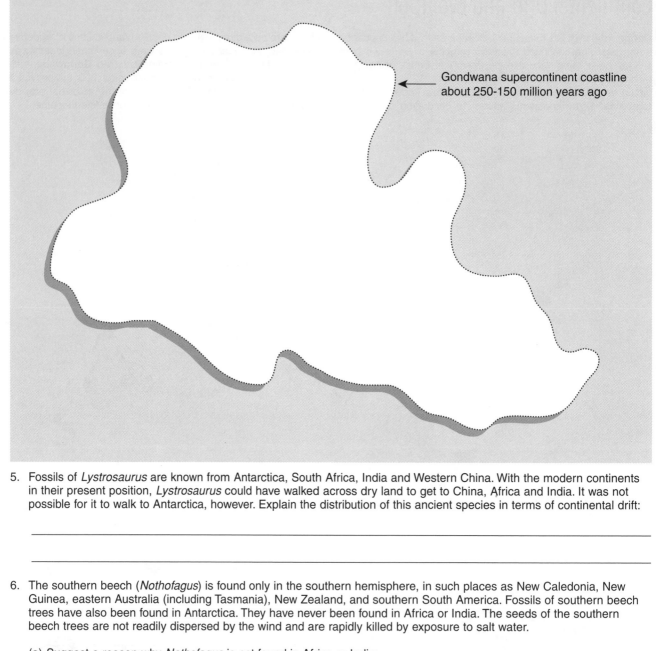

Gondwana supercontinent coastline about 250-150 million years ago

5. Fossils of *Lystrosaurus* are known from Antarctica, South Africa, India and Western China. With the modern continents in their present position, *Lystrosaurus* could have walked across dry land to get to China, Africa and India. It was not possible for it to walk to Antarctica, however. Explain the distribution of this ancient species in terms of continental drift:

_____

_____

6. The southern beech (*Nothofagus*) is found only in the southern hemisphere, in such places as New Caledonia, New Guinea, eastern Australia (including Tasmania), New Zealand, and southern South America. Fossils of southern beech trees have also been found in Antarctica. They have never been found in Africa or India. The seeds of the southern beech trees are not readily dispersed by the wind and are rapidly killed by exposure to salt water.

(a) Suggest a reason why *Nothofagus* is not found in Africa or India: _____

_____

(b) Use a colored pen to indicate the distribution of *Nothofagus* on the current world map (on the previous page) and on your completed map of Gondwana above.

(c) State how the arrangement of the continents into Gondwana explains this distribution pattern:

_____

_____

7. The Atlantic Ocean is currently opening up at the rate of 2 cm per year. At this rate in the past, calculate how long it would have taken to reach its current extent, with the distance from Africa to South America being 2300 km (assume the rate of spreading has been constant):

_____

8. Explain how continental drift provides evidence to support evolutionary theory: _____

_____

_____

_____

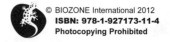

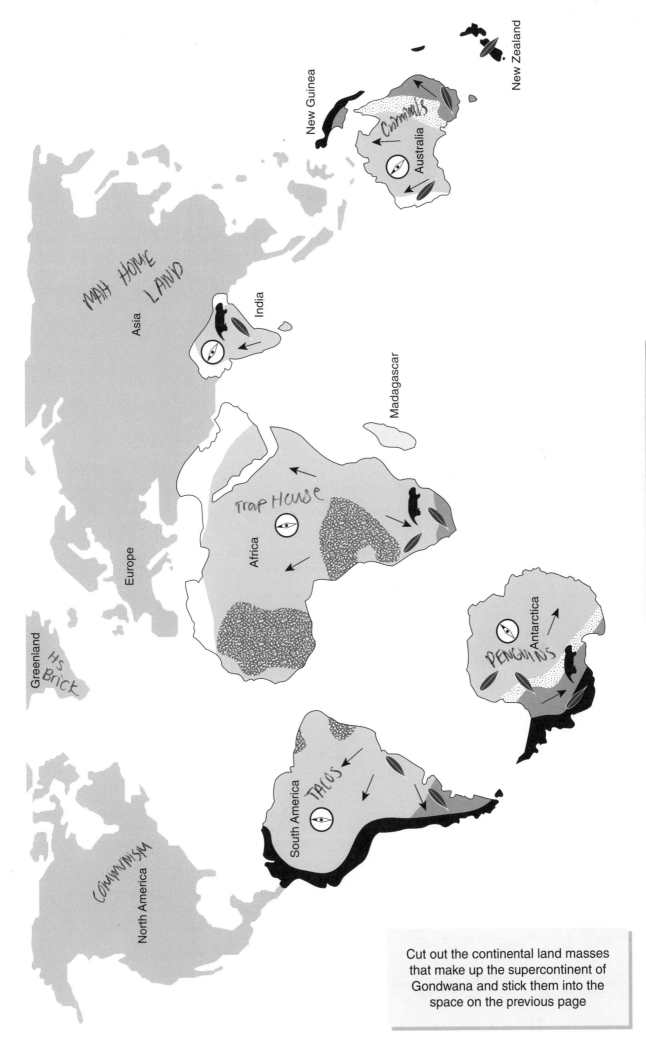

New Guinea

New Zealand

Australia

Cannibals

India

Madagascar

MAH HOME LAND

Asia

Europe

Trap House

Africa

Antarctica

PENGUINS

Greenland

Hs Brick

South America

TACOS

COMMUNISM

North America

Cut out the continental land masses
that make up the supercontinent of
Gondwana and stick them into the
space on the previous page

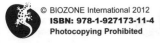

This page has been deliberately left blank

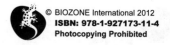

KEY TERMS: Flash Card Game

The cards below have a keyword or term printed on one side and its definition printed on the opposite side. The aim is to win as many cards as possible from the table. To play the game:

1) Cut out the cards and lay them definition side down on the desk. You will need one set of cards between two students.

2) Taking turns, choose a card and, BEFORE you pick it up, state your own best definition of the keyword to your opponent.

3) Check the definition on the opposite side of the card. If both you and your opponent agree that your stated definition matches, then keep the card. If your definition does not match then return the card to the desk.

4) Once your turn is over, your opponent may choose a card.

Fossil	Comparative anatomy	Radiometric dating
Molecular clock	Relative dating	Phylogeny
Chronometric dating	Biogeography	Homologous structures
Transitional fossil	Vestigial characteristic	Geologic time scale

Evidence for Biological Evolution

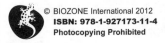
R 2

When you've finished the game keep these cutouts and use them as flash cards!

A technique used to date rocks and other material, based on comparing the relative abundances of a radioactive isotope and its decay products, using known decay rates.

The study of similarities and differences in the anatomy of organisms.

The preserved remains or traces (e.g. footprints) of past organisms.

The evolutionary history or genealogy of a group of organisms. Often represented as a 'tree' showing descent of new species from the ancestral one.

A method that determines the sequential order in which past events occurred, without necessarily determining their absolute age. Also called chronostatic dating.

A method, analogous to a timepiece, that uses molecular change to deduce the time in geologic history when two taxa diverged and so can be used to establish phylogenies.

Structures in different but related species that are derived from the same ancestral structure.

The study of how biodiversity is distributed over space and time.

The process of determining a specific date for an archeological or paleontological site or artifact, usually based on its physical or chemical properties. Also called absolute dating.

A system of chronological measurement relating stratigraphy to time. It is used by scientists to describe the timing and relationships between events in the history of the Earth.

Homologous characters (including anatomical structures, behaviors, and biochemical pathways) that have apparently lost all or most of their original function in a species through evolution.

The fossilized remains of organisms that illustrate an evolutionary transition in that they possess both primitive and derived characteristics.

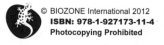

The Relatedness of Organisms

Key concepts

▶ The common ancestry of all organisms is reflected in biochemical similarities and shared core processes.

▶ Structural evidence at the subcellular level supports the relatedness of all eukaryotes.

▶ Phylogenetic trees and cladograms provide models of phylogeny and can be tested using new evidence.

▶ Organisms are assigned to taxonomic categories based on shared derived characteristics.

▶ Organisms are identified using a binomial naming system: genus and species.

Key terms

Archaea
Bacteria
binomial nomenclature
cladistics
cladogram
class
distinguishing feature
Eukarya
family
genus
kingdom
monophyletic
order
paraphyletic
phylogenetic tree
polyphyletic
phylogeny
phylum
shared derived characteristic (=synapomorphy)
Shared ancestral characteristic (=symplesiomorphy)
species
taxonomic category (taxon)

Essential Knowledge

☐ 1. Use the **KEY TERMS** to compile a glossary for this topic.

Structural and Functional Relatedness (1.B.1)   pages 290-291

☐ 2. Describe the biochemical evidence of structure and function that supports the relatedness of all domains (**Bacteria**, **Archaea**, and **Eukarya**), including:
  – DNA and RNA as carriers of genetic information through core processes of replication, transcription, and translation by highly conserved ribosomes.
  – Universality of major features of the genetic code.
  – Conservation of metabolic pathways such as glycolysis.
  – Use of ATP as energy currency by all extant life.

☐ 3. Describe the structural evidence supporting the relatedness of all eukaryotes:
  – Linear chromosomes
  – Membrane-bound organelles and endomembrane systems
  – Cytoskeletal elements

Cladograms and Phylogenetic Trees (1.B.2)   pages 292-306

☐ 4. Recognize a **cladogram** as a **phylogenetic tree** produced using **cladistic analysis**. Describe the use of cladistic analyses to produce **monophyletic phylogenies** based on **shared derived characteristics** (synapomorphies). Describe how cladistic analyses differ from traditional systematics and explain how and why the phylogenetic trees produced may differ.

☐ 5. Use an example to describe how the derivation or disappearance of traits can be represented using cladograms.

☐ 6. Explain how phylogenetic trees illustrate speciation events. Explain how, in a rooted phylogenetic tree, each node with descendants represents the inferred most recent common ancestor.

☐ 7. Describe how phylogenetic trees and cladograms are constructed using morphological similarities, and DNA and protein sequence similarities of fossil and living species.

☐ 8. Recognize the dynamic nature of phylogenetic analysis and explain how phylogenetic trees can change in the light of new information or improved technology. Use an example to illustrate your point.

☐ 9. Explain how, under a cladistic scheme, organisms are assigned to taxonomic categories on the basis of their shared derived characteristics. Recognize that classification schemes should (ideally) accurately reflect phylogeny.

☐ 10. Recognize at least seven major **taxonomic categories**: kingdom, phylum, class, order, family, genus, and species.

☐ 11. Explain how **binomial nomenclature** is used to classify organisms and demonstrate an understanding of what is meant by a distinguishing feature.

Periodicals:
Listings for this chapter are on page 384

Weblinks:
www.thebiozone.com/
weblink/AP1-3114.html

Descent and Common Ancestry

Our knowledge of how organisms are related has grown rapidly in recent decades. Traditional schemes for classifying the living world, which were based primarily on morphological comparisons, have been considerably revised in the light of new techniques in **molecular phylogenetics**. Such techniques compare the DNA, RNA, and proteins of organisms to establish evolutionary relationships. Molecular phylogenetics has enabled scientists to clarify the very earliest origins of eukaryotes and to recognize two prokaryote domains (rather than a single prokaryote superkingdom). Powerful evidence for the common ancestry of all life comes from the commonality in the genetic code, and from the similarities in the molecular machinery of all cells.

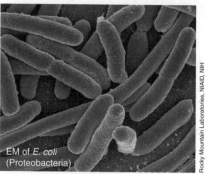

EM of *E. coli* (Proteobacteria)

Rocky Mountain Laboratories, NIAID, NIH

Xiangyux (PD)

LM of *Paramecium*

Barfooz and Josh Grosse (CC 3.0)

DOMAIN BACTERIA

Lack a distinct nucleus and cell organelles. Generally prefer less extreme environments than Archaea. Includes well-known pathogens, many harmless and beneficial species, and the cyanobacteria (photosynthetic bacteria containing the pigments chlorophyll a and phycocyanin). Endosymbiosis of ancestral bacteria during evolution gave rise to the eukaryotic organelles mitochondria and chloroplasts.

DOMAIN ARCHAEA

Closely resemble eubacteria in many ways but membrane and cell wall composition and aspects of metabolism are very different. Live in extreme environments similar to those on primeval Earth. They may utilize sulfur, methane, or halogens (chlorine, fluorine), and many tolerate extremes of temperature, salinity, or pH. The Korarchaeota are close to the stem group and live only in high temperature hydrothermal environments.

DOMAIN EUKARYA

Complex cells with organelles and a membrane-bound nucleus. This domain contains four of the kingdoms recognized under a traditional scheme. Note that Kingdom Protista is separated into distinct taxa, recognizing their diverse origins.

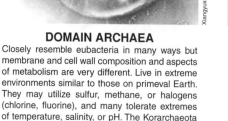

Animals   Fungi   Plants

Other bacteria   Cyanobacteria   Proteobacteria (many pathogens)   Sulfobacteria   Methanogens, extreme thermophiles, halobacteria   Algae

Bacteria that gave rise to chloroplasts

Ciliates

Bacteria that gave rise to mitochondria

Other single-celled eukaryotes

Korarchaeota

Hyperthermophillic bacteria

AN ARCHAEAN ORIGIN FOR EUKARYOTES
Archaeal RNA polymerase, which transcribes DNA, more closely resembles the equivalent molecule in eukaryotes than in bacteria. The protein components of archaeal ribosomes are also more like those in eukaryotes. These similarities in molecular machinery indicate that the eukaryotes diverged from the archaeans, not the bacteria.

Last Universal Common Ancestor (LUCA)

EUKARYOTES CARRY BACTERIAL GENES!
New evidence indicates the presence of bacterial genes in eukaryotes that are unrelated to photosynthesis or cellular respiration. These could be explained by gene transfers during evolution. A revised tree might include a complex network of connections to indicate single and multiple gene transfers across domains.

Adapted from: *Uprooting the tree of life (see Appendix)*

1. Explain the role of molecular phylogenetics in revising the traditional classification schemes (pre-1980):

   Compare the DNA, RNA and proteins of organisms to establish evolutionary relationships

2. Describe the evidence for the archaean origin of eukaryotic cells:

   Unrelated to photosynthesis or cellular respiration

3. What evidence is there for a Last Universal Common Ancestor?

   Common protein componets of archaeal ribosomes

Related activities: *The Origin of Eukaryotes*
Web links: *Types of Microbes, Introduction to the Archaea*

Periodicals:
Uprooting the tree of life,
To share and share alike

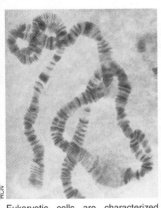

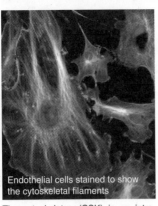
Endothelial cells stained to show the cytoskeletal filaments

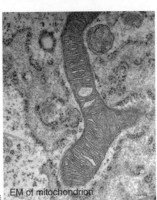
EM of mitochondrion

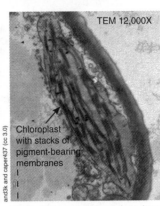
Chloroplast with stacks of pigment-bearing membranes

TEM 12,000X

Eukaryotic cells are characterized by large linear chromosomes within the cell's nucleus (above). These nuclear chromosomes are packaged by proteins into a condensed structure called chromatin. The evolution of linear chromosomes was related to the appearance of microtubule-dependent cell division (mitosis and meiosis).

The cytoskeleton (CSK) is an intra-cellular scaffold of protein filaments, important in intracellular movement of materials and cell division in eukaryotic cells. The CSK was once thought to be unique to eukaryotic cells, but homologs to all the major proteins of the eukaryotic cytoskeleton have also been found in prokaryotes.

The endosymbiotic origin of eukaryotic mitochondria is supported by the evidence from mitochondrial gene sequences, ribosomes, and protein synthesis that indicate a prokaryotic origin. Mitochondria are characteristic of eukaryotic cells and arose in a common ancestor to eukaryotic cells.

Chloroplasts are assumed to have been acquired, via endosymbiosis, after mitochondria and there is much evidence to suggest that they were acquired independently by more than one type of cyanobacterial cell (i.e. their origin is polyphyletic). The prokaryotic origin of chloroplasts is supported by ribosomal evidence.

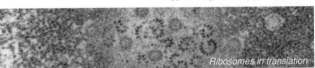
Ribosomes in translation

Most organisms share the same genetic code, i.e. the same combination of three DNA bases code for the same amino acid, although there are some minor variations (e.g. in mitochondria). Evidence suggests the code was subject to selection pressure which acted to minimize the effect of point mutations or errors in translation.

In all living systems, the genetic machinery consists of self-replicating DNA molecules. Some DNA is transcribed into RNA, some of which is translated into proteins. The machinery for translation (above) involves proteins and RNA. Ribsomal RNA analyses support a universal common ancestor.

Determining Phylogenetic Relationships

Increasingly, analyses to determine evolutionary relationships rely on cladistic analyses of character states. Cladism groups species according to their most recent common ancestor on the basis of shared derived characteristics or **synapomorphies**. All other characters are ignored. A phylogeny constructed using cladistics thus includes only **monophyletic groups**, i.e. the common ancestor and all of its descendants. It excludes both paraphyletic and polyphyletic groups (right). It is important to understand these terms when constructing cladograms. The cladist restriction to using only synapomorphies creates an unambiguous branching tree. One problem with this approach is that a strictly cladistic classification could theoretically have an impractically large number of taxonomic levels and may be incompatible with a Linnaean system.

Cladistic schemes have traditionally used morphological characteristics, with gain (or loss) of a character indicating a derived state. Increasingly, molecular comparisons are being used, particularly for highly conserved genes such as those coding for ribosomal RNA. For prokaryotes, molecular phylogeny studies have been the most important tool in revealing evolutionary relationships and revolutionizing traditional classification schemes.

Taxon 2 is polyphyletic as it includes organisms with different ancestors. The group "warm-blooded (endothermic) animals" is polyphyletic as it includes birds and mammals.

Taxon 3 is paraphyletic. It includes species A without including all of A's descendants. The traditional grouping of reptiles is paraphyletic because it does not include birds.

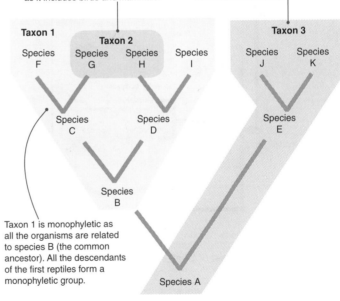

Taxon 1 is monophyletic as all the organisms are related to species B (the common ancestor). All the descendants of the first reptiles form a monophyletic group.

4. (a) Briefly describe how eukaryotic organelles, such as mitochondria and chloroplasts, are thought to have evolved:

acquired independently by more then one type of cyanobacteria cell

(b) What evidence is there to support this? _supperted by ribosomal evidence_

5. Discuss the benefits of cladistics for establishing evolutionary relationships. What useful groupings might it exclude?

The benefits of cladistics for establishing evolutronery relatronships are find common ancesters for organmisms.

© BIOZONE International 2012
ISBN: 978-1-927173-11-4
Photocopying Prohibited

Periodicals:
Evolution encoded

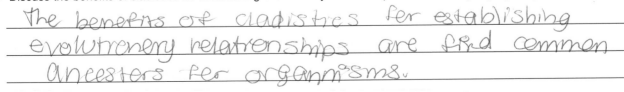
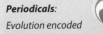

Relatedness of Organisms

The Phylogeny of Animals

The animals are a monophyletic group of multicellular eukaryotes that arose early in evolution from a group of protists called choanoflagellates. Sponges are the simplest animals, being sessile and without any body symmetry. They form a lineage separate to all other animal groups, and split off very early in animal evolution. The ctenophores and cnidarians form another lineage on the basis of having two (not three) tissue layers and radial symmetry. Bilateral symmetry, in which there are mirror-image left-right sides and an anterior and posterior end, is characteristic of all other animals (at least in embryonic stages). It arose later in animal evolution and is associated with the development of a head region and a coelom (an internal body cavity lined with mesoderm). Bilateral symmetry is considered to be an adaptation to the motile lifestyle of most animals.

Phylogeny of Animals

This phylogeny is based on DNA sequence data of a number of genes from various phyla. The short vertical bars indicate when novel traits arose. All animals more derived than the cnidarians and ctenophores are triploblastic.

Choanoflagellates

Sponges

Cnidaria

Ctenophora — TWO TISSUE LAYERS

Animal common ancestor

Acoels

Grow continuously
Rotifers

Flatworms

Annelids (segmented worms)

Mollusks — PROTOSTOMES

Grow by ecdysis
Nematodes

Arthropods

Coelom. Cephalization. Central nervous system

Echinoderms — DEUTEROSTOMES

Chordates

Cladogram adapted from Tudge (2000) with reference to Dunn et al (2008)

Radiata

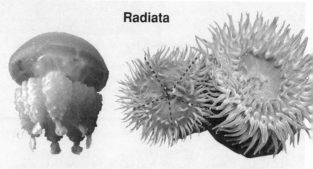

Cnidarians and ctenophores show **radial symmetry**, with similar body parts are arranged symmetrically around a central body axis. Adult echinoderms (e.g. sand dollar) are also radially symmetrical, but this is secondary as their larvae show bilateral symmetry.

Protostomes vs Deuterostomes

Protostomes and deuterostomes are divisions of the coelomate animals according to how the coelom develops and how the gut and embryonic germ layers form. Evolution has resulted in two ways to produce a bilateral, coelomate body plan. The origin of the coelom is uncertain, but probably arose in a Precambrian ancestor that gave rise to both deuterostome and protostome lineages.

Bilateria

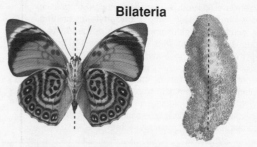

Most animals show **bilateral symmetry**, where the body can be divided evenly through only one plane. The division produces roughly equivalent, mirror-image halves. Bilateral animals also show triploblastic development (three tissue layers).

1. Suggest why the origin of **bilateral symmetry** is considered to be a great milestone in animal evolution:

   associated with the development of a head region and a coelom

2. (a) In the phylogenetic tree above, short vertical bars indicate when certain traits originated. Identify when each of the following traits arose by assigning each bar its corresponding letter, as follows (one is completed for you). Note that segmentation arises independently three times in three lineages. Radial symmetry in adults (**R**). Secondary loss of coelom (**L**). Segmentation (**S**). Multicellularity (**M**). Bilateral symmetry (**B**). Protostomy (**P**). Deuterostomy (**D**).

   (b) Place a vertical bar where you think ecdysis (molting of the cuticle or exoskeleton) arose in animal evolution.

3. (a) The free-swimming larvae of cnidarians are bilaterally symmetric. Also, new research shows that a few primitive cnidarians show bilateral symmetry. Where would you place the origin of bilateral symmetry based on this information?

   in fetal development

   (b) If the cnidarians were reclassified, on what basis would they remain distinct from other bilateral organisms?

   yes they would remain

Related activities: Phylogenetic Systematics, Features of Taxonomic Groups

Cladograms and Phylogenetic Trees

Phylogenetic systematics is a science in which the fields of taxonomy (the study of naming organisms) and phylogenetics (the study of evolutionary history) overlap. Traditional methods for establishing **phylogenetic trees** have emphasized morphological similarities in order to group species into genera and other higher level taxa. In contrast, **cladistics** is a methodology that relies on **shared derived characteristics** (**synapomorphies**) and ignores features that are not the result of shared ancestry. A cladogram is a phylogenetic tree constructed using cladistics.

Although cladistics has traditionally relied on morphological data, molecular data are increasingly being used to construct cladograms. Traditional and cladistic trees do not necessarily conflict, but there have been reclassifications of some taxa (notably primates, reptiles, and fish). Popular classifications will probably continue to reflect similarities and differences in appearance, rather than a strict evolutionary history. In this respect, they are a compromise between phylogeny and the need for a convenient filing system for species diversity.

Constructing a Simple Cladogram

A table listing the features for comparison allows us to identify where we should make branches in the **cladogram**. An outgroup (one which is known to have no or little relationship to the other organisms) is used as a basis for comparison.

Taxa

Comparative features	Jawless fish (outgroup)	Bony fish	Amphibians	Lizards	Birds	Mammals
Vertebral column	✔	✔	✔	✔	✔	✔
Jaws	✗	✔	✔	✔	✔	✔
Four supporting limbs	✗	✗	✔	✔	✔	✔
Amniotic egg	✗	✗	✗	✔	✔	✔
Diapsid skull	✗	✗	✗	✔	✔	✗
Feathers	✗	✗	✗	✗	✔	✗
Hair	✗	✗	✗	✗	✗	✔

The table above lists features shared by selected taxa. The outgroup (jawless fish) shares just one feature (vertebral column), so it gives a reference for comparison and the first branch of the cladogram (tree).

As the number of taxa in the table increases, the number of possible trees that could be drawn increases exponentially. To determine the most likely relationships, the rule of **parsimony** is used. This assumes that the tree with the least number of evolutionary events is most likely to show the correct evolutionary relationship.

Three possible cladograms are shown on the right. The top cladogram requires six events while the other two require seven events. Applying the rule of parsimony, the top cladogram must be taken as correct.

Parsimony can lead to some confusion. Some evolutionary events have occurred multiple times. An example is the evolution of the four chambered heart, which occurred separately in both birds and mammals. The use of fossil evidence and DNA analysis can help to solve problems like this.

Possible Cladograms

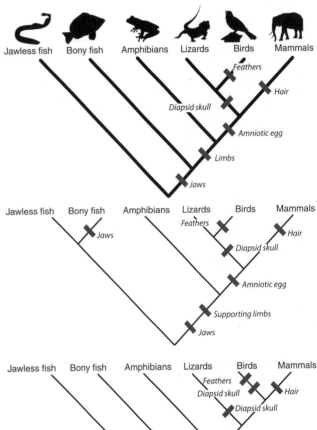

Using DNA Data

DNA analysis has allowed scientists to confirm many phylogenies and refute or redraw others. In a similar way to morphological differences, DNA sequences can be tabulated and analyzed. The ancestry of whales has been in debate since Darwin. The radically different morphologies of whales and other mammals makes it difficult work out the correct phylogenetic tree. However recently discovered fossil ankle bones, as well as DNA studies, show whales are more closely related to hippopotami than to any other mammal. Coupled with molecular clocks, DNA data can also give the time between each split in the lineage.

The DNA sequences on the right show part of the nucleotide subset 141-200 and some of the matching nucleotides used to draw the cladogram. Although whales were once thought most closely related to pigs, based on the DNA analysis the most parsimonious tree disputes this.

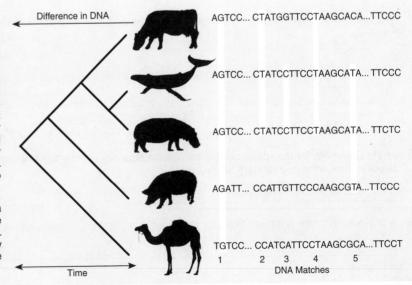

AGTCC... CTATGGTTCCTAAGCACA...TTCCC

AGTCC... CTATCCTTCCTAAGCATA... TTCCC

AGTCC... CTATCCTTCCTAAGCATA... TTCTC

AGATT... CCATTGTTCCCAAGCGTA...TTCCC

TGTCC... CCATCATTCCTAAGCGCA...TTCCT

1   2   3   4   5
DNA Matches

Difference in DNA

Time

© BIOZONE International 2012
ISBN: 978-1-927173-11-4
Photocopying Prohibited

Relatedness of Organisms

CHARACTER													
Taxon	1	2	3	4	5	6	7	8	9	10	11	12	13
Zebra-perch sea chub	0	0	0	0	0	0	0	0	0	0	0	0	0
Barred surfperch	1	0	0	0	0	0	0	0	0	1	1	0	0
Walleye surfperch	1	0	0	0	1	0	1	0	0	1	1	0	0
Black perch	1	1	1	0	0	0	0	0	0	0	0	1	0
Rainbow seaperch	1	1	1	0	0	0	0	0	0	0	0	1	0
Rubberlip surfperch	1	1	1	1	1	0	0	0	0	0	0	0	1
Pile surfperch	1	1	1	1	1	0	0	0	0	0	0	0	1
White seaperch	1	1	1	1	1	0	0	0	0	0	0	0	0
Shiner perch	1	1	1	1	1	1	0	0	0	0	0	0	0
Pink seaperch	1	1	1	1	1	1	1	1	0	0	0	0	0
Kelp perch	1	1	1	1	1	1	1	1	1	0	0	0	0
Reef perch	1	1	1	1	1	1	1	1	1	0	0	0	0

Juvenile surfperch

Steve Lonhart (SIMoN / MBNMS) PD NOAA

Surfperches are viviparous (live bearing) and the females give birth to relatively well developed young. Some of the characters (below, left) relate to adaptations of the male for internal fertilization. Others relate to deterring or detecting predators. In the matrix, characters are assigned a 0 or 1 depending on whether they represent the ancestral (0) or derived (1) state. This coding is common in cladistics because it allows the data to be analyzed by computer.

Data after Cailliel et al, 1986

Selected characters for cladogram assembly

		0	1
1.	Viviparity (live bearing)	0 No	1 Yes
2.	Males with flask organ	0 No	1 Yes
3.	Orbit without bony front wall	0 Yes	1 No
4.	Tail length	0 Short	1 Long
5.	Body depth	0 Deep	1 Narrow
6.	Body size	0 Large	1 Small
7.	Length of dorsal fin base	0 Long	1 Short
8.	Eye diameter	0 Moderate	1 Large
9.	Males with anal crescent	0 No	1 Yes
10.	Pectoral bone with process	0 No	1 Yes
11.	Length of dorsal sheath	0 Long	1 Short
12.	Body mostly darkish	0 No	1 Yes
13.	Flanks with large black bars	0 No	1 Yes

1. This activity provides the taxa and character matrix for 11 genera of marine fishes in the family of surfperches. The outgroup given is a representative of a sister family of rudderfishes (zebra-perch sea chub), which are not live-bearing. Your task is to create the most parsimonious cladogram from the matrix of character states provided. To help you, we have organized the matrix with genera having the smallest blocks of derived character states (1) at the top following the outgroup representative. Use a separate sheet of graph paper, working from left to right to assemble your cladogram. Identify the origin of derived character states with horizontal bars, as shown in the examples earlier in this activity. CLUE: You should end up with 15 steps. Two derived character states arise twice independently. Staple your cladogram to this page.

2. In the cladogram you have constructed for the surfperches, two characters have evolved twice independently:

   (a) Identify these two characters: _____

   (b) What selection pressures do you think might have been important in the evolution of these two derived states?

   _____

   _____

3. What assumption is made when applying the rule of parsimony in constructing a cladogram? _____

   _____

   _____

4. (a) Describe the contribution of biochemical evidence to clarifying some evolutionary relationships: _____

   _____

   (b) When might it not be advisable to rely on biochemical evidence alone? _____

   _____

   _____

5. In the DNA data for the whale cladogram (previous page) identify the DNA match that shows a mutation event must have happened twice in evolutionary history.

   _____

6. A phylogenetic tree is a hypothesis for an evolutionary history. How could you test it? _____

   _____

   _____

Features of Taxonomic Groups

In order to distinguish organisms, it is desirable to classify and name them (a science known as **taxonomy**). An effective classification system requires features that are distinctive to a particular group of organisms. The aim of classification is to organize species in a way that most accurately reflects their evolutionary history (**phylogeny**). Revised classification systems, recognizing three domains (rather than five or six kingdoms), do this rather better than traditional schemes, but for the purposes of describing the groups with which we are most familiar, the five kingdom system (used here) is still appropriate. The distinguishing features of some major **taxa** are provided in the following pages by means of diagrams and summaries. Recall that most animals show **bilateral symmetry**. **Radial symmetry** is a characteristic of cnidarians and ctenophores.

SUPERKINGDOM: PROKARYOTAE (Bacteria)

- Also known as prokaryotes. The term moneran is no longer in use.
- Two major bacterial lineages are recognized: the **Archaebacteria** (Archaea) and the more derived **Eubacteria** (Bacteria).
- All have a prokaryotic cell structure: they lack the nuclei and chromosomes of eukaryotic cells, and have smaller (70S) ribosomes.
- Have a tendency to spread genetic elements across species barriers by conjugation, viral transduction, and other processes.
- Asexual. Can reproduce rapidly by binary fission.

- Have evolved a wider variety of metabolism types than eukaryotes.
- Bacteria grow and divide or aggregate into filaments or colonies of various shapes. Colony type is often diagnostic.
- They are taxonomically identified by their appearance (form) and through biochemical differences.

Species diversity: 10,000+ Bacteria are rather difficult to classify to species level because of their relatively rampant genetic exchange, and because their reproduction is asexual.

Eubacteria

- Also known as 'true bacteria', they probably evolved from the more ancient Archaebacteria.
- Distinguished from Archaebacteria by differences in cell wall composition, nucleotide structure, and ribosome shape.
- Diverse group includes most bacteria.
- The **gram stain** is the basis for distinguishing two broad groups of bacteria. It relies on the presence of peptidoglycan in the cell wall. The stain is easily washed from the thin peptidoglycan layer of gram negative walls but is retained by the thick peptidoglycan layer of gram positive cells, staining them a dark violet color.

Gram Positive Bacteria

The walls of gram positive bacteria consist of many layers of peptidoglycan forming a thick, single-layered structure that holds the gram stain.

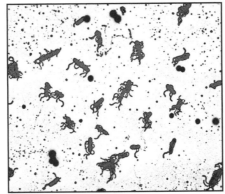

Photos: CDC

Bacillus alvei: a gram positive, flagellated bacterium. Note how the cells appear dark.

Gram Negative Bacteria

The cell walls of gram negative bacteria contain only a small proportion of peptidoglycan, so the dark violet stain is not retained by the organisms.

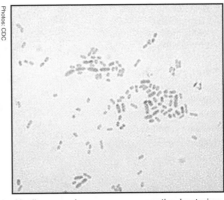

Alcaligenes odorans: a gram negative bacterium. Note how the cells appear pale.

SUPERKINGDOM: EUKARYOTAE
Kingdom: FUNGI

- Heterotrophic.
- Rigid cell wall made of chitin.
- Vary from single celled to large multicellular organisms.
- Mostly saprotrophic (ie. feeding on dead or decaying material).
- Terrestrial and immobile.

Examples:
Mushrooms/toadstools, yeasts, truffles, morels, molds, and lichens.

Species diversity: 80,000 +

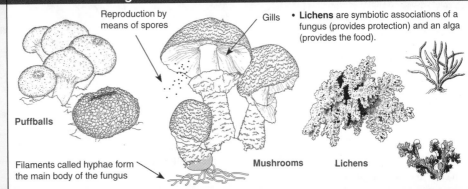

Reproduction by means of spores

Gills

- **Lichens** are symbiotic associations of a fungus (provides protection) and an alga (provides the food).

Puffballs

Filaments called hyphae form the main body of the fungus

Mushrooms

Lichens

Kingdom: PROTISTA

- A diverse group of organisms. They are polyphyletic and so better represented in the 3 domain system.
- Unicellular or simple multicellular.
- Widespread in moist or aquatic environments.

Examples of algae: green, red, and brown algae, dinoflagellates, diatoms.

Examples of protozoa: amoebas, foraminiferans, radiolarians, ciliates.

Species diversity: 55,000 +

Algae 'plant-like' protists

- Autotrophic (photosynthesis)
- Characterized by the type of chlorophyll present

Cell walls of cellulose, sometimes with silica

Diatom

Protozoa 'animal-like' protists

- Heterotrophic nutrition and feed via ingestion
- Most are microscopic (5 µm - 250 µm)

Move via projections called pseudopodia

Lack cell walls

Amoeba

Relatedness of Organisms

Kingdom: PLANTAE

- Multicellular organisms (the majority are photosynthetic and contain chlorophyll).
- Cell walls made of cellulose; food is stored as starch.
- Subdivided into two major divisions based on tissue structure: **Bryophytes** (non-vascular plants) and **Tracheophytes** (vascular plants).

Non-Vascular Plants:

- Non-vascular, lacking transport tissues (no xylem or phloem).
- Small and restricted to moist, terrestrial environments.
- Do not possess 'true' roots, stems, or leaves.

Phylum Bryophyta: Mosses, liverworts, and hornworts.

Species diversity: 18,600 +

Phylum: Bryophyta

Sexual reproductive structures

Flattened thallus (leaf like structure)

Sporophyte: reproduce by spores

Rhizoids anchor the plant into the ground

Liverworts

Mosses

Vascular Plants:

- Vascular: possess transport tissues.
- Possess true roots, stems, and leaves, as well as stomata.
- Reproduce via spores, not seeds.
- Clearly defined alternation of sporophyte and gametophyte generations.

Seedless Plants:

Spore producing plants, includes:
Phylum Filicinophyta: Ferns
Phylum Sphenophyta: Horsetails
Phylum Lycophyta: Club mosses
Species diversity: 13,000 +

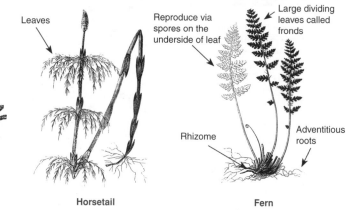

Phylum: Lycophyta

Leaves

Club moss

Phylum: Sphenophyta

Leaves

Horsetail

Phylum: Filicinophyta

Reproduce via spores on the underside of leaf

Large dividing leaves called fronds

Rhizome

Adventitious roots

Fern

Seed Plants:

Also called Spermatophyta. Produce seeds housing an embryo. Includes:

Gymnosperms

- Lack enclosed chambers in which seeds develop.
- Produce seeds in cones which are exposed to the environment.

Phylum Cycadophyta: Cycads
Phylum Ginkgophyta: Ginkgoes
Phylum Coniferophyta: Conifers
Species diversity: 730 +

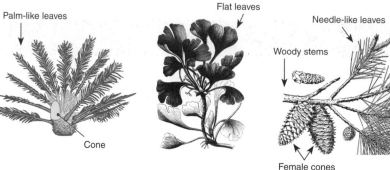

Phylum: Cycadophyta

Palm-like leaves

Cone

Cycad

Phylum: Ginkgophyta

Flat leaves

Ginkgo

Phylum: Coniferophyta

Needle-like leaves

Male cones

Woody stems

Female cones

Conifer

Angiosperms

Phylum: Angiospermophyta

- Seeds in specialised reproductive structures called flowers.
- Female reproductive ovary develops into a fruit.
- Pollination usually via wind or animals.

Species diversity: 260,000 +

The phylum Angiospermophyta may be subdivided into two classes:

Class Monocotyledoneae (Monocots)
Class Dicotyledoneae (Dicots)

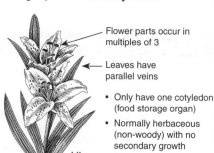

Angiosperms: **Monocotyledons**

Flower parts occur in multiples of 3

Leaves have parallel veins

- Only have one cotyledon (food storage organ)
- Normally herbaceous (non-woody) with no secondary growth

Lily

Examples: cereals, lilies, daffodils, palms, grasses.

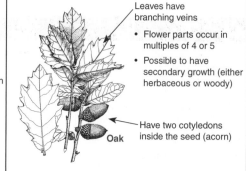

Angiosperms: **Dicotyledons**

Leaves have branching veins

- Flower parts occur in multiples of 4 or 5
- Possible to have secondary growth (either herbaceous or woody)

Have two cotyledons inside the seed (acorn)

Oak

Examples: many annual plants, trees and shrubs.

Kingdom: ANIMALIA

- Over 800,000 species described in 33 existing phyla.
- Multicellular, heterotrophic organisms.
- Animal cells lack cell walls.

- Further subdivided into major phyla on the basis of body symmetry, development of the coelom (protostome or deuterostome), and external and internal structures.

Phylum: Porifera

- Lack organs.
- All are aquatic (mostly marine).
- Asexual reproduction by budding.
- Lack a nervous system.

Examples: sponges.
Species diversity: 8000 +

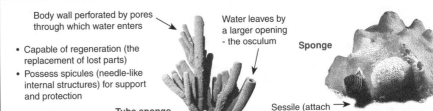

- Body wall perforated by pores through which water enters
- Water leaves by a larger opening - the osculum
- Sponge
- Capable of regeneration (the replacement of lost parts)
- Possess spicules (needle-like internal structures) for support and protection
- **Tube sponge**
- Sessile (attach to ocean floor)

Phylum: Cnidaria

- Diploblastic with two basic body forms:
  Medusa: umbrella shaped and free swimming by pulsating bell.
  Polyp: cylindrical, some are sedentary, others can glide, or somersault or use tentacles as legs.
- Some species have a life cycle that alternates between a polyp stage and a medusa stage.
- All are aquatic (most are marine).

Examples: Jellyfish, sea anemones, hydras, and corals.
Species diversity: 11,000 +

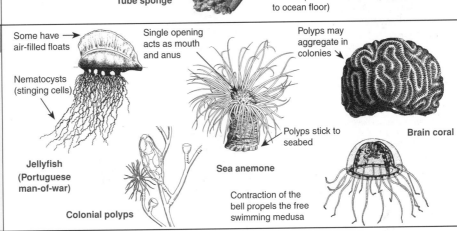

- Some have air-filled floats
- Single opening acts as mouth and anus
- Polyps may aggregate in colonies
- Nematocysts (stinging cells)
- Polyps stick to seabed
- **Brain coral**
- **Jellyfish (Portuguese man-of-war)**
- **Sea anemone**
- **Colonial polyps**
- Contraction of the bell propels the free swimming medusa

Phylum: Rotifera

- A diverse group of small, pseudocoelomates with sessile, colonial, and planktonic forms.
- Most freshwater, a few marine.
- Typically reproduce via cyclic parthenogenesis.
- Characterized by a wheel of cilia on the head used for feeding and locomotion, a large muscular pharynx (mastax) with jaw like trophi, and a foot with sticky toes.

Species diversity: 1500 +

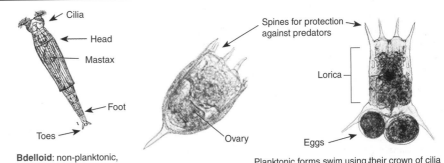

- Cilia
- Head
- Mastax
- Foot
- Toes
- Spines for protection against predators
- Lorica
- Ovary
- Eggs
- **Bdelloid**: non-planktonic, creeping rotifer
- Planktonic forms swim using their crown of cilia

Phylum: Platyhelminthes

- Unsegmented. Coelom has been lost.
- Flattened body shape.
- Mouth, but no anus.
- Many are parasitic.
Examples: Tapeworms, planarians, flukes.
Species diversity: 20,000 +

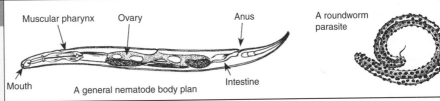

- Hooks
- *Detail of head (scolex)*
- **Liver fluke**
- **Tapeworm**
- **Planarian**

Phylum: Nematoda

- Tiny, unsegmented roundworms.
- Many are plant/animal parasites
Examples: Hookworms, stomach worms, lung worms, filarial worms
Species diversity: 80,000 - 1 million

- Muscular pharynx
- Ovary
- Anus
- A roundworm parasite
- Mouth
- Intestine
- A general nematode body plan

Phylum: Annelida

- Cylindrical, segmented body with chaetae (bristles).
- Move using hydrostatic skeleton and/or parapodia (appendages).

Examples: Earthworms, leeches, polychaetes (including tubeworms).
Species diversity: 15,000 +

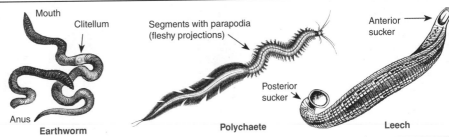

- Mouth
- Clitellum
- Segments with parapodia (fleshy projections)
- Anterior sucker
- Posterior sucker
- Anus
- **Earthworm**
- **Polychaete**
- **Leech**

Relatedness of Organisms

Kingdom: ANIMALIA (continued)

Phylum: Molluska

- Soft bodied and unsegmented.
- Body comprises head, muscular foot, and visceral mass (organs).
- Most have radula (rasping tongue).
- Aquatic and terrestrial species.
- Aquatic species possess gills.

Examples: Snails, mussels, squid.
Species diversity: 110,000 +

Class: Bivalvia

Radula lost in bivalves

Mantle secretes shell

Scallop

Two shells hinged together

Class: Gastropoda

Mantle secretes shell

Muscular foot for locomotion

Tentacles with eyes

Head

Land snail

Class: Cephalopoda

Well developed eyes

Squid

Foot divided into tentacles

Phylum: Arthropoda

- Exoskeleton made of chitin.
- Grow in stages after molting (ecdysis).
- Jointed appendages.
- Segmented bodies.
- Heart found on dorsal side of body.
- Open circulation system.
- Most have compound eyes.

Species diversity: 1 million +
Make up 75% of all living animals.

Arthropods are subdivided into the following classes:

Class: Crustacea (crustaceans)
- Mainly marine.
- Exoskeleton impregnated with mineral salts.
- Gills often present.
- Includes: Lobsters, crabs, barnacles, prawns, shrimps, isopods, amphipods
- **Species diversity:** 35,000 +

Class: Arachnida (chelicerates)
- Almost all are terrestrial.
- 2 body parts: cephalothorax and abdomen (except horseshoe crabs).
- Includes: spiders, scorpions, ticks, mites, horseshoe crabs.
- **Species diversity:** 57,000 +

Class: Insecta (insects)
- Mostly terrestrial.
- Most are capable of flight.
- 3 body parts: head, thorax, abdomen.
- Include: Locusts, dragonflies, cockroaches, butterflies, bees, ants, beetles, bugs, flies, and more
- **Species diversity:** 800,000 +

Myriapods (=many legs)
Class Diplopoda (millipedes)
- Terrestrial.
- Have a rounded body.
- Eat dead or living plants.
- **Species diversity:** 2000 +

Class Chilopoda (centipedes)
- Terrestrial.
- Have a flattened body.
- Poison claws for catching prey.
- Feed on insects, worms, and snails.
- **Species diversity:** 7000 +

Class: Crustacea

2 pairs of antennae

Cephalothorax (fusion of head and thorax)

Abdomen

My is burning

Crab

3 pairs of mouthparts

Cheliped (first leg)

Shrimp

Walking legs

Swimmerets

Amphipod

Class: Arachnida

Abdomen

Simple eyes

no eyes

4 pairs of walking legs

No antennae

2 pairs of feeding appendages

kohie

Cephalothorax

Spider

Scorpion

Carapace

Tick

me off

Abdomen

Telson (tail)

Horseshoe crab (exception to group: has 3 body parts)

Class: Insecta

1 pair of antennae

1 pair of compound eyes

Head

Thorax

Abdomen

2 pairs of wings

3 pairs of legs

Honey bee

Locust

Beetles are the largest group within the animal kingdom with more than 300,000 species.

Butterfly

Beetle

Class: Diplopoda

Body with many similar segments

Clearly defined head

1 pair of antennae

Each segment has 2 pairs of legs

1 pair of mouthparts

Class: Chilopoda

Body with many similar segments

1 pair of large antennae

Clearly defined head

1 pair of mouthparts

Each segment has 1 pair of legs

Phylum: Echinodermata

- Rigid body wall, internal skeleton made of calcareous plates.
- Many possess spines.
- Ventral mouth, dorsal anus.
- External fertilization.
- Unsegmented, marine organisms.
- Tube feet for locomotion.
- Water vascular system.

Examples: Starfish, brittlestars, feather stars, sea urchins, sea lilies.
Species diversity: 6000 +

Moveable spines

Starfish have central disc

Usually star shaped with 5 or more arms

Sea urchin

Starfish

Sand dollar

Numerous tube feet

Sea cucumber

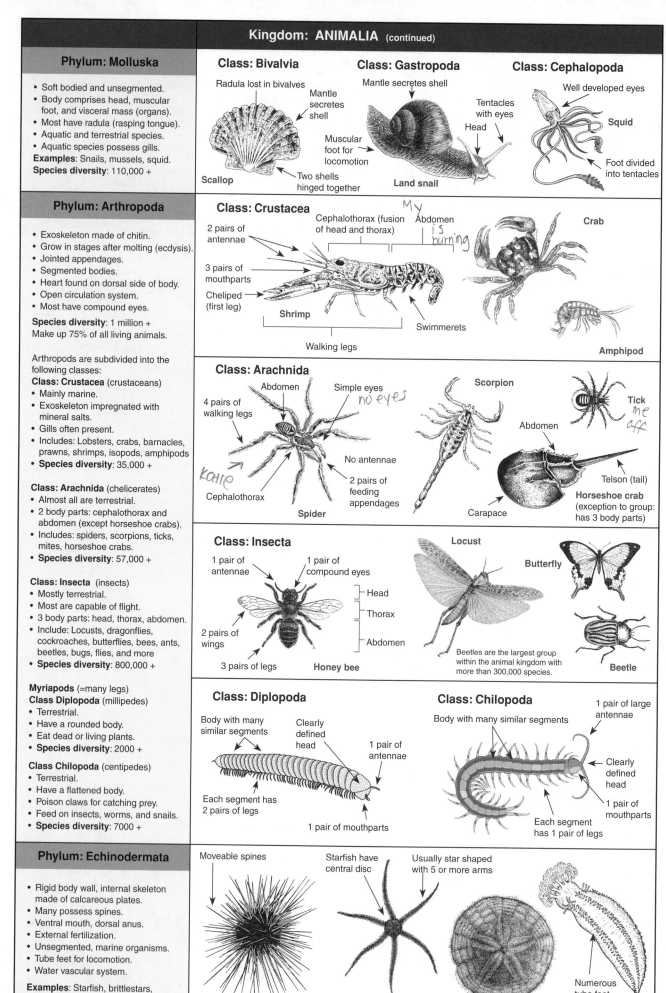

© BIOZONE International 2012
ISBN: 978-1-927173-11-4
Photocopying Prohibited

Kingdom: ANIMALIA (continued)

Phylum: Chordata

- Dorsal notochord (flexible, supporting rod) present at some stage in the life history.
- Post-anal tail present at some stage in their development.
- Dorsal, tubular nerve cord.
- Pharyngeal slits present.
- Circulation system closed in most.
- Heart positioned on ventral side.

Species diversity: 48,000 +

- A very diverse group with several sub-phyla:
  - Urochordata (sea squirts, salps)
  - Cephalochordata (lancelet)
  - Craniata (vertebrates)

Sub-Phylum Craniata (vertebrates)
- Internal skeleton of cartilage or bone.
- Well developed nervous system.
- Vertebral column replaces notochord.
- Two pairs of appendages (fins or limbs) attached to girdles.

Further subdivided into:

Class: Chondrichthyes (cartilaginous fish)
- Skeleton of cartilage (not bone).
- No swim bladder.
- All aquatic (mostly marine).
- Include: Sharks, rays, and skates.

Species diversity: 850 +

Class: Osteichthyes (bony fish)
- Swim bladder present.
- All aquatic (marine and fresh water).

Species diversity: 21,000 +

Class: Amphibia (amphibians)
- Lungs in adult, juveniles may have gills (retained in some adults).
- Gas exchange also through skin.
- Aquatic and terrestrial (limited to damp environments).
- Include: Frogs, toads, salamanders, and newts.

Species diversity: 3900 +

Class Reptilia (reptiles)
- Ectotherms with no larval stages.
- Teeth are all the same type.
- Eggs with soft leathery shell.
- Mostly terrestrial.
- Include: Snakes, lizards, crocodiles, turtles, and tortoises.

Species diversity: 7000 +

Class: Aves (birds)
- Terrestrial endotherms.
- Eggs with hard, calcareous shell.
- Strong, light skeleton.
- High metabolic rate.
- Gas exchange assisted by air sacs.

Species diversity: 8600 +

Class: Mammalia (mammals)
- Endotherms with hair or fur.
- Mammary glands produce milk.
- Glandular skin with hair or fur.
- External ear present.
- Teeth are of different types.
- Diaphragm between thorax/abdomen.

Species diversity: 4500 +
Subdivided into three subclasses:
Monotremes, marsupials, placentals.

Class: Chondrichthyes (cartilaginous fish)

- Lateral line sense organ
- Asymmetrical tail fin provides lift
- Skin with toothlike scales
- Ectotherms with endoskeleton made of cartilage
- Pelvic fin
- Pectoral fin
- No operculum (bony flap) over gills

Hammerhead shark

Stingray

Class: Osteichthyes (bony fish)

Eel

Seahorse

- Fins supported by bony rays
- Slippery skin with thin, bony scales
- Tail fin is symmetrical in shape
- Operculum (bony flap) over gills
- Sensory lateral line system
- Ectotherms with bony endoskeleton

Herring

Class: Amphibia

- Moveable eyelids
- Ectotherms with smooth (non-scaly), moist skin
- Well developed protrusible tongue
- Aquatic larvae undergo metamorphosis to adult
- Eardrum visible

Tadpole

Frog

Toad

Newt

Class: Reptilia

- Protective shell of horny plates
- Dry, watertight skin covered with overlapping scales
- Most with well developed eyes
- Limbs absent

Crocodile

Sea turtle

Rattlesnake

Class: Aves

Some birds are flightless

- Feathers
- Forelimbs modified as wings
- Horny beak with no teeth
- Horny scales on feet only

Penguin

Kiwi

chnzhne

Seagull

Class: Mammalia

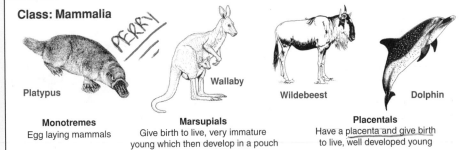

Platypus

PERRY

Wallaby

Wildebeest

Dolphin

Monotremes
Egg laying mammals

Marsupials
Give birth to live, very immature young which then develop in a pouch

Placentals
Have a placenta and give birth to live, well developed young

Relatedness of Organisms

Features of the Five Kingdoms

The classification of organisms into taxonomic groups is based on how biologists believe they are related in an evolutionary sense. Organisms in a taxonomic group share features which set them apart from other groups. By identifying these **distinguishing** **features**, it is possible to develop an understanding of the evolutionary history of the group. The focus of this activity is to summarize the distinguishing features of each of the five kingdoms in the five kingdom classification system.

1. Distinguishing features of Kingdom **Prokaryotae**:

   _____

   _____

   _____

   _____

   _____

   _____

2. Distinguishing features of Kingdom **Protista**:

   _____

   _____

   _____

   _____

   _____

   _____

3. Distinguishing features of Kingdom **Fungi**:

   _____

   _____

   _____

   _____

   _____

4. Distinguishing features of Kingdom **Plantae**:

   _____

   _____

   _____

   _____

   _____

   _____

   _____

5. Distinguishing features of Kingdom **Animalia**:

   _____

   _____

   _____

   _____

   _____

   _____

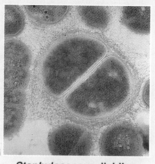

Staphylococcus dividing

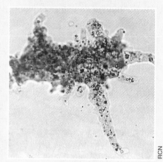

Helicobacter pylori

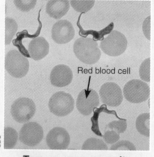

Red blood cell

Trypanosoma parasite

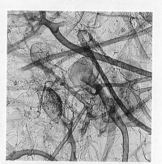

Amoeba

Mushrooms

Hyphae of *Rhizopus*

Moss

Pea plants

Cicada molting

Gibbon

© BIOZONE International 2012
ISBN: 978-1-927173-11-4
Photocopying Prohibited

Related activities: The New Tree of Life, Features of Taxonomic Groups

Features of Microbial Groups

A microorganism (or microbe) is literally a microscopic organism. The term is usually reserved for the organisms studied in microbiology: bacteria, fungi, microscopic protistans, and viruses. The first three of these represent three of the five kingdoms for which you described distinguishing features in an earlier activity (viruses are non-cellular and therefore not included in the five-kingdom classification). Most microbial taxa, but particularly the fungi, also have macroscopic representatives. The distinction between a macrofungus and a microfungus is an artificial but convenient one. Unlike microfungi, which are made conspicuous by the diseases or decay they cause, macrofungi are most likely to be observed with the naked eye. The microfungi include yeasts and pathogenic species. Macrofungi, e.g. mushrooms, toadstools, and lichens, are illustrated in *Features of Macrofungi and Plants*.

1. Describe aspects of each of the following for the bacteria and cyanobacteria (Kingdom Prokaryotae):

    (a) Environmental range: _____

    _____

    (b) Ecological role: _____

    _____

2. Identify an example within the bacteria of the following:

    (a) Photosynthetic: _____

    (b) Pathogen: _____

    (c) Decomposer: _____

    (d) Nitrogen fixer: _____

3. Describe aspects of each of the following for the microscopic protistans (Kingdom Protista):

    (a) Environmental range: _____

    _____

    (b) Ecological role: _____

    _____

4. Identify an example within the protists of the following:

    (a) Photosynthetic: _____

    (b) Pathogen: _____

    (c) Biological indicator: _____

5. Describe aspects of each of the following for the microfungi (Kingdom Fungi):

    (a) Environmental range: _____

    _____

    (b) Ecological role: _____

    _____

6. Identify examples within the microfungi of the following:

    (a) Animal pathogen: _____

    _____

    (b) Plant pathogens: _____

    _____

    _____

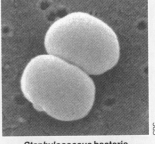

Spirillum bacterium

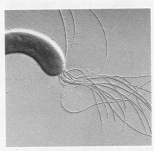

Staphylococcus bacteria

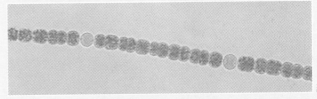

Anabaena cyanobacterium

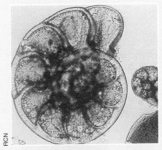

Foraminiferan

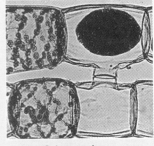

Spirogyra algae

Diatoms: *Pleurosigma*

Curvularia sp. conidiophore

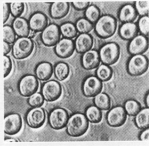

Yeast cells in solution

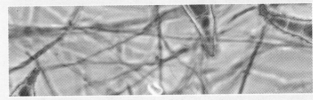

Microsporum distortum (a pathogenic fungus)

Relatedness of Organisms

Related activities: Features of Taxonomic Groups, Bacterial Cells

RA 1

Features of Macrofungi and Plants

Although plants and fungi are some of the most familiar organisms in our environment, their classification has not always been straightforward. We know now that the plant kingdom is monophyletic, meaning that it is derived from a common ancestor. The variety we see in plant taxa today is a result of their enormous diversification from the first plants. Although the fungi were once grouped together with the plants, they are unique organisms that differ from other eukaryotes in their mode of nutrition, structural organization, growth, and reproduction. The focus of this activity is to summarize the features of the fungal kingdom (**macrofungi**), the major divisions of the plant kingdom, and the two classes of flowering plants (angiosperms).

Lichen

Bracket fungus

Liverwort

Moss

Fern frond

Ground fern

Pine tree cone

Cycad

Coconut palms

Wheat plants

Deciduous tree

Flowering plant

1. **Macrofungi** features: _____

_____

_____

← Katip _____

_____

2. **Moss** and **liverwort** features: _____

_____

_____

_____

_____

3. **Fern** features: _____

_____

_____

_____

_____

4. **Gymnosperm** features: _____

_____

_____

_____

_____

5. **Monocot angiosperm** features: _____

_____

_____

_____

_____

6. **Dicot angiosperm** features: _____

_____

_____

_____

_____

Related activities: Features of Taxonomic Groups

© BIOZONE International 2012
ISBN: 978-1-927173-11-4
Photocopying Prohibited

Features of Animal Taxa

The animal kingdom is classified into about 35 major **phyla**. Representatives of the more familiar taxa are illustrated below: **cnidarians** (jellyfish, sea anemones, corals), **annelids** (segmented worms), **arthropods** (insects, crustaceans, spiders, scorpions, centipedes, millipedes), **mollusks** (snails, bivalve shellfish, squid, octopus), **echinoderms** (starfish, sea urchins),

vertebrates from the phylum **chordates** (fish, amphibians, reptiles, birds, mammals). The **arthropods** and the **vertebrates** have been represented in more detail, giving the **classes** for these **phyla**. This activity asks you to describe the **distinguishing features** of each of the taxa below. Underline the feature you think is most important in distinguishing the taxon from others.

Sea anemones

Jellyfish

Tubeworms

Earthworm

Long-horned beetle

Butterfly

Crab

Woodlouse

Scorpion

Spider

Centipede

Millipede

1. **Cnidarian** features: _____

_____

_____

_____

_____

2. **Annelid** features: _____

_____

_____

_____

_____

_____

3. **Insect** features: _____

_____

_____

_____

_____

4. **Crustacean** features: _____

_____

_____

_____

_____

5. **Arachnid** features: _____

_____

_____

_____

_____

6. **Myriopod** features (classes Chilopoda and Diplopoda):

_____

_____

_____

_____

_____

Relatedness of Organisms

Periodicals:
The family line

Related activities: Features of Taxonomic Groups

R 1

Chantine →

Nautilus

Abalone

Sea urchin

Starfish

Grouper

Shark

Frog

Salamander

← Katie

Iguana

Rattlesnake

Penguin

Pelican

Horse

Bear

7. **Mollusk** features: _____

_____

_____

_____

_____

8. **Echinoderm** features: _____

_____

_____

_____

_____

9. **Fish** features: _____

_____

_____

_____

_____

10. **Amphibian** features: _____

_____

_____

_____

_____

11. **Reptile** features: _____

_____

_____

_____

_____

12. **Bird** features: _____

_____

_____

_____

_____

13. **Mammal** features: _____

_____

_____

_____

_____

Classification System

The classification of organisms is designed to reflect how they are related to each other. The fundamental unit of classification of living things is the **species** (see *The Biological Species Concept* on page 310 for a discussion of what a species is). Its members are so alike genetically that they can interbreed. This genetic similarity also means that they are almost identical in their physical and other characteristics. Species are classified further into larger, more comprehensive categories (higher taxa). It must be emphasized that all such higher classifications are human inventions to suit a particular purpose.

1. The table below shows part of the classification for humans using the seven major levels of classification. For this question, use the example of the classification of the red kangaroo, on the next page, as a guide.

   (a) Complete the list of the taxonomic groupings on the left hand side of the table below:

Taxonomic Group	Human Classification
1. _____	_____
2. _____	_____
3. _____	_____
4. _____	_____
5. Family	Hominidae
6. _____	_____
7. _____	_____

   (b) Complete the classification for humans (*Homo sapiens*) on the table above.

2. Construct your own acronym or mnemonic to help you remember the principal taxonomic groupings in biology:

   _____

3. Describe the two-part scientific naming system (**binomial nomenclature**) which is used to name organisms:

   _____

4. Give two reasons why the classification of organisms is important:

   (a) _____

   (b) _____

5. Classification has traditionally been based on similarities in morphology but new biochemical methods are now widely used to determine species relatedness. Explain how these are being used to clarify the relationships between species:

   _____

   _____

   _____

   _____

   _____

6. Mammals have been divided into three major taxa: the monotremes, marsupials, and placentals. Describe the main morphological feature you would use to distinguish each taxon:

   (a) Monotreme (Prototheria): _____

   _____

   (b) Marsupial (Metatheria): _____

   _____

   (c) Placental (Eutheria): _____

   _____

Relatedness of Organisms

© BIOZONE International 2012
ISBN: 978-1-927173-11-4
Photocopying Prohibited

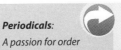
Periodicals:
A passion for order

Related activities: Phylogenetic Systematics

RA 2

Classification of the Ethiopian Hedgehog

Below is the classification for the **Ethiopian hedgehog**. Only one of each group is subdivided in this chart showing the levels that can be used in classifying an organism. Not all possible subdivisions have been shown here. For example, it is possible to indicate such categories as **super-class** and **sub-family**. The only natural category is the **species**, often separated into geographical **races**, or **sub-species**, which generally differ in appearance.

Kingdom: **Animalia**
Animals; one of five kingdoms

Phylum: **Chordata**
Animals with a notochord (supporting rod of cells along the upper surface)
tunicates, salps, lancelets, and vertebrates

23 other phyla

Sub-phylum: **Vertebrata**
Animals with backbones
fish, amphibians, reptiles, birds, mammals

Class: **Mammalia**
Animals that suckle their young on milk from mammary glands
placentals, marsupials, monotremes

Sub-class: **Eutheria or Placentals**
Mammals whose young develop for some time in the female's reproductive tract gaining nourishment from a placenta
placental mammals

Order: **Insectivora**
Insect eating mammals
An order of over 300 species of primitive, small mammals that feed mainly on insects and other small invertebrates.

17 other orders

Sub-order: **Erinaceomorpha**
The hedgehog-type insectivores. One of the three suborders of insectivores. The other suborders include the tenrec-like insectivores (*tenrecs and golden moles*) and the shrew-like insectivores (*shrews, moles, desmans, and solenodons*).

Family: **Erinaceidae**
The only family within this suborder. Comprises two subfamilies: the true or spiny hedgehogs and the moonrats (gymnures). Representatives in the family include the common European hedgehog, desert hedgehog, and the moonrats.

Genus: *Paraechinus*
One of eight genera in this family. The genus *Paraechinus* includes three species which are distinguishable by a wide and prominent naked area on the scalp.

7 other genera

Species: *aethiopicus*
The Ethiopian hedgehog inhabits arid coastal areas. Its diet consists mainly of insects, but includes small vertebrates and the eggs of ground nesting birds.

3 other species

The order Insectivora was first introduced to group together shrews, moles, and hedgehogs. It was later extended to include tenrecs, golden moles, desmans, tree shrews, and elephant shrews and the taxonomy of the group became very confused. Recent reclassification of the elephant shrews and tree shrews into their own separate orders has made the Insectivora a more cohesive group taxonomically.

Ethiopian hedgehog
Paraechinus aethiopicus

Speciation and Extinction

Key concepts

▶ Standard definitions of a species can be problematic.

▶ Reproductive isolation is essential for speciation. This is often preceded by allopatry.

▶ Larger scale patterns of evolution involve the diversification and extinction of species.

▶ Divergent evolution is frequently associated with the diversification of species into new niches.

▶ Current populations provide scientific evidence for the fact that populations continue to evolve.

Key terms

adaptation
adaptive radiation
allopatric (allopatry)
analogous structure
background extinction rate
biological species
coevolution
convergent evolution
divergent evolution
    (=cladogenesis)
evolution
extinction
gene flow
gene pool
genetic drift
homologous structure
homology
hybrid
mass extinction
natural selection
parallel evolution
phenotype
phyletic gradualism
    (=anagenesis)
phylogenetic species
phylogeny
polyploidy
population
postzygotic isolating
    mechanism
prezygotic isolating
    mechanism
reproductive isolation
ring species
speciation
species
sympatric (sympatry)

Essential Knowledge

☐ 1. Use the **KEY TERMS** to compile a glossary for this topic.

Species and Speciation (1.C.2) pages 310-323

☐ 2. Explain what is meant by a (biological) **species** and describe the limitations of its definition. Describe examples of **ring species** and their significance.

☐ 3. Describe the role of **natural selection**, **genetic drift**, and **isolation** in **speciation**. Use data to predict the effect of selection pressures on a population over time.

☐ 4. Explain **allopatric speciation** in terms of migration, isolation, and **adaptation** leading to reproductive (genetic) isolation of gene pools.

☐ 5. Describe and explain mechanisms of **reproductive isolation**, distinguishing between **prezygotic** and **postzygotic** reproductive isolating mechanisms.

☐ 6. Describe and explain **sympatric speciation**. Discuss the role of **polyploidy** in **instant speciation** events and in the development of some crops.

☐ 7. Justify your selection of data and examples to answer questions relating to **reproductive isolation** and **speciation**.

Patterns of Evolution (1.C.1) pages 308-309, 324, 329-340

☐ 8. Recognize patterns of species formation: **phyletic gradualism**, **coevolution**, **divergent evolution**, **adaptive radiation**. Analyze data related to speciation and extinction throughout Earth's history.

☐ 9. Describe **convergent evolution**. Explain how **analogous structures** (homoplasies) may arise through convergence. Distinguish between **homologies**, which indicate relatedness, and homoplasies, which do not.

☐ 10. Describe examples of **coevolution**. Evaluate the evidence for coevolution in species with close ecological relationships.

☐ 11. Explain how and why speciation rates vary. Distinguish between **punctuated equilibrium** and **gradualism** models for the pace of evolutionary change and evaluate the evidence for each model.

☐ 12. Describe the role of **extinction** in evolution. Recognize that extinction rates are rapid at times of ecological stress and relate this to the five major **mass extinctions** in Earth's history and in the current wave of extinctions.

Evolution Today (1.C.3) pages 325-328

☐ 13. Recall the evidence supporting the occurrence of evolution in all species.

☐ 14. Describe examples to support the current evolution of populations. Examples could include the evolution of chemical resistance in pathogens or observed phenotypic change in a population (e.g. the evolution of Galápagos finches).

Periodicals:
Listings for this
chapter are on page 384

Weblinks:
www.thebiozone.com/
weblink/AP1-3114.html

BIOZONE APP:
Student Review Series
Evolution

Small Flies and Giant Buttercups

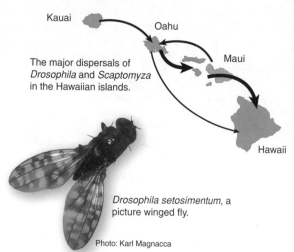

Kauai

Oahu

Maui

The major dispersals of *Drosophila* and *Scaptomyza* in the Hawaiian islands.

Hawaii

Drosophila setosimentum, a picture winged fly.

Photo: Karl Magnacca

Drosophilidae (commonly known as fruit flies) are a group of small flies found almost everywhere in the world. Two genera, *Drosophila* and *Scaptomyza*, are found in the Hawaiian islands and between them there are more than 800 species present on a land area of just 16,500 km^2; it is one of the densest concentrations of related species found anywhere. The flies range from 1.5 mm to 20 mm in length and display a startling range of wing forms and patterns, body shapes and colors, and head and leg shapes.

The diverse array of species and characteristics has made these flies the subject of much evolutionary and genetic research. Genetic analyses show that they are all related to a single species that may have arrived on the islands around 8 million years ago and diversified to exploit a range of unoccupied niches. Older species appear on the older islands and more recent species appear as one moves from the oldest to the newest islands. Such evidence points to numerous colonization events as new islands emerged from the sea. The volcanic nature of the islands means that newly isolated environments are a frequent occurrence. For example, forested areas may become divided by lava flows, so that flies in one region diverge rapidly from flies in another just tens of meters away. One such species is *D. silvestris*. Males have a series of hairs on their forelegs, which they brush against females during courtship. Males in the northeastern part of the island on which they are found have many more of these hairs than the males on the southwestern side of the island. While still the same species, the two demes are already displaying structural and behavioral isolation. Behavioral isolation is clearly an important phenomenon in drosophilid speciation. A second species, *D. heteroneura*, is closely related to *D. silvestris* and the two species live sympatrically. Although hybrid offspring are fully viable, hybridization rarely occurs because male courtship displays are very different.

Photo: Velela

The buttercups (*Ranunculus*) in alpine New Zealand are some of the largest in the world and are also the product of repeated speciation events. There are 14 species of *Ranunculus* in New Zealand; more than in the whole of North and South America combined. They occupy five distinct habitats ranging from snowfields and scree slopes to bogs. Genetic studies have shown that this diversity is the result of numerous isolation events following the growth and recession of glaciers. As the glaciers retreat, alpine habitat becomes restricted and populations are isolated at the tops of mountains. This restricts gene flow and provides the environment for species divergence. When the glaciers expand again, the extent of the alpine habitat increases, allowing isolated populations to come in contact and closely related species to hybridize.

1. Explain why so many drosophilidae are present in Hawaii: _____

_____

_____

_____

2. Explain why these flies are of interest: _____

_____

_____

3. Describe the relationship between the age of the islands and the age of the fly species: _____

_____

_____

_____

4. Explain why New Zealand has so many alpine buttercups: _____

_____

_____

_____

_____

_____

Related activities: Allopatric Speciation

Periodicals:

Evolution in New Zealand

© BIOZONE International 2012
ISBN: 978-1-927173-11-4
Photocopying Prohibited

Patterns of Evolution

The diversification of an ancestral group into two or more species in different habitats is called **divergent evolution**. This process is shown below, where two species have diverged from a **common ancestor**. Note that another species budded off, only to become extinct. Divergence is common in evolution. When divergent evolution involves the formation of a large number of species to occupy different niches, this is called an **adaptive radiation**. The example below (right) describes the radiation of the mammals that occurred after the extinction of the dinosaurs; an event that made niches available. Note that the evolution of species may not necessarily involve branching: a species may accumulate genetic changes that, over time, result in the emergence of what can be recognised as a different species. This is known as **sequential evolution** (also called **phyletic gradualism** or anagenesis).

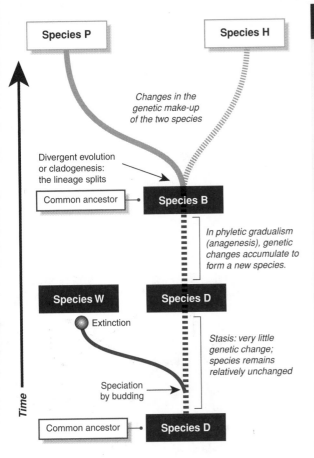

Species P — **Species H**

Changes in the genetic make-up of the two species

Divergent evolution or cladogenesis: the lineage splits

Common ancestor — **Species B**

In phyletic gradualism (anagenesis), genetic changes accumulate to form a new species.

Species W — **Species D**

Extinction

Stasis: very little genetic change; species remains relatively unchanged

Speciation by budding

Common ancestor — **Species D**

Time

Mammalian Adaptive Radiation

Marine predator niche

Arboreal herbivore niche

Underground herbivore niche

Terrestrial predator niche

Freshwater predator niche

Browsing/ grazing niche

Flying predator/ frugivore niche

Megazostrodon: one of the first mammals

The earliest true mammals evolved about 195 million years ago, long before they underwent their major adaptive radiation some 65-50 million years ago. These ancestors to the modern forms were very small (12 cm), many were nocturnal and fed on insects and other invertebrate prey. *Megazostrodon* (above) is a typical example. This shrew-like animal is known from fossil remains in South Africa and first appeared in the Early Jurassic period (about 195 million years ago).

It was climatic change as well as the extinction of the dinosaurs (and their related forms) that suddenly left many niches vacant for exploitation by such adaptable 'generalists'. All modern mammal orders developed very quickly and early.

1. In the hypothetical example of divergent evolution illustrated above, left:

(a) Describe the type of evolution that produced species B from species D: *divergent evolution*

(b) Describe the type of evolution that produced species P and H from species B: *sequential evolution*

(c) Name all species that evolved from: **Common ancestor D:** *B;W* **Common ancestor B:** *P;H*

(d) Explain why species B, P, and H all possess a physical trait not found in species D or W: *W diverged by speciation by budding*

2. (a) Explain the distinction between **divergence** and **adaptive radiation**: *In adaptive radiation involves the formation of a large number of species to occupy different niches*

(b) Discuss the differences between **sequential evolution** and **divergent evolution**: *Sequential evolution involves branching a species may accumulate genetic changes that over time result in the emergence of what can be recognised as a different species.*

© BIOZONE International 2012
ISBN: 978-1-927173-11-4
Photocopying Prohibited

Periodicals:
Dinosaurs take wing ,
Evolution: five big questions

Related activities: *Adaptive Radiation in Mammals*
Weblinks: *Macroevolution*

A 2

The Biological Species Concept

The **species** is the basic unit of taxonomy. A **biological species** is defined as a group of organisms capable of interbreeding to produce fertile offspring. However, there are some difficulties in applying the biological species concept (BSC) in practice. Morphologically identical but reproductively isolated cryptic species and closely related species that interbreed to produce fertile hybrids (e.g. *Canis* species), indicate that the boundaries of a species gene pool can be unclear. The BSC is also more successfully applied to animals than to plants. Plants hybridize easily and can reproduce vegetatively. For some, e.g. cotton and rice, F_1 hybrids are fertile but hybrid breakdown in subsequent generations stops hybrids proliferating and maintains the parental types in the wild. In addition, the BSC cannot be applied to asexually reproducing or extinct organisms. Increasingly, biologists are using DNA analyses to clarify relationships between the related populations that we regard as one species.

Distribution of *Canis* species

The global distribution of most species of *Canis* (dogs and wolves) is shown on the map, right. The gray wolf inhabits the forests of North America, northern Europe, and Siberia. The red wolf and Mexican wolf (original distributions shown) were once distributed more widely, but are now extinct in the wild except for reintroductions. In contrast, the coyote has expanded its original range and is now found throughout North and Central America. The range of the three jackal species overlap in the open savannah of eastern Africa. The dingo is distributed throughout the Australian continent.

Distribution of the domesticated dog is global as a result of the spread of human culture. The dog has been able to interbreed with all other members of the genus listed here to form fertile hybrids. Contrast this with members of the horse family, in which hybrid offspring are viable but sterile.

Interbreeding between *Canis* species

The ability of many *Canis* species to interbreed to produce fertile hybrids illustrates one of the problems with the traditional species concept. Red wolves, gray wolves, Mexican wolves, and coyotes will all interbreed to produce fertile offspring. Red wolves are very rare, and it is possible that hybridization with coyotes has been a factor in their decline. By contrast, no interbreeding occurs between the three distinct species of jackal, even though their ranges overlap in the Serengeti of eastern Africa. These animals are highly territorial, but simply ignore members of the other jackal species.

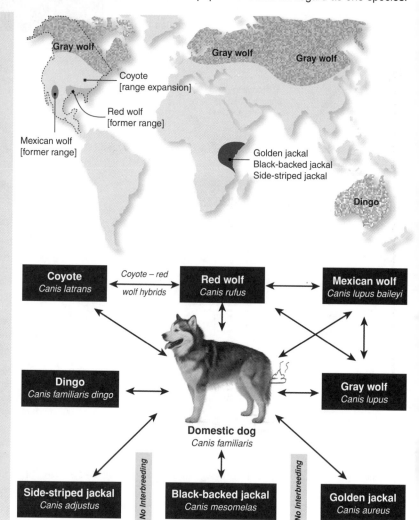

1. Describe the type of barrier that prevents the three species of jackal from interbreeding: _____

_____

2. Describe the factor that has prevented the dingo from interbreeding with other *Canis* species (apart from the dog):

_____

3. Describe a possible contributing factor to the occurrence of interbreeding between the coyote and red wolf:

_____

4. The gray wolf is a widely distributed species. Explain why the North American population is considered to be part of the same species as the northern European and Siberian populations:

_____

_____

5. Explain some of the limitations of using the biological species concept to assign species: _____

_____

_____

Related activities: The Phylogenetic Species Concept
Weblinks: Species and Speciation

Periodicals:
Species and species
formation

© BIOZONE International 2012
ISBN: 978-1-927173-11-4
Photocopying Prohibited

The Phylogenetic Species Concept

Although the biological species concept is useful, there are many situations in which it is difficult to apply, e.g. for asexual populations (including bacteria) or extinct organisms. In such situations, the phylogenetic species concept (PSC) can be more useful. It not reliant on the criterion of successful interbreeding and can be applied to asexually or sexually reproducing organisms, and to extinct organisms. Phylogenetic species are defined on the basis of their shared evolutionary ancestry, which is determined on the basis of **shared derived characteristics**, which may be morphological, especially for higher taxonomic ranks, or biochemical (e.g. DNA differences). The PSC defines a species as the smallest group that all share a derived character state. It is widely applicable in paleontology because biologists can compare both living and extinct organisms. While the PSC solves some difficulties, it creates others. It does not apply well to morphologically different species that are connected by gene flow. Similarly, the ability to distinguish genetically distinct but morphologically identical cryptic species on the basis of DNA analyses can lead to a proliferation of extant species that is not helpful in establishing a phylogeny.

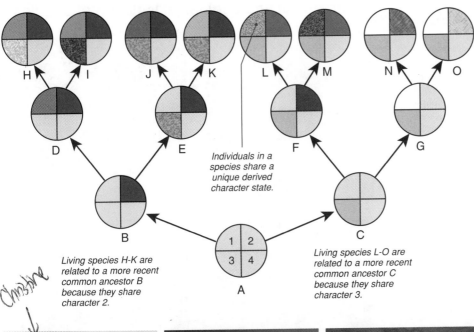

Individuals in a species share a unique derived character state.

Living species H-K are related to a more recent common ancestor B because they share character 2.

Living species L-O are related to a more recent common ancestor C because they share character 3.

Christine

This simplified phylogenetic tree traces four characters among 15 species (8 present and 7 ancestral). The 8 modern species (species H-O) share a character (4) derived from a distant common ancestor (A). Although the primitive character unites all 8 species, the branching of the tree is based on characters derived from the ancestral ones. Classification on the basis of shared derived characters defines the species as the smallest group diagnosable by a unique combination of characters. If large numbers of characters are included in the analysis, it is easy to see how this method results in a proliferation of species that may or may not be meaningful. Under the PSC model, there are no subspecies; either a population is a phylogenetic species or it is not taxonomically distinguishable.

Tree sparrows (*P. montanus*) are ~10% smaller than the similar house sparrow but the two species hybridize freely.

House sparrows (*P. domesticus*) are widespread with many intermediate "subspecies" of unknown status.

Mallards are infamous for their ability to hybridize freely with a large number of other duck "species".

True sparrows all belong to the genus *Passer*. There are a large number of species distinguished on the basis of song, plumage, and size. A vestigial dorsal outer primary feather and an extra bone in the tongue are ancestral characters. Many populations are not good biological species in that they hybridize freely to produce fertile off-spring. A similar situation exists within the genus *Anas* of dabbling ducks (which includes the mallards). Many birds are best described using the PSC rather than the BSC.

1. (a) Explain the basis by which species are assigned under the PSC: _____

_____

_____

(b) Describe one problem with the use of the PSC: _____

_____

(c) Describe situations where the use of the PSC might be more appropriate than the BSC: _____

_____

_____

2. Suggest how genetic techniques could be used to elucidate the phylogeny of a cluster of related phylogenetic species:

_____

_____

Periodicals:
What is a species?

Related activities: Isolating Mechanisms
Weblinks: Species and Speciation

Isolation and Species Formation

Isolating mechanisms are barriers to successful interbreeding between species. Reproductive isolation is fundamental to the **biological species concept**, which defines a species by its inability to breed with other species to produce fertile offspring. Prezygotic isolating mechanisms act before fertilization occurs, preventing species ever mating, whereas postzygotic barriers take effect after fertilization. **Geographical barriers** are not regarded as reproductive isolating mechanisms because they are not part of the species' biology, although they are often a necessary precursor to reproductive isolation in sexually reproducing populations. Ecological isolating mechanisms are those that isolate gene pools on the basis of ecological preferences, e.g habitat selection. Although ecological and geographical isolation are sometimes confused, they are quite distinct, as ecological isolation involves a component of the species biology. Similarly, the **temporal isolation** of species, through differences in the timing of important life cycle events, effectively prevents potentially interbreeding species from successfully reproducing.

Geographical Isolation

Geographical isolation describes the isolation of a species population (gene pool) by some kind of physical barrier, e.g. mountain range, water body, isthmus, desert, or ice sheet. Geographical isolation is a frequent first step in the subsequent reproductive isolation of a species. For example, geologic changes to the lake basins have been instrumental in the subsequent proliferation of cichlid fish species in the rift lakes of East Africa (right). Similarly, many Galápagos Island species (e.g. iguanas, finches) are now quite distinct from the Central and South American species from which they arose after isolation from the mainland.

Ecological (Habitat) Isolation

Ecological isolation describes the existence of a **prezygotic reproductive barrier** between two species (or sub-species) as a result of them occupying or breeding in different habitats within the same general geographical area. Ecological isolation includes small scale differences (e.g. ground or tree dwelling) and broad differences (e.g. desert vs grasslands). The red-browed and brown **treecreepers** (*Climacteris* spp.) are sympatric in south-eastern Australia and both species feed largely on ants. However the brown spends most of its time foraging on the ground or on fallen logs while the red-browed forages almost entirely in the trees.

Ecological isolation often follows geographical isolation, but in many cases the geographical barriers may remain in part. For example, five species of **antelope squirrels** occupy different habitat ranges throughout the southwestern United States and northern Mexico, a region divided in part by the Grand Canyon. The white tailed antelope squirrel is widely distributed in desert areas to the north and south of the canyon, while the smaller Harris' antelope squirrel has a much more limited range only to the south in southern Arizona. The Grand Canyon still functions as a barrier to dispersal but the species are now ecologically isolated as well.

Geographical and Ecological Isolation of Species

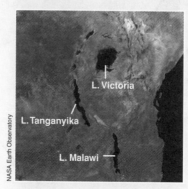

Malawi cichlid species

L. Victoria
L. Tanganyika
L. Malawi

Red-browed treecreeper

Brown treecreeper

White-tailed antelope squirrel

The Grand Canyon - a massive rift in the Colorado Plateau

Harris' antelope squirrel

1. Describe the role of isolating mechanisms in maintaining the integrity of a species: _Create differene_ _In the timing of important life cycle events, effectvely prevents potentionally interbreeding species from successfully reprodue_

2. (a) Why is geographical isolation not regarded as a reproductive isolating mechanism? _The isolateon of the genepool by some kind of physical barror_

   (b) Explain why, despite this, it often precedes reproductive isolation: _Species are unable to reproduce because their separated_

3. Distinguish between geographical and ecological isolation: _ecological isolation results from Species ocupying or breeding in different habitats withih the same general area_

Related activities: Reproductive Isolation, Speciation in Australia

Periodicals: Cichlids of the Rift Lakes

© BIOZONE International 2012
ISBN: 978-1-927173-11-4
Photocopying Prohibited

Reproductive Isolation

Reproductive isolation prevents interbreeding (and therefore gene flow) between species. Any factor that impedes two species from producing viable, fertile hybrids contributes to reproductive isolation. Single barriers may not completely stop gene flow, so most species commonly have more than one type of barrier.

Single barriers to reproduction (including geographical barriers) often precede the development of a suite of reproductive isolating mechanisms (RIMs). Most operate before fertilization (prezygotic RIMs) with postzyotic RIMs being important in preventing offspring between closely related species.

Prezygotic Isolating Mechanisms

Temporal Isolation

Individuals from different species do not mate because they are active during different times of the day, or in different seasons. Plants flower at different times of the year or even at different times of the day to avoid hybridization (e.g. members of the orchid genus *Dendrobium*, which occupy the same location and flower on different days). Closely related animal species may have quite different breeding seasons or periods of emergence. **Periodical cicadas** (right) of the genus *Magicicada* are so named because members of each species in a particular region are developmentally synchronized, despite very long life cycles. Once their underground period of development (13 or 17 years depending on the species) is over, the entire population emerges at much the same time to breed.

Gamete Isolation

The gametes from different species are often incompatible, so even if they meet they do not survive. For animals where fertilization is internal, the sperm may not survive in the reproductive tract of another species. If the sperm does survive and reach the ovum, chemical differences in the gametes prevent fertilization. Gamete isolation is particularly important in aquatic environments where the gametes are released into the water and fertilized externally, such as in reproduction in frogs. Chemical recognition is also used by flowering plants to recognize pollen from the same species.

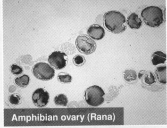
Amphibian ovary (Rana)

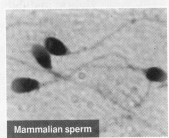
Mammalian sperm

Behavioral (ethological) Isolation

Behavioral isolation operates through differences in species courtship behaviors. Courtship is a necessary prelude to mating in many species and courtship behaviors are species specific. Mates of the same species are attracted with distinctive, usually ritualized, dances, vocalizations, and body language. Because they are not easily misinterpreted, the courtship behaviors of one species will be unrecognized and ignored by individuals of another species. Birds exhibit a remarkable range of courtship displays. The use of song is widespread but ritualized movements, including nest building, are also common. For example, the elaborate courtship bowers of bowerbirds are well known, and Galápagos frigatebirds have an elaborate display in which they inflate a bright red gular pouch (right). Amongst insects, empid flies have some of the most elaborate courtship displays. They are aggressive hunters, so ritualized behavior involving presentation of a prey item facilitates mating. The sexual organs of the flies are also like a lock and key, providing mechanical reproductive isolation as well (see below).

Male frigatebird courtship display

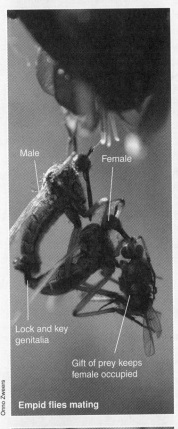
Male
Female
Lock and key genitalia
Gift of prey keeps female occupied
Empid flies mating

Male tree frog calling

Mechanical (morphological) Isolation

Structural differences (incompatibility) in the anatomy of reproductive organs prevents sperm transfer between individuals of different species. This is an important isolating mechanism preventing breeding between closely related species of arthropods. Many flowering plants have coevolved with their animal pollinators and have flower structures to allow only that insect access. Structural differences in the flowers and pollen of different plant species prevents cross breeding because pollen transfer is restricted to specific pollinators and the pollen itself must be species compatible.

Wing beating in male sage grouse

Damselflies mating

Complex flowers in orchids

Periodicals:
Listen, we're different

Related activities: Isolation and Species Formation

Postzygotic Isolating Mechanisms

Hybrid sterility

Even if two species mate and produce hybrid offspring that are vigorous, the species are still reproductively isolated if the hybrids are sterile (genes cannot flow from one species' gene pool to the other). Such cases are common among the horse family (such as the zebra and donkey shown on the right). One cause of this sterility is the failure of meiosis to produce normal gametes in the hybrid. This can occur if the chromosomes of the two parents are different in number or structure (see the "**zebronkey**" karyotype on the right). The **mule**, a cross between a donkey stallion and a horse mare, is also an example of **hybrid vigor** (they are robust) as well as **hybrid sterility**. Female mules sometimes produce viable eggs but males are infertile.

Zebra stallion (2N = 44) **X** Donkey jenny (2N = 62)

Karyotype of '**Zebronkey**' offspring (2N = 53)

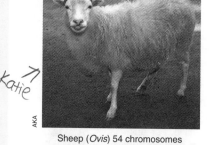

Chromosomes contributed by zebra stallion · Y

Chromosomes contributed by donkey jenny · X

Hybrid inviability

Mating between individuals of two species may produce a zygote, but genetic incompatibility may stop development of the zygote. Fertilized eggs often fail to divide because of mis-matched chromosome numbers from each gamete. Very occasionally, the hybrid zygote will complete embryonic development but will not survive for long. For example, although sheep and goats seem similar and can be mated together, they belong to different genera. Any offspring of a sheep-goat pairing is generally stillborn.

Sheep (*Ovis*) 54 chromosomes

Goat (*Capra*) 60 chromosomes

Hybrid breakdown

Hybrid breakdown is common feature of some plant hybrids. The first generation (F₁) may be fertile, but the second generation (F₂) are infertile or inviable. Examples include hybrids between species of cotton (near right), species of *Populus*, and strains of the cultivated rice *Oryza* (far right).

1. In the following examples, classify the reproductive isolating mechanism as either **prezygotic** or **postzygotic** and describe the mechanisms by which the isolation is achieved (e.g. structural isolation, hybrid sterility, etc.):

   (a) Some different cotton species can produce fertile hybrids, but breakdown of the hybrid occurs in the next generation when the offspring of the hybrid die in their seeds or grow into defective plants:

   Prezygotic / ~~postzygotic~~ (delete one)   Mechanism of isolation: ___hybrid breakdown___

   (b) Many plants have unique arrangements of their floral parts that stop transfer of pollen between plants:

   ~~Prezygotic~~ / postzygotic (delete one)   Mechanism of isolation: ___mechanical___

   (c) Two skunk species do not mate despite having habitats that overlap because they mate at different times of the year:

   ~~Prezygotic~~ / postzygotic (delete one)   Mechanism of isolation: ___temporal___

   (d) Several species of the frog genus *Rana* live in the same regions and habitats, where they may occasionally hybridize. The hybrids generally do not complete development, and those that do are weak and do not survive long:

   Prezygotic / ~~postzygotic~~ (delete one)   Mechanism of isolation: ___hybrid inviability___

2. Postzygotic isolating mechanisms are said to reinforce prezygotic ones. Explain why this is the case:

   ___If a prezygotic mechanism doesn't stop the breeding of a new species than a postzygotic mechanism will___

Are Ring Species Real?

A **ring species** is a connected series of closely related populations, distributed around a geographical barrier, in which the adjacent populations in the ring are able to interbreed, but those at the extremes of the ring are reproductively isolated. The ring species concept was proposed by Ernst Mayr in 1942 to account for the circumpolar distribution of species of herring gulls (*Larus* species). The idea of a ring species is attractive to biologists because it appears to show speciation in action (i.e.

incipient species). However, such examples are rare, and rigorous analysis of supposed ring species, such as the herring gull complex, have shown that they do not meet all the necessary criteria to be ring species as defined. Although ring species are rare, the concept is still helpful because it can allow us to reconstruct the divergence of populations from an ancestral species. Ring species also provide evidence that speciation can occur without complete geographical isolation.

What is a Ring Species?

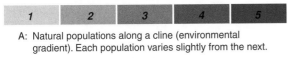

A: Natural populations along a cline (environmental gradient). Each population varies slightly from the next.

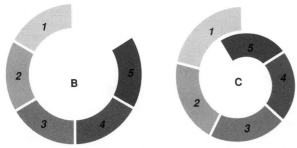

The variation in populations may occur in a geographical ring, e.g. around a continental shoreline (B). Adjacent populations in the cline can interbreed. If the ring closes (C), the populations at the extremes of the ring may meet but are too different to interbreed.

Criteria for a Ring Species

▶ Ring species develop from a single ancestral population with isolation by distance.

▶ They show expansion of their range around a geographic barrier, such a mountain range or desert.

▶ Adjacent populations in the ring can interbreed to produce fertile offspring (gene flow).

▶ The terminal populations are reproductively isolated.

The circumpolar distribution of *Larus* subspecies (still often cited in many texts) inspired Mayr to propose the ring species hypothesis. However, mtDNA studies (Liebers *et al.* 2004) have indicated that there were two ancestral gull populations (not one) and most of Mayr's subspecies in fact deserve species status. What is more, the *Larus* complex includes several species, excluded by Mayr, whose taxonomy is unclear. Ring species do appear to be a very rare phenomenon if they exist at all. In contrast, cryptic (hidden) species, which are morphologically identical but behave as (reproductively isolated) true species, appear to be quite common.

The herring gull (front) and black-backed gull (rear) do not interbreed at the ends of the circumpolar ring where they coexist. However, genetic analyses do not support a ring species.

Populations of *Ensatina* in the USA occupy a ring around California's Central Valley. While they show some of the characteristics of ring species, many of the adjacent populations are in fact genetically isolated and do not interbreed. What is more, the yellow-eyed *Ensatina* (above) has evolved to be a mimic of the toxic California newt, and this has probably driven its genetic isolation from adjacent *Ensatina* populations.

Greenish warbler populations occupy a ring around the Tibetan Plateau. Eastern and western populations meet in Siberia but do not interbreed. Analyses support their being a ring species.

1. Why is the phenomenon of ring species interesting to evolutionary biologists? *The adjacent populations in the ring are able to interbreed but those at the extremes of the ring are reproductively isolated.*

2. Discuss how modern genetic analyses are changing the way we view species and determine species status: *It shows how speciation can occur without complete geographical isolation.*

3. What implications might the existence of ring species have for conservation: *Allows us to reconstruct the divergence of populations from an ancestral species.*

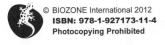

© BIOZONE International 2012
ISBN: 978-1-927173-11-4
Photocopying Prohibited

Related activities: Ring Species: The Greenish Warbler

A 2

Ensatina in North America: Ring Species or Cryptic Species?

Ensatina eschscholtzii is a species of lungless salamander found on the west coast of North America from the Pacific North-west of the USA to Baja California in Mexico. *E. eschscholtzii* has long been considered a ring species, which probably expanded southwards from an ancestral population in Oregon along either side of California's Central Valley. However, molecular analyses are now indicating that the story of *Ensatina* is more complicated than first supposed. Geographically adjacent populations within the ring may be genetically isolated or comprise morphologically identical but genetically distinct **cryptic species**. Regardless of the conclusions drawn from the evidence (below), species such as *E. eschscholtzii* give us reason to reevaluate how we define species and quantify biodiversity.

Oregon Ensatina,
E. e. oregonensis

The ancestral Oregon population spread south along either side of California's Central Valley. Coastal and inland populations diverged.

Oregon

Inland populations occupy the Sierra Nevada range. This inland flank of the distribution is not geographically continuous (note the gap in the ring) and most of the *Ensatina* 'subspecies' are genetically isolated from geographically adjacent 'subspecies'. In addition, the inland populations include two (or more) morphologically undistinguishable 'cryptic' species.

Central Valley of California

Painted Ensatina,
E. e. picta

Sierra Nevada,
E. e. platensis

Yellow-eyed,
E. e. xanthoptica

Yellow blotched,
E. e. croceater

The yellow-eyed *Ensatina* has crossed the central valley to overlap in a narrow contact zone with the Sierra Nevada form. They occasionally interbreed to produce fertile offspring, but mostly the populations remain distinct.

Gap in the ring
(Mojave desert)

	Criteria for a ring species	Ensatina?
1	A single ancestral population with isolation by distance.	Yes
2	Range expansion around both sides of an area of inhospitable habitat.	Yes
3	Lack of gene flow at the terminus of the ring.	Yes
4	Continued gene flow around the rest of the ring.	Not entirely

Monterey Ensatina,
E. e. eschscholtzii

In southern California, the ranges of the coastal Monterey form and the inland large-blotched form overlap, but little or no gene flow occurs between them. If they interbreed, the hybrids are infertile or have extremely reduced fitness. Analysis of enzymes and DNA indicates that they are different species.

Large-blotched,
E. e. klauberi

4. The *Ensatina* species complex fulfils two of the three criteria necessary to define a ring species (table, above left) yet does not fit comfortably with Mayr's definition of a biological species. Describe the aspects of *Ensatina* that:

(a) Supports the idea that they are a single species: _Genetically identical but different phenotypes_

(b) Does not agree with the standard definition of a biological species: _Adjacent populations within the ring may be genetical isolated_

5. Yellow-eyed *Ensatina* is a mimic of the toxic California newt. What might this suggest about the selection pressures on this subspecies and their influence on the rate at which the population becomes genetically distinct?

This suggests that the selection pressure of the Yellow-eyed Ensatina is high and they become genetically distinct fast.

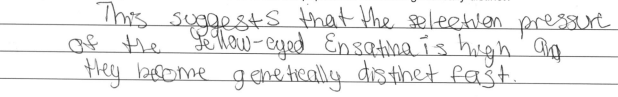

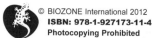

Ring Species: The Greenish Warbler

Greenish warblers (*Phylloscopus trochiloides*) are found in forests across much of northern and central Asia. They inhabit the ring of mountains surrounding the large area of desert which includes the Tibetan Plateau, and Taklamakan and Gobi deserts, and extends into Siberia. In Siberia, two distinct subspecies coexist and do not interbreed, but are apparently connected by gene flow around the Himalayas to the south. The greenish warblers may thus form a rare example of a ring species.

2 Populations spread both east and west along the Himalayas. Populations developed unique characteristics, but adjacent populations remained able to breed together.

3 East and west populations eventually rejoined in Siberia, but because of morphological, behavioral, and genetic differences they do not interbreed.

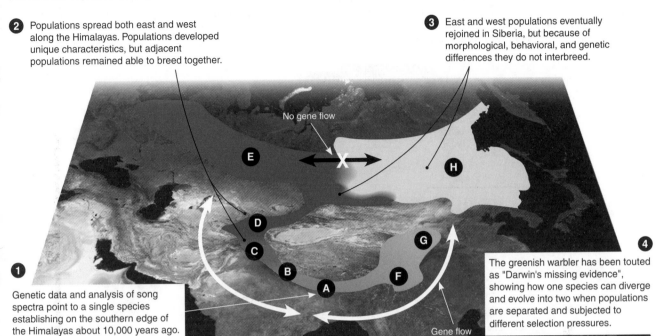

No gene flow

1 Genetic data and analysis of song spectra point to a single species establishing on the southern edge of the Himalayas about 10,000 years ago.

4 The greenish warbler has been touted as "Darwin's missing evidence", showing how one species can diverge and evolve into two when populations are separated and subjected to different selection pressures.

Gene flow

Song Spectra of the Greenish Warbler

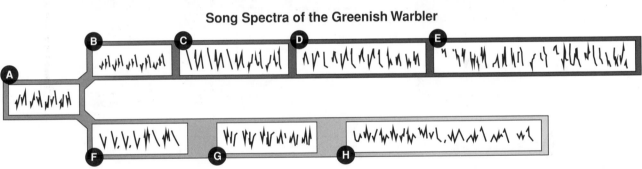

The two coexisting subspecies of greenish warblers can be distinguished by their songs and the number of bars on the wings. The warbler in western Siberia has one light bar across the top of the wing, while the warbler in eastern Siberia has two. Analysis of the songs around the ring show that all songs can be traced to the population labeled A above. Songs become progressively different moving east or west around the ring. The songs of the eastern warblers (H) and western warblers (E) in Siberia are so different that neither recognizes the other. Eastern and western forms have **subspecies status**.

JM Garg, Wikipedia CC 3.0

1. How do the eastern and western Siberian populations of greenish warblers differ? _____

_____

2. Explain how these differences occurred: _____

_____

_____

3. Explain why the greenish warbler has been touted as "evolution in action": _____

_____

_____

Related activities: Are Ring Species Real?

A 2

Stages in Species Development

The diagram below represents a possible sequence of genetic events involved in the origin of two new species from an ancestral population. As time progresses (from top to bottom of the diagram) the amount of genetic variation increases and each group becomes increasingly isolated from the other. The mechanisms that operate to keep the two gene pools isolated from one another may begin with **geographical barriers**. This may be followed by **prezygotic** mechanisms which protect the gene pool from unwanted dilution by genes from other pools. A longer period of isolation may lead to **postzygotic** mechanisms (see the pages on reproductive isolating mechanisms). As the two gene pools become increasingly isolated and different from each other, they are progressively labelled: population, race, and subspecies. Finally they attain the status of separate species.

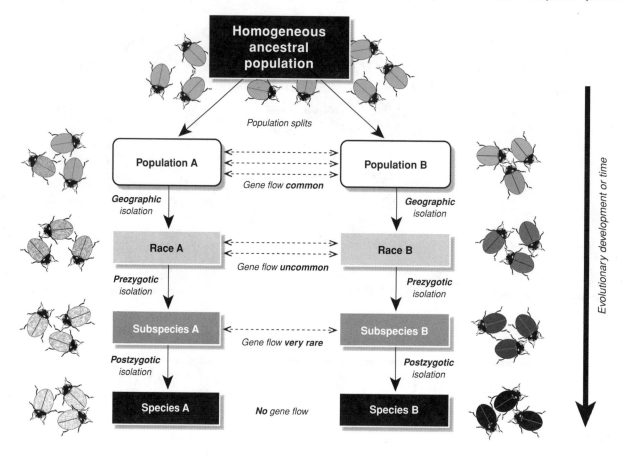

1. What happens to the amount of gene flow between two diverging populations as they gradually attain species status?

_____

2. Early human populations about 500,000 years ago were scattered across the continents of Africa, Europe, and Asia. This was a time of many regional variants, collectively called archaic *Homo sapiens* (or *Homo heidelbergensis*). The fossil skulls from different regions showed odd mixtures of characteristics, some modern and some considered 'primitive'. These regional populations are generally given the status of subspecies. Suggest reasons why gene flow may have been rare between these populations, but still occasionally took place:

_____

_____

_____

_____

3. In the USA, the species status of several duck species, including the black duck (*Anas rubripes*) and the mottled duck in Florida (*A. fulvigula*) is threatened by interbreeding with the now widespread and very adaptable mallard duck (*A. platyrhynchos*). Similar threatened extinction though hybridization has occurred in New Zealand and Australia, where the native Pacific duck species have been virtually eliminated as a result of interbreeding with the introduced mallard.

(a) Suggest why these hybrids threaten the species status of some native duck species: _____

_____

(b) What factor(s) may still deter interbreeding between native duck species and mallards:_____

_____

Related activities: Isolation and Species Formation, Reproductive Isolation

Allopatric Speciation

Allopatric speciation refers to speciation as a result of populations of a species becoming isolated from each other. It is probably the most common mechanism of speciation and has been important in regions where there have been cycles of geographical fragmentation. Such cycles can occur as the result of ice expansion and retreat during glacial and interglacial periods. Such events are also accompanied by sea level changes which can isolate populations within small geographical regions.

Stage 1: Moving into new environments

There are times when the range of a species expands for a variety of different reasons. A single population in a relatively homogeneous environment will move into new regions of their environment when they are subjected to intense competition (whether it is interspecific or intraspecific). The most severe form of competition is between members of the same species since they are competing for identical resources in the habitat. In the diagram on the right there is a 'parent population' of a single species with a common gene pool with regular 'gene flow' (theoretically any individual has access to all members of the opposite sex for mating purposes).

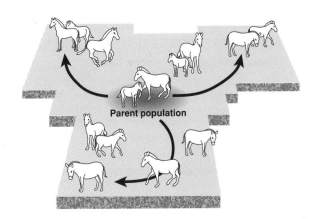

Parent population

Stage 2: Geographical isolation

Isolation of parts of the population may occur due to the formation of **physical barriers**. These barriers may cut off those parts of the population that are at the extremes of the species range and gene flow is prevented or rare. The rise and fall of the sea level has been particularly important in functioning as an isolating mechanism. Climatic change can leave 'islands' of habitat separated by large inhospitable zones that the species cannot traverse.

Example: In mountainous regions, alpine species are free to range widely over extensive habitat during cool climatic periods. During warmer periods, however, they may become isolated because their habitat is reduced to 'islands' of high ground surrounded by inhospitable lowland habitat.

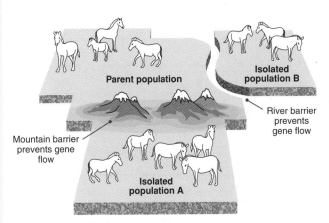

Parent population

Isolated population B

River barrier prevents gene flow

Mountain barrier prevents gene flow

Isolated population A

Stage 3: Different selection pressures

The isolated populations (A and B) may be subjected to quite different selection pressures. These will favor individuals with traits that suit each particular environment. For example, population A will be subjected to selection pressures that relate to drier conditions. This will favor those individuals with phenotypes (and therefore genotypes) that are better suited to dry conditions. They may for instance have a better ability to conserve water. This would result in improved health, allowing better disease resistance and greater reproductive performance (i.e. more of their offspring survive). Finally, as allele frequencies for certain genes change, the population takes on the status of a subspecies. Reproductive isolation is not yet established but the **subspecies** are significantly different genetically from other related populations.

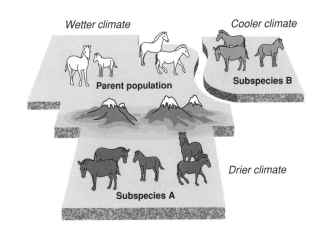

Wetter climate

Cooler climate

Parent population

Subspecies B

Drier climate

Subspecies A

Stage 4: Reproductive isolation

The separated populations (isolated subspecies) will often undergo changes in their genetic makeup as well as their behavior patterns. These ensure that the gene pool of each population remains isolated and 'undiluted' by genes from other populations, even if the two populations should be able to remix (due to the removal of the geographical barrier). Gene flow does not occur. The arrows (in the diagram to the right) indicate the zone of overlap between two species after the new Species B has moved back into the range inhabited by the parent population. Closely-related species whose distribution overlaps are said to be **sympatric species**. Those that remain geographically isolated are called a**llopatric species**.

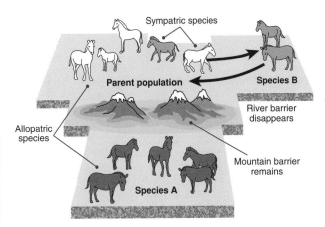

Sympatric species

Parent population

Species B

River barrier disappears

Allopatric species

Mountain barrier remains

Species A

Related activities: Stages in Species Development, Reproductive Isolation
Weblinks: Allopatric Speciation, Species and Speciation

RA 2

1. Explain why some animals, given the opportunity, move into new environments: _____

_____

_____

2. (a) Plants are unable to move. How might plants disperse to new environments? _____

_____

(b) Describe the amount of **gene flow** within the parent population prior to and during this range expansion:

_____

_____

_____

3. What **process** causes the formation of new **mountain ranges**?_____

4. What event can cause large changes in **sea level** (up to 200 meters)? _____

5. Describe six **physical barriers** that could isolate different parts of the same population: _____

(a) _____   (d) _____

(b) _____   (e) _____

(c) _____   (f) _____

6. Describe the effect of physical barriers on **gene flow**: _____

_____

_____

_____

7. (a) Describe four types of **selection pressure** that could affect a gene pool: _____

_____

_____

(b) Describe the possible effect of these selection pressures on the **allele frequencies** of an isolated gene pool:

_____

_____

_____

8. Describe two types of **prezygotic** and two types of **postzygotic** reproductive isolating mechanisms:

(a) Prezygotic: _____

_____

(b) Postzygotic: _____

_____

9. Distinguish between **allopatry** and **sympatry** in populations: _____

_____

_____

_____

_____

_____

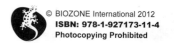

Sympatric Speciation

New species may be formed even where there is no separation of the gene pools by physical barriers. This is called **sympatric speciation**, and it is rarer than allopatric speciation, although it is not uncommon in plants which form **polyploids**. There are two situations where sympatric speciation is thought to occur. These are described below:

Speciation Through Niche Differentiation

Niche isolation

In a heterogeneous environment (one that is not the same everywhere), a population exists within a diverse collection of **microhabitats**. Some organisms prefer to occupy one particular type of 'microhabitat' most of the time, only rarely coming in contact with fellow organisms that prefer other microhabitats. Some organisms become so dependent on the resources offered by their particular microhabitat that they never meet up with their counterparts in different microhabitats.

Reproductive isolation

The individual groups that have remained genetically isolated because of their microhabitat preferences become reproductively isolated. They become new species, with subtle differences in behavior, structure, and physiology. Gene flow (via sexual reproduction) is limited to organisms that share a similar microhabitat preference (as shown in the diagram on the right).

Example: Some beetles prefer to find plants identical to the species they grew up on, when it is time for them to lay eggs. Individual beetles of the same species have different preferences.

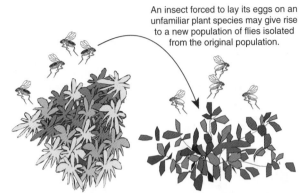

An insect forced to lay its eggs on an unfamiliar plant species may give rise to a new population of flies isolated from the original population.

Original host plant species   New host plant species

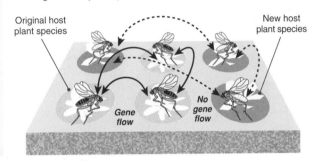

Original host plant species

New host plant species

Gene flow

No gene flow

Instant Speciation by Polyploidy

Polyploidy may result in the formation of a new species without isolation from the parent species. This event, occurring during meiosis, produces sudden reproductive isolation for the new group. Because the sex-determining mechanism is disturbed, animals are rarely able to achieve new species status this way (they are effectively sterile, e.g. tetraploid XXXX). Many plants, on the other hand, are able to reproduce vegetatively, or carry out self pollination. This ability to reproduce on their own enables such polyploid plants to produce a breeding population.

Speciation by allopolyploidy

This type of polyploidy usually arises from the doubling of chromosomes in a hybrid between two different species. The doubling often makes the hybrid fertile.

Examples: Modern wheat. Swedes are polyploid species formed from a hybrid between a type of cabbage and a type of turnip.

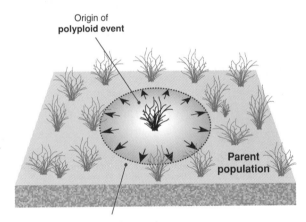

Origin of **polyploid event**

Parent population

New polyploid plant species spreads outwards through the existing parent population

---

1. Explain what is meant by **sympatric speciation** (do not confuse this with sympatric species):

_____

_____

2. Explain how **polyploidy** can result in the formation of a new species: _____

_____

_____

3. Identify an example of a species that has been formed by polyploidy: _____

4. Explain how **niche differentiation** can result in the formation of a new species: _____

_____

_____

Related activities: Breeding Modern Wheat
Weblinks: Sympatric Speciation

Speciation and Dispersal of the Camelidae

The camel family, Camelidae, consists of six modern-day species that have survived on three continents: Asia, Africa and South America. They are characterised by having only two functional toes, supported by expanded pads for walking on sand or snow. The slender snout bears a cleft upper lip. The recent distribution of the camel family is fragmented. Geophysical phenomena such as plate tectonics and the ice age cycles have controlled the extent of their distribution. South America, for example, was separated from North America until the end of the Pliocene, about 2.5 million years ago. Three general principles about the dispersal and distribution of land animals are:

- When very closely related animals (as shown by their anatomy) were present at the same time in widely separated parts of the world, it is highly probable that there was no barrier to their movement in one or both directions between the localities in the past.
- The most effective barrier to the movement of land animals (particularly mammals) was a sea between continents (as was caused by changing sea levels during the ice ages).
- The scattered distribution of modern species may be explained by the movement out of the area they originally occupied, or extinction in those regions between modern species.

Origin and Dispersal of the Camel Family

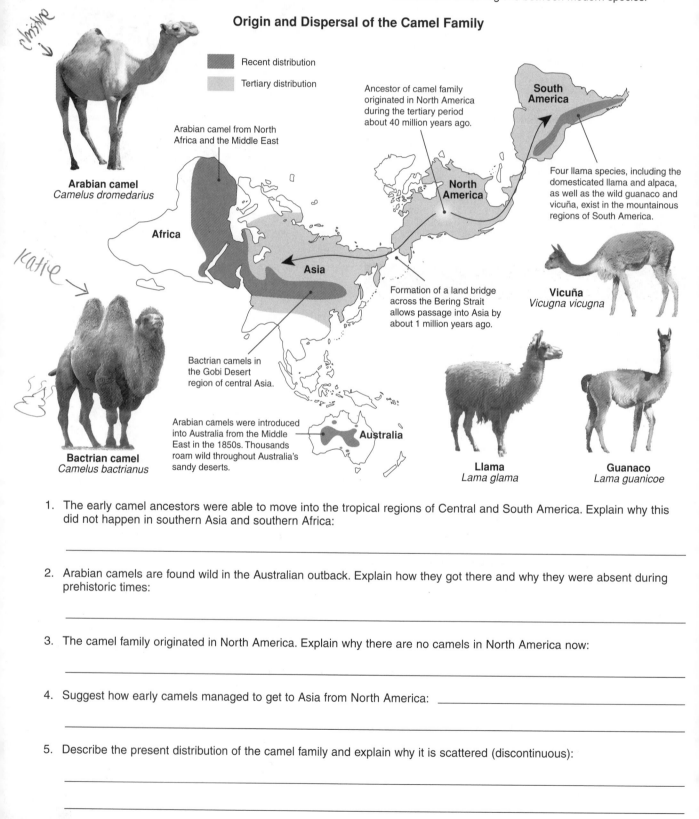

Recent distribution

Tertiary distribution

Ancestor of camel family originated in North America during the tertiary period about 40 million years ago.

Arabian camel from North Africa and the Middle East

Arabian camel
Camelus dromedarius

Africa

Asia

North America

South America

Four llama species, including the domesticated llama and alpaca, as well as the wild guanaco and vicuña, exist in the mountainous regions of South America.

Vicuña
Vicugna vicugna

Formation of a land bridge across the Bering Strait allows passage into Asia by about 1 million years ago.

Bactrian camels in the Gobi Desert region of central Asia.

Bactrian camel
Camelus bactrianus

Arabian camels were introduced into Australia from the Middle East in the 1850s. Thousands roam wild throughout Australia's sandy deserts.

Australia

Llama
Lama glama

Guanaco
Lama guanicoe

1. The early camel ancestors were able to move into the tropical regions of Central and South America. Explain why this did not happen in southern Asia and southern Africa:

_____

2. Arabian camels are found wild in the Australian outback. Explain how they got there and why they were absent during prehistoric times:

_____

3. The camel family originated in North America. Explain why there are no camels in North America now:

_____

4. Suggest how early camels managed to get to Asia from North America: _____

_____

5. Describe the present distribution of the camel family and explain why it is scattered (discontinuous):

_____

_____

Speciation of Australian Treecreepers

Speciation on continents occurs with geographical isolating mechanisms that are somewhat different to those on islands. In Australia, geographical barriers exist in the form of regions of inhospitable habitat. These create 'islands' of preferred habitat cut off from one another. Australian treecreeper birds (genus *Climacteris*) comprise several species and subspecies, each distinguishable by variations in the color patterns of their plumage. They probably evolved from a single **common ancestor**. Their distribution is restricted to savannah woodland

areas. A distinctive form is associated with each of the major woodland areas (shaded in different patterns on the map of Australia). Barriers to expansion of the distribution of these populations are indicated by a circle; each barrier comprises a district with unsuitable dry habitat. The distribution of two of the treecreeper species has undergone a secondary range expansion (see arrows), where they have extended their range beyond their region of origin, into new neighboring habitat.

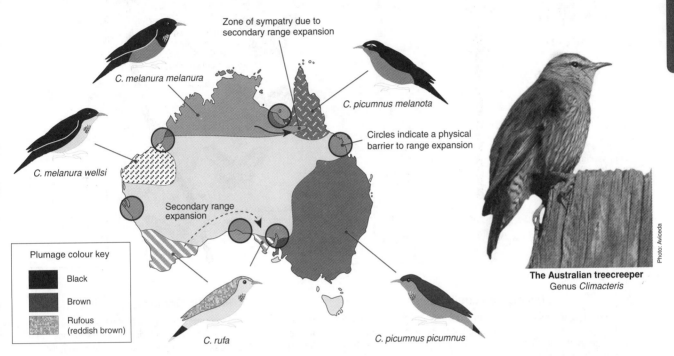

Zone of sympatry due to secondary range expansion

C. melanura melanura

C. picumnus melanota

C. melanura wellsi

Circles indicate a physical barrier to range expansion

Secondary range expansion

Plumage colour key

- ■ Black
- ■ Brown
- ▨ Rufous (reddish brown)

C. rufa

C. picumnus picumnus

The Australian treecreeper
Genus *Climacteris*

Photo: Aviceda

1. (a) How many **species** are illustrated above? Explain your answer: _____

_____

   (b) Describe the distribution of these treecreeper populations in Australia: _____

_____

_____

2. Explain why there are no treecreeper populations in the central region of Australia: _____

_____

3. Two species in the NE of Australia, *C. melanura melanura* and *C. picumnus melanota*, exhibit sympatric distribution.

   (a) What is meant by the zone of sympatry in this context? _____

_____

   (b) What mechanisms are most likely to prevent interbreeding between these two species? _____

_____

4. What is meant by **secondary range expansion** in the two populations above: _____

_____

5. Describe the physical barriers that have prevented the neighboring populations from mixing (in all but one case):

_____

6. Predict a likely outcome to the distribution of these species, should the climate change to produce more coastal rainfall:

_____

Related activities: Isolation and Species Formation, Allopatric Speciation

The Rate of Evolutionary Change

The pace of evolution is much debated, with two basic models proposed: **phyletic gradualism** and **punctuated equilibrium**. Some scientists believe that both mechanisms may operate at different times and in different circumstances. Interpretations of the fossil record vary depending on the time scales involved. During

its formative millenia, a species may have accumulated changes gradually (e.g. over 50,000 years). If that species survives for 5 million years, the evolution of its defining characteristics would have occurred in just 1% of its time on Earth. In the fossil record, the species would appear quite suddenly.

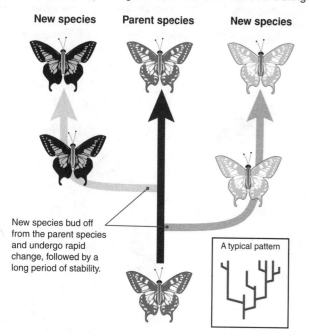

New species **Parent species** **New species**

New species bud off from the parent species and undergo rapid change, followed by a long period of stability.

A typical pattern

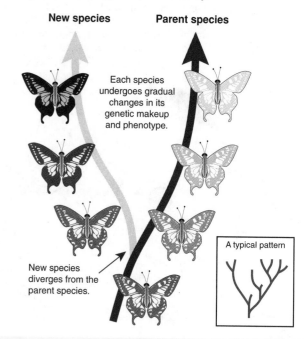

New species **Parent species**

Each species undergoes gradual changes in its genetic makeup and phenotype.

New species diverges from the parent species.

A typical pattern

Punctuated Equilibrium

There is abundant evidence in the fossil record that, instead of gradual change, species stay much the same for long periods of time (called stasis). These periods are punctuated by short bursts of evolution which produce new species quite rapidly. According to the punctuated equilibrium theory, most of a species' existence is spent in **stasis** and little time is spent in active evolutionary change. The stimulus for evolution occurs when some crucial aspect of the environment changes.

Phyletic Gradualism

Phyletic gradualism assumes that populations slowly diverge by accumulating adaptive characteristics in response to different selective pressures. If species evolve by gradualism, there should be transitional forms seen in the fossil record, as is seen with the evolution of the horse. Trilobites, an extinct marine arthropod, are another group of animals that have exhibited gradualism. In a study in 1987 a researcher found that they changed gradually over a three million year period.

1. Suggest the kinds of environments that would support the following paces of evolutionary change:

   (a) Punctuated equilibrium: _The kind of environment that would support a punctuated equilibrium is a constant environment_

   (b) Gradualism: _The kind of environment that would support a punctuated equilibrium is an environment that doesn't undergo drastic chan_

2. In the fossil record of early human evolution, species tend to appear suddenly, then linger for often very extended periods before disappearing suddenly. There are few examples of smooth intergradations from one species to the next. Explain which of the above models best describes the rate of human evolution:

   Gradualism describes the rate of human evolution

3. Some species apparently show little evolutionary change over long periods of time (hundreds of millions of years).

   (a) Name two examples of such species: _humans,_

   (b) State the term given to this lack of evolutionary change: _Gradalism_

   (c) Suggest why such species have changed little over evolutionary time: _Their environment haven't gone under drastic change._

Related activities: Patterns of Evolution
Weblinks: The Big Issues in Evolution

Periodicals:
Evolution in the fast lane

© BIOZONE International 2012
ISBN: 978-1-927173-11-4
Photocopying Prohibited

Chloroquine Resistance in Protozoa

Chloroquine is an antimalarial drug discovered in 1934. It was first used clinically to prevent malaria in 1946. Chloroquine was widely used because it was cheap to produce, safe, and very effective. Chloroquine resistance in *Plasmodium falciparum* first appeared in the late 1950s, and the subsequent spread of resistance has significantly decreased chloroquine's effectiveness. The WHO regularly updates the global status of anti-malarial drug efficacy and drug resistance. Their 2010 report shows that when chloroquine is used as a monotherapy, it is still effective at preventing malaria in Central American countries (chloroquine resistance has not yet developed there). In 30 other countries, chloroquine failure rates ranged between 20 and 100%. In some regions, chloroquine used in combination with other anti-malarial drugs is still an effective treatment.

Global Spread of Chloroquine Resistance

1980s
1970s
1960s
1957
1960s
1980s
1959
1978
1970s
1960s
1980s

Areas of chloroquine resistance in *P. falciparum*.

Malaria in humans is caused by various species of *Plasmodium*, a protozoan parasite transmitted by *Anopheles* mosquitoes. The inexpensive antimalarial drug **chloroquine** was used successfully to treat malaria for many years, but its effectiveness has declined since resistance to the drug was first recorded in the 1950s. Chloroquine resistance has spread steadily (above) and now two of the four *Plasmodium* species, *P. falciparum* and *P. vivax*, are chloroquine-resistant. *P. falciparum* alone accounts for 80% of all human malarial infections and 90% of the deaths, so this rise in resistance is of global concern. New anti-malarial drugs have been developed, but are expensive and often have undesirable side effects. Resistance to even these newer drugs is already evident, especially in *P. falciparum*, although this species is currently still susceptible to artemisinin, a derivative of the medicinal herb *Artemisia annua*.

Recent studies have demonstrated a link between mutations in the chloroquine resistance transporter (PfCRT) gene, and resistance to chloroquine in *P. falciparum*. PfCRT is a membrane protein involved in drug and metabolite transport.

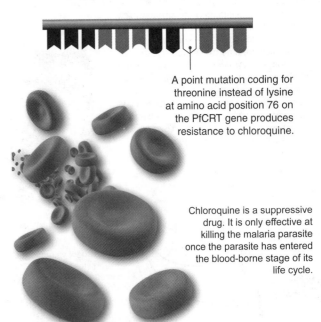

A point mutation coding for threonine instead of lysine at amino acid position 76 on the PfCRT gene produces resistance to chloroquine.

Chloroquine is a suppressive drug. It is only effective at killing the malaria parasite once the parasite has entered the blood-borne stage of its life cycle.

The use of chloroquine in many African countries was halted during the 1990s because resistance developed in *P. falciparum*. Recent studies in Malawi and Kenya have revealed a significant decrease in chloroquine resistance since the drug was withdrawn. There may be a significant fitness cost to the PfCRT mutants in the absence of anti-malaria drugs, leading to their decline in frequency once the selection pressure of the drugs is removed. This raises the possibility of re-introducing chloroquine as an anti-malarial treatment in the future.

1. Describe the benefits of using chloroquine to prevent malaria: _____

_____

_____

2. With reference to *Plasmodium falciparum*, explain how chloroquine resistance arises: _____

_____

_____

_____

3. Describe two strategies to reduce the spread of chloroquine resistance while still treating malaria:

(a) _____

_____

(b) _____

_____

© BIOZONE International 2012
ISBN: 978-1-927173-11-4
Photocopying Prohibited

Periodicals:
Beating the
bloodsuckers

Related activities: Malaria, Drug Resistance in HIV

RA 3

Drug Resistance in HIV

Although many diseases are treated very effectively with drugs, the emergence of drug resistant pathogens is increasingly undermining the ability to treat and control diseases such as HIV/AIDS. HIV's high mutation rate and short generation times contribute to the rapid spread of drug resistance. Rapid evolution in pathogens is exacerbated by the strong selection pressures created by the wide use and misuse of antiviral drugs, the poor quality of available drugs in some sectors of the population, and lack of education on drug use. The most successful treatment for several diseases, including HIV/AIDS, appears to be a multi-pronged attack using a cocktail of drugs to target the pathogen in several different ways.

Drug Resistance in HIV

Strains of drug-resistant HIV arise when the virus mutates during replication. Resistance may develop as a result of a single mutation, or through a step-wise accumulation of specific mutations. These mutations may alter drug binding capacity or increase viral fitness, or they may be naturally occurring polymorphisms (which occur in untreated patients). Drug resistance is likely to develop in patients who do not follow their treatment schedule closely, as the virus has an opportunity to adapt more readily to a "non-lethal" drug dose. The best practice for managing the HIV virus is to treat it with a cocktail of anti-retroviral drugs with different actions to minimize the number of viruses in the body. This minimizes the replication rate, and also the chance of a drug resistant mutation being produced.

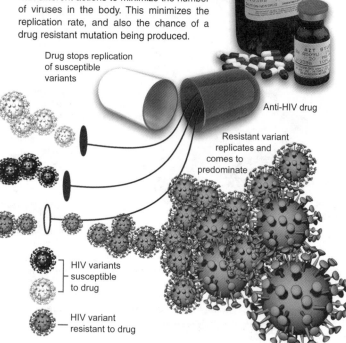

Drug stops replication of susceptible variants

Anti-HIV drug

Resistant variant replicates and comes to predominate

HIV variants susceptible to drug

HIV variant resistant to drug

Causes of Drug Resistance

- Poor drug compliance: patients stop taking their medication or do not follow the treatment as directed.

- Low levels of drug absorption: drug absorption is reduced if a patient has diarrhoea, is vomiting, or has an intestinal infection.

- Individual variation: the effectiveness of the body to absorb, distribute, metabolise, and eliminate a drug varies between patients.

- Toxicity: the side effects of the medication may make the patient very sick, so they stop taking it.

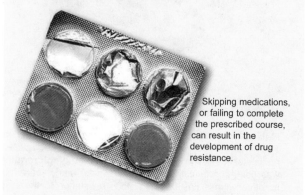

Skipping medications, or failing to complete the prescribed course, can result in the development of drug resistance.

Superinfection arises when a person already infected with HIV acquires a second strain of the virus. Superinfection increases the body's viral load, and can speed up disease progression or result in the patient acquiring a drug resistant HIV strain. The phenomenon of superinfection has implications for the development of a successful HIV vaccine because it shows that the body does not develop an immunological memory to HIV. This means that future HIV vaccines may be ineffective at preventing HIV infection.

1. Describe factors contributing to the rapid spread of drug resistance in pathogens: _____
_____
_____

2. With reference to HIV/AIDS, explain how drug resistance arises in a pathogen population: _____
_____
_____
_____
_____

3. Explain the implications of HIV superinfection on the development of a successful HIV vaccine: _____
_____
_____
_____

The Evolution of Darwin's Finches

The Galápagos Islands, off the west coast of Ecuador, comprise 16 main islands and six smaller islands. They are home to a unique range of organisms, including 13 species of finches, each of which has evolved from a single species of grassquit. After colonizing the islands, the grassquits underwent adaptive radiation in response to the availability of unexploited feeding niches. This radiation is most evident in the present beak shape of each species. The beaks are adapted for different purposes such as crushing seeds, pecking wood, or probing flowers for nectar. Current consensus recognizes ground finches, cactus finches, tree finches, and warbler finches. Between them, the 13 species of this endemic group fill the roles of seven different families of South American mainland birds. Darwin speculated that the finches shared a common ancestor and DNA analyses have confirmed this, showing that all 13 species evolved from a flock of about 30 birds arriving one million years ago.

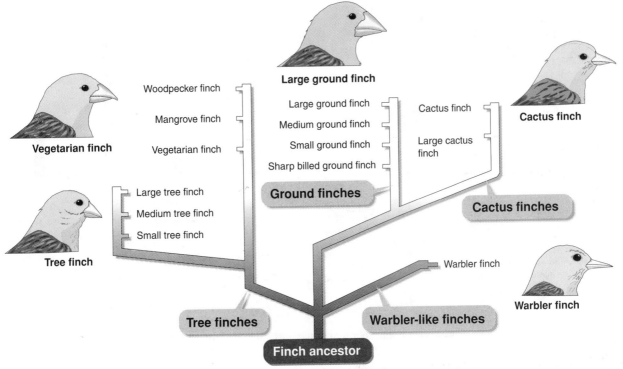

Vegetarian finch

Woodpecker finch
Mangrove finch
Vegetarian finch

Large ground finch

Large ground finch
Medium ground finch
Small ground finch
Sharp billed ground finch

Ground finches

Cactus finch

Large cactus finch

Cactus finches

Cactus finch

Tree finch

Large tree finch
Medium tree finch
Small tree finch

Tree finches

Warbler finch

Warbler-like finches

Warbler finch

Finch ancestor

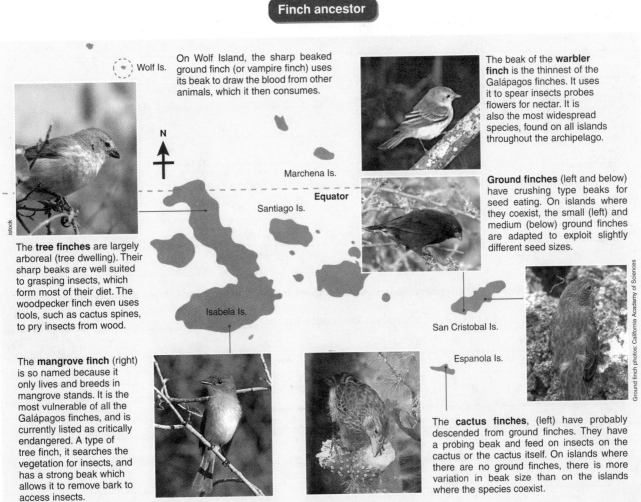

Wolf Is.

On Wolf Island, the sharp beaked ground finch (or vampire finch) uses its beak to draw the blood from other animals, which it then consumes.

The beak of the **warbler finch** is the thinnest of the Galápagos finches. It uses it to spear insects probes flowers for nectar. It is also the most widespread species, found on all islands throughout the archipelago.

N

Marchena Is.

Equator

Santiago Is.

Ground finches (left and below) have crushing type beaks for seed eating. On islands where they coexist, the small (left) and medium (below) ground finches are adapted to exploit slightly different seed sizes.

The **tree finches** are largely arboreal (tree dwelling). Their sharp beaks are well suited to grasping insects, which form most of their diet. The woodpecker finch even uses tools, such as cactus spines, to pry insects from wood.

Isabela Is.

San Cristobal Is.

Espanola Is.

The **mangrove finch** (right) is so named because it only lives and breeds in mangrove stands. It is the most vulnerable of all the Galápagos finches, and is currently listed as critically endangered. A type of tree finch, it searches the vegetation for insects, and has a strong beak which allows it to remove bark to access insects.

The **cactus finches**, (left) have probably descended from ground finches. They have a probing beak and feed on insects on the cactus or the cactus itself. On islands where there are no ground finches, there is more variation in beak size than on the islands where the species coexist.

Related activities: Adaptations and Fitness, Allopatric Speciation
Weblinks: Darwin's Finches

A 2

328

Adaptation in response to resource competition on bill size in small and medium ground finches

Abingdon, Bindloe, James, Jervis Islands

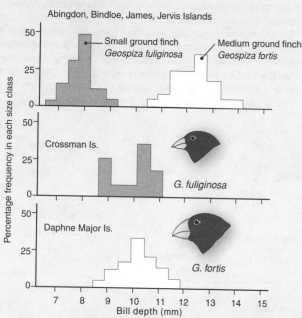

Small ground finch
Geospiza fuliginosa

Medium ground finch
Geospiza fortis

Crossman Is.

G. fuliginosa

Daphne Major Is.

G. fortis

Percentage frequency in each size class

Bill depth (mm)

Proportions in the diet of various seed sizes in three species of ground finch

Small ground finch
G. fuliginosa

Medium ground finch
G. fortis

Large ground finch
G. magnirostris

Daphne Major Island

Diet proportions

Seed depth (mm)

Two species of ground finch (*G. fuliginosa* and *G. fortis*) are found on a number of islands in the Galápagos. On islands where the species occur together, the bill sizes of the two species are quite different and they feed on different sized seeds, thus avoiding direct competition. On islands where each of these species occurs alone, and there is no competition, the bill sizes of both species move to an intermediate range.

Data based on an adaptation by Strickberger (2000)

Ground finches feed on seeds, but the upper limit of seed size they can handle is constrained by the bill size. Even though small seeds are accessible to all, the birds concentrate on the largest seeds available to them because these provide the most energy for the least handling effort. For example, the large ground finch can easily open smaller seeds, but concentrates on large seeds for their high energy rewards.

1. Describe the main factors that have contributed to the adaptive radiation of Darwin's finches: _____

_____

_____

_____

2. (a) What evidence is there to indicate that species of *Geospiza* compete for the same seed sizes?_____

_____

_____

(b) How have adaptations in bill size enabled coexisting species of *Geospiza* to avoid resource competition?

_____

_____

_____

3. The range of variability shown by a phenotype in response to environmental variation is called **phenotypic plasticity**.

(a) Discuss the evidence for phenotypic plasticity in Galápagos finches: _____

_____

_____

_____

(b) Explain what this suggests about the biology of the original finch ancestor: _____

_____

_____

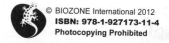

Adaptive Radiation in Mammals

Adaptive radiation is diversification (both structural and ecological) among the descendants of a single ancestral group to occupy different niches. The mammals underwent a spectacular adaptive radiation immediately following the sudden extinction of the non-avian dinosaurs. Most of the modern mammalian taxa became established very early. The diagram below shows the divergence of the mammals into major orders with many occupying niches left vacant by the dinosaurs. The vertical extent of each gray shape shows the time span for which that particular mammal order has existed (note that the scale for the geologic time scale in the diagram is not linear). Those that reach the top of the chart have survived to the present day. The width of a gray shape indicates how many species were in existence at any given time (narrow means there were few, wide means there were many). The dotted lines indicate possible links between the various mammal orders for which there is no direct fossil evidence.

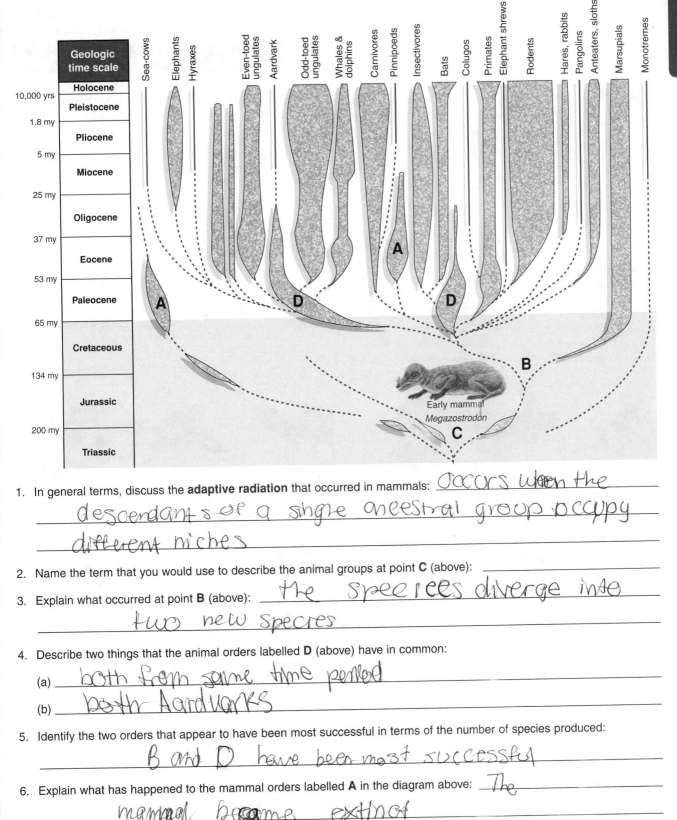

1. In general terms, discuss the **adaptive radiation** that occurred in mammals: _Occurs when the descendants of a single ancestral group occupy different niches_

2. Name the term that you would use to describe the animal groups at point **C** (above): _____

3. Explain what occurred at point **B** (above): _the species diverge into two new species_

4. Describe two things that the animal orders labelled **D** (above) have in common:

   (a) _both from same time period_

   (b) _both Aardvarks_

5. Identify the two orders that appear to have been most successful in terms of the number of species produced: _B and D have been most successful_

6. Explain what has happened to the mammal orders labelled **A** in the diagram above: _The mammal became extinct_

7. Identify the **epoch** during which there was the most adaptive radiation: _C_

8. Describe two key features that distinguish mammals from other vertebrates:

   (a) _have mamary glands_   (b) _three middle ear bones_

9. Describe the principal reproductive features that distinguish each of the major mammalian lines (sub-classes):

   (a) Monotremes: _lay eggs_

   (b) Marsupials: _have pouches_

   (c) Placentals: _give birth to a fully developed live young_

10. There are 18 orders of placental mammals (or 17 in schemes that include the pinnipeds within the Carnivora). Their names and a brief description of the type of mammal belonging to each group is provided below. Identify and label each of the diagrams with the correct name of their order:

Orders of Placental Mammals

Order	Description
Insectivora	Insect-eating mammals
Macroscelidae	Elephant shrews (formerly classified with insectivores)
Chiroptera	Bats
Cetacea	Whales and dolphins
Pholidota	Pangolins
Rodentia	Rodents
Proboscidea	Elephants
Sirenia	Sea-cows (manatees)
Artiodactyla	Even-toed hoofed mammals
Dermoptera	Colugos
Primates	Primates
Xenarthra	Anteaters, sloths, and armadillos
Lagomorpha	Pikas, hares, and rabbits
Carnivora	Flesh-eating mammals (canids, raccoons, bears, cats)
Pinnipedia	Seals, sealions, walruses (often now included as a sub-order of Carnivora)
Tubulidentata	Aardvark
Hyracoidea	Hyraxes
Perissodactyla	Odd-toed hoofed mammals

1 _proboscidea_ 2 _rodentia_ 3 _xenarthra_

4 _Chiroptera_ 5 _perissodactyla_ 6 _carnivora_

7 _pinnipedia_ 8 _lagomorpha_ 9 _primates_ 10 _artiodactyla_ 11 _dermoptera_ 12 _sirenia_

13 _macroscelidae_ 14 _insectivora_ 15 _Tubulidentata_ 16 _hyraxes_ 17 _pholidota_ 18 _cetacea_

11. For each of three **orders** of placental mammal, describe one **adaptive feature** that allows it to exploit a different niche from other placentals, and describe a **biological advantage** conferred by the adaptation:

   (a) Order: _insectivora_   Adaptive feature: _spikey back_

   Biological advantage: _protect itself against predaters_

   (b) Order: _lagomorpha_   Adaptive feature: _short limbs_

   Biological advantage: _allows animal to move faster_

   (c) Order: _carnivora_   Adaptive feature: _claws_

   Biological advantage: _do harm to pray_

Adaptive Radiation in Ratites

The **ratites** evolved from a single common ancestor. They are a monophyletic group of birds that lost the power of flight very early on in their evolutionary development. Ratites possess two features distinguishing them from other birds: a flat breastbone (instead of the more usual keeled shape) and a primitive palate (roof of the mouth). Flightlessness in itself is not unique to this group. There are other examples of birds that have lost the power of flight, particularly on remote, predator-free islands. Fossil evidence indicates that the ancestors of ratites were flying birds living about 80 million years ago. These ancestors also had a primitive palate, but they possessed a keeled breastbone.

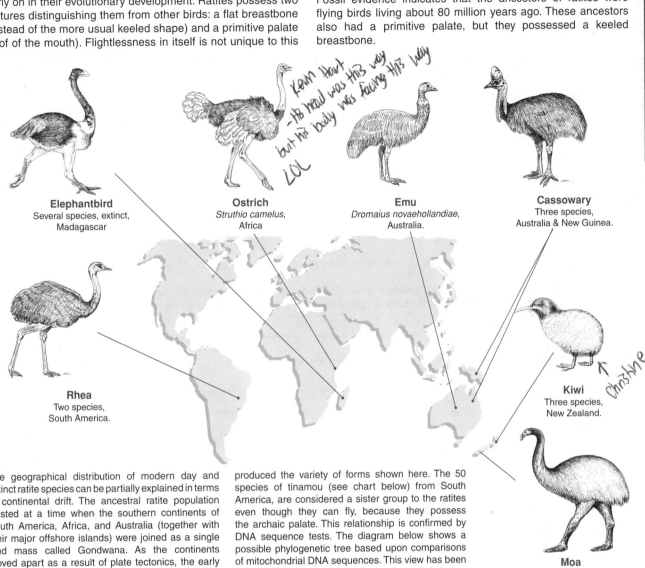

Elephantbird
Several species, extinct, Madagascar

Ostrich
Struthio camelus, Africa

Emu
Dromaius novaehollandiae, Australia.

Cassowary
Three species, Australia & New Guinea.

Rhea
Two species, South America.

Kiwi
Three species, New Zealand.

Moa
Eleven species (Lambert *et al.* 2004*), all extinct, New Zealand.

The geographical distribution of modern day and extinct ratite species can be partially explained in terms of continental drift. The ancestral ratite population existed at a time when the southern continents of South America, Africa, and Australia (together with their major offshore islands) were joined as a single land mass called Gondwana. As the continents moved apart as a result of plate tectonics, the early ratite populations were carried with them. Subsequent speciation on each continent and some of the islands produced the variety of forms shown here. The 50 species of tinamou (see chart below) from South America, are considered a sister group to the ratites even though they can fly, because they possess the archaic palate. This relationship is confirmed by DNA sequence tests. The diagram below shows a possible phylogenetic tree based upon comparisons of mitochondrial DNA sequences. This view has been supported by the extensive comparison of skeletons from the different ratite species.

Mesozoic Era

Birds evolved from a saurischian (small theropod) dinosaur ancestor about 150 million years ago (below)

Ratites diverge from the line to the rest of the birds about 100 million years ago.

Cenozoic Era

Fossil evidence suggests that **ratite ancestors** possessed a keeled breastbone and an archaic palate (roof of mouth)

Ratites

All other living birds
Moa 1: *Anomalopteryx*
Moa 2: *Pachyornis*
Moa 3: *Dinornis*
Moa 4: *Megalapteryx*
Little spotted kiwi
Great spotted kiwi
Brown kiwi
Emu
Cassowary
Ostrich
Rhea 1
Rhea 2
Tinamou (can fly)

A Letters indicate common ancestors

* Lambert *et al.* 2004. "Ancient DNA solves sex mystery of moa." Australasian Science, 25(8), Sept. 2004, pp. 14-16.

1. (a) Describe three physical features distinguishing all ratites from most other birds: _____

_____

(b) Identify the primitive feature shared by ratites and tinamou: _____

_____

2. Describe two anatomical changes, common to all ratites, which have evolved as a result of flightlessness. For each, describe the selection pressures for the anatomical change:

(a) Anatomical change: _____

Selection pressure: _____

_____

(b) Anatomical change: _____

Selection pressure: _____

_____

3. Name the ancient supercontinent that the ancestral ratite population inhabited: _____

4. (a) The extinct elephantbird from Madagascar is thought to be very closely related to another modern ratite. Based purely on the **geographical distribution** of ratites, identify the modern species that is the most likely relative:

_____

(b) Explain why you chose the modern ratite in your answer to (a) above: _____

_____

(c) Draw lines on the diagram at the bottom of the previous page to represent the divergence of the elephantbird from the modern ratite you have selected above.

5. (a) Name two other flightless birds that are not ratites: _____

(b) Explain why these other flightless species are not considered part of the ratite group: _____

_____

_____

6. Eleven species of moa is an unusually large number compared to the species diversity of the kiwis, the other ratite group found in New Zealand. The moas are classified into at least four genera, whereas kiwis have only one genus. The diets of the moas and the kiwis are thought to have had a major influence on each group's capacity to diverge into separate species and genera. The moas were herbivorous, whereas kiwis are nocturnal feeders, feeding on invertebrates in the leaf litter. Explain why, on the basis of their diet, moas diverged into many species, whereas kiwis diverged little:

_____

_____

_____

7. The DNA evidence suggests that New Zealand had two separate invasions of ratites, an early invasion from the moas (before the breakup of Gondwana) followed by a second invasion of the ancestors of the kiwis. Describe a possible sequence of events that could account for this:

_____

_____

_____

_____

8. The common ancestors of divergent groups are labelled (A-L) on the diagram at the bottom of the previous page. State the **letter** identifying the **common ancestor** for:

(a) The kiwis and the Australian ratites: _____   (b) The kiwis and the moas: _____

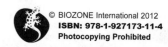

Convergent Evolution

Not all similarities between species are the result of common ancestry. Species from different evolutionary lineages may come to resemble each other if they have similar ecological roles and natural selection has shaped similar adaptations. This is called **convergent evolution** or **convergence**. It can be difficult to distinguish convergent and parallel evolution, as both produce similarity of form. The distinction is somewhat arbitrary and relates to how recently the taxa shared a common ancestor. Generally, similarity of form arising in closely related lineages (e.g. within marsupial mice) is regarded as parallelism, whereas similarity arising in more distantly related taxa is convergence (e.g. similarities between marsupial mice and placental mice).

Convergence in Swimming Form

Selection pressures to solve similar problems in particular environments may result in similarity of form and function in unrelated species. The development of succulent forms in unrelated plant groups (*Euphorbia* and the cactus family) is an example of **convergence** in plants. In the example (right), the selection pressures of the aquatic environment have produced a similar **streamlined** body shape in unrelated vertebrate groups. Icthyosaurs, penguins, and dolphins each evolved from terrestrial species that took up an aquatic lifestyle. Their general body form has evolved to become similar to that of the shark, which has always been aquatic. Note that flipper shape in mammals, birds, and reptiles is a result of convergence, but its origin from the pentadactyl limb is an example of **homology**.

Analogous Structures

Analogous structures (or **homoplasies**) are those that have the same function and often the same appearance, but quite different origins. The example on the right illustrates how the **eye** has developed independently in two unrelated taxa. The appearance of the eye is similar, but there is no genetic relatedness between the two groups (mammals and cephalopod mollusks).

The **wings** of birds and insects are also analogous structures. The wings have the same function, but the two taxa do not share a common ancestor. *Longisquama*, a lizard-like creature that lived about 220 mya, also had 'wings' that probably allowed gliding between trees. These 'wings' were not a modification of the forearm (as in birds), but highly modified long scales or feathers extending from its back.

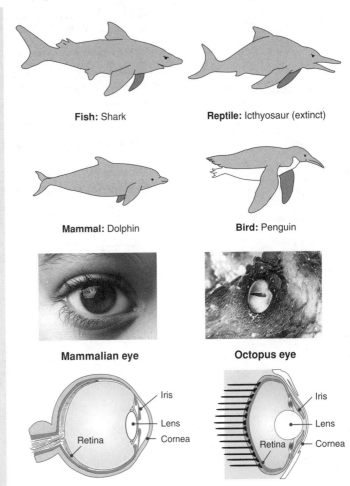

Fish: Shark   **Reptile:** Icthyosaur (extinct)

Mammal: Dolphin   **Bird:** Penguin

Mammalian eye   **Octopus eye**

Iris
Lens
Retina
Cornea

Iris
Lens
Retina
Cornea

---

1. In the example above illustrating convergence in swimming form, describe two ways in which the body form has evolved in response to the particular selection pressures of the aquatic environment:

   (a) _Dolphins evolved from terrestrial species that took up an aquatic lifestyle_

   (b) _Flipper shape in mammals, birds and reptiles_

2. Describe two of the selection pressures that have influenced the body form of the swimming animals above:

   (a) _Aquatic lifestyle_

   (b) _Prey vs. Predater_

3. When early taxonomists encountered new species in the Pacific region and the Americas, they were keen to assign them to existing taxonomic families based on their apparent similarity to European species. In recent times, many of the new species have been found to be quite unrelated to the European families they were assigned to. Explain why the traditional approach did not reveal the true evolutionary relationships of the new species:

   The traditional approach did not reveal the true evolutionary relationships of the new species because they based the comparison on phenotype instead of genetics

Periodicals: Which came first?

Related activities: Adaptations and Fitness
Weblinks: Convergent Evolution

RA 2

4. For each of the paired examples, briefly describe the adaptations of body shape, diet and locomotion that appear to be similar in both forms, and the likely selection pressures that are acting on these mammals to produce similar body forms:

Convergence Between Marsupials and Placentals

Australia

Marsupial and **placental** mammals diverged very early in mammalian evolution (about 120 mya), probably in what is now the Americas. Marsupials were widespread throughout the ancient supercontinent of Gondwana as it began to break up through the Cretaceous, but then became isolated on the southern continents, while the placentals diversified in the Americas and elsewhere, displacing the marsupials in most habitats around the world. Australia's isolation from other landmasses in the Eocene meant that the Australian marsupials escaped competition with the placentals and diversified into a wide variety of forms, ecologically equivalent to the North American placental species.

North America

Some older sources cite this example as one of parallelism, rather than convergence. However, a greater degree of morphological difference than now once separated the ancestors of the placental and marsupial lineages being compared, making this a case of convergence, not parallelism.

Marsupial Mammals

Placental Mammals

Marsupial		Placental
Wombat	(a) Adaptations: Rodent-like teeth, eat roots and above ground plants, and can excavate burrows. Selection pressures: Diet requires chisel-like teeth for gnawing. The need to seek safety from predators on open grassland.	Woodchuck
Flying phalanger	(b) Adaptations: Selection pressures:	Flying squirrel
Marsupial mole	(c) Adaptations: Selection pressures:	Mole
Marsupial mouse	(d) Adaptations: Selection pressures:	Mouse
Tasmanian wolf (tiger)	(e) Adaptations: Selection pressures:	Wolf
Long-eared bandicoot	(f) Adaptations: Selection pressures:	Jack rabbit

Coevolution

The term **coevolution** is used to describe cases where two (or more) species reciprocally affect each other's evolution. Each party in a coevolutionary relationship exerts selective pressures on the other and, over time, the species develop a relationship that may involve mutual dependency. Coevolution is a likely consequence when different species have close ecological interactions with one another. These ecological relationships include predator-prey and parasite-host relationships and mutualistic relationships such as those between plants and their pollinators. There are many examples of coevolution amongst parasites or pathogens and their hosts, and between predators and their prey, as shown on the following page.

Swollen-thorn *Acacia* lack the cyanogenic glycosides found in related *Acacia* spp. and the thorns are large and hollow, providing living space for the aggressive, stinging *Pseudomyrmex* ants which patrol the plant and protect it from browsing herbivores. The *Acacia* also provides the ants with protein rich food.

Hummingbirds are important pollinators in the tropics. Their needle-like bills and long tongues can take nectar from flowers with deep tubes. Their ability to hover enables them to feed quickly from dangling flowers. As they feed, their heads are dusted with pollen, which is efficiently transferred between flowers.

Butterflies find flowers by vision and smell them after landing to judge their nectar source. Like bees, they can remember characteristics of desirable flowers and so exhibit constancy, which benefits both pollinator and plant. Butterfly flowers are very fragrant and are blue, purple, deep pink, red, or orange.

Bees are excellent pollinators; they are strong enough to enter intricate flowers and have medium length tongues which can collect nectar from many flower types. They have good color vision, which extends into the UV, but they are red-blind, so bee pollinated flowers are typically blue, purplish, or white and they may have nectar guides that are visible as spots.

Beetles represent a very ancient group of insects with thousands of modern species. Their high diversity has been attributed to extensive coevolution with flowering plants. Beetles consume the ovules as well as pollen and nectar and there is evidence that ovule consumption by beetles might have driven the evolution of protective carpels in angiosperms.

Neotropical fruit bats (*Artibeus*) pollinate flowers while feeding on nectar and pollen

Bats are nocturnal and color-blind but have an excellent sense of smell and are capable of long flights. Flowers that have coevolved with bat pollinators are open at night and have light colors that do not attract other pollinators. Bat pollinated flowers also produce strong fragrances that mimic the smell of bats and have a wide bell shape for easy access.

1. Using examples, explain what you understand by the term coevolution: _Coevolution is exerts selective pressures on the other species and over time the species develop a relationship that may involve mutual dependency._

2. Describe some of the strategies that have evolved in plants to attract pollinators: _The plants have evolved to attract pollinators by producing strong fragrances that mimic the pollinator._

Weblinks: The Evolutionary Arms Race, The Coevolutionary Arms Race, Toxic Newts

RA 3

Predators, Parasites, and Coevolution

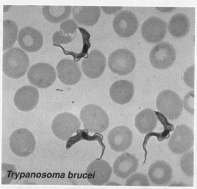
Trypanosoma brucei

Predators have obviously evolved to exploit their prey, and have evolved effective offensive weapons and hunting ability. Prey have evolved numerous strategies to protect themselves from predators, including large size and strength, protective coverings, defensive weapons, and toxicity. Lions have evolved the ability to hunt cooperatively to increase their chance of securing a kill from swift herding species such as zebra and gazelles.

Female *Helicornius* butterflies will avoid laying their eggs on plants already occupied by eggs, because their larvae are highly cannibalistic. Passionfruit plants (*Passiflora*) have exploited this by creating fake, yellow eggs on leaves and buds. *Passiflora* has many chemical defenses against herbivory, but these have been breached by *Heliconius*. It has thus counter-evolved new defenses against this herbivory by this genus.

Trypanosomes provide a good example of **host-parasite coevolution**. Trypanosomes must evolve strategies to evade their host's defenses, but their virulence is constrained by needing to keep their host alive. Molecular studies show that *Trypanosoma brucei* coevolved in Africa with the first hominids around 5 mya, but *T. cruzi* contact with human hosts occurred in South America only after settlements were made by nomadic cultures.

3. Explain how coevolution could lead to an increase in biodiversity: _Coevolution can lead to an increase in biodiversity because it can cause interaction between species which can cause a change in the gene pool._

4. Discuss some of the possible consequences of species competition: _The possible consequences of species competion can cause mutation or the weaker species to become extinct._

5. The analogy of an "arms race" is often used to explain the coevolution of exploitative relationships such as those of a parasite and its host. Form a small group to discuss this idea and then suggest how the analogy is flawed:
Adaption follows adaption in something of a long term 'arms race' between interacting populations of different populations; organisms must consiently adapt, evolve and proliferate not merely to gain reproductive advantage but to survive.

Extinction

Extinction is an important process in evolution as it provides opportunities, in the form of vacant niches, for the development of new species. Most species that have ever lived are now extinct. The species alive today make up only a fraction of the total list of species that have lived on Earth throughout its history. Extinction is a natural process. Background extinction is the steady rate of species turnover in a taxonomic group (a group of related species). The duration of a species is thought to range from as little as 1 million years for complex larger organisms to as long as 10-20 million years for simpler organisms. Superimposed on this constant background extinction are catastrophic events or **mass extinctions** that wipe out vast numbers of species in relatively brief periods of time in geologic terms. The diagram below shows how the number of species has varied over the history of life on Earth. The number of species is indicated on the graph by families: a taxonomic group comprising many genera and species. There have been five major extinction events and a sixth event, which began in the Late Pleistocene and continues today.

Major Mass Extinctions

The Permian extinction
(250 million years ago)

This was the most devastating mass extinction of all. Nearly all life on Earth perished, with 90% of marine species and probably many terrestrial ones also, disappearing from the fossil record. This extinction event marks the **Paleozoic-Mesozoic** boundary.

The Cretaceous extinction
(65 million years ago)

This extinction event marks the boundary between the Mesozoic and Cenozoic eras. More than half the marine species and many families of terrestrial plants and animals became extinct, including nearly all the dinosaur species (the birds are now known to be direct descendants of the dinosaurs).

Quaternary extinction
(15,000-10,000 years ago)

This first phase of the sixth extinction is marked by the extinction of many species of giant mammal, although there were few extinctions of entire families. It is known as the **Pleistocene overkill** because it was associated with the hunting activities of prehistoric humans. Many large marsupials in Australia and placental species elsewhere became extinct.

The sixth extinction
(second phase occurring now)

The current phase of the sixth extinction is largely due to human destruction of habitats (e.g. coral reefs, tropical forests) and pollution. It is considered far more serious and damaging than some earlier mass extinctions because of the speed at which it is occurring. The increasing human impact is making biotic recovery difficult.

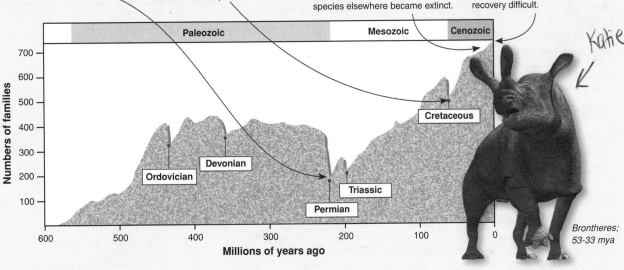

Brontheres;
53-33 mya

1. Describe the main features (scale and type of organisms killed off) of each of the following major extinction events:

   (a) Permian extinction: _almost all marine animals and terrestrial vertebrate_

   (b) Cretaceous extinction: _mass extinction of plants, animals and non-avian dinosaurs_

   (c) Megafaunal extinction: _Large - bodied mammals became extinct_

2. Explain how human activity has contributed to the most recent mass extinction: _Taken over land to create shelter and cut down trees and use as resour_

3. In general terms, describe the effect that past mass extinctions had on the way the surviving species **further evolved**:
   The surviving species after a mass extinction probably evolve faster

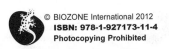
Periodicals:
Mass extinctions,
The sixth extinction

Related activities: Causes of Mass Extinctions
Weblinks: What Killed the Dinosaurs?

RA 2

Causes of Mass Extinctions

There have been about 24 mass extinctions over the last 540 million years. Five of these were major and 19 were minor (based on the percentage of genera that became extinct). Some extinction peaks coincided with known comet or asteroid impacts, implying that they may have been causal. Environmental deterioration associated with climate change or a major cosmic, geologic, or biological event, is deemed the main cause of extinctions. A sixth extinction, which began in the Late Pleistocene, continues today. The ability of the biosphere to recover from such crises is evident by the fact that life continues to exist today.

Large asteroid/comet impacts

Large solar flares

Volcanism

Possible causes of mass extinction	Effects on the extinction rate	Examples of extinction events and their likely causes
Impacts by large asteroid/comet or 'showers': Shock waves, heat-waves, wildfires, impact 'winters' caused by global dust clouds, super-acid rain, toxic oceans, superwaves and superfloods from an oceanic impact.	Global extinction of much of the planet's biodiversity. Smaller comet showers could cause stepwise, regional extinctions.	**Quaternary** (15,000-10,000 ya) *Extinction of:* Many large mammals and large flightless birds. *Probable cause:* Warming of the global climate after the last ice age plus predation (hunting) by humans.
Supernovae radiation: Direct exposure to X-rays and cosmic rays. Ozone depletion and subsequent exposure to excessive UV radiation from the sun.	Causes mutations and kills organisms. Selective mass extinctions, particularly of animals (but not plants) exposed to the atmosphere, as well as shallow-water aquatic forms.	**Late Cretaceous** (65 mya) *Extinction of:* Dinosaurs, plesiosaurs, icthyosaurs, mosasaurs, pterosaurs, ammonites and belemnites (squid-like animals), and many other groups.
Large solar flares: Exposure to large doses of X-rays, and UV radiation. Ozone depletion.	Mass extinctions.	*Probable cause:* An asteroid impact (probably Yucatan peninsula) produced catastrophic environmental disturbance.
Geomagnetic reversals: Increased flux of cosmic rays.	Mass extinctions.	**Late Permian** (250 mya) *Extinction of:* 90% of marine species and 70% of land species. Coral reefs, trilobites, some amphibians, and mammal-like reptiles, were eliminated (the Great Dying).
Continental drift: Climatic changes, such as glaciations and droughts, occur when continents move towards or away from the poles.	Global cooling due to changes in the pattern of oceans currents caused by shifting land masses. Extinctions as species find themselves in inhospitable climates.	
Intense volcanism: Cold conditions caused by volcanic dust reducing solar input. Volcanic gases causing acid rain and reduced alkalinity of oceans. Toxic trace elements.	Stepwise mass extinctions.	*Probable cause:* An asteroid impact, followed by furious volcanic activity, a rapid heating of the atmosphere, and depletion of life-giving oxygen from the oceans.
Sea level change: Loss of habitat.	Mass extinctions of susceptible species (e.g. marine reptiles, coral reefs, coastal species).	**Late Devonian** (360 mya) *Extinction of:* Many corals, bivalves, fish, sponges (21% of all marine families). Collapse of tropical reef communities.
Arctic spill over: Release of cold fresh water or brackish water from an isolated Arctic Ocean. Ocean temperature falls by 10°C, resulting in atmospheric cooling and drought.	Mass extinctions in marine ecosystems. Change of vegetation with drastic effect on large reptiles.	*Probable cause:* Global cooling associated with (or causing?) widespread oxygen deficiency of shallow seas.
Anoxia: Shortage of oxygen.	Mass extinctions in the oceans.	
Spread of disease/predators: Direct effects due to changing geographic distribution.	Mass extinctions.	

Source: *Environmental Change – The Evolving Ecosphere (1997)*, by R.J. Huggett

Source: *Evolution: A Biological and Paleontological Approach*, Skelton, P. (ed.), Addison-Wesley (1993)

1. Describe how each of the following events might have caused mass extinctions in the past:

   (a) Large asteroid/comet impact: ___Super waves, super floods, super-acid rain, heat waves, shock waves___

   (b) Continental drift: ___Glaciations and droughts, earthquakes___

   (c) Volcanism: ___Cold conditions caused by volcanic dush reducing solar input; toxic trace elements; acid rain___

2. Explain how the arrival of a new plant or animal species onto a continent may cause the demise of other species there:
   ___Increase species competition___

The Sixth Extinction

Human activity dominates Earth. Humans can be found almost everywhere on the globe, even at the South Pole. As humans have spread across the planet, from Africa into Europe and Asia, and across into the Americas and beyond, they have changed the environment around them to suit their needs. How these changes have occurred has varied according to the technology available and the general social environment and attitudes at the time. These changes have had a profound impact on the globe's physical and biological systems. Only in the past century have we begun to fully evaluate the impact of human activity on the Earth. Humans have caused the deliberate or accidental extinction of numerous species and brought many more to the brink. So many species have been lost as a direct or indirect result of human activity that this period in history has been termed the Sixth Extinction. There is debate over when this extinction began, its extent, and the degree of human involvement. However, it is clear that many species are being lost and many that existed before humans appeared in their domains, no longer do.

What is the Sixth Extinction?

There have been five previous mass extinctions. The last, 65 mya, saw the end of the dinosaurs and most marine reptiles. The Sixth Extinction refers to the apparent human-induced loss of much of the Earth's biodiversity. Twenty years ago, Harvard biologist E.O. Wilson estimated that as many as 30,000 species a year were being lost, one every 20 minutes. The extinction dates for some examples are given below.

Once Here, Now Gone...

Moa (New Zealand)
~1500 AD

Reunion giant tortoise
(Reunion Is.)~1800 AD

Passenger Pigeon
(United States) 1914 AD

Thylacine (Australia)
1930 AD

Golden toad (Costa Rica) 1989 AD

Caribbean monk seal
(Caribbean) 2008 AD

Estimating Extinction Rates

Estimates of the rate of species loss can be made by using the **background extinction rate** as a reference. It is estimated that one species per million species per year becomes extinct. By totalling the number of extinctions known or suspected over a time period, we can compare the current rate of loss to the background extinction rate.

Birds provide one of the best examples. There are about 10,000 living or recently extinct species of birds. In the last 500 years an estimated 150 or more have become extinct. From the background extinction rate we would expect one species to become extinct every 100 years (10,000/1,000,000 = 0.01 extinctions per year = 1 extinction per 100 years). It then becomes apparent that 150/500 = 0.3 extinctions per year or 30 extinctions per century, 30 times greater than the background rate. The same can be calculated for most other groups of animals and plants.

Organism*	Total number of species (approx)*	Known extinction (since ~1500)*
Mammals	5487	87
Birds	9975	150
Reptiles	10,000	22
Amphibians	6700	39
Plants	300,000	114

* These numbers vastly underestimate the true numbers because so many species are undescribed.

The Carolina Parakeet

The Carolina parakeet was the only parakeet (or parrot) native to the eastern United States. It was a very social bird, forming large flocks in upland forests and forest edges, from the Ohio Valley to the Gulf of Mexico. It was a spectacular bird, with bright green plumage, a yellow head, and orange face and bill. Its natural diet was wild fruit, seeds, and nuts, but when settlers began planting crops, it quickly exploited them and was treated as a pest (despite a useful role in destroying invasive weed species). Entire flocks were easily targeted and shot because the parakeets' behavioral strategy was to rally around distressed or injured birds. The last confirmed sighting of the parakeet in the wild was in 1904 and the last captive bird died in 1918.

The reasons for the Carolina parakeet's extinction are not fully known. Although large numbers were killed by hunting or habitat destruction, the last, apparently healthy, populations were left more or less alone. It appears disease may have been a major factor in the final extinction of the last populations, probably from introduced domestic fowl.

1. Define the **Sixth Extinction**: _____

_____

_____

_____

The Lighthouse Keeper's Cat
and the extinction of the Stephens Island wren

The tale of the lighthouse keeper's cat has become a well known story in New Zealand where 42% of the country's native bird species became extinct following the arrival of humans. The story occurs on Stephens Island, an offshore island located on the northern tip of the South Island of New Zealand, with an area of just 150 hectares. In 1894, a lighthouse was built on the island and kept by three keepers, including assistant keeper David Lyall who brought along a cat. Over the next year, the cat brought 10 tiny birds back to the lighthouse. Lyall sent them to be studied and it was found the bird - called the **Stephens Island wren** - was new to science. Unfortunately within a short time of its discovery, the Stephens Island wren had been exterminated by the cat.

The tale is not as straightforward as is presented in the popular view. Evidence suggests that there were many cats on the island; the litter of a female which may have arrived in 1894. Certainly, by 1895 there was more than one cat and by 1898 the island was effectively infested with cats.

The Stephens Island wren was once common throughout New Zealand but, by the time Europeans settled, it was restricted to Stephens Island. It was first described in early 1894 and was seen alive perhaps just twice by Europeans, despite a number of searches. It was effectively extinct by mid 1895. The last cats on the island were finally exterminated in 1925.

2. Use the data in the table to calculate (1) the **rate of species extinction per century** for each of the groups and (2) how many times greater this is than the background extinction rate:

(a) Mammals: _____

_____

(b) Reptiles: _____

_____

(c) Amphibians: _____

_____

(d) Plants _____

_____

3. Humans are frequently not directly responsible for the extinction of a species, yet recently extinct species have often become so after humans arrive in their habitat. Discuss the reasons for these extinctions:

_____

_____

_____

_____

4. Explain how the story of the lighthouse keeper's cat shows what can happen when new species are introduced to new and isolated habitats:

_____

_____

_____

5. Identify two species, other than those mentioned earlier, that have recently become extinct. Identify the date and possible reasons for their extinction:

_____

_____

KEY TERMS: Flash Card Game

The cards below have a keyword or term printed on one side and its definition printed on the opposite side. The aim is to win as many cards as possible from the table. To play the game.....

1) Cut out the cards and lay them definition side down on the desk. You will need one set of cards between two students.

2) Taking turns, choose a card and, BEFORE you pick it up, state

your own best definition of the keyword to your opponent.

3) Check the definition on the opposite side of the card. If both you and your opponent agree that your stated definition matches, then keep the card. If your definition does not match then return the card to the desk.

4) Once your turn is over, your opponent may choose a card.

Adaptation	Adaptive radiation	Allopatric
Background extinction rate	Convergent evolution	Divergent evolution
Evolution	Extinction	Homology
Natural selection	Phenotype	Phyletic gradualism
Phylogeny	Population	Reproductive isolation
Ring species	Speciation	Species

When you've finished the game keep these cutouts and use them as flash cards!

Speciation in which the populations are physically separated. A term for physically separated populations.

A form of divergent evolution in which there is rapid speciation of one ancestral species to fill many different ecological niches.

The evolutionary process by which a species becomes better suited to its niche. Also a feature that results from this process.

Evolutionary process in which a species or related species follow different evolutionary pathways to eventually become less related.

The evolution of two or more lineages towards similar morphologies or adaptations so that the lineages appear similar despite their different ancestry.

The normal rate of extinction of species during the Earth's geological and biological history, excluding major extinction events when this rate increased.

Similarity between species that results from common ancestry.

The complete dying out of a species so that there are no representatives of the species remaining anywhere.

Changes in gene pools over time.

Evolution characterised by gradual change within a lineage, the change in any single generation being extremely small.

Observable characteristics in an organism.

The process by which traits become more or less common in a population through differential survival and reproduction.

The situation in which members of a group of organisms breed with each other but not with members of other groups.

The total number of individuals of a species within a set habitat or area.

The evolutionary history or genealogy of a group of organisms. Often represented as a 'tree' showing descent of new species from the ancestral one.

Group or population of individuals that can interbreed to produce viable offspring.

The division of one species, during evolution, into two or more separate species.

A population sharing the same gene pool where groups next to each other are able to interbreed but groups at the extreme ends of the population can not.

The Origin of Living Systems

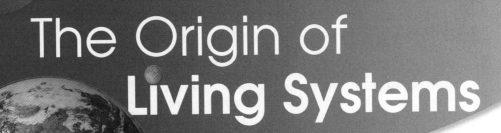

Key concepts

▶ Scientific evidence supports models for the formation of the Earth's first organic molecules.

▶ RNA could have served as both a template for information and an enzyme.

▶ RNAs were the first self-replicating molecules.

▶ Molecular and genetic evidence supports a common ancestry of all living things.

Key terms

endosymbiosis

endosymbiotic theory

Last Universal Common
  Ancestor

Miller-Urey experiments

panspermia

primordial environment

RNA world

stromatolite

Essential Knowledge

☐ 1. Use the **KEY TERMS** to compile a glossary for this topic.

Hypotheses for the Origin of Life *(1.D.1)*   pages 344-347

☐ 2. Describe the **primordial environment** and the likely events that led to the formation of life on Earth, including:
  − The role of inorganic precursors, available free energy, and lack of oxygen
  − The formation of monomers as building blocks for larger molecules
  − The production of self-replicating polymers.

☐ 3. Describe the likely role of RNA as the first self-replicating molecule. Outline RNA's role as an enzyme and in the origin of the first self-replicating cells.

Scientific Evidence for the Origin of Life *(1.D.2)* pages 290-291, 348-350

☐ 4. Identify major stages in the evolution of life on Earth. Summarize the main ideas related to where life originated: ocean surface, extraterrestrial (**panspermia**), and deep sea thermal vents.

☐ 5. Outline the experiments of **Miller** and **Urey** that attempted to simulate the prebiotic environment on Earth. Describe their importance in our understanding of the probable origin of organic compounds.

☐ 6. Describe the evidence for the first aquatic prokaryotes. Discuss the importance of these early organisms to the later evolution of diversity.

☐ 7. Explain what the current ecology of some bacterial groups (Bacteria and Archaea) tells us about the probable conditions of early life on Earth.

☐ 8. Discuss the molecular and genetic evidence from living and extinct organisms that indicates common ancestry (a **Last Universal Common Ancestor**).

☐ 9. Discuss the **endosymbiotic theory** for the evolution of mitochondria, chloroplasts, and other eukaryotic organelles. Summarize the evidence in support of this theory.

Periodicals:
Listings for this chapter are on page 384

Weblinks:
www.thebiozone.com/
weblink/AP1-3114.html

BIOZONE APP:
Student Review Series
The Origin of Life

Life in the Universe

Complex organic molecules (as are found in living things) have been detected beyond Earth in interstellar dust clouds and in meteorites that have landed on Earth. Scientists hypothesize that more than 4 billion years ago, one such dust cloud collapsed into a swirling **accretion disk** that gave rise to the sun and planets in our solar system. Some of the fragile molecules survived the heat of solar system formation by sticking together in comets at the disk's fringe where temperatures were freezing. Later, the comets and other cloud remnants carried the molecules to Earth.

The formation of these organic molecules and their significance to the origin of life on Earth are currently being investigated experimentally (see below). The study of the origin of life on Earth is closely linked to the search for life elsewhere in our solar system. There are further plans to send solar and lunar orbiters to other planets and their moons, and even to land on a comet in November 2014. Their objective will be to look for signs of life or its chemical precursors. If detected, such a discovery would suggest that life may be widespread in the universe.

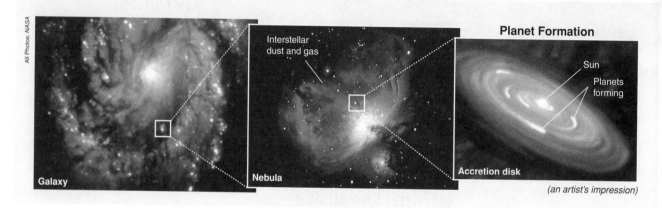

Galaxy

Nebula

Interstellar dust and gas

Planet Formation

Sun

Planets forming

Accretion disk

(an artist's impression)

All Photos: NASA

How Organic Molecules Might Form in Space

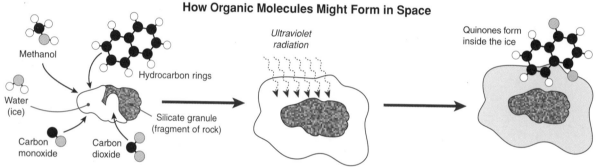

Methanol

Hydrocarbon rings

Water (ice)

Carbon monoxide

Carbon dioxide

Silicate granule (fragment of rock)

Ultraviolet radiation

Quinones form inside the ice

Interstellar ice begins to form when molecules such as methanol, water, and hydrocarbons freeze onto sandlike granules of silicate drifting in dense interstellar clouds.

Ultraviolet radiation from nearby stars causes some of the chemical bonds of the frozen compounds in the ice to break.

The broken down molecules recombine into structures such as quinones, which would never form if the fragments were free to float away.

Two **Mars Exploration rovers** landed on Mars in early 2004. Each rover carried sophisticated instruments, which were used to determine the history of climate and water at two sites where conditions may once have been favorable for life.

NASA

Organic Molecules Detected in Space

In a simple cloud-chamber experiment with simulated space ice (frozen water, methanol, and ammonia), complex compounds were yielded, including: ketones, nitriles, ethers, alcohols, and quinones. These same organic molecules are found in carbon-rich meteorites. A six-carbon molecule (known as HMT) was also created. In warm, acidic water it is known to produce amino acids.

In another investigation into compounds produced in this way, some of the molecules displayed a tendency to form capsule-like droplets in water. These capsules were similar to those produced using extracts of a meteorite from Murchison, Australia in 1989. When organic compounds from the meteorite were mixed with water, they spontaneously assembled into spherical structures similar to cell membranes. These capsules were found to be made up of a host of complex organic molecules.

Source: *Life's far-flung raw materials*, Scientific American, July 1999, pp. 26-33

1. Suggest how sampling the chemical makeup of a comet might assist our understanding of life's origins:

   Can help assit when organism started to develop

2. Explain the significance of molecules from space that naturally form capsule-like droplets when added to water:

   Organic molecules Show the origin of life on Earth

3. How are scientists able to know about the existence of complex organic molecules in space? A 6-carbon molecule can produce amino acids

Related activities: The Origin of Life on Earth
Weblinks: Life in the Universe

Periodicals:
The ice of life

The Origin of Life on Earth

Recent discoveries of **prebiotic** conditions on other planets and their moons has rekindled interest in the origin of life on primeval Earth. Experiments demonstrate that both peptides and nucleic acids may form polymers naturally in the conditions that are thought to have existed in a primitive terrestrial environment. RNA has also been shown to have enzymatic properties (**ribozymes**) and is capable of self-replication. These discoveries have removed some fundamental obstacles to creating a plausible scientific model for the origin of life from a prebiotic soup. Much research is now underway and space probes have been sent to Mercury, Venus, Mars, Pluto, and Pluto's moon, Charon. They will search for evidence of prebiotic conditions or primitive microorganisms. The study of life in such regions beyond our planet is called **exobiology**.

Steps Proposed in the Origin of Life

The appearance of life on our planet may be understood as the result of evolutionary processes involving the following major steps:

1. Formation of the Earth (4600 mya) and its acquisition of volatile organic chemicals by collision with comets and meteorites, which provided the precursors of biochemical molecules.

2. Prebiotic synthesis and accumulation of amino acids, purines, pyrimidines, sugars, lipids, and other organic molecules in the primitive terrestrial environment.

3. Condensation reactions involving the synthesis of polymers of peptides (proteins) and nucleic acids (probably RNA) with self-replicating and catalytic abilities.

4. Synthesis of lipids, their self-assembly into double-layered membranes and liposomes, and the 'capturing' of prebiotic (self-replicating and catalytic) molecules within their boundaries.

5. Formation of a **protobiont**; this is an immediate precursor to the first living systems. Such protobionts would exhibit cooperative interactions between small catalytic peptides, replicative molecules, proto-tRNA, and protoribosomes.

Photo: Ron Lind

These living **stromatolites** from a beach in Western Australia are created by mats of cyanobacteria. Similar, fossilized stromatolites have been found in rocks dating back to 3500 million years ago.

Where Might Life Have Originated?

Scientists continue to speculate about where life might have originated. Three alternative views of how the key processes occurred are illustrated below.

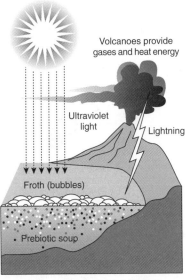

Ocean Surface (Tidal Pools)
This popular theory suggests that life arose in a tidepool, pond or on moist clay on the primeval Earth. Gases from volcanoes would have been energized by UV light or electrical discharges to form the prebiotic molecules in froth.

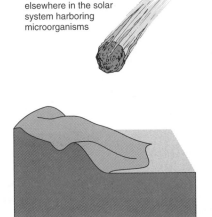

Comet or meteorite from elsewhere in the solar system harboring microorganisms

Panspermia
Cosmic ancestry (panspermia) is a serious scientific theory that proposes living organisms were 'seeded' on Earth as 'passengers' aboard comets or meteors. Such incoming organisms would have to survive the heat of re-entry.

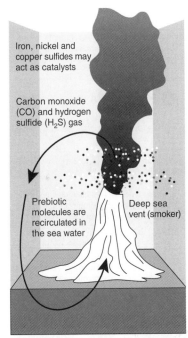

Iron, nickel and copper sulfides may act as catalysts

Carbon monoxide (CO) and hydrogen sulfide (H_2S) gas

Prebiotic molecules are recirculated in the sea water

Deep sea vent (smoker)

Undersea Thermal Vents
A recently proposed theory suggests that life may have arisen at ancient volcanic vents (called smokers). This environment provides the necessary gases, energy, and a possible source of catalysts (metal sulfides).

1. Summarize the steps hypothesized for the appearance of life on Earth: _peptides and nucleic acids may form polymers naturally in the conditions that are thought to have existed in a primitive terrestrial environment_

2. Explain why ocean surfaces and undersea thermal vents are likely environments for the origin of life on Earth: _The environment provides the necessary gases, energy and a possible source of catalysts_

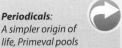

Periodicals:
A simpler origin of life, Primeval pools

Related activities: Prebiotic Experiments
Weblinks: 4 Billion Years of Evolution

An RNA World

A key problem in understanding how life began is how biological information was first stored and how it was copied and replicated. Modern life requires many complex proteins and molecules for replication, but these did not exist in life's early history. The discovery of ribozymes in 1982 (more than fifteen years after they were first hypothesized) helped to solve this problem, at least in part. Ribozymes are enzymes formed from RNA, which itself can store biological information. The ribozymes can catalyze the replication of the original RNA molecule. This mechanism for self replication has lead to the theory of an "**RNA world**".

❶ Pre RNA World

The individual ribonucleotides of RNA are difficult to assemble without enzymes.

This has led to proposals of a pre-RNA world, where polymers similar to RNA formed and acted as the very first catalysts.

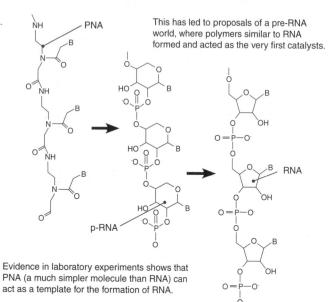

Evidence in laboratory experiments shows that PNA (a much simpler molecule than RNA) can act as a template for the formation of RNA.

❷ RNA World

RNA is able to act as a vehicle for both information storage and catalysis. It therefore provides a way around the problem that genes require enzymes to form and enzymes require genes to form. The first stage of evolution may have proceeded by RNA molecules folding up to form ribozymes which could then catalyze the replication of other similar RNAs.

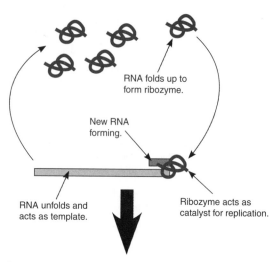

RNA folds up to form ribozyme.

New RNA forming.

RNA unfolds and acts as template.

Ribozyme acts as catalyst for replication.

❸ Mutation and Competition

From the establishment of self replicating RNA molecules there would have been competition, of a sort. Incorrect copies of the original RNA produced new varieties of RNA.

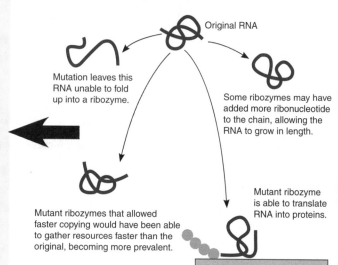

Original RNA

Mutation leaves this RNA unable to fold up into a ribozyme.

Some ribozymes may have added more ribonucleotide to the chain, allowing the RNA to grow in length.

Mutant ribozymes that allowed faster copying would have been able to gather resources faster than the original, becoming more prevalent.

Mutant ribozyme is able to translate RNA into proteins.

❹ Proto-cells

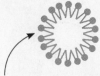

Large groups of micelles can interact to form vesicles large enough to contain other molecules.

Certain types of organic molecule (such as fatty acids) spontaneously form micelles when placed in an aqueous solution.

Mutually cooperating RNAs and proteins trapped inside a vesicle would be able to replicate without moving away from each other.

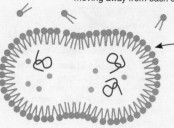

Vesicle growth by the attraction of micelles would eventually cause it to be unstable and split in two. Each vesicle would take a random number of RNAs with it.

1. Why did the discovery of ribozymes add weight to the RNA world hypothesis? _The ribozymes can catalyze the replication of the original RNA molecule_

2. Explain how mutations in RNA templates led to the first form of evolution: _mutations in the RNA template can lead to a mutation in the DNA causing evolutionary change._

Related activities: The Origin of Life on Earth
Weblinks: Exploring Life's Origins, The Discovery of Ribozymes

© BIOZONE International 2012
ISBN: 978-1-927173-11-4
Photocopying Prohibited

Prebiotic Experiments

In the 1950s, Stanley Miller and Harold Urey attempted to recreate the conditions of primitive Earth. They hoped to produce, in an experiment, to the biological molecules that preceded the development of the first living organisms. Researchers at the time believed that the Earth's early atmosphere was made up of methane, water vapor, ammonia, and hydrogen gas, so these were the components included in the experiments. Many variations on this experiment have produced similar results (below). It seems that the building blocks of life are relatively easy to create. Many types of organic molecules have even been detected in deep space.

The Miller-Urey Experiment

The experiment (right) was run for a week after which samples were taken from the collection trap for analysis. Up to 4% of the carbon (from the methane) had been converted to amino acids. In this and subsequent experiments, it has been possible to form all 20 amino acids commonly found in organisms, along with nucleic acids, several sugars, lipids, adenine, and even ATP (if phosphate is added to the flask). Researchers now believe that the early atmosphere may be similar to the vapors given off by modern volcanoes: carbon monoxide (CO), carbon dioxide (CO_2), and nitrogen (N_2). Note the absence of free atmospheric oxygen.

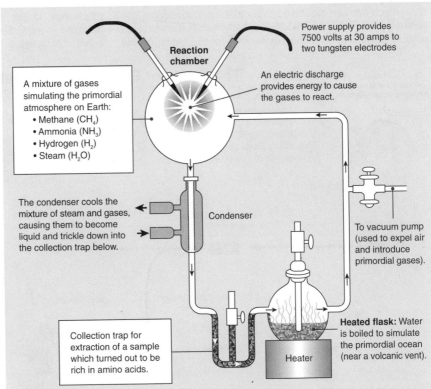

Power supply provides 7500 volts at 30 amps to two tungsten electrodes

Reaction chamber

A mixture of gases simulating the primordial atmosphere on Earth:
• Methane (CH_4)
• Ammonia (NH_3)
• Hydrogen (H_2)
• Steam (H_2O)

An electric discharge provides energy to cause the gases to react.

The condenser cools the mixture of steam and gases, causing them to become liquid and trickle down into the collection trap below.

Condenser

To vacuum pump (used to expel air and introduce primordial gases).

Collection trap for extraction of a sample which turned out to be rich in amino acids.

Heated flask: Water is boiled to simulate the primordial ocean (near a volcanic vent).

Heater

Iron pyrite (fool's gold) (above) has been proposed as a possible stabilizing surface for the synthesis of organic compounds in the prebiotic world.

The Origin of Living Systems

Some scientists envisage a global winter scenario for the formation of life. Organic compounds are more stable in colder temperatures and could combine in a lattice of ice. This frozen 'world' could be thawed later.

Lightning is a natural phenomenon associated with volcanic activity. It may have supplied a source of electrical energy for the formation of new compounds (e.g. oxides of nitrogen) which were incorporated into organic molecules.

The early Earth was subjected to volcanism everywhere. At volcanic sites such as deep sea hydrothermal vents and geysers (like the one above), gases delivered vital compounds to the surface, where reactions took place.

1. In the Miller-Urey experiment simulating the conditions on primeval Earth, identify parts of the apparatus equivalent to:

   (a) Primeval atmosphere: ___reactch chember___   (c) Lightning: ___electric discharge___

   (b) Primeval ocean: ___Heated flask___   (d) Volcanic heat: ___vacum pump___

2. What organic molecules were created by this experiment? ___amino acids___

3. (a) Why do you think the Miller-Urey experiment is not an accurate model of what happened on the primeval Earth?
   ___Absence of free atmospheric oxygen___

   (b) What changes to the experiment could help it to better fit our understanding of the Earth's primordial conditions?
   ___Absense of free Oxygen___

© BIOZONE International 2012
ISBN: 978-1-927173-11-4
Photocopying Prohibited

Periodicals:
Earth in the beginning

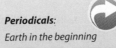

Related activities: The Origin of Life on Earth
Weblinks: An Interview with Stanley Miller

A 3

The Origin of Eukaryotes

The first eukaryotes were unicellular and occur only rarely in microfossils. The first fossil evidence dates to 2.1 bya, but molecular evidence suggests that the eukaryotic lineage is much more ancient and closer to the origin of life. The original endosymbiotic theory (Margulis, 1970) proposed that eukaryotes arose as a result of an endosymbiosis between two prokaryotes, one of which was aerobic and gave rise to the mitochondrion. The hypothesis has since been modified to recognize that eukaryotes probably originated with the appearance of the nucleus and flagella, with later acquisition of mitochondria and chloroplasts by endosymbiosis. Primitive eukaryotes probably acquired mitochondria by engulfing purple bacteria. Similarly, chloroplasts may have been acquired by engulfing primitive cyanobacteria. In both instances, the organelles produced became dependent on the nucleus of the host cell to direct some of their metabolic processes. Unlike mitochondria, chloroplasts were probably acquired independently by more than one organism, so their origin is polyphyletic.

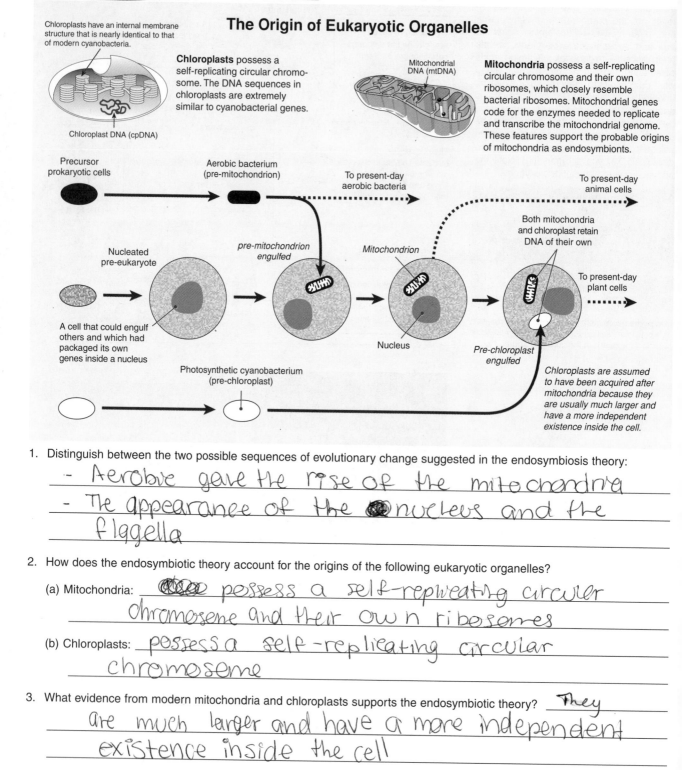

The Origin of Eukaryotic Organelles

Chloroplasts have an internal membrane structure that is nearly identical to that of modern cyanobacteria.

Chloroplasts possess a self-replicating circular chromosome. The DNA sequences in chloroplasts are extremely similar to cyanobacterial genes.

Chloroplast DNA (cpDNA)

Mitochondrial DNA (mtDNA)

Mitochondria possess a self-replicating circular chromosome and their own ribosomes, which closely resemble bacterial ribosomes. Mitochondrial genes code for the enzymes needed to replicate and transcribe the mitochondrial genome. These features support the probable origins of mitochondria as endosymbionts.

Precursor prokaryotic cells

Aerobic bacterium (pre-mitochondrion)

To present-day aerobic bacteria

To present-day animal cells

Both mitochondria and chloroplast retain DNA of their own

Nucleated pre-eukaryote

pre-mitochondrion engulfed

Mitochondrion

To present-day plant cells

A cell that could engulf others and which had packaged its own genes inside a nucleus

Nucleus

Pre-chloroplast engulfed

Photosynthetic cyanobacterium (pre-chloroplast)

Chloroplasts are assumed to have been acquired after mitochondria because they are usually much larger and have a more independent existence inside the cell.

1. Distinguish between the two possible sequences of evolutionary change suggested in the endosymbiosis theory:

   - Aerobic gave the rise of the mitochondria
   - The appearance of the nucleus and the flagella

2. How does the endosymbiotic theory account for the origins of the following eukaryotic organelles?

   (a) Mitochondria: possess a self-replicating circular chromosome and their own ribosomes

   (b) Chloroplasts: possess a self-replicating circular chromosome

3. What evidence from modern mitochondria and chloroplasts supports the endosymbiotic theory? They are much larger and have a more independent existence inside the cell

4. Comment on how the fossil evidence of early life supports or contradicts the endosymbiotic theory: The fossils of early life supports the endosymbiotic theory

Related activities: The Origin of Life of Earth
Weblinks: Endosymbiosis and the Origin of Eukaryotes
Periodicals: The rise of life on Earth

© BIOZONE International 2012
ISBN: 978-1-927173-11-4
Photocopying Prohibited

Landmarks in Earth's History

The scientific explanation of the origin of life on Earth is based soundly on the extensive fossil record, as well as the genetic comparison of modern life forms. Together they clearly indicate that modern life forms arose from ancient ancestors that have long since become extinct. These ancient life forms themselves originally arose from primitive cells living some 3500 million years ago in conditions quite different from those on Earth today. The earliest fossil records of living things show only simple cell types. It is believed that the first cells arose as a result of evolution at the chemical level in a 'primordial soup' (a rich broth of chemicals in a warm pool of water, perhaps near a volcanic vent). Life appears very early in Earth's history, but did not evolve beyond the simple cell stage until much later (about 600 mya). This would suggest that the evolution of complex life forms required more difficult evolutionary hurdles to be overcome. One hypothesis is that fluctuating nutrient levels and rising oxygen following the end of the snowball glaciations may have been important in the evolution of multicellularity.

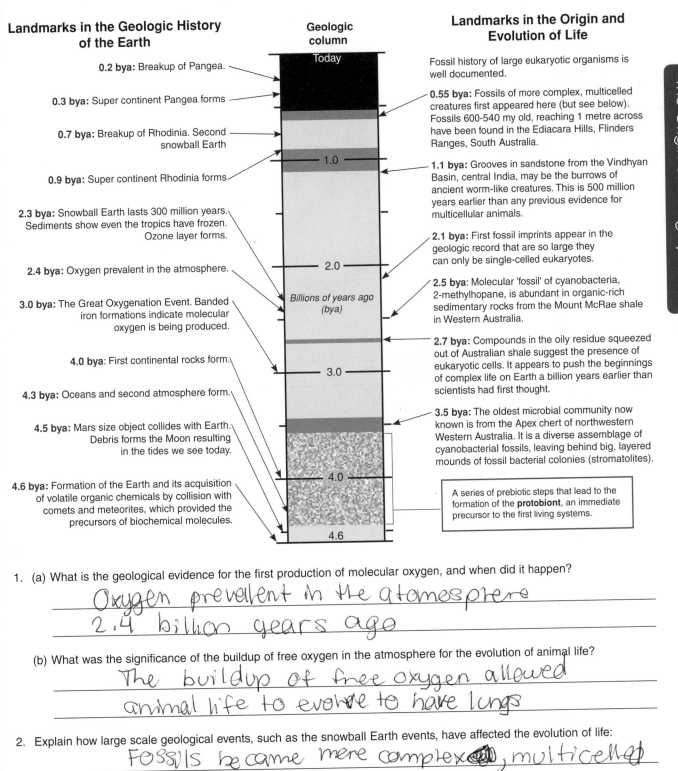

Landmarks in the Geologic History of the Earth

0.2 bya: Breakup of Pangea.

0.3 bya: Super continent Pangea forms

0.7 bya: Breakup of Rhodinia. Second snowball Earth

0.9 bya: Super continent Rhodinia forms

2.3 bya: Snowball Earth lasts 300 million years. Sediments show even the tropics have frozen. Ozone layer forms.

2.4 bya: Oxygen prevalent in the atmosphere.

3.0 bya: The Great Oxygenation Event. Banded iron formations indicate molecular oxygen is being produced.

4.0 bya: First continental rocks form.

4.3 bya: Oceans and second atmosphere form.

4.5 bya: Mars size object collides with Earth. Debris forms the Moon resulting in the tides we see today.

4.6 bya: Formation of the Earth and its acquisition of volatile organic chemicals by collision with comets and meteorites, which provided the precursors of biochemical molecules.

Geologic column

Today

1.0

2.0

Billions of years ago (bya)

3.0

4.0

4.6

Landmarks in the Origin and Evolution of Life

Fossil history of large eukaryotic organisms is well documented.

0.55 bya: Fossils of more complex, multicelled creatures first appeared here (but see below). Fossils 600-540 my old, reaching 1 metre across have been found in the Ediacara Hills, Flinders Ranges, South Australia.

1.1 bya: Grooves in sandstone from the Vindhyan Basin, central India, may be the burrows of ancient worm-like creatures. This is 500 million years earlier than any previous evidence for multicellular animals.

2.1 bya: First fossil imprints appear in the geologic record that are so large they can only be single-celled eukaryotes.

2.5 bya: Molecular 'fossil' of cyanobacteria, 2-methylhopane, is abundant in organic-rich sedimentary rocks from the Mount McRae shale in Western Australia.

2.7 bya: Compounds in the oily residue squeezed out of Australian shale suggest the presence of eukaryotic cells. It appears to push the beginnings of complex life on Earth a billion years earlier than scientists had first thought.

3.5 bya: The oldest microbial community now known is from the Apex chert of northwestern Western Australia. It is a diverse assemblage of cyanobacterial fossils, leaving behind big, layered mounds of fossil bacterial colonies (stromatolites).

A series of prebiotic steps that lead to the formation of the **protobiont**, an immediate precursor to the first living systems.

The Origin of Living Systems

1. (a) What is the geological evidence for the first production of molecular oxygen, and when did it happen?

Oxygen prevellent in the atomesphere 2.4 billion years ago

(b) What was the significance of the buildup of free oxygen in the atmosphere for the evolution of animal life?

The buildup of free oxygen allowed animal life to evolve to have lungs

2. Explain how large scale geological events, such as the snowball Earth events, have affected the evolution of life:

Fossils became more complex, multicelled creatures.

Periodicals:
A cool early life

Related activities: The Origin of Life on Earth
Weblinks: Deep Time, A Brief History of Life

A 2

The Clock of History

The visualization of Earth's history as a 24 hour clock enables almost unimaginable lengths of time to be compressed into a somewhat more meaningful, or at least imaginable, scale. Some of the key events in Earth's history are listed below:

Cenozoic Era

0.01 mya: Modern history begins.

0.2 mya: Modern humans evolve. Their activities, starting at the most recent ice age, are implicated in the extinction of the megafauna.

3 mya: Early humans arise from ape-like ancestors.

~65 mya: Major shifts in climate. Major adaptive radiations of angiosperms (flowering plants), insects, birds, and mammals. Start of Cenozoic Era.

Mesozoic Era

~65 mya: Apparent asteroid impact implicated in mass extinctions of many marine species and all dinosaurs. End of Mesozoic Era.

181 mya: Major radiations of dinosaurs and evolution of birds.

230 mya: Origin of mammals Gymnosperms become dominant land plants.

Paleozoic Era

240 mya: Mass extinction of nearly all species on land and in the sea.

340 mya: Reptiles evolve.

370 mya: Amphibians evolve.

435 mya: Major adaptive radiations of early fishes. Plants colonize the land.

550 mya: Cambrian explosion: Animals with hard parts appear. Simple marine communities indicated by early Cambrian fossil forms.

Precambrian Era (Hadean, Archean, and Proterozoic)

1100 mya: Multicellular life evolves.

3500 mya: First fossils of bacteria.

4300 mya: Oceans form.

4500 mya: Moon forms from planetary collision.

4600 mya: Origin of Earth.

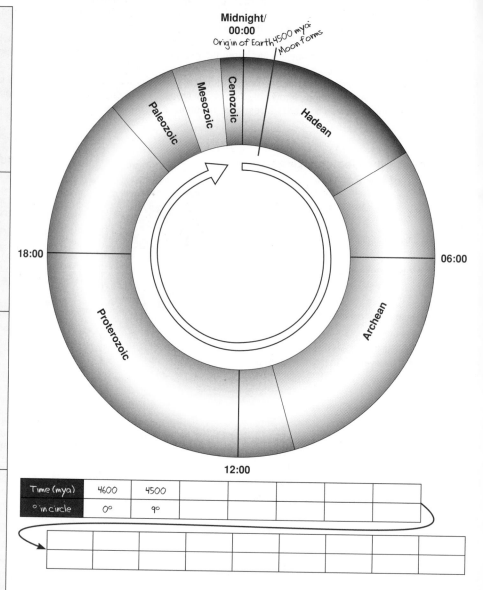

Time (mya)	4600	4500					
° in circle	0°	9°					

3. (a) Using the times given in the table above left, calculate the degrees each time represents and write it in the space provided. The first two are done for you. (Hint use the formula: 360 - (mya x 0.078)):

(b) Using the degrees you have calculated, mark on the clock the events listed. The first two are done for you:

4. Discuss the effects of changing selection pressures on the evolution of life on Earth: _____

_____

_____

_____

_____

_____

_____

_____

_____

_____

_____

_____

Appendix A: Science **Practices**

Key concepts

▸ The basis of all science is observation, hypothesis, and investigation.

▸ Scientists collect and analyze data to test their hypotheses.

▸ The design of an experiment should enable you to test your hypothesis.

▸ Data can be analyzed and presented in various ways, including in graphs and tables.

Key terms

accuracy

Chi-squared

control

controlled variable

dependent variable

graph

hypothesis

independent variable

mean

median

model (scientific)

observation

precision

prediction

rate

scientific method

table

trend

variable

Essential Knowledge

☐ 1. Use the **KEY TERMS** to compile a glossary for this topic.

The Scientific Method   pages 352-356

☐ 2. Describe and explain the principles of the **scientific method**, including the role of **observations** as a prelude to making **hypotheses** from which testable **predictions** are generated. [SP3]

☐ 3. Describe how scientific **models** can be used to illustrate biological processes and concepts, communicate information, make predictions, and describe systems. [SP1]

Planning an Investigation & Collecting Data   pages 357-360

☐ 4. Demonstrate an ability to plan and collect data appropriate for a particular scientific question [SP4]:

(a) identify the data to be collected.

(b) Describe the **control** and its role in the experiment.

(c) Identify **dependent** and **independent variables**, and how they are measured.

(d) identify the **controlled variables**, their significance, and how they will be controlled.

☐ 5. Demonstrate an ability to **record** your data systematically and accurately.

☐ 6. Distinguish between **accuracy** and **precision** and explain the significance of these when collecting quantitative data.

Transforming and Analyzing Data   pages 361-379

☐ 7. Use the flow chart provided in this section to determine the appropriate analysis for your data. [SP2]

☐ 8. Demonstrate appropriate application of mathematical routines to data., e.g. determining **mean** and **median**, calculating **rates**, and **Chi-square** analysis. [SP2]

☐ 9. identify and describe any **trends**, patterns, or relationships (e.g. correlations) in your data. [SP5]

☐ 10. Demonstrate an ability to organize different types of data appropriately in a **table**, including any calculated values.

☐ 11. Demonstrate and ability to construct a **graph** based on your collected data. Identify outliers, and explain their significance. [SP5]

Weblinks:

www.thebiozone.com/
weblink/AP1-3114.html

Teacher Resource
CD-ROM: Spreadsheets
and Statistics

The Scientific Method

Scientific knowledge grows through a process called the **scientific method**. This process involves observation and measurement, hypothesizing and predicting, and planning and executing investigations designed to test formulated **hypotheses**. A scientific hypothesis is a tentative explanation for an observation, which is capable of being tested by experimentation. Hypotheses lead to **predictions** about the system involved and they are accepted or rejected on the basis of the investigation's findings. Acceptance of the hypothesis is not necessarily permanent: explanations may be rejected later in light of new findings.

The Scientific Method

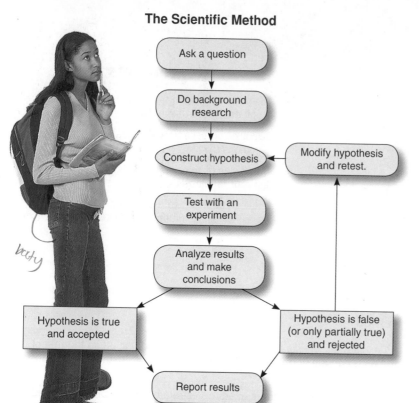

- Ask a question
- Do background research
- Construct hypothesis
- Modify hypothesis and retest.
- Test with an experiment
- Analyze results and make conclusions
- Hypothesis is true and accepted
- Hypothesis is false (or only partially true) and rejected
- Report results

betty

Forming a Hypothesis

Features of a sound hypothesis:
- It is based on observations and prior knowledge of the system.
- It offers an explanation for an observation.
- It refers to only one independent variable.
- It is written as a definite statement and not as a question.
- It is testable by experimentation.
- It leads to predictions about the system.

Testing a Hypothesis

Features of a sound method:
- It tests the validity of the hypothesis.
- It is repeatable.
- It includes a control which does not receive treatment.
- All variables are controlled where possible.
- The method includes a dependent and independent variable.
- Only the independent variable is changed (manipulated) between treatment groups.

Hypothesis involving manipulation

Used when the effect of manipulating a variable on a biological entity is being investigated. **Example:** The composition of applied fertilizer influences the rate of growth of plant A.

Hypothesis of choice

Used when investigating species preference, e.g. for a particular habitat type or microclimate. **Example:** Woodpeckers (species A) show a preference for tree type when nesting.

Hypothesis involving observation

Used when organisms are being studied in their natural environment and conditions cannot be changed. **Example:** Fern abundance is influenced by the degree to which the canopy is established.

1. Why might an accepted hypothesis be rejected at a later date? _A hypothesis can be proven wrong_

2. Explain why a method must be repeatable: _In order to see if it can have a different effect in different environments_

3. In which situation(s) is it difficult, if not impossible, to control all the variables? _____

Related activities: Planning A Quantitative Investigation
Weblinks: Hypothesis

© BIOZONE International 2012
ISBN: 978-1-927173-11-4
Photocopying Prohibited

Systems and Models

Systems are assemblages of interrelated components working together by way of a driving force. A simple example of a system is our eight-planet solar system. Each of the planet's orbits represents a single component of the system. The driving force of the system is gravity from the Sun. Modelling systems helps to understand how they work. A **model** is a representation of an object or system that shares important characteristics with the object or system being studied. A model does not necessarily have to incorporate all the characteristics or be fully accurate to be useful. It depends in the level of understanding required.

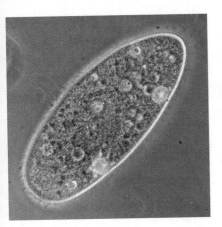

Open systems are able to exchange matter, energy and information with their surroundings. This causes them to be constantly changing, although the overall processes and outcomes remain relatively constant. Open systems are the most common type in natural systems. Examples include ecosystems, living organisms, and the ocean.

Closed systems exchange energy with their surroundings, but not matter. Closed systems are uncommon on Earth, although the cycling of certain materials, such as water and nitrogen, approximates them. The Earth itself is essentially a closed system. It receives energy from the Sun but exchanges virtually no matter with the universe (apart from the occasional meteorite).

Isolated systems exchange no energy, information or matter with their surroundings. No such systems are known to exist (with the possible exception of the entire universe). Some natural systems approximate isolated systems, at least for certain lengths of time. The solar system is essentially isolated, as is the Milky Way galaxy if gravity from nearby stars or galaxies is ignored.

Models are extremely important when trying to understand how a system operates. Models are useful for breaking complex systems or organizations down into manageable parts and often only part of a system is modelled at a time. As understanding of the system progresses, more and more data can be built into the model so that it more closely represents the real world system or object. A common example is the use of models to represent atoms. The three illustrations (right) become more complex from left to right.

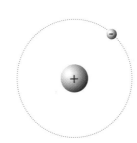

Atomic model showing position of charge

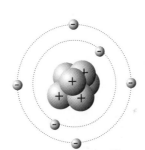

Atomic model showing 2D electron orbitals

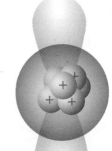

Atomic model showing 3D electron clouds

Feedback Loops

A feedback loop is a system in which the output is fed back into the system as input.

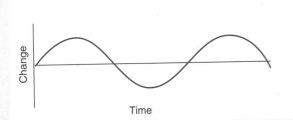

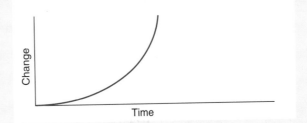

Negative Feedback Loops

Negative feedback loops are probably the most common in natural systems and counteract change in an equilibrium. As one component of the equilibrium changes, it causes changes in a second component, which returns the first component to its original state. Examples include predator-prey cycles, population changes in K-selected species, and the regulation of body systems in living organisms.

Positive feedback loops

These are less common in natural systems than negative feedback loops because they accelerate or exaggerate change and are therefore destabilizing. A change in one part of the system causes change in a second part which increases the magnitude of the first. Examples include exponential population growth, certain possible aspects of global warming, and some physiological systems in animals, with the intent of achieving a purpose, e.g. childbirth.

1. Identify each of the following as either an open, closed, or isolated system:

   (a) Reef ecosystem: _____   (e) Solar system: _____

   (b) Nitrogen cycle: _____   (f) Digestive system: _____

   (c) Earth: _____   (g) A National Park: _____

   (d) Biosphere: _____   (h) A large lake: _____

2. Identify each of the following examples as a negative or positive feedback loop:

   (a) Alarm leading to panic: _____   (d) x = (x+1): _____

   (b) Temperature regulation: _____   (e) Predator prey oscillation: _____

   (c) Hormonal changes during childbirth: _____   (f) Exponential population growth: _____

3. Explain in why the water cycle approximates a closed system: _____

   _____

   _____

   _____

   _____

4. Explain why there are no known isolated systems: _____

   _____

   _____

   _____

5. (a) Explain why there are generally few positive feedback loops in the physiological systems of living organisms:

   _____

   _____

   (b) Explain the purpose of positive feedback in physiological systems: _____

   _____

   _____

6. Explain why models are never 100% accurate representations of the system being studied: _____

   _____

   _____

   _____

7. Discuss the advantages and disadvantages of using models to explain a system: _____

   _____

   _____

   _____

   _____

   _____

   _____

   _____

Hypotheses and Predictions

A hypothesis offers a tentative explanation to questions generated by observations and leads to one or more **predictions** about the way a biological system will behave. Experiments are constructed to test these predictions. For every hypothesis, there is a corresponding **null hypothesis**; a hypothesis of no difference or no effect. Creating a null hypothesis enables a hypothesis to be tested in a meaningful way using statistical tests. If the results of an experiment are statistically significant, the null hypothesis can be rejected. If a hypothesis is accepted, anyone should be able to test the predictions with the same methods and get a similar result each time. Scientific hypotheses may be modified as more information becomes available.

Observations, Hypotheses, and Predictions

Observation is the basis for formulating hypotheses and making predictions. An observation may generate a number of plausible hypotheses, and each hypothesis will lead to one or more predictions, which can be tested by further investigation.

Observation 1: Some caterpillar species are brightly colored and appear to be conspicuous to predators such as insectivorous birds. Predators appear to avoid these species. These caterpillars are often found in groups, rather than as solitary animals.

Observation 2: Some caterpillar species are cryptic in their appearance or behavior. Their camouflage is so convincing that, when alerted to danger, they are difficult to see against their background. Such caterpillars are usually found alone.

Assumptions

Any biological investigation requires you to make **assumptions** about the biological system you are working with. Assumptions are features of the system (and your investigation) that you assume to be true but do not (or cannot) test. Possible assumptions about the biological system described above include:

- Insectivorous birds have color vision.
- Caterpillars that look bright or cryptic to us, also appear that way to insectivorous birds.
- Insectivorous birds can learn about the palatability of prey by tasting them.

1. Study the example above illustrating the features of cryptic and conspicuous caterpillars, then answer the following:

   (a) Generate a hypothesis to explain the observation that some caterpillars are brightly colored and conspicuous while others are cryptic and blend into their surroundings:

   Hypothesis: _____

   _____

   _____

   (b) State the null form of this hypothesis: _____

   _____

   (c) Describe one of the **assumptions** being made in your hypothesis: _____

   _____

   _____

   (d) Based on your hypothesis, generate a **prediction** about the behavior of insectivorous birds towards caterpillars:

   _____

   _____

© BIOZONE International 2012
ISBN: 978-1-927173-11-4
Photocopying Prohibited

Related activities: The Scientific Method

Weblinks: Hypothesis

A 2

2. During the course of any investigation, new information may arise as a result of observations unrelated to the original hypothesis. This can lead to the generation of further hypotheses about the system. For each of the incidental observations described below, formulate a prediction, and an outline of an investigation to test it. *The observation described in each case was not related to the hypothesis the experiment was designed to test:*

(a) Bacterial cultures

Prediction: _____

_____

_____

_____

_____

_____

_____

Outline of the investigation:

Observation: During an experiment on bacterial growth, these girls noticed that the cultures grew at different rates when the dishes were left overnight in different parts of the laboratory.

_____

_____

_____

_____

_____

(b) Plant cloning

Prediction: _____

_____

_____

_____

_____

_____

_____

Outline of the investigation:

Observation: During an experiment on plant cloning, a scientist noticed that the root length of plant clones varied depending on the concentration of a hormone added to the agar.

_____

_____

_____

_____

_____

_____

Accuracy and Precision

The terms accuracy and precision are often confused, or used interchangeably, but their meanings are different. In any study, **accuracy** refers to how close a measured or derived value is to its true value. Simply put, it is the correctness of the measurement. It can sometimes be a feature of the sampling equipment or its calibration. **Precision** refers to the closeness of repeated measurements to each other, i.e. the ability to be exact. A balance with a fault in it could give very precise (i.e. repeatable) but inaccurate (untrue) results. Using the analogy of a target, repeated measurements are compared to arrows being shot at a target. This analogy can be useful when thinking about the difference between accuracy and precision.

Accurate but imprecise	Precise but inaccurate	Inaccurate and imprecise	Accurate and precise

Accurate but imprecise

The measurements are all close to the true value but quite spread apart.

Analogy: The arrows are all close to the bullseye.

Precise but inaccurate

The measurements are all clustered close together but not close to the true value.

Analogy: The arrows are all clustered close together but not near the bullseye.

Inaccurate and imprecise

The measurements are all far apart and not close to the true value.

Analogy: The arrows are spread around the target.

Accurate and precise

The measurements are all close to the true value and also clustered close together.

Analogy: The arrows are clustered close together near the bullseye.

The accuracy of a measurement refers to how close the measured (or derived) value is to the true value. The precision of a measurement relates to its repeatability. In most laboratory work, we usually have no reason to suspect a piece of equipment is giving inaccurate measurements (is biased), so making precise measures is usually the most important consideration. We can test the precision of our measurements by taking repeated measurements from individual samples.

Population studies present us with an additional problem. When a researcher makes measurements of some variable in a study (e.g. fish length), they are usually trying to obtain an estimate of the true value for a parameter of interest (e.g. the mean size, therefore age, of fish). Populations are variable, so we can more accurately estimate a population parameter if we take a large number of random samples from the population.

A digital device such as this pH meter (above left) will deliver precise measurements, but its accuracy will depend on correct calibration. The precision of measurements taken with instruments such as callipers (above) will depend on the skill of the operator.

1. Distinguish between accuracy and precision: _____

   _____

   _____

2. Describe why it is important to take measurements that are both accurate and precise: _____

   _____

3. A researcher is trying to determine at what temperature enzyme A becomes denatured. Their temperature probe is incorrectly calibrated. Discuss how this might affect the accuracy and precision of the data collected:

   _____

   _____

   _____

   _____

Related activities: Terms and Notation

A 2

Variables and Data

When planning any kind of biological investigation, it is important to consider the type of data that will be collected. It is best, whenever possible, to collect quantitative or numerical data, as these data lend themselves well to analysis and statistical testing. Recording data in a systematic way as you collect it,

e.g. using a table or spreadsheet, is important, especially if data manipulation and transformation are required. It is also useful to calculate summary, descriptive statistics (e.g. mean, median) as you proceed. These will help you to recognize important trends and features in your data as they become apparent.

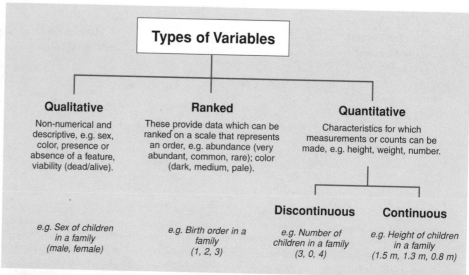

Types of Variables

Qualitative
Non-numerical and descriptive, e.g. sex, color, presence or absence of a feature, viability (dead/alive).

Ranked
These provide data which can be ranked on a scale that represents an order, e.g. abundance (very abundant, common, rare); color (dark, medium, pale).

Quantitative
Characteristics for which measurements or counts can be made, e.g. height, weight, number.

Discontinuous

Continuous

e.g. Sex of children in a family (male, female)

e.g. Birth order in a family (1, 2, 3)

e.g. Number of children in a family (3, 0, 4)

e.g. Height of children in a family (1.5 m, 1.3 m, 0.8 m)

The values for monitored or measured variables, collected during the course of the investigation, are called data. Like their corresponding variables, data may be quantitative, qualitative, or ranked.

A: Leaf shape

B: Number per litter

C: Fish length

1. For each of the photographic examples (A – C above), classify the variables as quantitative, ranked, or qualitative:

   (a) Leaf shape: _____qualitative_____

   (b) Number per litter: _____ranked_____

   (c) Fish length: _____quantitative_____

2. Why it is desirable to collect quantitative data where possible in biological studies? _____

   _____

   _____

   _____

   _____

3. How you might measure the color of light (red, blue, green) quantitatively? _____

   _____

4. (a) Give an example of data that could not be collected in a quantitative manner, explaining your answer:

   _____

   _____

   (b) Sometimes, ranked data are given numerical values, e.g. rare = 1, occasional = 2, frequent = 3, common = 4, abundant = 5. Suggest why these data are sometimes called **semi-quantitative**:

   _____

   _____

Related activities: Descriptive Statistics

Planning a Quantitative Investigation

The middle stage of any investigation (following the planning) is the practical work when the data are collected. Practical work may be laboratory or field based. Typical laboratory based experiments involve investigating how a biological response is affected by manipulating a particular **variable**, e.g. temperature. The data collected for a quantitative practical task should be recorded systematically, with due attention to safe practical techniques, a suitable quantitative method, and accurate measurements to a an appropriate degree of precision. If your quantitative practical task is executed well, and you have taken care throughout, your evaluation of the experimental results will be much more straightforward and less problematic.

Carrying Out Your Practical Work

Preparation

Familiarize yourself with the equipment and how to set it up. If necessary, calibrate equipment to give accurate measurements.

Read through the methodology and identify key stages and how long they will take.

Execution

Know how you will take your measurements, how often, and to what degree of precision.

If you are working in a group, assign tasks and make sure everyone knows what they are doing.

Recording

Record your results systematically, in a hand-written table or on a spreadsheet.

Record your results to the appropriate number of significant figures according to the precision of your measurement.

Identifying Variables

A variable is any characteristic or property able to take any one of a range of values. Investigations often look at the effect of changing one variable on another. It is important to identify all variables in an investigation: independent, dependent, and controlled, although there may be nuisance factors of which you are unaware. In all fair tests, only one variable is changed by the investigator.

Dependent variable

- Measured during the investigation.
- Recorded on the y axis of the graph.

Controlled variables

- Factors that are kept the same or controlled.
- List these in the method, as appropriate to your own investigation.

Independent variable

- Set by the experimenter.
- Recorded on the graph's x axis.

Experimental Controls

A **control** refers to standard or reference treatment or group in an experiment. It is the same as the experimental (test) group, except that it lacks the one variable being manipulated by the experimenter. Controls are used to demonstrate that the response in the test group is due a specific variable (e.g. temperature). The control undergoes the same preparation, experimental conditions, observations, measurements, and analysis as the test group. This helps to ensure that responses observed in the treatment groups can be reliably interpreted.

The experiment above tests the effect of a certain nutrient on microbial growth. All the agar plates are prepared in the same way, but the control plate does not have the test nutrient applied. Each plate is inoculated from the same stock solution, incubated under the same conditions, and examined at the same set periods. The control plate sets the baseline; any growth above that seen on the control plate is attributed to the presence of the nutrient.

Examples of Investigations

Aim		Variables	
Investigating the effect of varying...	**on the following...**	**Independent variable**	**Dependent variable**
Temperature	Leaf width	Temperature	Leaf width
Light intensity	Activity of woodlice	Light intensity	Woodlice activity
Soil pH	Plant height at age 6 months	pH	Plant height

Related activities: Variables and Data, Terms and Notation

A 2

Appendices

In order to write a sound method for your investigation, you need to determine how the independent, dependent, and controlled variables will be set and measured (or monitored). A good understanding of your methodology is crucial to a successful investigation. You need to be clear about how much data, and what type of data, you will collect. You should also have a good idea about how you plan to analyze those data. Use the example below to practise identifying this type of information.

Case Study: Catalase Activity

Catalase is an enzyme that converts hydrogen peroxide (H_2O_2) to oxygen and water. An experiment investigated the effect of temperature on the rate of the catalase reaction. Small (10 cm³) test tubes were used for the reactions, each containing 0.5 cm³ of enzyme and 4 cm³ of hydrogen peroxide. Reaction rates were assessed at four temperatures (10°C, 20°C, 30°C, and 60°C). For each temperature, there were two reaction tubes (e.g. tubes 1 and 2 were both kept at 10°C). The height of oxygen bubbles present after one minute of reaction was used as a measure of the reaction rate; a faster reaction rate produced more bubbles. The entire experiment, involving eight tubes, was repeated on two separate days.

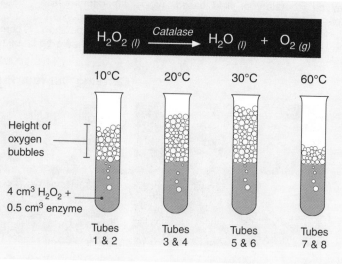

1. Write a suitable aim for this experiment: _____

   _____

2. Write a suitable hypothesis for this experiment: _____

   _____

3. (a) Identify the **independent variable:** _____

   (b) State the range of values for the independent variable: _____

   (c) Name the unit for the independent variable: _____

   (d) List the equipment needed to set the independent variable, and describe how it was used: _____

   _____

4. (a) Identify the **dependent variable**: _____

   (b) Name the unit for the dependent variable: _____

   (c) List the equipment needed to measure the dependent variable, and describe how it was used: _____

   _____

5. (a) Each temperature represents a treatment/sample/trial (circle one):

   (b) State the number of tubes at each temperature: _____

   (c) State the sample size for each treatment: _____

   (d) State how many times the whole investigation was repeated: _____

6. Explain why it would have been desirable to have included an extra tube containing no enzyme: _____

   _____

7. Identify three variables that might have been controlled in this experiment, and how they could have been monitored:

   (a) _____

   (b) _____

   (c) _____

8. Explain why controlled variables should be monitored carefully: _____

   _____

Constructing Tables

Tables provide a convenient way to systematically record and condense a large amount of information for later presentation and analysis. The protocol for creating tables for recording data during the course of an investigation is provided elsewhere, but tables can also provide a useful summary in the results section of a finished report. They provide an accurate record of numerical values and allow you to organize your data in a way that allows you to clarify the relationships and trends that are apparent. Columns can be provided to display the results of any data transformations such as rates. Some basic descriptive statistics (such as mean or standard deviation) may also be included prior to the data being plotted. For complex data sets, graphs tend to be used in preference to tables, although the latter may be provided as an appendix.

Presenting Data in Tables

Tables should have an accurate, descriptive title. Number tables consecutively through the report.

Heading and subheadings identify each set of data and show units of measurement.

Independent variable in the left column.

Table 1: Length and growth of the third internode of bean plants receiving three different hormone treatments (data are given ± standard deviation).

Treatment	Sample size	Mean rate of internode growth (mm day^{-1})	Mean internode length (mm)	Mean mass of tissue added (g day^{-1})
Control	50	0.60 ± 0.04	32.3 ± 3.4	0.36 ± 0.025
Hormone 1	46	1.52 ± 0.08	41.6 ± 3.1	0.51 ± 0.030
Hormone 2	98	0.82 ± 0.05	38.4 ± 2.9	0.56 ± 0.028
Hormone 3	85	2.06 ± 0.19	50.2 ± 1.8	0.68 ± 0.020

Control values (if present) should be placed at the beginning of the table.

Each row should show a different experimental treatment, organism, sampling site etc.

Columns for comparison should be placed alongside each other. Show values only to the level of significance allowable by your measuring technique.

Organize the columns so that each category of like numbers or attributes is listed vertically.

Tables can be used to show a calculated measure of spread of the values about the mean.

1. Describe two advantages of using a table format for data presentation:

   (a) _____

   _____

   (b) _____

   _____

2. Why might you tabulate data before you presented it in a graph? _____

   _____

   _____

3. (a) What is the benefit of tabulating basic descriptive statistics rather than raw data? _____

   _____

   _____

   (b) Why would you include a measure of spread (dispersion) for a calculated statistic in a table?

   _____

   _____

4. Why should you place control values at the beginning of a table? _____

   _____

   _____

Appendices

Constructing Graphs

Presenting results in a graph format provides a visual image of trends in data in a minimum of space. The choice between graphing or tabulation depends on the type and complexity of the data and the information that you are wanting to convey. Presenting graphs properly requires attention to a few basic details, including correct orientation and labelling of the axes, and accurate plotting of points. Common graphs include scatter plots and line graphs (for continuous data), and bar charts and histograms (for categorical data). Where there is an implied trend, a line of best fit can be drawn through the data points, as indicated in the figure below.

Presenting Data in Graph Format

A key identifies symbols. This information sometimes appears in the title.

Fig. 1: Cumulative water loss in μL from a geranium shoot in still and moving air.

Graphs (called figures) should have a concise, explanatory title. If several graphs appear in your report they should be numbered consecutively.

Label both axes and provide appropriate units of measurement if necessary.

Place the dependent variable e.g. biological response, on the vertical (Y) axis (if you are drawing a scatter graph it does not matter).

Plot points accurately. Different responses can be distinguished using different symbols, lines or bar colors.

Two or more sets of results can be plotted on the same figure and distinguished by a key. For time series it is appropriate to join the plotted points with a line.

Each axis should have an appropriate scale. Decide on the scale by finding the maximum and minimum values for each variable.

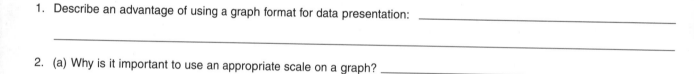

Place the independent variable e.g. treatment, on the horizontal (X) axis

1. Describe an advantage of using a graph format for data presentation: _____

_____

2. (a) Why is it important to use an appropriate scale on a graph? _____

_____

   (b) Scales on X and Y axes may sometimes be "floating" (not meeting in the lower left corner), or they may be broken using a double slash and recontinued. When and why would you use these techniques?

_____

_____

3. (a) What is wrong with the graph plotted to the right:

_____

_____

_____

_____

   (b) Describe the graph's appearance if it were plotted correctly:

_____

_____

_____

_____

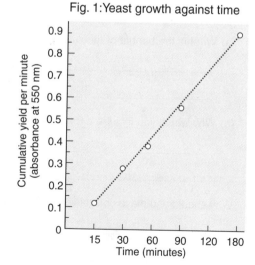

Fig. 1: Yeast growth against time

Related activities: Interpreting Sample Variability

© BIOZONE International 2012
ISBN: 978-1-927173-11-4
Photocopying Prohibited

Manipulating Raw Data

The data collected by measuring or counting in the field or laboratory are called **raw data.** They often need to be changed (**transformed**) into a form that makes it easier to identify important features of the data (e.g. trends). Some basic calculations, such as totals (the sum of all data values for a variable), are made as a matter of course to compare replicates or as a prelude to other transformations. The calculation of **rate** (amount per unit time) is another example of a commonly performed calculation,

and is appropriate for many biological situations (e.g. measuring growth or weight loss or gain). For a line graph, with time as the independent variable plotted against the values of the biological response, the slope of the line is a measure of the rate. Biological investigations often compare the rates of events in different situations (e.g. the rate of photosynthesis in the light and in the dark). Other typical transformations include frequencies (number of times a value occurs) and percentages (fraction of 100).

Tally Chart

Records the number of times a value occurs in a data set

HEIGHT (cm)	TALLY	TOTAL
0-0.99	|||	3
1-1.99	++++		6
2-2.99	++++ ++++	10
3-3.99	++++ ++++ ||	12
4-4.99	|||	3
5-5.99	||	2

- A useful first step in analysis; a neatly constructed tally chart doubles as a simple histogram.
- Cross out each value on the list as you tally it to prevent double entries. Check all values are crossed out at the end and that totals agree.

Example: Height of 6d old seedlings

Percentages

Expressed as a fraction of 100

Women	Body mass (kg)	Lean body mass (kg)	% lean body mass
Athlete	50	38	76.0
Lean	56	41	73.2
Normal weight	65	46	70.8
Overweight	80	48	60.0
Obese	95	52	54.7

- Percentages provide a clear expression of what proportion of data fall into any particular category, e.g. for pie graphs.
- Allows meaningful comparison between different samples.
- Useful to monitor change (e.g. % increase from one year to the next).

Example: Percentage of lean body mass in women

Rates

Expressed as a measure per unit time

Time (minutes)	Cumulative sweat loss (mL)	Rate of sweat loss (mL min⁻¹)
0	0	0
10	50	5
20	130	8
30	220	9
60	560	11.3

- Rates show how a variable changes over a standard time period (e.g. one second, one minute, or one hour).
- Rates allow meaningful comparison of data that may have been recorded over different time periods.

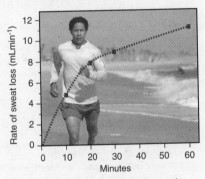
Example: Rate of sweat loss in exercise

1. (a) Explain what it means to transform data: _____

(b) Briefly explain the general purpose of transforming data: _____

2. For each of the following examples, state a suitable transformation, together with a reason for your choice:

(a) Determining relative abundance from counts of four plant species in two different habitat areas:

Suitable transformation: _____

Reason: _____

(b) Determining the effect of temperature on the production of carbon dioxide by respiring seeds:

Suitable transformation: _____

Reason: _____

Related activities: Variables and Data

DA

3. Complete the transformations for each of the tables on the right. The first value is provided in each case.

(a) TABLE: Incidence of cyanogenic clover in different areas:

Working: 124 ÷ 159 = 0.78 = 78%

> This is the number of cyanogenic clover out of the total.

Incidence of cyanogenic clover in different areas

Clover plant type	Frost free area		Frost prone area		Totals
	Number	%	Number	%	
Cyanogenic	124	78	26		
Acyanogenic	35		115		
Total	159				

(b) TABLE: Plant water loss using a bubble potometer

Working: (9.0 – 8.0) ÷ 5 min = 0.2

> This is the distance the bubble moved over the first 5 minutes. Note that there is no data entry possible for the first reading (0 min) because no difference can be calculated.

Plant water loss using a bubble potometer

Time (min)	Pipette arm reading (cm^3)	Plant water loss ($cm^3\ min^{-1}$)
0	9.0	–
5	8.0	0.2
10	7.2	
15	6.2	
20	4.9	

(c) TABLE: Photosynthetic rate at different light intensities:

Working: 1 ÷ 15 = 0.067

> This is time taken for the leaf to float. A reciprocal gives a per minute rate (the variable measured is the time taken for an event to occur).

NOTE: In this experiment, the flotation time is used as a crude measure of photosynthetic rate. As oxygen bubbles are produced as a product of photosynthesis, they stick to the leaf disc and increase its buoyancy. The faster the rate, the sooner they come to the surface. The rates of photosynthesis should be measured over similar time intervals, so the rate is transformed to a 'per minute' basis (the reciprocal of time).

Photosynthetic rate at different light intensities

Light intensity (%)	Average time for leaf disc to float (min)	Reciprocal of time (min^{-1})
100	15	0.067
50	25	
25	50	
11	93	
6	187	

(d) TABLE: Frequency of size classes in a sample of eels:

Working: (7 ÷ 270) x 100 = 2.6 %

> This is the number of individuals out of the total that appear in the size class 0-50 mm. The relative frequency is rounded to one decimal place.

Frequency of size classes in a sample of eels

Size class (mm)	Frequency	Relative frequency (%)
0-50	7	2.6
50-99	23	
100-149	59	
150-199	98	
200-249	50	
250-299	30	
300-349	3	
Total	270	

Taking the Next Step

By this stage, you will have completed many of the early stages of your investigation. Now is a good time to review what you have done and reflect on the biological significance of what you are investigating. Review the first page of this flow chart in light of your findings so far. You are now ready to begin a more in-depth analysis of your results. Never under-estimate the value of plotting your data, even at a very early stage. This will help you decide on the best type of data analysis (following page).

Photos courtesy of Pasco

Observation

Something...

- Changes or affects something else.
- Is more abundant, etc. along a transect, at one site, temperature, concentration, etc. than others.
- Is bigger, taller, or grows more quickly.

Pilot study

Lets you check...

- Equipment, sampling sites, sampling interval.
- How long it takes to collect data.
- Problems with identification or other unforeseen issues.

Research

To find out...

- Basic biology and properties.
- What other biotic or abiotic factors may have an effect.
- Its place within the broader biological context.

Analysis

Are you looking for a...

- **Difference**.
- **Trend** or relationship.
- **Goodness of fit** (to a theoretical outcome).

GO TO NEXT PAGE

Be prepared to revise your study design in the light of the results from your pilot study

Variables

Next you need to...

- Identify the key variables likely to cause the effect.
- Identify variables to be controlled in order to give the best chance of showing the effect that you want to study.

Hypothesis

Must be...

- Testable
- Able to generate predictions

so that in the end you can say whether your data supports or allows you to reject your hypothesis.

*Related activities: Chi-Squared Test, Chi-Squared Exercise in Ecology
Chi-Squared Exercise in Genetics*

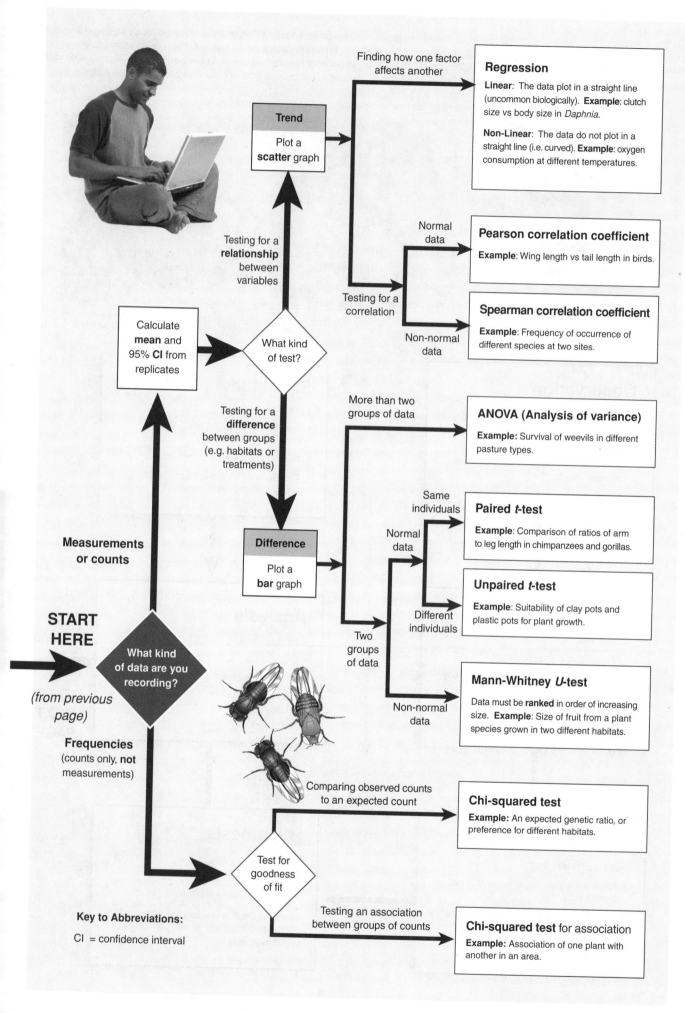

Finding how one factor affects another

Regression

Linear: The data plot in a straight line (uncommon biologically). **Example**: clutch size vs body size in *Daphnia*.

Non-Linear: The data do not plot in a straight line (i.e. curved). **Example**: oxygen consumption at different temperatures.

Trend

Plot a **scatter** graph

Testing for a **relationship** between variables

Normal data

Pearson correlation coefficient

Example: Wing length vs tail length in birds.

Testing for a correlation

Spearman correlation coefficient

Example: Frequency of occurrence of different species at two sites.

Non-normal data

Calculate **mean** and 95% **CI** from replicates

What kind of test?

More than two groups of data

ANOVA (Analysis of variance)

Example: Survival of weevils in different pasture types.

Testing for a **difference** between groups (e.g. habitats or treatments)

Same individuals

Paired *t*-test

Example: Comparison of ratios of arm to leg length in chimpanzees and gorillas.

Normal data

Difference

Plot a **bar** graph

Unpaired *t*-test

Example: Suitability of clay pots and plastic pots for plant growth.

Different individuals

Measurements or counts

Two groups of data

Mann-Whitney *U*-test

Data must be **ranked** in order of increasing size. **Example**: Size of fruit from a plant species grown in two different habitats.

Non-normal data

START HERE

What kind of data are you recording?

(from previous page)

Frequencies (counts only, **not** measurements)

Comparing observed counts to an expected count

Chi-squared test

Example: An expected genetic ratio, or preference for different habitats.

Test for goodness of fit

Testing an association between groups of counts

Chi-squared test for association

Example: Association of one plant with another in an area.

Key to Abbreviations:

CI = confidence interval

Descriptive Statistics

For most investigations, measures of the biological response are made from more than one sampling unit. The sample size (the number of sampling units) will vary depending on the resources available. In lab based investigations, the sample size may be as small as two or three (e.g. two test-tubes in each treatment). In field studies, each individual may be a sampling unit, and the sample size can be very large (e.g. 100 individuals). It is useful to summarize the data collected using **descriptive statistics.**

Descriptive statistics, such as mean, median, and mode, can help to highlight trends or patterns in the data. Each of these statistics is appropriate to certain types of data or distributions, e.g. a mean is not appropriate for data with a skewed distribution (see below). Frequency graphs are useful for indicating the distribution of data. Standard deviation and standard error are statistics used to quantify the amount of spread in the data and evaluate the reliability of estimates of the true (population) mean.

Variation in Data

Whether they are obtained from observation or experiments, most biological data show variability. In a set of data values, it is useful to know the value about which most of the data are grouped; the centre value. This value can be the mean, median, or mode depending on the type of variable involved (see schematic below). The main purpose of these statistics is to summarize important trends in your data and to provide the basis for statistical analyses.

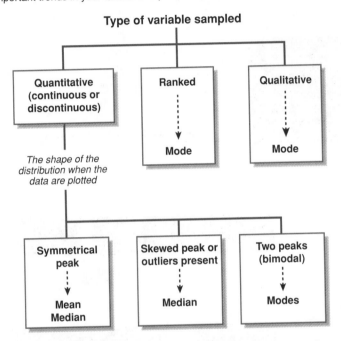

The shape of the distribution will determine which statistic (mean, median, or mode) best describes the central tendency of the sample data.

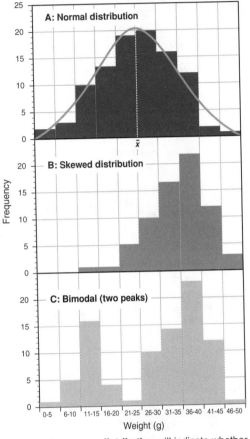

A **frequency distribution** will indicate whether the data are normal, skewed, or bimodal.

Case Study: Height of Swimmers

Data (below) and descriptive statistics (left) from a survey of the height of 29 members of a male swim squad.

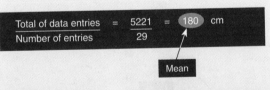

1. Give a reason for the difference between the mean, median, and mode for the swimmers' height data:

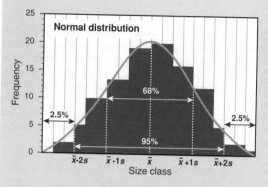

Measuring Spread

The **standard deviation** is a frequently used measure of the variability (spread) in a set of data. It is usually presented in the form $\bar{x} \pm s$. In a normally distributed set of data, 68% of all data values will lie within one standard deviation (s) of the mean ($\bar{x}$) and 95% of all data values will lie within two standard deviations of the mean (left).

Two different sets of data can have the same mean and range, yet the distribution of data within the range can be quite different. In both the data sets pictured in the histograms below, 68% of the values lie within the range $\bar{x} \pm 1s$ and 95% of the values lie within $\bar{x} \pm 2s$. However, in B, the data values are more tightly clustered around the mean.

Histogram A has a larger standard deviation; the values are spread widely around the mean.

Both plots show a normal distribution with a symmetrical spread of values about the mean.

Histogram B has a smaller standard deviation; the values are clustered more tightly around the mean.

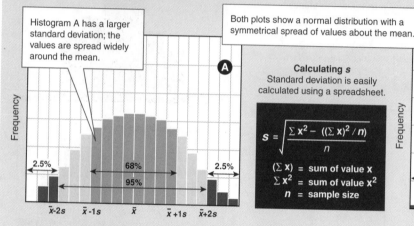

Calculating s
Standard deviation is easily calculated using a spreadsheet.

$$s = \sqrt{\frac{\sum x^2 - ((\sum x)^2 / n)}{n}}$$

$(\sum x)$ = sum of value x
$\sum x^2$ = sum of value x^2
n = sample size

Case Study: Fern Reproduction

Raw data (below) and descriptive statistics (right) from a survey of the number of sori found on the fronds of a fern plant.

Fern spores

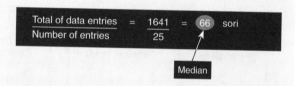

$$\frac{\text{Total of data entries}}{\text{Number of entries}} = \frac{1641}{25} = 66 \text{ sori}$$

Median

Raw data: Number of sori per frond

64	60	64	62	68	66	66	63
69	70	63	70	70	63	63	62
71	69	59	70	66	61	61	70
67	64	63	64				

Number of sori per frond (in rank order)		Sori per frond	Tally	Total
59	66	59	✔	1
60	66	60	✔	1
61	67	61	✔	1
62	68	62	✔✔	2
62	69	63	✔✔✔✔	4
63	69	64	✔✔✔✔	4
63	70	65		0
63	70	66	✔✔	2
63	70	67	✔	1
64	70	68	✔	1
64	70	69	✔✔	2
64	71	70	✔✔✔✔✔	5
64		71	✔	1

Median
Mode

2. Give a reason for the difference between the mean, median, and mode for the fern sori data:

_____

_____

_____

_____

3. Calculate the mean, median, and mode for the data on ladybird masses below. Draw up a tally chart and show all calculations:

Ladybird mass (mg)		
10.1	8.2	7.7
8.0	8.8	7.8
6.7	7.7	8.8
9.8	8.8	8.9
6.2	8.8	8.4

Interpreting Sample Variability

Measures of central tendency, such as mean, attempt to identify the most representative value in a set of data, but the description of a data set also requires that we know something about how far the data values are spread around that central measure. As we have seen in the previous activity, the **standard deviation** (*s*) gives a simple measure of the spread or **dispersion** in data. The **variance** (*s²*) is also a measure of dispersion, but the standard deviation is usually preferred because it is expressed in the original units. Two data sets could have exactly the same mean values, but very different values of dispersion. If we were simply to use the central tendency to compare these data sets, the results would (incorrectly) suggest that they were alike. The assumptions we make about a population will be affected by what the sample data tell us. This is why it is important that sample data are unbiased (e.g. collected by **random sampling**) and that the sample set is as large as practicable. This exercise will help to illustrate how our assumptions about a population are influenced by the information provided by the sample data.

Complete sample set
n = 689 (random)

Length in mm	Freq
25	1
26	0
27	0
28	0
29	0
30	0
31	0
32	2
33	3
34	3
35	4
36	5
37	10
38	23
39	22
40	33
41	39
42	41
43	41
44	36
45	49
46	32
47	14
48	32
49	27
50	25
51	24
52	17
53	18
54	27
55	21
56	20
57	11
58	18
59	16
60	22
61	13
62	8
63	10
64	5
65	7
66	2
67	3
68	3
69	1
70	0
71	1

Random Sampling, Sample Size, and Dispersion in Data

Sample size and sampling bias can both affect the information we obtain when we sample a population. In this exercise you will calculate some descriptive statistics for some sample data.

The complete set of sample data we are working with comprises 689 length measurements of year zero (young of the year) perch (column left). Basic descriptive statistics for the data have bee calculated for you below and the frequency histogram has also been plotted.

Look at this data set and then complete the exercise to calculate the same statistics from each of two smaller data sets (tabulated right) drawn from the same population. This exercise shows how random sampling, large sample size, and sampling bias affect our statistical assessment of variation in a population.

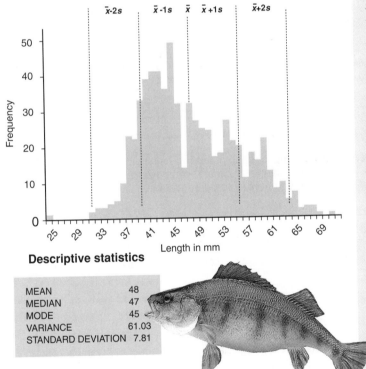

Length of year zero perch

Descriptive statistics

MEAN	48
MEDIAN	47
MODE	45
VARIANCE	61.03
STANDARD DEVIATION	7.81

Small sample set
n = 30 (random)

Length in mm	Freq
25	1
26	0
27	0
28	0
29	0
30	0
31	0
32	0
33	0
34	0
35	2
36	0
37	0
38	3
39	2
40	1
41	3
42	0
43	0
44	0
45	0
46	1
47	0
48	2
49	0
50	0
51	1
52	3
53	0
54	0
55	0
56	0
57	1
58	0
59	3
60	2
61	2
62	0
63	0
64	0
65	0
66	0
67	2
68	1

This population was sampled randomly to obtain this data set

This column records the number of fish of each size

30 — Number of fish in the sample

Small sample set
n = 50 (bias)

Length in mm	Freq
46	1
47	0
48	0
49	1
50	0
51	0
52	1
53	1
54	1
55	1
56	0
57	2
58	2
59	4
60	1
61	0
62	8
63	10
64	13
65	2
66	0
67	2

50

The person gathering this set of data was biased towards selecting larger fish because the mesh size on the net was too large to retain small fish

1. For the complete data set (*n* = 689) calculate the percentage of data falling within:

   (a) ± one standard deviation of the mean: _____

   (b) ± two standard deviations of the mean: _____

   (c) Explain what this information tells you about the distribution of year zero perch from this site: _____

   _____

2. Give another reason why you might reach the same conclusion about the distribution: _____

   _____

© BIOZONE International 2012
ISBN: 978-1-927173-11-4
Photocopying Prohibited

Related activities: Descriptive Statistics

DA 3

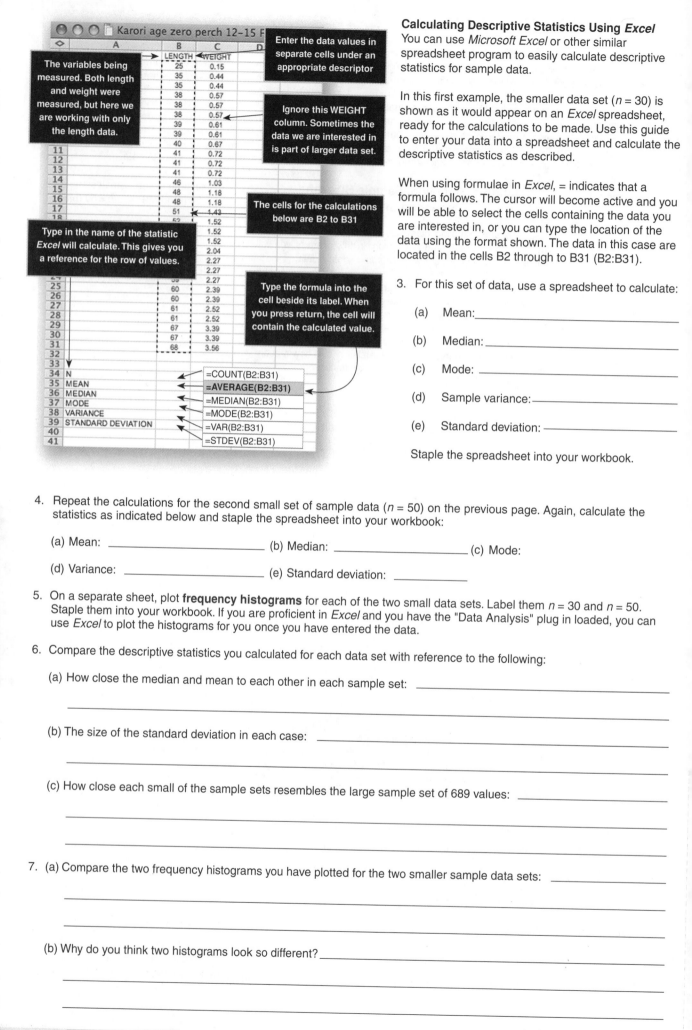

Karori age zero perch 12-15 F

	A	B	C	D
		LENGTH	WEIGHT	
		25	0.15	
		35	0.44	
		35	0.44	
		38	0.57	
		38	0.57	
		38	0.57	
		39	0.61	
		39	0.61	
		40	0.67	
11		41	0.72	
12		41	0.72	
13		41	0.72	
14		46	1.03	
15		48	1.18	
16		48	1.18	
17		51	1.43	
18		52	1.52	
			1.52	
			1.52	
			2.04	
			2.27	
			2.27	
			2.27	
25		60	2.39	
26		60	2.39	
27		61	2.52	
28		61	2.52	
29		67	3.39	
30		67	3.39	
31		68	3.56	
32				
33				
34	N			
35	MEAN			
36	MEDIAN			
37	MODE			
38	VARIANCE			
39	STANDARD DEVIATION			
40				
41				

The variables being measured. Both length and weight were measured, but here we are working with only the length data.

Type in the name of the statistic *Excel* will calculate. This gives you a reference for the row of values.

Enter the data values in separate cells under an appropriate descriptor

Ignore this WEIGHT column. Sometimes the data we are interested in is part of larger data set.

The cells for the calculations below are B2 to B31

Type the formula into the cell beside its label. When you press return, the cell will contain the calculated value.

```
=COUNT(B2:B31)
=AVERAGE(B2:B31)
=MEDIAN(B2:B31)
=MODE(B2:B31)
=VAR(B2:B31)
=STDEV(B2:B31)
```

Calculating Descriptive Statistics Using *Excel*

You can use *Microsoft Excel* or other similar spreadsheet program to easily calculate descriptive statistics for sample data.

In this first example, the smaller data set (n = 30) is shown as it would appear on an *Excel* spreadsheet, ready for the calculations to be made. Use this guide to enter your data into a spreadsheet and calculate the descriptive statistics as described.

When using formulae in *Excel*, = indicates that a formula follows. The cursor will become active and you will be able to select the cells containing the data you are interested in, or you can type the location of the data using the format shown. The data in this case are located in the cells B2 through to B31 (B2:B31).

3. For this set of data, use a spreadsheet to calculate:

    (a)   Mean:_____

    (b)   Median: _____

    (c)   Mode: _____

    (d)   Sample variance:_____

    (e)   Standard deviation: _____

Staple the spreadsheet into your workbook.

4. Repeat the calculations for the second small set of sample data (n = 50) on the previous page. Again, calculate the statistics as indicated below and staple the spreadsheet into your workbook:

    (a) Mean: _____   (b) Median: _____   (c) Mode: _____

    (d) Variance: _____   (e) Standard deviation: _____

5. On a separate sheet, plot **frequency histograms** for each of the two small data sets. Label them n = 30 and n = 50. Staple them into your workbook. If you are proficient in *Excel* and you have the "Data Analysis" plug in loaded, you can use *Excel* to plot the histograms for you once you have entered the data.

6. Compare the descriptive statistics you calculated for each data set with reference to the following:

    (a) How close the median and mean to each other in each sample set: _____

    _____

    (b) The size of the standard deviation in each case: _____

    _____

    (c) How close each small of the sample sets resembles the large sample set of 689 values: _____

    _____

    _____

7. (a) Compare the two frequency histograms you have plotted for the two smaller sample data sets: _____

    _____

    _____

    (b) Why do you think two histograms look so different? _____

    _____

    _____

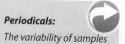

Periodicals:
The variability of samples

The Reliability of the Mean

You have already seen how to use the **standard deviation** (s) to quantify the spread or **dispersion** in your data. The **variance** (s²) is another such measure of dispersion, but the standard deviation is usually the preferred of these two measures because it is expressed in the original units. Usually, you will also want to know how good your sample mean (x̄) is as an estimate of the true population mean (μ). This can be indicated by the standard error of the mean (or just **standard error** or SE). SE is often used as an error measurement simply because it is small, rather than for any good statistical reason. However, it is does allow you

calculate the **95% confidence interval (95% CI)**. The calculation and use of 95% CIs is outlined below and the following page. By the end of this activity you should be able to:
- Enter data and calculate descriptive statistics using a spreadsheet program such as *Microsoft Excel*. You can follow this procedure for any set of data.
- Calculate standard error and 95% confidence intervals for sample data and plot these data appropriately with error bars.
- Interpret the graphically presented data and reach tentative conclusions about the findings of the experiment.

Reliability of the Sample Mean

When we take measurements from samples of a larger population, we are using those samples as indicators of the trends in the whole population. Therefore, when we calculate a sample mean, it is useful to know how close that value is to the true population mean (μ). This is not merely an academic exercise; it will enable you to make inferences about the aspect of the population in which you are interested. For this reason, statistics based on samples and used to estimate population parameters are called **inferential statistics**.

Ladybird-population

When we measure a particular attribute from a sample of a larger population and calculate a mean for that attribute, we can calculate how closely our sample mean (the statistic) is to the true population mean for that attribute (the parameter).
Example: If we calculated the mean number of carapace spots from a sample of six ladybird beetles, how reliable is this statistic as an indicator of the mean number of carapace spots in the whole population? We can find out by calculating the **95% confidence interval**.

The Standard Error (SE)

The standard error (SE) is simple to calculate and is usually a small value. Standard error is given by:

$$SE = \frac{s}{\sqrt{n}}$$

where s = the standard deviation, and n = sample size.

Standard errors are sometimes plotted as error bars on graphs, but it is more meaningful to plot the **95% confidence intervals** (see box below). All calculations are easily made using a spreadsheet (see opposite).

The 95% Confidence Interval

SE is required to calculate the 95% confidence interval (CI) of the mean. This is given by:

$$95\% \text{ CI} = SE \times t_{P(n-1)}$$

Do not be alarmed by this calculation; once you have calculated the value of the SE, it is a simple matter to multiply this value by the value of t at P = 0.05 (from the t table) for the appropriate degrees of freedom (df) for your sample (n – 1).

For example: where the SE = 0.6 and the sample size is 10, the calculation of the 95% CI is:

$$95\% \text{ CI} = 0.6 \times 2.262 = \boxed{1.36}$$

Part of the t table is given to the right for P = 0.05. Note that, as the sample becomes very large, the value of t becomes smaller. For very large samples, t is fixed at 1.96, so the 95% CI is slightly less than twice the SE

All these statistics, including a plot of the data with Y error bars, can be calculated using a program such as *Microsoft Excel*.

Critical values of Student's t distribution at P = 0.05.

df	P 0.05
1	12.71
2	4.303
3	3.182
4	2.776
5	2.571
6	2.447
7	2.365
8	2.306
9	2.262
10	2.228
20	2.086
30	2.042
40	2.021
60	2.000
120	1.980
>120	1.960

Value of t at n–1 = 9

Maximum value of t at this level of P

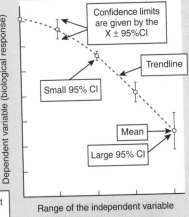

Relationship of Y against X (± 95% confidence intervals, n = 10)

Confidence limits are given by the X ± 95%CI. Trendline. Small 95% CI. Mean. Large 95% CI. Dependent variable (biological response). Range of the independent variable

Plotting your confidence intervals

Once you have calculated the 95% CI for the means in your data set, you can plot them as error bars on your graph. Note that the **95% confidence limits** are given by the value of the **mean ± 95%CI**. A 95% confidence limit (i.e. P = 0.05) tells you that, on average, 95 times out of 100, the limits will contain the true population mean.

© BIOZONE International 2012
ISBN: 978-1-927173-11-4
Photocopying Prohibited

Related activities: Descriptive Statistics, Taking the Next Step
Weblinks: Introduction to Descriptive Statistics

DA

Comparing Treatments Using Descriptive Statistics

In an experiment, the growth of newborn rats on four different feeds was compared by weighing young rats after 28 days on each of four feeding regimes. The suitability of each food type for maximizing growth in the first month of life was evaluated by comparing the means of the four experimental groups. Each group comprised 10 individual rats. All 40 newborns were born to sibling mothers with the same feeding history. For this activity, follow the steps outlined below and reproduce them yourself.

1 Calculating Descriptive Statistics

Entering your data and calculating descriptive statistics.

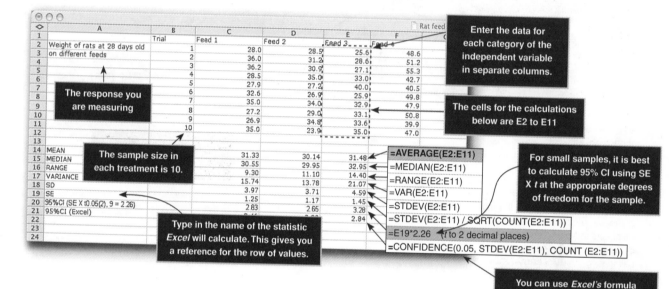

Enter the data for each category of the independent variable in separate columns.

The response you are measuring

The cells for the calculations below are E2 to E11

The sample size in each treatment is 10.

	Trial	Feed 1	Feed 2	Feed 3	Feed 4		
Weight of rats at 28 days old on different feeds	1	28.0	28.5	25.6	48.6		
	2	36.0	31.2	28.6	51.2		
	3	36.2	30.9	27.1	55.3		
	4	28.5	35.0	33.0	42.7		
	5	27.9	27.2	40.0	40.5		
	6	32.6	26.9	25.9	49.8		
	7	35.0	34.0	32.9	47.9		
	8	27.2	29.0	33.1	50.8		
	9	26.9	34.8	33.6	39.9		
	10	35.0	23.9	35.0	47.0		
MEAN		31.33	30.14	31.48		=AVERAGE(E2:E11)	
MEDIAN		30.55	29.95	32.95		=MEDIAN(E2:E11)	
RANGE		9.30	11.10	14.40		=RANGE(E2:E11)	
VARIANCE		15.74	13.78	21.07		=VAR(E2:E11)	
SD		3.97	3.71	4.59		=STDEV(E2:E11)	
SE		1.25	1.17	1.45		=STDEV(E2:E11) / SQRT(COUNT(E2:E11))	
95%CI (SE X t0.05(2), 9 = 2.26)		2.83	2.65	3.28		=E19*2.26 (to 2 decimal places)	
95%CI (Excel)				2.84		=CONFIDENCE(0.05, STDEV(E2:E11), COUNT (E2:E11))	

Type in the name of the statistic *Excel* will calculate. This gives you a reference for the row of values.

For small samples, it is best to calculate 95% CI using SE X *t* at the appropriate degrees of freedom for the sample.

You can use *Excel's* formula to calculate this statistic, but it approximates *t* to 1.96, which is inaccurate for small sample sizes.

2 Drawing the Graph

To plot the graph, you will need to enter the data values you want to plot in a format that *Excel* can use (above). To do this, enter the values in columns under each category.

▶ Each column will have two entries: mean and 95% CI. In this case, we want to plot the mean weight of 28 day rats fed on different foods and add the 95% confidence intervals as error bars.

▶ The independent variable is categorical, so the correct graph type is a column chart. Select the row of mean values (including column headings).

Mean and 95% confidence interval

Mean values and column headings

Rat feeding CI.xls

	Feed 1	Feed 2	Feed 3	Feed 4	M
Mean	31.33	30.14	31.48	47.37	
95% CI	2.83	2.65	3.28	3.55	

Enter the values for means and 95% CI in columns under the categories.

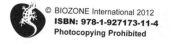

▶ **STEP 1**:
From the menu bar choose: **Insert > Chart > Column**. This is **Step 1** in the Chart Wizard. Click **Next**.

▶ **STEP 2**:
Click **Next**.

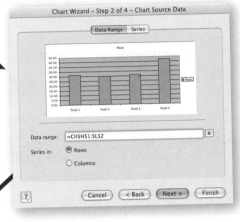

▶ **STEP 3**:
You have the option to add a title, labels for your X and Y axes, turn off gridlines, and add (or remove) a legend (key). When you have added all the information, click **Next**.

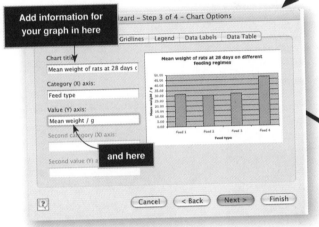

▶ **STEP 4**:
Specify the chart location. It should appear "as object in" Sheet 1 by default. Click on the chart and move it to reveal the data.

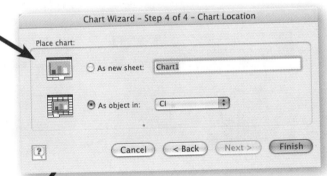

❸ **Formatting the graph**

A chart will appear on the screen. Right click (Ctrl-click on Mac) on any part of any column and choose **Format data series**. To add error bars, select the Y error bars tab, and click on the symbol that shows Display both. Click on Custom, and use the data selection window to select the row of 95% CI data for "+" and "–" fields.

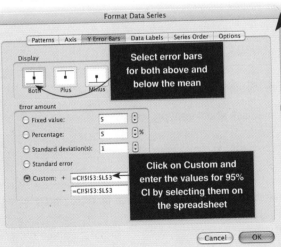

Click on OK and your chart will plot with error bars.

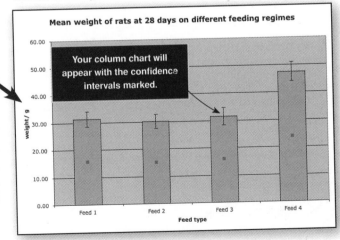

Appendices

Further Data Transformations

Raw data usually needs some kind of processing so that trends in the data and relationships between variables can be easily identified and tested statistically. The simplest and most powerful statistical tests generally require data to exhibit a normal distribution, yet many biological variables are not distributed in this way. It is possible to get around this apparent problem by transforming the data. Data transformation can help to account for differences between sample sizes in different treatments. It is also a perfectly legitimate way to normalize data so that its distribution meets the criteria for analysis. It is not a way to manipulate data to get the result you want. Your choice of transformation is based on the type of data you have and how you propose to analyze it. Some experimental results may be so clear, a complex statistical analysis is unnecessary.

Reciprocals

1 / x is the reciprocal of x.

Enzyme concentration ($\mu g\ mL^{-1}$)	Reaction time (min)	Reciprocal value
6	25	0.04
12.5	20	0.05
25	14	0.07
50	5	0.20
100	2.5	0.40
150	1.75	0.57

- Reciprocals of time (1/data value) can provide a crude measure of rate in situations where the variable measured is the total time taken to complete a task.

Problem: Responses are measured over different time scales.

Example: Time taken for color change in an enzyme reaction.

Square Root

A square root is a value that when multiplied by itself gives the original number.

Sampling site	No. of Woodlice	Square root
1	10	3.16
2	7	2.65
3	5	2.24
4	3	1.73
5	1	1
6	0	0
7	1	1
8	1	1

- Applied to data that counts something.
- The square root of a negative number cannot be taken. Negative numbers are made positive by the addition of a constant value.
- Helps to normalize skewed data.

Problem: Skewed data.

Example: The number of woodlice distributed across a transect.

Log$_{10}$

A log transformation has the effect of normalizing data.

- Log transformations are useful for data where there is an exponential increase in numbers (e.g. cell growth).
- Log transformed data will plot as a straight line and the numbers are more manageable.
- To find the $\log_{10}$ of a number, e.g. 32, using a calculator, key in log 32 = ___.

The answer should be 1.51.

Problem: Exponential increases.

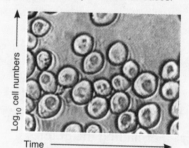

Example: Cell growth in a yeast culture

1. Why might a researcher transform skewed or non-normal data prior to statistical analysis? _____

_____

_____

_____

2. For each of the following examples, state a suitable transformation, together with a reason for your choice:

(a) Comparing the time taken for chemical precipitation to occur in a flask at different pH values:

Suitable transformation: _____

Reason: _____

(b) Analyzing the effect of growth environment on the number of bacterial colonies developing in 50 agar plates:

Suitable transformation: _____

Reason: _____

Related activities: Manipulating Raw Data

Using the Chi-Squared Test in Genetics

The **chi-squared test**, χ^2, is frequently used for testing the outcome of dihybrid crosses against an expected (predicted) Mendelian ratio, and it is appropriate for use in this way. When using the chi-squared test for this purpose, the null hypothesis predicts the ratio of offspring of different phenotypes according to the expected Mendelian ratio for the cross, assuming independent assortment of alleles (no linkage). Significant departures from the predicted Mendelian ratio indicate linkage of the alleles in question. Raw counts should be used and a large sample size is required for the test to be valid.

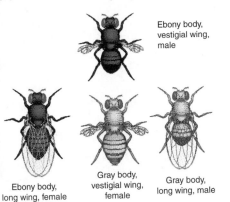

Ebony body, vestigial wing, male

Ebony body, long wing, female

Gray body, vestigial wing, female

Gray body, long wing, male

Images of *Drosophila* courtesy of **Newbyte Educational Software**: *Drosophila* Genetics Lab (**www.newbyte.com**)

Using χ^2 in Mendelian Genetics

In a *Drosophila* genetics experiment, two individuals were crossed (the details of the cross are not relevant here). The predicted Mendelian ratios for the offspring of this cross were 1:1:1:1 for each of the four following phenotypes: gray body-long wing, gray body-vestigial wing, ebony body-long wing, ebony body-vestigial wing. The observed results of the cross were not exactly as predicted. The following numbers for each phenotype were observed in the offspring of the cross:

Observed results of the example *Drosophila* cross

Gray body, long wing	98	Ebony body, long wing	102
Gray body, vestigial wing	88	Ebony body, vestigial wing	112

Using χ^2, the probability of this result being consistent with a 1:1:1:1 ratio could be tested. Worked example as follows:

Step 1: Calculate the expected value (E)

In this case, this is the sum of the observed values divided by the number of categories (see note below)

$$\frac{400}{4} = 100$$

Step 2: Calculate O – E

The difference between the observed and expected values is calculated as a measure of the deviation from a predicted result. Since some deviations are negative, they are all squared to give positive values. This step is usually performed as part of a tabulation (right, darker blue column).

Category	O	E	O – E	$(O – E)^2$	$\dfrac{(O – E)^2}{E}$
Gray, long wing	98	100	–2	4	0.04
Gray, vestigial wing	88	100	–12	144	1.44
Ebony, long wing	102	100	2	4	0.04
Ebony, vestigial wing	112	100	12	144	1.44
Total = 400				χ^2	$\Sigma = 2.96$

Step 3: Calculate the value of χ^2

$$\chi^2 = \sum \frac{(O – E)^2}{E}$$

Where: O = the observed result
E = the expected result
Σ = sum of

The calculated χ^2 value is given at the bottom right of the last column in the tabulation.

Step 5a: Using the χ^2 table

On the χ^2 table (part reproduced in Table 1 below) with 3 degrees of freedom, the calculated value for χ^2 of 2.96 corresponds to a probability of between 0.2 and 0.5 (see arrow). *This means that by chance alone a χ^2 value of 2.96 could be expected between 20% and 50% of the time.*

Step 4: Calculating degrees of freedom

The probability that any particular χ^2 value could be exceeded by chance depends on the number of degrees of freedom. This is simply **one less than the total number of categories** (this is the number that could vary independently without affecting the last value). **In this case: 4–1 = 3.**

Step 5b: Using the χ^2 table

The probability of between 0.2 and 0.5 is higher than the 0.05 value which is generally regarded as significant. The null hypothesis cannot be rejected and we have no reason to believe that the observed results differ significantly from the expected (at $P = 0.05$).

Footnote: Many Mendelian crosses involve ratios other than 1:1. For these, calculation of the expected values is not simply a division of the total by the number of categories. Instead, the total must be apportioned according to the ratio. For example, for a total of 400 as above, in a predicted 9:3:3:1 ratio, the total count must be divided by 16 (9+3+3+1) and the expected values will be 225: 75: 75: 25 in each category.

Table 1: Critical values of χ^2 at different levels of probability. By convention, the critical probability for rejecting the null hypothesis (H_0) is 5%. If the test statistic is less than the tabulated critical value for $P = 0.05$ we cannot reject H_0 and the result is not significant. If the test statistic is greater than the tabulated value for $P = 0.05$ we reject H_0 in favor of the alternative hypothesis.

Degrees of freedom	Level of probability (P)									
	0.98	0.95	0.80	0.50	0.20	0.10	0.05	0.02	0.01	0.001
1	0.001	0.004	0.064	0.455	1.64	2.71	3.84	5.41	6.64	10.83
2	0.040	0.103	0.466	1.386	3.22	4.61	5.99	7.82	9.21	13.82
3	0.185	0.352	1.005	2.366	4.64	6.25	7.82	9.84	11.35	16.27
4	0.429	0.711	1.649	3.357	5.99	7.78	9.49	11.67	13.28	18.47
5	0.752	0.145	2.343	4.351	7.29	9.24	11.07	13.39	15.09	20.52

χ^2 (for row 1, 0.50 column): 1.64 ← Do not reject H_0 Reject H_0 →

Appendices

Chi-Squared Exercise in Genetics

The following problems examine the use of the chi-squared (χ^2) test in genetics. A worked example illustrating the use of the chi-squared test for a genetic cross is provided on the previous page.

1. In a tomato plant experiment, two heterozygous individuals were crossed (the details of the cross are not relevant here). The predicted Mendelian ratios for the offspring of this cross were 9:3:3:1 for each of the **four following phenotypes**: purple stem-jagged leaf edge, purple stem-smooth leaf edge, green stem-jagged leaf edge, green stem-smooth leaf edge.

The observed results of the cross were not exactly as predicted. The numbers of offspring with each phenotype are provided below:

Observed results of the tomato plant cross			
Purple stem-jagged leaf edge	12	Green stem-jagged leaf edge	8
Purple stem-smooth leaf edge	9	Green stem-smooth leaf edge	0

(a) State your null hypothesis for this investigation (H$_0$): _____

_____

(b) State the alternative hypothesis (H$_A$): _____

_____

2. Use the chi-squared (χ^2) test to determine if the differences observed between the phenotypes are significant. The table of critical values of χ^2 at different P values is provided on the previous page.

(a) Enter the observed values (number of individuals) and complete the table to calculate the χ^2 value:

Category	O	E	O — E	(O — E)2	$\frac{(O - E)^2}{E}$
Purple stem, jagged leaf					
Purple stem, smooth leaf					
Green stem, jagged leaf					
Green stem, smooth leaf					
	Σ				Σ

(b) Calculate χ^2 value using the equation:

$$\chi^2 = \sum \frac{(O - E)^2}{E} \qquad \chi^2 = \underline{\quad\quad}$$

(c) Calculate the degrees of freedom: _____

(d) Using the χ^2 table, state the P value corresponding to your calculated χ^2 value:

_____

(e) State your decision: reject H$_0$ / do not reject H$_0$

(circle one)

3. Students carried out a pea plant experiment, where two heterozygous individuals were crossed. The predicted Mendelian ratios for the offspring were 9:3:3:1 for each of the **four following phenotypes**: round-yellow seed, round-green seed, wrinkled-yellow seed, wrinkled-green seed.

The observed results were as follows:

Round-yellow seed	441	Wrinkled-yellow seed	143
Round-green seed	159	Wrinkled-green seed	57

Use a separate piece of paper to complete the following:

(a) State the null and alternative hypotheses (H$_0$ and H$_A$).

(b) Calculate the χ^2 value.

(c) Calculate the degrees of freedom and state the P value corresponding to your calculated χ^2 value.

(d) State whether or not you reject your null hypothesis: reject H$_0$ / do not reject H$_0$ (circle one)

4. Comment on the whether the χ^2 values obtained above are similar. Suggest a reason for any difference:

_____

_____

Related activities: Using the Chi-Squared Test in Genetics

© BIOZONE International 2012
ISBN: 978-1-927173-11-4
Photocopying Prohibited

Using the Chi-Squared Test in Ecology

The **chi-squared test** (χ^2), like the student's *t* test, is a test for difference between data sets, but it is used when you are working with frequencies (counts) rather than measurements. It is a simple test to perform but the data must meet the requirements of the test. These are as follows:

- It can only be used for data that are raw counts (not measurements or derived data such as percentages).
- It is used to compare an experimental result with an expected theoretical outcome (e.g. an expected Mendelian ratio or a theoretical value indicating "no preference" or "no difference" between groups in a response such as habitat preference).
- It is not a valid test when sample sizes are small (<20).

Like all statistical tests, it aims to test the null hypothesis; the hypothesis of no difference between groups of data. The following exercise is a worked example using chi-squared for testing an ecological study of habitat preference. As with most of these simple statistical tests, chi-squared is easily calculated using a spreadsheet.

Using χ^2 in Ecology

Pneumatophores

In an investigation of the ecological niche of the mangrove, *Avicennia marina var. resinifera*, the density of pneumatophores was measured in regions with different substrate. The mangrove trees were selected from four different areas: mostly sand, some sand, mostly mud, and some mud. Note that the variable, substrate type, is categorical in this case. Quadrats (1 m by 1 m) were placed around a large number of trees in each of these four areas and the numbers of pneumatophores were counted. Chi-squared was used to compare the observed results for pneumatophore density (as follows) to an expected outcome of no difference in density between substrates.

Mangrove pneumatophore density in different substrate areas

Mostly sand	85	Mostly mud	130
Some sand	102	Some mud	123

Using χ^2, the probability of this result being consistent with the expected result could be tested. Worked example as follows:

Step 1: Calculate the expected value (E)

In this case, this is the sum of the observed values divided by the number of categories.

$$\frac{440}{4} = 110$$

Step 2: Calculate O – E

The difference between the observed and expected values is calculated as a measure of the deviation from a predicted result. Since some deviations are negative, they are all squared to give positive values. This step is usually performed as part of a tabulation (right, darker blue column).

Category	O	E	O–E	(O–E)^2	$\dfrac{(O-E)^2}{E}$
Mostly sand	85	110	-25	625	5.68
Some sand	102	110	-8	64	0.58
Mostly sand	130	110	20	400	3.64
Some sand	123	110	13	169	1.54

Total = 440

χ^2 → $\Sigma = 11.44$

Step 3: Calculate the value of χ^2

$$\chi^2 = \sum \frac{(O - E)^2}{E}$$

Where:
O = the observed result
E = the expected result
Σ = sum of

The calculated χ^2 value is given at the bottom right of the last column in the tabulation.

Step 4: Calculating degrees of freedom

The probability that any particular χ^2 value could be exceeded by chance depends on the number of degrees of freedom. This is simply **one less than the total number of categories** (this is the number that could vary independently without affecting the last value). *In this case: 4–1 = 3.*

Step 5a: Using the χ^2 table

On the χ^2 table (part reproduced in Table 1 below) with 3 degrees of freedom, the calculated value for χ^2 of 11.44 corresponds to a probability of between 0.01 and 0.001 (see arrow). *This means that by chance alone a χ^2 value of 11.44 could be expected between 1% and 0.1% of the time.*

Step 5b: Using the χ^2 table

The probability of between 0.1 and 0.01 is lower than the 0.05 value which is generally regarded as significant. The null hypothesis can be rejected and we have reason to believe that the observed results differ significantly from the expected (at P = 0.05).

Table 1: Critical values of χ^2 at different levels of probability. By convention, the critical probability for rejecting the null hypothesis (H_0) is 5%. If the test statistic is less than the tabulated critical value for P = 0.05 we cannot reject H_0 and the result is not significant. If the test statistic is greater than the tabulated value for P = 0.05 we reject H_0 in favour of the alternative hypothesis.

Degrees of freedom	Level of probability (P)									
	0.98	0.95	0.80	0.50	0.20	0.10	0.05	0.02	0.01	0.001
1	0.001	0.004	0.064	0.455	1.64	2.71	3.84	5.41	6.64	10.83
2	0.040	0.103	0.466	1.386	3.22	4.61	5.99	7.82	9.21	13.82
3	0.185	0.352	1.005	2.366	4.64	6.25	7.82	9.84	11.35	16.27
4	0.429	0.711	1.649	3.357	5.99	7.78	9.49	11.67	13.28	18.47
5	0.752	0.145	2.343	4.351	7.29	9.24	11.07	13.39	15.09	20.52

χ^2 10.83 (column 0.01 → 0.001, with arrow pointing down to 16.27)

← Do not reject H_0

Reject H_0 →

Appendices

Chi-Squared Exercise in Ecology

The following exercise illustrates the use of chi-squared (χ^2) in ecological studies of habitat preference. In the first example, it is used for determining if the flat periwinkle *(Littorina littoralis)* shows significant preference for any of the four species of seaweeds with which it is found. Using quadrats, the numbers of periwinkles associated with each seaweed species were recorded. The data from this investigation are provided for you in Table 1. In the second example, the results of an investigation into habitat preference in woodlice (also called pillbugs, sowbugs, or slaters) are presented for analysis (Table 2).

1. (a) State your null hypothesis for this investigation (H_0):

_____

_____

_____

(b) State the alternative hypothesis (H_A): _____

_____

_____

Table 1: Number of periwinkles associated with different seaweed species

Seaweed species	Number of periwinkles
Spiral wrack	9
Bladder wrack	28
Toothed wrack	19
Knotted wrack	64

2. Use the chi-squared test to determine if the differences observed between the samples are significant or if they can be attributed to chance alone. The table of critical values of χ^2 is provided in "*Using The Chi-Squared Test in Ecology*".

(a) Enter the observed values (no. of periwinkles) and complete the table to calculate the χ^2 value:

(b) Calculate χ^2 value using the equation:

$$\chi^2 = \sum \frac{(O - E)^2}{E} \qquad \chi^2 = \underline{\hphantom{aaaa}}$$

(c) Calculate the degrees of freedom: _____

(d) Using the χ^2, state the P value corresponding to your calculated χ^2 value:

_____

(e) State whether or not you reject your null hypothesis:

reject H_0 / do not reject H_0 (*circle one*)

Category	O	E	O – E	(O – E)²	$\frac{(O - E)^2}{E}$
Spiral wrack					
Bladder wrack					
Toothed wrack					
Knotted wrack					
Σ					Σ

3. Students carried out an investigation into habitat preference in woodlice. In particular, they were wanting to know if the woodlice preferred a humid atmosphere to a dry one, as this may play a part in their choice of habitat. They designed a simple investigation to test this idea. The woodlice were randomly placed into a choice chamber for 5 minutes where they could choose between dry and humid conditions (atmosphere). The investigation consisted of five trials with ten woodlice used in each trial. Their results are shown on Table 2 (right):

(a) State the null and alternative hypotheses (H_0 and H_A) :

_____

_____

Table 2: Habitat preference in woodlice

Trial	Atmosphere	
	Dry	Humid
1	2	8
2	3	7
3	4	6
4	1	9
5	5	5

Use a separate piece of paper (or a spreadsheet) to calculate the chi-squared value and summarize your answers below:

(b) Calculate χ^2 value: _____

(c) Calculate the degrees of freedom and state the P value corresponding to your calculated χ^2 value: _____

(d) State whether or not you reject your null hypothesis: reject H_0 / do not reject H_0 (*circle one*)

Spearman Rank Correlation

The Spearman rank correlation is a test used to determine if there is a statistical dependence (correlation) between two variables. The test is appropriate for data that have a non-normal distribution (or where the distribution is not known) and assesses the degree of association between the X and Y variables. For the test to work, the values used must be **monotonic** i.e. the values must increase or decrease together or one increases while the other decreases. A value of 1 indicates a perfect correlation; a value of 0 indicates no correlation between the variables. The example below examines the relationship between the frequency of the drumming sound made by male frigatebirds (Y) and the volume of their throat pouch (X).

Spearman's Rank Data for Frigate Bird Pouch Volume and Drumming Frequency

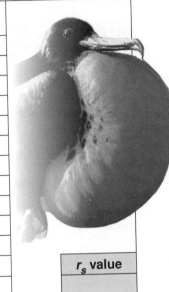

Bird	Volume of pouch (cm³)	Rank (R₁)	Frequency of drumming sound (Hz)	Rank (R₂)	Difference (D) (R₁-R₂)	D²
1	2550		461			
2	2440	I	473	6	-5	25
3	2740		532			
4	2730		465			
5	3010		485			
6	3370		488			
7	3080		527			
8	4910		478			
9	3740		485			
10	5090		434			
11	5090		468			
12	5380		449			
				Σ(Sum)		

Based on Madsen et al 2004

r_s value

Analyzing the Data

Step one: Rank the data for each variable. For each variable, the numbers are ranked in descending order, e.g. for the variable, volume, the highest value 5380 cm³ is given the rank of 12 while its corresponding frequency value is given the rank of 2. Fill in the rank columns in the table above in the same way. If two numbers have the same rank value, then use the mean rank of the two values (e.g. 1+2 = 3. 3/2= 1.5).

Step two: Calculate the difference (D) between each pair of ranks (R₁-R₂) and enter the value in the table (as a check, the sum of all differences should be 0).

Step three: Square the differences and enter them into the table above (this removes any negative values).

Step four: Sum all the D² values and enter the total into the table.

Step five: Use the formula below to calculate the Spearman Rank Correlation Coefficient (r_s). Enter the r_s value in the box above.

$$r_s = 1 - \left(\frac{6\Sigma D^2}{n(n^2-1)} \right)$$

Spearman Rank Correlation Coefficient

Step six: Compare the r_s value to the table of critical values (right) for the appropriate number of pairs. If the r_s value (ignoring sign) is greater than or equal to the critical value then there is a significant correlation. If r_s is positive then there is a positive correlation. If r_s is negative then there is a negative value correlation.

Number of pairs of measurements	Critical value
5	1.00
6	0.89
7	0.79
8	0.74
9	0.68
10	0.65
12	0.59
14	0.54
16	0.51
18	0.48
20	0.45

1. State the null hypothesis for the data set. _____

2 (a) Identify the critical value for the frigate bird data: _____

   (b) State is the correlation is positive of negative: _____

   (c) State whether the correlation is significant: _____

3. Explain why the data collected must be monotonic if a Spearman rank correlation is to be used: _____

Appendices

Appendix B

TERMS AND NOTATION

The definitions for some commonly encountered terms related to making biological investigations are provided below. Use these as you would use a biology dictionary when planning your investigation and writing up your report. It is important to be consistent with the use of terms i.e. use the same term for the same procedure or unit throughout your study. Be sure, when using a term with a specific statistical meaning, such as sample, that you are using the term correctly.

General Terms

Data: Facts collected for analysis.

Qualitative: Not quantitative. Described in words or terms rather than by numbers. Includes subjective descriptions in terms of variables such as color or shape.

Quantitative: Able to be expressed in numbers. Numerical values derived from counts or measurements.

The Design of Investigations

Hypothesis: A tentative explanation of an observation, capable of being tested by experimentation. Hypotheses are written as clear statements, not as questions.

Control treatment (control): A standard (reference) treatment that helps to ensure that responses to other treatments can be reliably interpreted. There may be more than one control in an investigation.

Dependent variable: A variable whose values are determined by another variable (the independent variable). In practice, the dependent variable is the variable representing the biological response.

Independent variable: A variable whose values are set, or systematically altered, by the investigator.

Controlled variables: Variables that may take on different values in different situations, but are controlled (fixed) as part of the design of the investigation.

Experiment: A contrived situation designed to test (one or more) hypotheses and their predictions. It is good practice to use sample sizes that are as large as possible for experiments.

Investigation: A very broad term applied to scientific studies; investigations may be controlled experiments or field based studies involving population sampling.

Parameter: A numerical value that describes a characteristic of a population (e.g. the mean height of all 17 year-old males).

Prediction: The prediction of the response (Y) variable on the basis of changes in the independent (X) variable.

Random sample: A method of choosing a sample from a population that avoids any subjective element. It is the equivalent to drawing numbers out of a hat, but using random number tables. For field based studies involving quadrats or transects, random numbers can be used to determine the positioning of the sampling unit.

Repeat / Trial: The entire investigation is carried out again at a different time. This ensures that the results are reproducible. Note that repeats or trials are not **replicates** in the true sense unless they are run at the same time.

Replicate: A duplication of the entire experimental design run at the same time.

Sample: A sub-set of a whole used to estimate the values that might have been obtained if every individual or response was measured. A sample is made up of **sampling units**, In lab based investigations, the sampling unit might be a test-tube, while in field based studies, the sampling unit might be an individual organism or a quadrat.

Sample size (*n*): The number of samples taken. In a field study, a typical sample size may involve 20-50 individuals or 20 quadrats. In a lab based investigation, a typical sample size may be two to three sampling units, e.g. two test-tubes held at 10°C.

Sampling unit: Sampling units make up the sample size. Examples of sampling units in different investigations are an individual organism, a test tube undergoing a particular treatment, an area (e.g. quadrat size), or a volume. The size of the sampling unit is an important consideration in studies where the area or volume of a habitat is being sampled.

Statistic: An estimate of a parameter obtained from a sample (e.g. the mean height of all 17 year-old males in your class). A precise (reliable) statistic will be close to the value of the parameter being estimated.

Treatments: Well defined conditions applied to the sample units. The response of sample units to a treatment is intended to shed light on the hypothesis under investigation. What is often of most interest is the comparison of the responses to different treatments.

Variable: A factor in an experiment that is subject to change. Variables may be controlled (fixed), manipulated (systematically altered), or represent a biological response.

Precision and Significance

Accuracy: The correctness of the measurement (the closeness of the measured value to the true value). Accuracy is often a function of the calibration of the instrument used for measuring.

Measurement errors: When measuring or setting the value of a variable, there may be some difference between your answer and the 'right' answer. These errors are often as a result of poor technique or poorly set up equipment.

Objective measurement: Measurement not significantly involving subjective (or personal) judgment. If a second person repeats the measurement they should get the same answer.

Precision (of a measurement): The repeatability of the measurement. As there is usually no reason to suspect that a piece of equipment is giving inaccurate measures, making precise measurements is usually the most important consideration. You can assess or quantify the precision of any measurement system by taking repeated measurements from individual samples.

Precision (of a statistic): How close the statistic is to the value of the parameter being estimated. Also called **reliability**.

The Expression of Units

The value of a variable must be written with its units where possible. Common ways of recording measurements in biology are: volume in liters, mass in grams, length in meters, time in seconds. The following example shows different ways to express the same term. Note that ml and cm³ are equivalent.

Oxygen consumption (milliliters per gram per hour)

Oxygen consumption ($ml\ g^{-1}\ h^{-1}$) or ($mL\ g^{-1}\ h^{-1}$)

Oxygen consumption ($ml/g/h$) or ($mL/g/h$)

Oxygen consumption ($cm^3\ g^{-1}\ h^{-1}$)

Statistical significance: An assigned value that is used to establish the probability that an observed trend or difference represents a true difference that is not due to chance alone. If a level of significance is less than the chosen value (usually 1-10%), the difference is regarded as statistically significant. Remember that in rigorous science, it is the hypothesis of no difference or no effect (the null hypothesis, H_0) that is tested. The alternative hypothesis (your tentative explanation for an observation) can only be accepted through statistical rejection of H_0.

Validity: Whether or not you are truly measuring the right thing.

Appendix C

CALCULATIONS, CONVERSIONS, AND MULTIPLES

Carrying out calculations:

When answering computational questions, it is important to show all the working associated with calculating the answer. Some examples of the calculations and conversions you may encounter are described below.

1. Converting between multiples:

(a) Convert the following to kilometers:

   (i) 5 millimeters: 5 mm = 0.000 005 km = 5×10^{-6} km.

   (ii) 10 000 centimeters: 10 000 cm = 0.1 km.

   (iii) 8000 meters: 8000 m = 8 km.

(b) Convert 12 ms^{-1} to kmhr^{-1}: $12 \times 60 \times 60 = 43\ 200$ mhr^{-1}. 43 200/1000 = 43.2 kmhr^{-1}.

2. Sampling:

(a) A study of a bear population discovers that there were 5 bears living within a 4 km^2 area of a forest:

   (i) Calculate the total population if the forest is 100 km^2.
5/4 = 1.25 bears per 1 km^2. $1.25 \times 100 = 125$ bears.

   (ii) It is estimated that the bear population may be 20% larger than the sample suggests. Calculate the new population: $125 + (125 \times 0.20) = 150$ bears.

   (iii) It is estimated that the population has an annual growth rate of 1.3 percent. Calculate the bear population in a further five years:
$Pop_{future\ bears} = Pop_{present\ bears} \times (1 + 0.013)^5 = 150 \times (1.013)^5 = 160$ bears in 5 years time.

(b) A coal deposit is estimated at 2000 tonnes (t). If the coal is extracted at a rate of 30 thr^{-1} calculate how long the deposit will last: 2000/30 = 66.67 hours.

3. Reading off a graph:

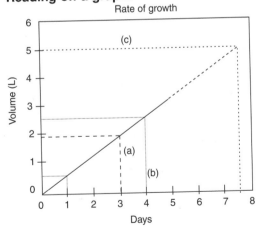

Rate of growth

(a) State the volume after 3 days: 1.95 L

(b) Calculate the rate of growth per day:
Day 1 = 0.5 L Day 4 = 2.5 L Change = 2.5-0.5 = 2.0 over 3 days. 2.0/3 = 0.667 L per day.

(c) Extrapolate the graph to determine how long it will take to reach a volume of 5 L: 7.55-7.65 days. (confirm from calculation 5/0.667 = 7.5).

INTERNATIONAL SYSTEM OF UNITS (SI)

Examples of SI derived units

DERIVED QUANTITY	NAME	SYMBOL
area	square meter	m^2
volume	cubic meter	m^3
speed, velocity	meter per second	ms^{-1}
acceleration	meter per second squared	ms^{-2}
mass density	kilogram per cubic meter	kgm^{-3}
specific volume	cubic meter per kilogram	m^{3}kg^{-1}
amount-of-substance/ concentration	mole per cubic meter mole per liter	molm^{-3} molL^{-1}
luminance	candela per square meter	cdm^{-2}

MULTIPLES

MULTIPLE	PREFIX	SYMBOL	EXAMPLE
10^9	giga	G	gigawatt (GW)
10^6	mega	M	megawatt (MW)
10^3	kilo	k	kilogram (kg)
10^2	hecto	h	hectare (ha)
10^{-1}	deci	d	decimeter (dm)
10^{-2}	centi	c	centimeter (cm)
10^{-3}	milli	m	milliimeter (mm)
10^{-6}	micro	μ	microsecond (μs)
10^{-9}	nano	n	nanometer (nm)
10^{-12}	pico	p	picosecond (ps)

CONVERSION FACTORS FOR COMMON UNITS OF MEASURE

For all conversions multiply by the factor shown

LENGTH
Centimeters to inches:	0.393
Meters to feet:	3.280
Kilometers to miles:	0.621

VOLUME
Milliliters to fluid ounces:	0.034
Liters to gallons:	0.264
Cubic meters to gallons:	264.1

AREA
Square meters to square feet:	10.76
Hectares to acres:	2.471
Square kilometers to square miles	0.386

TEMPERATURE
°C to °F:	0 °C = 32 °F
	100 °C = 212 °F
Formula °C to °F:	°F = °C x 1.8 + 32

Appendix D

THE BIOCHEMISTRY OF LIFE

▶ **Water, Life, and Hydrogen Bonding**
Biol. Sci. Rev., 21(2) Nov. 2008, pp. 18-20. *The molecules of life and the important role of hydrogen bonding.*

▶ **What is Tertiary Structure?**
Biol. Sci. Rev., 21(1) Sept. 2008, pp. 10-13. *How amino acid chains fold into the functional shape of a protein.*

▶ **Glucose & Glucose-Containing Carbohydrates**
Biol. Sci. Rev., 19(1) Sept. 2006, pp. 12-15. *The structure of glucose and its polymers.*

▶ **Designer Starches**
Biol. Sci. Rev., 19(3) Feb. 2007, pp. 18-20. *The composition of starch, and an excellent account of its properties and functions.*

▶ **Getting in and Out**
Biol. Sci. Rev., 20(3), Feb. 2008, pp. 14-16. *Diffusion: some adaptations and some common misunderstandings.*

THE INTERNAL STRUCTURE OF CELLS

▶ **Size Does Matter**
Biol. Sci. Rev., 17 (3) February 2005, pp. 10-13. *Measuring the size of organisms and calculating magnification and scale.*

▶ **Bacteria**
National Geographic, 184(2) August 1993, pp. 36-61. *The structure and diversity of bacteria; our most abundant and useful organisms.*

▶ **Cellular Factories**
New Scientist, 23 November 1996 (Inside Science). *The role of different organelles in plant and animal cells.*

▶ **Light Microscopy**
Biol. Sci. Rev., 13(1) Sept. 2000, pp. 36-38. *The basis and various techniques of light microscopy.*

CELL MEMBRANES AND TRANSPORT

▶ **Border Control**
New Scientist, 15 July 2000 (Inside Science). *The role of the plasma membrane in cell function: membrane structure, transport processes, and the role of receptors on the cell membrane.*

▶ **The Fluid-Mosaic Model for Membranes**
Biol. Sci. Rev., 22(2), Nov. 2009, pp. 20-21. *Diagrammatic revision of membrane structure and function.*

▶ **Getting in and Out**
Biol. Sci. Rev., 20(3), Feb. 2008, pp. 14-16. *Diffusion: some adaptations and some common misunderstandings.*

▶ **How Biological Membranes Achieve Selective Transport**
Biol. Sci. Rev., 21(4), April 2009, pp. 32-36. *The structure of the plasma membrane and the proteins that enable the selective transport of molecules.*

▶ **What is Endocytosis?**
Biol. Sci. Rev., 22(3), Feb. 2010, pp. 38-41. *The mechanisms of endocytosis and the role of membrane receptors in concentrating important molecules before ingestion.*

CELLULAR COMMUNICATION

▶ **How and Why Bacteria Communicate**
Scientific American, 276 (2) Feb. 1997, pp. 52-57. *Chemical communication is involved in triggering a number of critical biological events in bacteria.*

▶ **New Jobs for Ancient Chaperones**
Scientific American, 299 (1) July 2008, pp. 32-37. *Guardian proteins are found in all forms of life and keep critical cellular processes running smoothly.*

DNA AND RNA

▶ **Control Centre**
New Scientist, 17 July 1999, (Inside Science). *An easy to read account of the organisation of DNA in eukaryotic cells, the origin of the nucleus, how genes code for proteins, and the role of ribosomes and RNA in translation.*

▶ **DNA: 50 Years of the Double Helix**
New Scientist, 15 March 2003, pp. 35-51. *A special issue on DNA: structure and function, repair, the new-found role of histones, and the functional significance of chromosome position in the nucleus.*

▶ **DNA Polymerase**
Biol. Sci. Rev., 22(4) April 2010, pp. 38-41. *How DNA polymerase operates during DNA replication in the cell and how the enzyme is used in PCR.*

▶ **What is a Gene?**
Biol. Sci. Rev., 15(2) Nov. 2002, pp. 9-11. *A synopsis of genes, mutations, and transcriptional control of gene expression.*

▶ **Gene Structure and Expression**
Biol. Sci. Rev., 12 (5) May 2000, pp. 22-25. *The structure and function of genes, and the basis of gene regulation in eukaryotes and prokaryotes.*

▶ **The Alternative Genome**
Scientific American, April, 2005, pp. 40-47. *The old axiom "one gene, one protein" no longer holds true. The more complex an organism, the more likely it became that way by extracting multiple protein meanings from individual genes.*

▶ **The Hidden Genetic Program**
Scientific American, Oct. 2004, pp. 30-37. *Large portions of the DNA of complex organisms may encode RNAs with important regulatory functions.*

▶ **Transfer RNA**
Biol. Sci. Rev., 15(3) Feb. 2003, pp. 26-29. *A good account of the structure and role of tRNA in protein synthesis.*

▶ **Genetics of Sickle Cell Anaemia**
Biol. Sci. Rev., 20(4) April 2008, pp. 14-17. *The molecular and physiological basis of sickle cell disease.*

▶ **Tailor-Made Proteins**
Biol. Sci. Rev., 13(4) March 2001, pp. 2-6. *Recombinant proteins and their uses in industry and medicine.*

▶ **Rice, Risk and Regulations**
Biol. Sci. Rev., 20(2) Nov. 2007, pp. 17-20. *The genetic engineering of one of the world's most important cereal crops is an ethical concern for many.*

▶ **The Engineering of Crop Plants**
Biol. Sci. Rev., 20(4) April 2008, pp. 30-36. *Crop plants can be engineered to increase the nutritional value of foods and to improve non-food crops as sources of raw materials for industry.*

CHROMOSOMES AND CELL DIVISION

▶ **The Cell Cycle and Mitosis**
Biol. Sci. Rev., 14(4) April 2002, pp. 37-41. *Cell growth and division, stages in the cell cycle, and the complex control over different stages of mitosis.*

▶ **Mechanisms of Meiosis**
Biol. Sci. Rev., 15(4), April 2003, pp. 20-24. *A clear and thorough account of the events and mechanisms of meiosis.*

THE CHROMOSOMAL BASIS OF INHERITANCE

▶ **Mendel's Legacy**
Biol. Sci. Rev., 18(4), April 2006, pp. 34-37. *Explores the accuracy of Mendel's laws in light of today's knowledge.*

▶ **The Y Chromosome: It's a Man Thing**
Biol. Sci. Rev., 20(4) April 2008, pp. 2-6. *The Y chromosome is at the root of sex determination. This account discusses the nature of the Y chromosome, non-disjunction and Y chromosome disorders and the inheritance of Y linked diseases*

▶ **Strange Inheritance**
New Scientist, 12 July 2008, pp. 28-33.

Appendix D

Studies of epigenetic inheritance is forcing us to rethink what we know about genetics and evolution.

▶ **Secrets of The Gene**

National Geographic, 196(4) Oct. 1999, pp. 42-75. *The nature of genes and the inheritance of particular genetic traits through certain populations.*

▶ **The Color Code**

New Scientist, 10 March 2002, pp. 34-37. *Researchers are uncovering the five to ten genes responsible for skin pigmentation.*

▶ **What is Variation?**

Biol. Sci. Rev., 13(1) Sept. 2000, pp. 30-31. *The nature of continuous and discontinuous variation. The distribution pattern of traits that show continuous variation is discussed.*

REGULATION OF GENE EXPRESSION

▶ **Cell Differentiation**

Biol. Sci. Rev., 20(4), April 2008, pp. 10-13. *How tissues arise through the control of cellular differentiation during development. The example provided is the differentiation of blood cells.*

▶ **Regulating Evolution**

Sci. American, May 2008, pp. 34-45. *Mutations in the DNA switches controlling body-shaping genes, rather than the genes themselves, have been significant in the evolution of morphological differences.*

▶ **What is a Mutation?**

Biol. Sci. Rev., 20(3) Feb. 2008, pp. 6-9. *The nature of mutations: causes, timing, and effects. Sickle cell disease is described.*

▶ **What is Cell Suicide?**

Biol. Sci. Rev., 20(1) Sept. 2007, pp. 17-20. *An account of the mechanisms behind cell suicide and its role in normal growth and development.*

▶ **Gene Structure and Expression**

Biol. Sci. Rev., 12 (5) May 2000, pp. 22-25. *The structure and function of genes, and the basis of gene regulation in eukaryotes and prokaryotes.*

▶ **Living with the Enemy**

New Scientist, 25 Oct. 2008, pp. 26-33. *The sheer diversity of mutations that can turn cells cancerous and drive tumor growth gives endless opportunities to outwit our defense. How can we protect ourselves effectively?*

SOURCES OF VARIATION

▶ **What is Variation?**

Biol. Sci. Rev., 13(1) Sept. 2000, pp. 30-31. *The nature of continuous and discontinuous variation. The distribution pattern of traits that show continuous variation is discussed.*

▶ **What is a Mutation?**

Biol. Sci. Rev., 20(3) Feb. 2008, pp. 6-9. *The nature of mutations: causes, timing, and effects. Sickle cell disease is described.*

▶ **The Price of Silent Mutations**

Scientific American, June 2009, pp. 34-41. *Small changes to DNA that were once considered silent are proving important in human disease, acting through a variety of mechanisms.*

▶ **Genetics of Sickle Cell Anaemia**

Biol. Sci. Rev., 20(4) April 2008, pp. 14-17. *The molecular and physiological basis of sickle cell disease.*

▶ **What is a Chromosome Mutation?**

Biol. Sci. Rev., 22(1) Sept. 2009, pp. 14-18. *An account of chromosomal disorders and how they arise. Some common human aneuploidies and their karyotypes are described.*

GENETIC CHANGE IN POPULATIONS

▶ **Was Darwin Wrong?**

National Geographic, 206(5) Nov. 2004, pp. 2-35. *An excellent account of the overwhelming scientific evidence for evolution. A good starting point for reminding students that the scientific debate around evolutionary theory is associated with the mechanisms by which evolution occurs, not the fact of evolution itself.*

▶ **Animal Attraction**

National Geographic, July 2003, pp. 28-55. *An engaging and expansive account of mating in the animal world.*

▶ **The Hardy-Weinberg Principle**

Biol. Sci. Rev., 15(4), April 2003, pp. 7-9. *A succinct explanation of the basis of the Hardy-Weinberg principle, and its uses in estimating genotype frequencies and predicting change in populations.*

▶ **Evolution at a Snail's Pace**

Biol. Sci. Rev., 23(2), Nov. 2010, pp. 26-29. *The banded snail provides a perfect case study in phenotypic variation and shifts in allele frequencies in response to changing environments.*

▶ **The Moths of War**

New Scientist, 8 Dec. 2007, pp 46-49. *New research into the melanism of the peppered moth reaffirms it as an example of evolution, reclaiming it back from Creationists.*

▶ **Skin Deep**

Scientific American, October 2002, pp. 50-57. *This article examines the evolution of skin color in humans and presents powerful evidence for skin color ("race") being the end result of opposing selection forces (the need for protection of folate from UV vs the need*

to absorb vitamin D). Clearly written and of high interest, this is a must for student discussion and a perfect vehicle for examining natural selection.

▶ **Fair Enough**

New Scientist, 12 Oct. 2002, pp. 34-37. *Skin color in humans: this article examines the argument for there being a selective benefit to being dark or pale in different environments.*

▶ **Black Squirrels**

Biol. Sci. Rev., 21(2), Nov. 2008, pp. 39-41. *A look at how squirrel types have changed in Britain over time, and the selection pressures acting on the pigmentation and melanism.*

▶ **Polymorphism**

Biol. Sci. Rev., 14(1) Sept. 2001, pp. 19-21. *A good account of genetic polymorphism. Examples include the carbonaria gene (Biston), the sickle cell gene, and aphids.*

▶ **Genetics of Sickle Cell Anemia**

Biol. Sci. Rev., 20(4) April 2008, pp. 14-17. *This account includes explanation of how a mutation is retained in the population as a result of heterozygous advantage.*

▶ **The Enemy Within**

Scientific American, April 2011, pp. 26-33. *Antibiotic resistance is spreading in the transfer of genes that confer resistance in a new pattern globally. New medications are not being developed quickly enough to treat gram-negative bacteria.*

▶ **The Cheetah: Losing the Race?**

Biol. Sci. Rev., 14(2) Nov. 2001, pp. 7-10. *The inbred status of cheetahs and its evolutionary consequences.*

▶ **Taming the Wild**

National Geographic, 219(3) March 2011, pp. 34-59. *Species that have been successfully domesticated and live with human contact have different genetic traits to those that are not domesticated. An experiment in fox breeding begun in the 1960s showed after nine generations a collection of genes conferring tame behavior.*

EVIDENCE FOR BIOLOGICAL EVOLUTION

▶ **A Cool Early Life**

Scientific American, Oct. 2005, pp. 40-47. *Evidence suggests that the Earth cooled as early as 4.4 bya. These cooler, wet surroundings were necessary for life to evolve.*

▶ **How Old is...**

National Geographic, 200(3) Sept. 2001, pp. 79-101. *A discussion of dating methods and their application.*

Appendix D

▶ **The Quick and the Dead**
New Scientist, 5 June 1999, pp. 44-48. *The formation of fossils: fossil types and preservation in different environments.*

▶ **The Accidental Discovery of a Feathered Giant Dinosaur**
Biol. Sci. Rev., 20(4), April 2008, pp. 18-20. *How scientists piece together and interpret confusing fossil evidence.*

▶ **A Fin is a Limb is a Wing**
National Geographic, 210(5) Nov. 2006, pp. 110-135. *An excellent account of the role of developmental genes in the evolution of complex organs and structures in animals. Beautifully illustrated, compelling evidence for the mechanisms of evolutionary change.*

▶ **A Waste of Space**
New Scientist, 25 April 1998, pp. 38-39. *Vestigial organs: how they arise in an evolutionary sense and the role they might play.*

▶ **Uprooting the Tree of Life**
Scientific American Feb. 2000, pp. 72-77. *Using molecular techniques to redefine phylogeny and divulge the path of evolution.*

THE RELATEDNESS OF ORGANISMS

▶ **Uprooting the Tree of Life**
Scientific American, 282(2) Feb. 2000, pp. 72-77. *Additions to the consensus view of the three domain tree of life. What part did gene transfers play in the evolution of eukaryotic groups.*

▶ **To Share and Share Alike**
Scientific American, 304(4) April 2011, p. 13. *Markers in archaean and bacterial genomes show that 88-98% of new genes are acquired by horizontal gene transfer, which may be the dominant force in prokaryote evolution.*

▶ **Evolution Encoded**
Scientific American, 290 (4) April 2004, pp. 56-63. *Natural selection has produced and maintained the genetic code to minimize harmful errors.*

▶ **The Family Line - The Human-Cat Connection**
National Geographic, 191(6) June 1997, pp. 77-85. *An examination of the genetic diversity and lineages within the felidae. A good context within which to study classification.*

▶ **A Passion for Order**
National Geographic, 211(6) June 2007, pp. 73-87. *The history of Carl Linnaeus and the classification of plant species.*

SPECIATION AND EXTINCTION

▶ **Evolution in New Zealand**
Biol. Sci. Rev., 21(3) Feb. 2009, pp. 33-37. *NZ offers a unique suite of case studies in evolution*

▶ **Dinosaurs take Wing**
National Geographic, 194(1) July 1998, pp. 74-99. *An account of the evolution of birds from small theropod dinosaurs, including an exploration of the homology between the typical dinosaur limb and the wing of the modern bird.*

▶ **Evolution: Five Big Questions**
New Scientist, 14 June 2003, pp. 32-39, 48-51. *A synopsis of the five most common points of discussion regarding evolution and the mechanisms by which it occurs.*

▶ **Species and Species Formation**
Biol. Sci. Rev., 20(3), Feb. 2008, pp. 36-39. *A feature covering the definition of species and how new species come into being through speciation.*

▶ **What is a Species?**
Scientific American June 2008, pp. 48-55. *The science of classification; modern and traditional approaches, the value of each, and the importance of taxonomy to identifying and recognising diversity. Excellent.*

▶ **Cichlids of the Rift Lakes**
Scientific American, February 1999, pp. 44-49. *An excellent account of the recent speciation events in cichlid fishes in Lake Victoria, as revealed by mtDNA.*

▶ **Listen, We're Different**
New Scientist, 17 July 1999, pp. 32-35. *An account of speciation in periodic cicadas as a result of behavioural and temporal isolating mechanisms.*

▶ **Evolution in the Fast Lane**
New Scientist, 2 April 2011, pp. 32-36. *The evolution of stickleback armour provides an example of rapid evolution. Fluctuating selection pressures produce unexpected patterns of evolution in algal and rotifer populations also. It seems that rapid evolution may not be the exception, but the norm.*

▶ **The Rise of Mammals**
National Geographic, 203(4), pp. April 2003, p. 2-37. *An account of the adaptive radiation of mammals and the significance of the placenta in mammalian evolution.*

▶ **Beating the Bloodsuckers**
Biol. Sci. Rev., 16(3) Feb. 2004, pp. 31-35. *The global distribution of malaria, the current state of malaria research, and an account of the biology of the Plasmodium parasite and the body's immune response to it.*

▶ **Search for a Cure**
National Geographic, 201(2) February 2002, pp. 32-43. *An account of the status of the AIDS epidemic and the measures to stop it.*

▶ **Which Came First?**
Scientific American, Feb. 1997, pp. 12-14. *Shared features among fossils; convergence or common ancestry?*

▶ **Mass Extinctions**
New Scientist, 5 March 2011, pp i-iv. *An 'instant expert' article covering the nature of mass extinction using two important mass extinction events as examples: the loss of the dinosaurs and the end Permian extinction which resulted in the loss of 80-90% of species. It also describes how life rebounds following extinction events.*

▶ **The Sixth Extinction**
National Geographic, 195(2) Feb. 1999, pp. 42-59. *High extinction rates have occurred five times in the past. Human impact is driving the sixth extinction.*

THE ORIGIN OF LIVING SYSTEMS

▶ **Earth in the Beginning**
National Geographic, 210(6) Dec. 2006, pp. 58-67. *Modern landscapes offer glimpses of the way Earth may have looked billions of years ago.*

▶ **The Ice of Life**
Scientific American, August 2001, pp. 37-41. *Space ice may promote organic molecules and may have seeded life on Earth (teacher's reference).*

▶ **A Simpler Origin of Life**
Scientific American, June 2007, pp. 24-31. *Two conflicting theories on how a complicated molecule such as RNA formed (teacher's reference).*

▶ **Primeval Pools**
New Scientist, 2 July 2005, pp. 40-43. *Ecosystems where microbes dominate as they did millions of years in the past.*

▶ **The Rise of Life on Earth**
National Geographic, 193(3) March 1998, pp. 54-81. *A series of articles covering the theories for life's origins, the evolution of life's diversity, and the origin of eukaryotic cells.*

▶ **A Cool Early Life**
Scientific American, Oct. 2005, pp. 40-47. *Evidence suggests that the Earth cooled as early as 4.4 bya. These cooler, wet surroundings were necessary for life to evolve.*

Index

Index